ACS SYMPOSIUM SERIES 949

Current Fluoroorganic Chemistry

New Synthetic Directions, Technologies, Materials, and Biological Applications

Vadim A. Soloshonok, Editor
The University of Oklahoma

Koichi Mikami, Editor
Tokyo Institute of Technology

Takashi Yamazaki, Editor
Tokyo University of Agriculture and Technology

John T. Welch, Editor
State University of New York at Albany

John F. Honek, Editor
The University of Waterloo

Sponsored by the
ACS Divisions of Fluorine Chemistry and Medicinal Chemistry

American Chemical Society, Washington, DC

Library of Congress Cataloging-in-Publication Data

Current fluoroorganic chemistry : new synthetic directions, technologies, materials, and biological applications / Vadim Anatol′evich Soloshonok, editor ... [et al.] ; sponsored by the ACS divisions of fluorine chemistry and medicinal chemistry.

p. cm.—(ACS symposium series ; 949)

Includes bibliographical references and index.

ISBN 13: 978–0–8412–7403–7 (alk. paper)

ISBN 10: 0–8412–7403–7 (alk. paper)

1. Organofluorine compounds. 2. Organic compounds—Synthesis.

I. Soloshonok, V. A. II. American Chemical Society. Division of Fluorine Chemistry. III. Division of Medicinal Chemistry.

QD412.F1C87 2006
547′.02—dc22 2006050085

The paper used in this publication meets the minimum requirements of American National Standard for Information Sciences—Permanence of Paper for Printed Library Materials, ANSI Z39.48–1984.

Distributed by Oxford University Press

PRINTED IN THE UNITED STATES OF AMERICA

Current Fluoroorganic Chemistry

9955

Foreword

The ACS Symposium Series was first published in 1974 to provide a mechanism for publishing symposia quickly in book form. The purpose of the series is to publish timely, comprehensive books developed from ACS sponsored symposia based on current scientific research. Occasionally, books are developed from symposia sponsored by other organizations when the topic is of keen interest to the chemistry audience.

Before agreeing to publish a book, the proposed table of contents is reviewed for appropriate and comprehensive coverage and for interest to the audience. Some papers may be excluded to better focus the book; others may be added to provide comprehensiveness. When appropriate, overview or introductory chapters are added. Drafts of chapters are peer-reviewed prior to final acceptance or rejection, and manuscripts are prepared in camera-ready format.

As a rule, only original research papers and original review papers are included in the volumes. Verbatim reproductions of previously published papers are not accepted.

ACS Books Department

Contents

Preface xi

New Synthetic Directions

1. **Trifluoromethylation of Metal Enolates and Theoretical Guideline** 2
   Koichi Mikami and Yoshimitsu Itoh

2. **Syntheses of Organofluorine Derivatives by a New Radical Process** 25
   Samir Z. Zard

3. **An Issue of Synthesis Strategy: To Make or Rather to Transform Organofluorine Compounds** 39
   Manfred Schlosser, Qian Wang, and Fabrice Cottet

4. **Nitrogen-Containing Organofluorine Compounds through Metathesis Reactions** 54
   Santos Fustero, Juan F. Sanz-Cervera, Carlos del Pozo, and José Luis Aceña

5. **Highly Diastereoselective Anodic Fluorination** 69
   Toshio Fuchigami and Toshiki Tajima

6. **A Stereospecific Entry to Functionalized *cis*-1,2-Difluoroalkenes** 83
   Anilkumar Raghavanpillai and Donald J. Burton

7. **Stereoselective Synthesis of Polyfluorinated *Exo*-Tricyclononenes and Norbornenes** 113
   Viacheslav A. Petrov

8. **Catalytic In-Situ Generation of Trifluoroacetaldehyde from Its Hemiacetal and Successive Direct Aldol Reaction with Ketones** 141
   Kazumasa Funabiki and Masaki Matsui

9. **Cationic Cyclizations of Fluoro Alkenes: Fluorine as *a Controller* and *an Activator* 155**
Junji Ichikawa

New Technologies and Materials

10. **A Hydro–Fluoro Hybrid Compound, Hermaphrodite, Is Fluorophilic or Fluorophobic? A New Concept for the Fluorine-Based Crystal Engineering 170**
T. Ono, H. Hayakawa, N. Yasuda, H. Uekusa, and Y. Ohashi

11. **Fluorous Nanoflow Microreactor: Nanoflow Microreactor with Fluorous Lanthanide Catalysts for Increase in Reactivity and Selectivity 190**
Koichi Mikami and Takayuki Tonoi

12. **Fluorous Chemistry for Synthesis and Purification of Biomolecules: Peptides, Oligosaccharides, Glycopeptides, and Oligonucleotides 207**
Wei Zhang

13. **Functional Compounds Based on Hypervalent Sulfur Fluorides 221**
Peer Kirsch and Gerd-Volker Röschenthaler

14. **Synthesis of Macrolactam Marine Natural Products Using Fluorous Protecting Groups 244**
Yutaka Nakamura and Seiji Takeuchi

15. **Carbohydrate Microarrays and Fluorous-Phase Synthesis: Interfacing Fluorous-Phase Tags with the Direct Formation of Glycoarrays 261**
Nicola L. Pohl

16. **Fluorous Carboxylates as Useful Metal Ligands for Catalysts–Products Recovery Procedures and Sensing 271**
Mounir El Bakkari and Jean-Marc Vincent

New Biological Applications

17. Organofluorine Chemistry at the Biomedical Interface: A Case Study on Fluoro-Taxoid Anticancer Agents 288
Iwao Ojima, Larissa V. Kuznetsova, and Liang Sun

18. Synthesis of *gem*-Difluoromethylated Sugar Nucleosides 305
Feng-Ling Qing and Xiao-Long Qiu

19. Polyfluoropyridyl Glycosyl Donors 323
Christopher A. Hargreaves, Graham Sandford, and Benjamin G. Davis

20. New 10-Trifluoromethyl Monomers, Dimers, and Chimeras of Artemisinin from a Key Allyl Bromide Precursor 337
Danièle Bonnet-Delpon, Jean-Pierre Bégué, Benoit Crousse, Fabienne Grellepois, Constance Chollet, Fatima Chorki, Guillaume Magueur, Michèle Ourévitch, Nguyen Thi Ngoc Tam, Philippe Grellier, Nguyen Van Hung, Truong Van Nhu, Doan Hanh Nhan, and Nguyen Thi Minh Thu

21. Synthesis of Novel *gem*-Difluorinatedcyclopropane Hybrids: Applications for Material and Medicinal Sciences 352
Toshiyuki Itoh

22. The Synthesis of an Antiviral Fluorinated Purine Nucleoside: 3′-α-Fluoro-2′, 3′-dideoxyguanosine 363
Kunisuke Izawa, Takayoshi Torii, Tomoyuki Onishi, and Tokumi Maruyama

23. Solvent-Peptide Interactions in Fluoroalcohol–Water Mixtures 379
C. Chatterjee, G. Hovagimyan, and J. T. Gerig

24. Fluorinated Methionines as Probes in Biological Chemistry 393
John F. Honek

25. Synthesis and Conformational Analysis of Fluorine-Containing Oligopeptides 409
Takashi Yamazaki, Takamasa Kitamoto, and Shunsuke Marubayashi

26. **Fluorinated Inhibitors of Matrix Metalloproteinases** 420
Fiorenza Viani, Luca Bruché, Gabriele Candiani, Florent Huguenot, Monika Jagodzinska, Raffaella Maffezzoni, Nathalie Moussier, Monica Sani, Roberta Sinisi, and Matteo Zanda

27. **Using Fluorinated Amino Acids for Structure Analysis of Membrane-Active Peptides by Solid-State ^{19}F-NMR** 431
Parvesh Wadhwani, Pierre Tremouilhac, Erik Strandberg, Sergii Afonin, Stephan Grage, Marco Ieronimo, Marina Berditsch, and Anne S. Ulrich

28. **Asymmetric Synthesis of β-Trifluoromethylated β-Amino Carbonyl Compounds Based on the 1,2-Addition to Trifluoroacetaldehyde SAMP- or RAMP-Hydrazones** 447
Kazumasa Funabiki, Masashi Nagamori, Masaki Matsui, Gerhard Raabe, and Dieter Enders

29. **A Novel Strategy for the Synthesis of Trifluoroalanine Oligopeptides** 462
Kenji Uneyama, Yong Guo, Kana Fujiwara, and Yumi Komatsu

30. **Hyperstable Collagen Based on 4-Fluoroproline Residues** 477
Ronald T. Raines

31. **Fluorinated Amino Acids and Reagents in Protein Design and Biomolecule Separation** 487
He Meng, Venkateshwarlu Kalsani, and Krishna Kumar

Indexes

Author Index 503

Subject Index 505

Preface

Selective fluorination of a lead compound can dramatically improve the performance of that material whether used in liquid crystals or pharmacologically active substances. The task confronting the synthetic chemist is how to construct such materials: whether to employ fluorinated building blocks or fluorination of a suitably functionalized precursor. The use of either strategy introduces a second conundrum: What are the effects of fluorination on reactivity and how can those effects be employed for a desirable result? While both these topics have been the subject of innumerable publications and symposia over the last half-century, the questions remain.

This symposium series volume is derived from three independent symposia held on organofluorine chemisty. The first of these, sponsored jointly by the American Chemical Society (ACS) Divisions of Fluorine Chemistry and Medicinal Chemistry, was entitled *New Frontiers of Fluoroorganic Chemistry* held at the 230th National ACS Meeting in Washington, D.C. (August 28 to September 1, 2005) and was organized by Vadim Soloshonok and Koichi Mikami. At the same occasion, a symposium entitled *Fluorous Chemistry* was organized by Dennis Curran. The third symposium, held at the International Chemical Congress of Pacific Basin Societies, in Honolulu, Hawaii (December 15–20, 2005) focused on *Fluorine-Containing Amino Acids: Preparation and Application to Biological Systems*, was convened by Takashi Yamazaki, John T. Welch, and John F. Honek. Both meetings were supported by the Petroleum Research Fund of the ACS and the ACS Division of Fluorine Chemistry. Additional financial support for the Honolulu meeting came from the Japanese Society of Fluorine Chemistry.

From the 55 presentations at these three symposia this volume includes 31 contributions. These chapters are organized into three groups; New Synthetic Directions, New Technologies, and New Biological Applications. The first section describes new methods for the incorporation of fluorinated building blocks and the application of electrochemical fluorination in synthesis. The section on new technologies presents several new applications of fluorous technology or the applications of extensively fluorinated compounds in preparative reactions. The final section has a very strong emphasis on fluorinated amino acids, peptides, and proteins but also discusses fluorinated anti-cancer and antiviral compounds in addition to carbohydrates and nucleosides. This volume, therefore, addresses in part the questions raised earlier.

It is our pleasure to acknowledge the contributions of all the speakers to the symposia and to the authors for their contributions to this volume. Organofluorine chemistry remains a challenging and exciting discipline and promises to remain so far into the future.

Vadim A. Soloshonok
Department of Chemistry and Biochemistry
University of Oklahoma
Norman, OK 73019–3051

Koichi Mikami
Department of Applied Chemistry
Tokyo Institute of Technology
Tokyo, Japan

Takashi Yamazaki
Department of Applied Chemistry
Tokyo University of Agriculture and Technology
Tokyo 152–8552, Japan

John T. Welch
Department of Chemistry
State University of New York at Albany
1400 Washington Avenue
Albany, NY 12222

John F. Honek
Department of Chemistry
University of Waterloo

Current Fluoroorganic Chemistry

New Synthetic Directions

Chapter 1

Trifluoromethylation of Metal Enolates and Theoretical Guideline

Koichi Mikami and Yoshimitsu Itoh

Department of Applied Chemistry, Tokyo Institute of Technology, Meguro-ku, Tokyo 152–8552, Japan

The difficulty as well as the significance of the direct generation of metal enolates of α-CF_3-ketones cannot be easily understood by chemists unfamiliar with fluorine. In sharp contrast to the great success in non-fluorine, hydrocarbon chemistry, fluorine chemistry has long been overshadowed. Metal enolates of carbonyl compounds are synthetically important in carbon-carbon bond forming reactions. However, the metal enolates of fluorinated carbonyl compounds have been severely limited to α-F metal enolates. α-CF_3 Metal enolates have generally been recognized as unstable and difficult to prepare because of the facile β-M-F elimination. However, we have developed a direct generation and synthetic application of α-CF_3 metal enolates. Therefore, the present results regarding the direct generation and synthetic use of metal enolates of α-CF_3-ketones might be recognized as a paradigm shift to the general use of metal enolates in fluorine chemistry.

Introduction

Organofluorine compounds have attracted much attention because of their rapidly increasing applications in the biological and material sciences (*1*). In particular, CF_3-compounds of specific absolute configurations are employed as novel lead analogues that exhibit particularly high physiological activity and remarkable physical properties (*2*). In sharp contrast to normal alkyl halides (R_h-X), however, perfluoroalkyl halides (R_f-X) can not react as a R_f^+ reagent (*eg.* nucleophilic alkylation) because the electronegativities of the perfluoroalkyl groups are higher than those of the halogens. The polarization of perfluoroalkyl halides is as $R_f^{\delta-}$-$X^{\delta+}$ and treatment with a nucleophile could not produce R_f-Nu (*3*).

Therefore, we report our research on the radical trifluoromethylation of metal enolates of ketones with trifluoromethyl iodide.

Direct Generation of Titanium Enolate of α-CF_3 Ketone

α-CF_3 metal enolates have generally been recognized to be unstable and difficult to prepare because of the rapid β-M-F elimination (Scheme 1) (*4*).

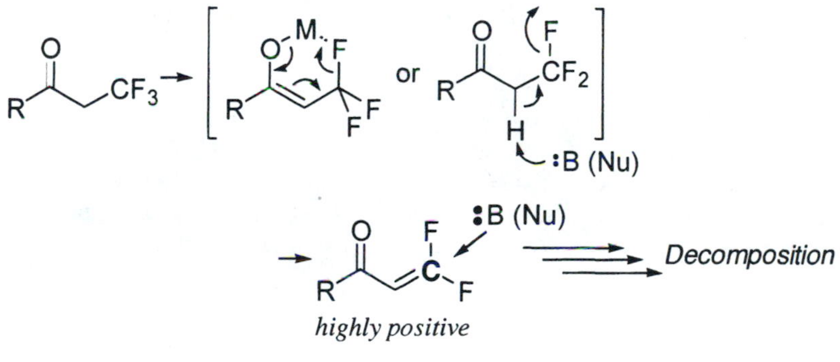

R CF3 O → [O M F R F F or O F CF2 R H :B (Nu)] → O F :B (Nu) R C F Decomposition *highly positive*

Scheme 1. Decomposition of α-CF_3 metal enolates.

We have already reported (*5*) that the enolate can be formed quantitatively by using $TiCl_4$ and Et_3N (Table 1) (*6*). α-Deuteriated ketone (**D-1a**) was obtained in quantitative yield using 35 wt% DCl/D_2O for quenching the enolate (run 2) (*7*). This result clearly shows that $TiCl_3$-enolates can be directly generated quantitatively from α-CF_3-ketones. In sharp contrast, the lithium enolate was formed only in low yield using LDA, and the total recovery of **1a** and **D-1a** was only 35% (run 3). This result presumes that defluorination took place in the case of lithium enolate (*8*).

Table 1. Formation of $TiCl_3$-enolate of α-CF_3 ketone.

$R = C(CH_3)_2Bn$ **1a** — Metal / Base, -78 °C / 30 min → [metal enolate, O–M] — D_2O, M = Ti → **D-1a**

run		% yield[a] 1a	D-1a
1	$TiCl_4$ (1.2) / Et_3N (1.4)	16	81
2[b]		trace	>99
3[c]	LDA (1.0)	31	4

[a] Determined by ^{19}F NMR. [b] 35wt% DCl / D_2O was used.
[c] THF was used as a solvent.

Theoretical Rationale for the Stability of the Titanium Enolate

According to crystallographic data, titanium, rather than lithium, has a strong affinity for fluorine (*9*). Therefore, it seems strange that titanium enolates of α-CF_3 ketones could be formed without defluorination. In order to clarify this anomaly, the structures of titanium and lithium enolates of α-CF_3-acetone were optimized at the B3LYP/631LAN (LANL2DZ for titanium, 6-31G* for others) level (*10, 11*). The optimized structures are shown in Figure 1.

In the lithium enolate, strong interaction between lithium and fluorine is clearly observed and it can be easily understood that defluorination readily takes place. In contrast, the Ti-O-C angle is almost straight (170.2°) and Ti-F interaction cannot be observed in the titanium enolate (*12*). The stability of the $TiCl_3$-enolate of α-CF_3 ketone can be explained by the linearity of the Ti-O-C bond, which suppresses any Ti-F interaction. The linearity of the Ti-O-C angle stems from the multiple bonding nature of the Ti-O bond. Donation of the lone electron pair of the oxygen to the empty d-orbital of titanium causes the multiple bonding. This is supported by the X-ray structure of titanium complexes and by computational studies (*13*).

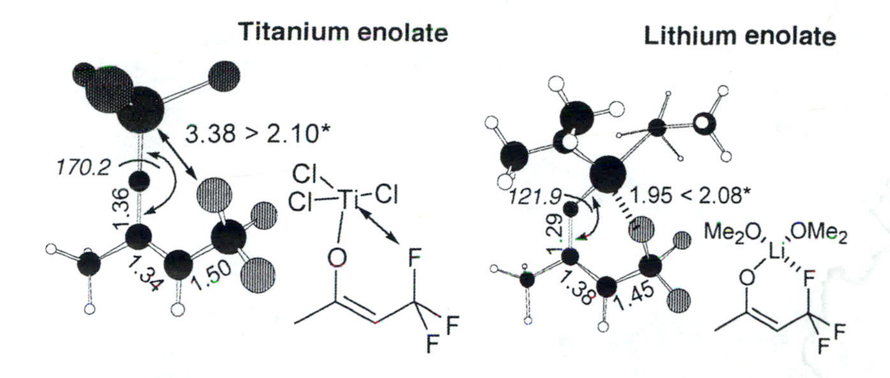

*Figure 1. 3D structures of titanium and lithium enolates optimized at the B3LYP/631LAN level, with the bond lengths in Å and the bond angles (italics) in degrees. * Sum of vdW radii of F and ionic radii of the corresponding metal.*

Radical Trifluoromethylation

Radical Trifluoromethylation of Titanium Ate Enolates

The synthesis of α-CF_3 carbonyl compounds itself has not been fully established. In sharp contrast to normal CH_3-X, CF_3-X cannot undergo alkylation of ketone enolates (14) because the electronegativity of the CF_3 group is higher than those of halogens (3). Therefore, radical trifluoromethylation of enolates is, in principle, one of the simplest ways to introduce a CF_3 unit at the α position of a carbonyl group. However, there are only a few examples especially in the case of ketone (15, 16, 17). It has been widely reported that the synthetic difficulty is due to defluorination of the α-CF_3-ketone product by the parent enolate or base during the reaction (16). However, we have succeeded in the efficient generation of titanium enolates of α-CF_3 ketones and high yielding aldol reactions (5). The stability of the titanium enolates of α-CF_3-ketones stems from the linearity of the Ti-O-C bonds caused by the donation of the lone electron pair of the oxygen to the empty d-orbital of titanium to suppress Ti-F interaction (9) and successive defluorination. Because titanium enolates of α-CF_3-ketones are stable to defluorination, it occurred to us that titanium enolates could be applied to radical trifluoromethylation for the synthesis of α-CF_3 ketones (Scheme 2) (18).

First, several titanium enolates of cyclohexanone were examined with CF_3 radical, which was generated by CF_3I (ca. 5 eq.) and Et_3B (1.0 eq.) (19). The reaction was carried out at -78 °C for 2 hours. The yields were determined by ^{19}F NMR using BTF (benzotrifluoride) as an internal standard (Scheme 3). In the case of $TiCl_3$-enolate (formed by $TiCl_4$ and Et_3N in CH_2Cl_2 at -78°C), no α-

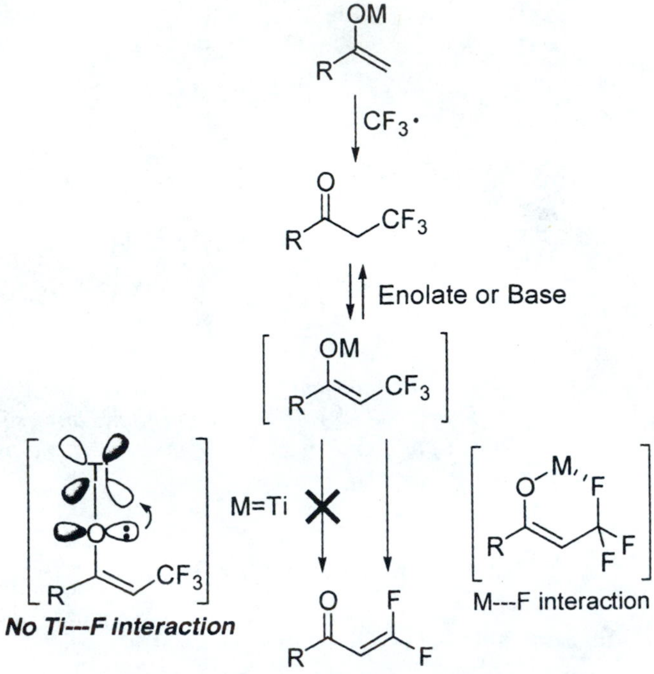

Scheme 2. Radical trifluoromethylation of titanium enolates.

CF_3-ketone (**1b**) was obtained. In the case of $Ti(O^iPr)_3$ enolate (formed by the addition of $Ti(O^iPr)_3Cl$ to the corresponding lithium enolate in THF at -78°C), α-CF_3 ketone (**1b**) was formed, however, in low yield (23%). In order to increase the reactivity of the enolate, the titanium ate [20, 21] enolate was examined. Titanium ate enolates could be easily formed just by adding $Ti(O^iPr)_4$ to lithium enolate at low temperature [20]. Upon treatment of titanium ate enolate (**3b**) with CF_3 radical, the α-CF_3-ketone was obtained in an increased yield (56%).

Therefore, radical trifluoromethylation of titanium ate enolate (**3b**) was further investigated and excess amounts of LDA and $Ti(O^iPr)_4$ were found to be important in increasing the yield (Table 2). When the enolate (**3b**) was formed by 1.0 eq. of LDA and 1.0 eq. of $Ti(O^iPr)_4$, the product (**1b**) was formed in only 56% yield (run 1). When 1.6 eq. of LDA and 1.6 eq. of $Ti(O^iPr)_4$ were used, the yield was increased up to 81% (run 3). Using 1.0 eq. of LDA and 1.6 eq. of $Ti(O^iPr)_4$ gave the α-CF_3 ketone (**1b**) in almost the same yield as in run 1 (52%, run 5). Therefore, excess amounts of both LDA and $Ti(O^iPr)_4$ should be employed.

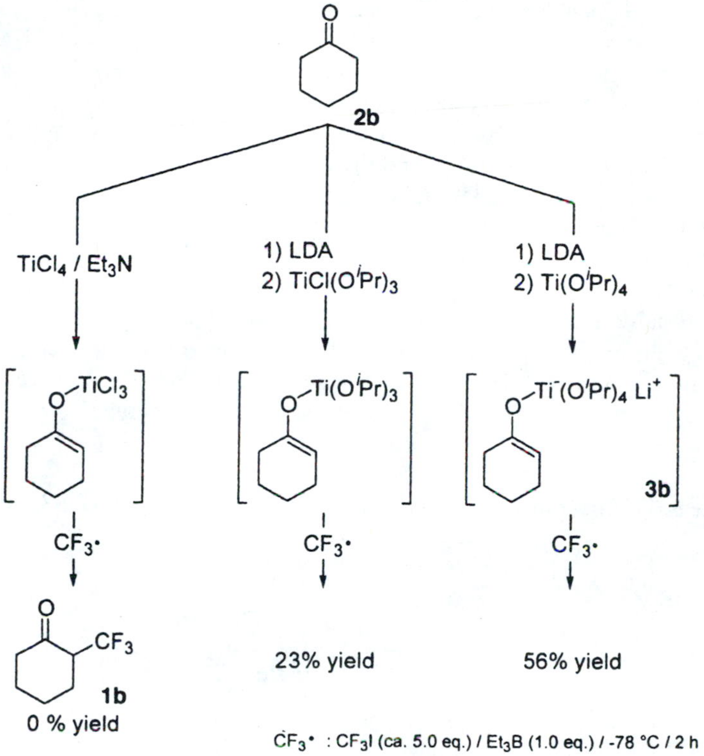

2b
TiCl4 / Et3N
1) LDA
2) TiCl(OiPr)3
1) LDA
2) Ti(OiPr)4
TiCl3
Ti(OiPr)3
Ti-(OiPr)4 Li+
3b
CF3•
CF3
1b
0 % yield
23% yield
56% yield
CF3• : CF3I (ca. 5.0 eq.) / Et3B (1.0 eq.) / -78 °C / 2 h

Scheme 3. Various titanium enolates for radical trifluoromethylation.

Table 2. Trifluoromethylation of titanium ate enolates.

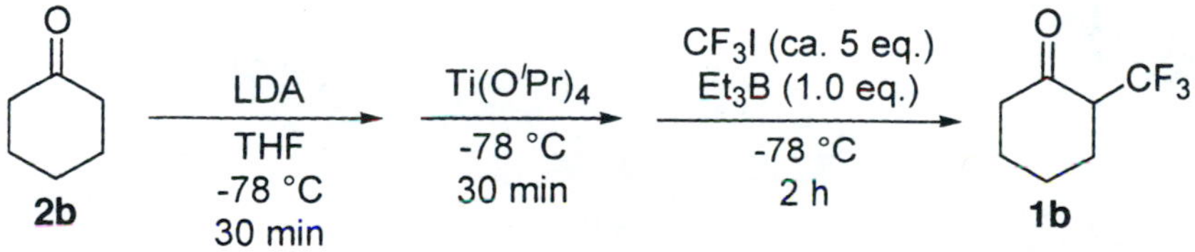

2b
LDA
THF
-78 °C
30 min
Ti(OiPr)4
-78 °C
30 min
CF3I (ca. 5 eq.)
Et3B (1.0 eq.)
-78 °C
2 h
CF3
1b

run	LDA (eq.)	$Ti(O^iPr)_4$ (eq.)	% yield[a]
1	1.0	1.0	56
2	1.3	1.3	72
3	1.6	1.6	81
4	2.0	2.0	80
5	1.0	1.6	52

[a] Determined by ^{19}F NMR using BTF as an internal standard.

The titanium ate enolate is prepared from the corresponding lithium enolate. When LDA was used for the preparation of lithium enolate, 1 eq. of $^{i}Pr_2NH$ was formed simultaneously. In order to investigate the effect of $^{i}Pr_2NH$, $^{n}BuLi$ was added to the corresponding silyl enol ether (22). By this silyl-to-lithium method, lithium enolate could be generated without the formation of $^{i}Pr_2NH$ and the amount of $^{i}Pr_2NH$ could be controlled at will (Table 3). When the reaction was carried out without the addition of $^{i}Pr_2NH$, the yields did not change significantly even by increasing the amount of $^{n}BuLi$ and/or $Ti(O^{i}Pr)_4$ (Table 3, runs 1-3). On the contrary, when three reagents ($^{n}BuLi$, $^{i}Pr_2NH$, and $Ti(O^{i}Pr)_4$) were used in 1.0 eq. each, the yield was decreased (run 4 (vs. run 1)). In the case that these three reagents were used in 1.6 eq., the yield was increased (run 5). Although the yields in runs 4 and 5 in Table 3 were slightly decreased than those in runs 1 and 3 in Table 2, a similar tendency was observed in the relationships of the yields and the amounts of reagents employed.

Table 3. Trifluoromethylation of titanium ate enolates starting from silyl enol ether.

OTMS
4b —[$^{n}BuLi$, THF, 0 °C, 20~30 min]→ —[R_2NH, 0 °C, 20 min]→ —[$Ti(O^{i}Pr)_4$, -78 °C, 30 min]→ —[CF_3I (ca. 5 eq.), Et_3B (1.0 eq.), -78 °C, 2 h]→ **1b** (CF_3)

run	$^{n}BuLi$ (eq.)	R_2NH (eq.)	$Ti(O^{i}Pr)_4$ (eq.)	% yield[a]
1	1.0	-	1.0	63
2	1.0	-	1.6	62
3	1.6	-	1.6	68
4	1.0	$^{i}Pr_2NH$ (1.0)	1.0	49
5	1.6	(1.6)	1.6	74
6	1.0	2,2,6,6-Me_4-piperidine (1.0)	1.0	57
7	1.6	(1.6)	1.6	72
8	1.0	Et_2NH (1.0)	1.0	6
9	1.6	(1.6)	1.6	11

[a] Determined by ^{19}F NMR using BTF as an internal standard.

From these results, the effect of $^{i}Pr_2NH$ can rationalized as follows (Scheme 4). $^{i}Pr_2NH$, which is formed by using LDA in the preparation of titanium ate enolate (**3b**), would exchange with the $^{-}O^{i}Pr$ ligand to form $^{i}PrOH$. $^{i}PrOH$ would protonate the enolate to form the parent ketone and the titanium amide complex (LDA/$Ti(O^{i}Pr)_4$) (23). This mechanism rationalizes not only the decreased yield upon the addition of $^{i}Pr_2NH$ (Table 3, run 4) (protonation of the enolate to reduce the amount of the reactive enolate species) but also the increased yield using an excess amount of LDA and $Ti(O^{i}Pr)_4$ (the equilibrium shifts to titanium ate enolate). In order to support the proposal, 2,2,6,6-tetramethylpiperidine and Et_2NH were investigated. 2,2,6,6-Tetramethylpiperidine is bulkier than $^{i}Pr_2NH$ and its coordinating ability is lower than $^{i}Pr_2NH$. Thus, it could be expected that the proposed equilibrium (Scheme 4) would shift to the titanium ate enolate (**3b**). In fact, when 2,2,6,6-tetramethylpiperidine was used in 1.0 eq. (run 6), the decreased yield was not so significant relative to $^{i}Pr_2NH$ (run 4), and, when 2,2,6,6-tetramethylpiperidine was used in 1.6 eq. (run 7), the yield increased as with $^{i}Pr_2NH$ (run 5). In contrast, Et_2NH is less bulky than $^{i}Pr_2NH$ and its coordination ability is higher than $^{i}Pr_2NH$. Thus, it could be expected that the proposed equilibrium (Scheme 4) would shift to the ketone (**2b**). In fact, when Et_2NH was used in 1.0 eq. (run 8) and 1.6 eq. (run 9), the yield decreased. It should be noted that the LDA/$Ti(O^{i}Pr)_4$ complex, which can act as a base to decompose the α-CF_3 product, works not for decreasing the yield but for increasing the yield of the α-CF_3 products.

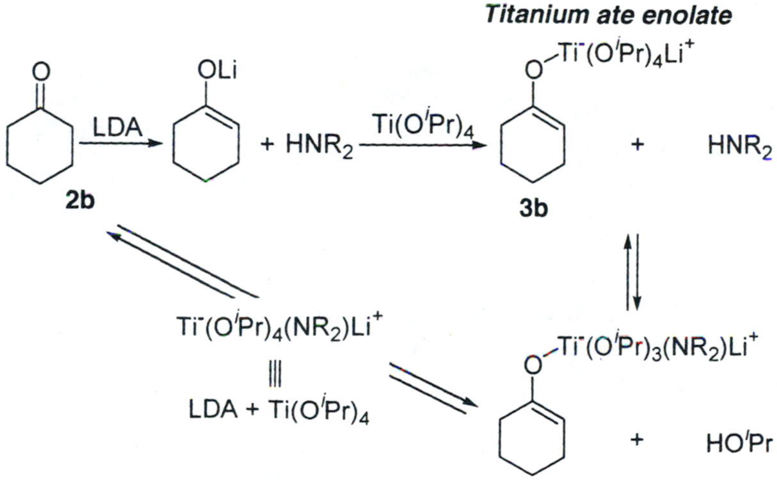

Scheme 4. Equilibrium between the titanium ate enolate and the parent ketone.

Several ketonic substrates were investigated (Table 4). Although cyclohexanone gave the α-CF_3 product (**1b**) in 81% yield with ketone:LDA:$Ti(O^iPr)_4$=1.0:1.6:1.6 conditions (run 2), cyclopentanone gave poor yield (**1c**) with both ketone:LDA:$Ti(O^iPr)_4$=1.0:1.0:1.0 and =1.0:1.6:1.6 conditions (38% in run 3 and 33% in run 4). In the case of cycloheptanone, the α-CF_3 product (**1d**) was obtained in good yield (61%) by ketone:LDA:$Ti(O^iPr)_4$

Table 4. Trifluoromethylation of various titanium ate enolates.

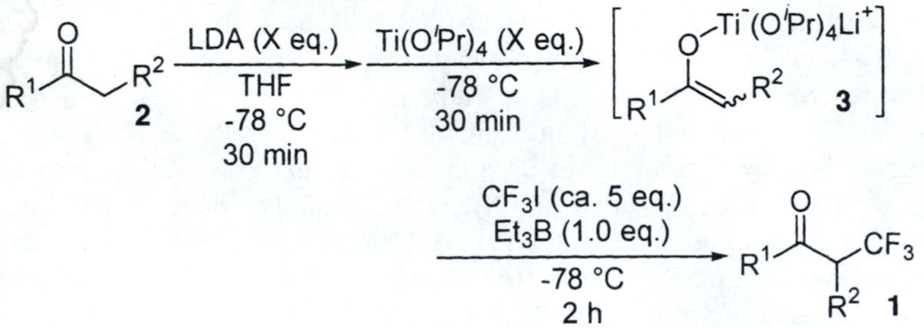

$R^1C(O)CH_2R^2$ (**2**) → LDA (X eq.), THF, -78 °C, 30 min → $Ti(O^iPr)_4$ (X eq.), -78 °C, 30 min → [$R^1C(O^-Ti(O^iPr)_4Li^+)=CHR^2$] (**3**) → CF_3I (ca. 5 eq.), Et_3B (1.0 eq.), -78 °C, 2 h → $R^1C(O)CH(R^2)CF_3$ (**1**)

run	product	X (eq.)	% yield[a]
1	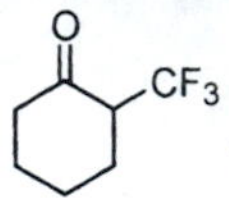 **1b**	1.0	63
2	**1b**	1.6	81
3	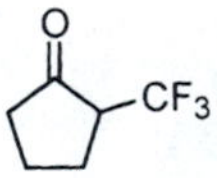 **1c**	1.0	38
4	**1c**	1.6	33
5	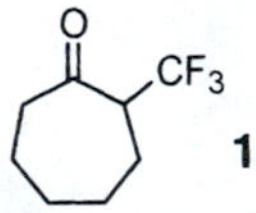 **1d**	1.0	49
6	**1d**	1.6	61
7	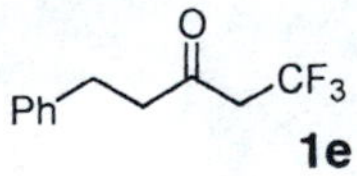 **1e**	1.0	38
8	**1e**	1.6	14
9	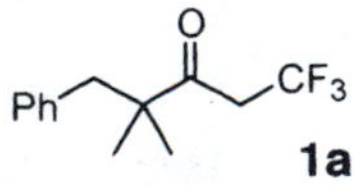 **1a**	1.0	(50)
10	**1a**	1.6	69 (65)
11	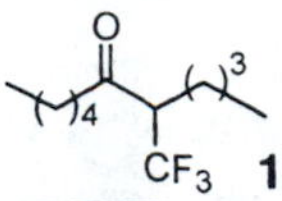 **1f**	1.0	32
12	**1f**	1.6	64

[a] Determined by ^{19}F NMR using BTF as an internal standard
The number in parentheses refer to the yield of the isolated products.

=1.0:1.6:1.6 condition (run 6). For acyclic substrates, **1a** and **1f** gave the products in good yield (65 and 64% in runs 10 and 12, respectively). The yield of **1e** was poor in the case that ketone:LDA:Ti(O^{i}Pr)$_4$=1.0:1.6:1.6 (run 8, 14%). Interestingly however, when the number of equivalents were decreased (ketone:LDA:Ti(O^{i}Pr)$_4$=1.0:1.0:1.0), the result was better (38%, run 7) than that under 1.0:1.6:1.6 conditions for this acyclic substrates, yet the reason is still unclear. The yields of **1e** (runs 7 and 8) were decreased compared to **1a** (runs 8 and 10), which has similar structure to **1e**. Steric bulkiness could account for the difference since less hindered product (**1e**) could easily be enolized leading to the decomposition. When acetophenone, ester, and amide were used as substrates, the α-CF_3 product was not obtained at all.

Table 5. Trifluoromethylation of various titanium ate enolates prepared from silyl enol ether.

Scheme: **4** (R¹C(OTMS)=CR²R³) → ^{n}BuLi (X eq.), THF, 0 °C, 20~30 min → ^{i}Pr$_2$NH (X eq.), 0 °C, 20 min → Ti(O^{i}Pr)$_4$ (X eq.), -78 °C, 30 min → [R¹C(O–Ti$^-$(O^{i}Pr)$_4$Li$^+$)=CR²R³] **3** → CF_3I (ca. 5 eq.), Et_3B (1.0 eq.), -78 °C, 2 h → R¹C(=O)C(R²)(R³)CF_3 **1**

run	substrate	product	X (eq.)	% yield[a]
1	4b	1b	1.0	49
2			1.6	74
3	4g	1g	1.0	30
4			1.6	42
5	4h	1h	1.0	32 (35)
6			1.6	44 (43)

[a] Determined by ^{19}F NMR using BTF as an internal standard.
The number in parentheses refer to the yields of the isolated products.

Although LDA could generate only the kinetic lithium enolate, the thermodynamic enolate could also be prepared from silyl enol ethers. Therefore, the thermodynamic titanium ate enolate of an α-substituted ketone could be generated by the silyl-to-lithium transmetalation method to construct a quaternary carbon center attached with the CF_3 substituent. In the case of α-Me (**4g**) and α-Ph (**4h**) substituted cyclohexanones, the products were obtained in reasonable yields (42 and 43% yield, respectively) (Table 5).

Radical Trifluoromethylation of Lithium Enolates

During the course of our exploration of the radical trifluoromethylation of titanium ate enolates, we discovered that the lithium enolate could in fact be applied to radical trifluoromethylation and that the reaction proceeded at an extremely rapid reaction rate.

Table 6. Radical trifluoromethylation of lithium enolate.

O
2b
LDA
THF
-78 °C
30 min
O-Li
5b
$Ti(O^iPr)_4$
-78 °C
30 min
$O-Ti^-(O^iPr)_4Li^+$
3b
CF_3I (ca. 5 eq.)
Et_3B (1.0 eq.)
-78 °C
2 h
O
CF_3
1b

run	LDA (eq.)	$Ti(O^iPr)_4$ (eq.)	% yield[a]
1	1.6	1.6	81
2	1.6	0	41
3	1.0	0	63

[a] Determined by ^{19}F NMR using BTF as an internal standard.

Radical trifluoromethylation of titanium ate enolate gave 81% of α-CF_3-cyclohexanone (**1b**) (Table 6, run 1). However, the yield greatly decreased without $Ti(O^iPr)_4$ (run 2). This might be caused by the decomposition of the α-

CF_3 product (**1b**) by an excess amount of LDA. Surprisingly, the use of just one equivalent of LDA gave a 63% yield of the α-CF_3 product. This indicates that the lithium enolate (**5b**) itself is less likely to rapidly decompose the α-CF_3 product. Therefore, we further investigated the radical trifluoromethylation of the lithium enolate.

First, the preparation time of the lithium enolate (**5b**) was investigated (Table 7). In the case of the trifluoromethylation of titanium ate enolate, the preparation time of the lithium enolate (**5b**), which is a precursor of the titanium ate enolate, took only 30 min. However, in the case of the trifluoromethylation of lithium enolate (**5b**), a preparation time of 60 min was needed to attain sufficient yield (run 2). A preparation time of more than 1 h was not necessary (run 3). The titanium ate enolate (**3b**) is in equilibrium between the ate enolate and the ate complex (LDA/$Ti(O^iPr)_4$) with the ketone (**2b**) (Scheme 4). Therefore, even the enolization was not completed in 30 min, enolization by titanium ate complex (LDA/$Ti(O^iPr)_4$) would proceed under the reaction conditions to compensate the incomplete enolization by LDA.

Table 7. Preparation time of lithium enolate.

2b → LDA (1.0 eq.), THF, -78 °C, X min → [5b (O-Li enolate)] → CF_3I (ca. 5 eq.), Et_3B (1.0 eq.), -78 °C, 2 h → 1b (CF_3)

run	X (min)	% yield[a]
1	30	63
2	60	73
3	120	72

[a] Determined by ^{19}F NMR using BTF as an internal standard.

Next, the radical reaction time was investigated using the lithium enolate (**5b**) prepared over 60 min (Table 8). When the reaction was carried out for 13 h, the yield was 62% (run 1). A shorter reaction time increased the yield; the α-CF_3 product (**1b**) was obtained in 80% yield when the reaction was quenched at 1 h (run 3). Surprisingly, the yield did not change when the reaction time was shortened (run 4-7). Finally, ~1 sec of the reaction time gave the α-CF_3 product (**1b**) in 81% yield (run 7). Compared with the radical trifluoromethylation of titanium ate enolate (Table 6, run 1), which took 2 h to give the same yield, the reaction of lithium enolate is extremely fast!

Table 8. Investigation of reaction time.

2b → LDA (1.0 eq.), THF, -78 °C, 60 min → [5b] → CF_3I (ca. 5 eq.), Et_3B (1.0 eq.)[a], -78 °C → 1b

run	reaction time	% yield[b]
1	13 h	62
2	2 h	73
3	1 h	80
4	15 min	81
5	1 min	83
6	15 sec	82
7	~1 sec	81

[a] Et_3B was added at the time of 15 sec.
[b] Determined by ^{19}F NMR using BTF as an internal standard.

A variety of ketonic substrates were investigated using LDA to generate lithium enolate (Table 9). In the case of cyclohexanone (run 1), and 4-^{t}Bu (run 2), 2-Me (run 3), and 2-Ph (run 4) cyclohexanones, the reactions proceeded with extremely fast reaction rates and provided the α-CF_3 products in fair to good yields. The reaction rates of cyclopentanone (run 5) and cycloheptanone (run 6) were relatively slow (5 min). For acyclic substrates, the yield was only poor (ca. 20~30% yield). Ester and amide were also investigated but did not give the α-CF_3 product at all. From the results described above, cyclohexanone derivatives are the most suitable substrate for this reaction system.

Table 9. Trifluoromethylation of various lithium enolates.

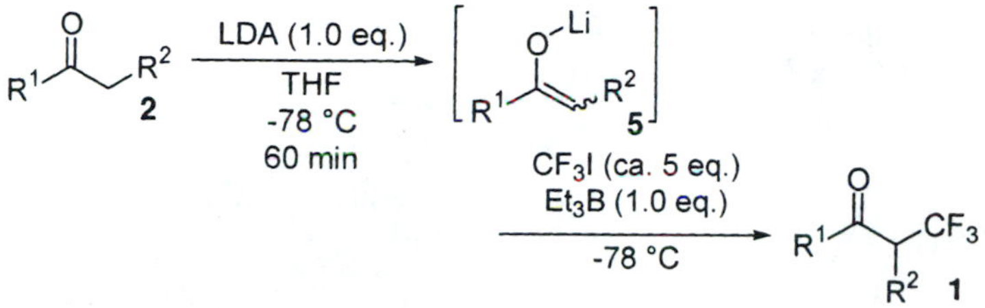

$R^1C(O)CH_2R^2$ **2** → LDA (1.0 eq.), THF, -78 °C, 60 min → [lithium enolate **5**] → CF_3I (ca. 5 eq.), Et_3B (1.0 eq.), -78 °C → $R^1C(O)CH(R^2)CF_3$ **1**

run	substrate	product	reaction time	% yield[b]
1	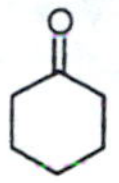 **2b**	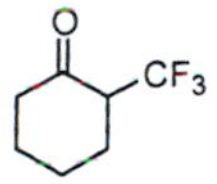 **1b**	~1 s	81
2	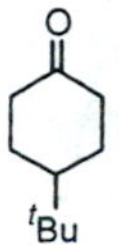 **2i**	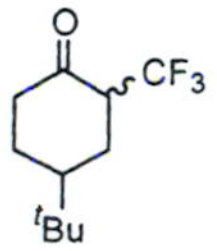 **1i**	~1 s	71 (67) [73:27]
3	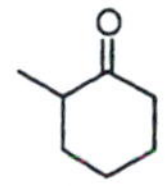 **2j**	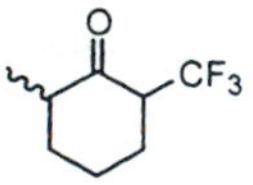 **1j**	~1 s	74 [57:43]
4	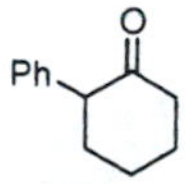 **2k**	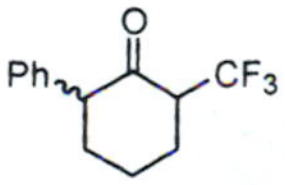 **1k**	~1 s	43 (40) [57:43]
5	 **2c**	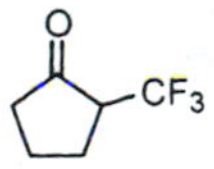 **1c**	5 min	40
6	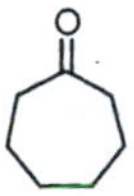 **2d**	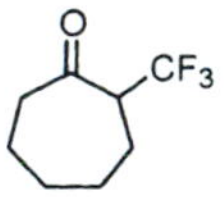 **1d**	5 min	48

[a] Et_3B was added in flat 15 s

[b] Determined by ^{19}F NMR using BTF as an inernal standard. The values in ( ) refer to the yields of isolated products The values in [ ] are the diastereomeric ratio and determined by ^{19}F NMR

As mentioned before, the thermodynamic lithium enolate could be generated by treatment of the corresponding silyl enol ether with 1.0 eq. of ^{n}BuLi to give the regioisomeric α-CF_3 product. Lithium enolates prepared by the silyl-to-lithium transmetallation method were also investigated for the most suitable cyclohexanone derivatives (Table 10). α-Me and α-Ph substituted substrates provided the products, which bear quaternary carbon centers attached with CF_3, in fair yields (runs 2 and 3).

Table 10. Trifluoromethylation of various lithium enolates prepared from silyl enol ether.

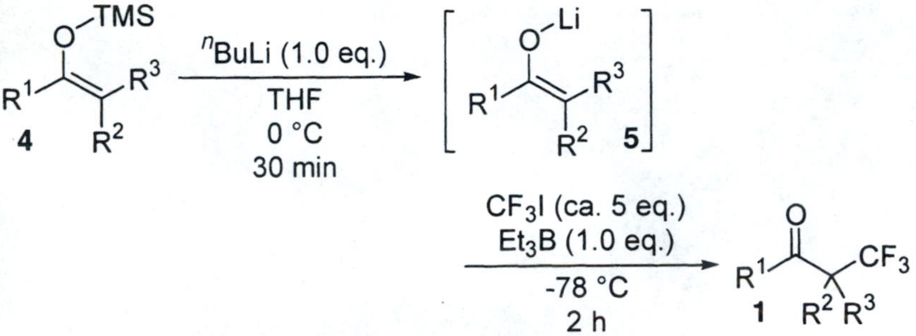
TMS
R1
R3
4
R2
nBuLi (1.0 eq.)
THF
0 °C
30 min
Li
R1
R3
R2
5
CF3I (ca. 5 eq.)
Et3B (1.0 eq.)
-78 °C
2 h
R1
CF3
1
R2 R3

run	substrate	product	% yield[b]
1	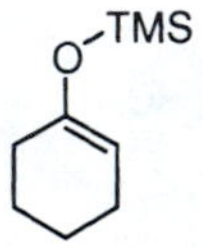 **4b**	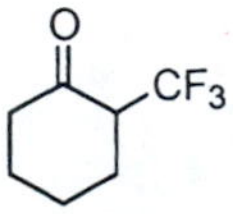 **1b**	77
2[c]	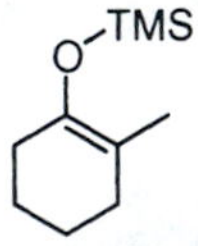 **4g**	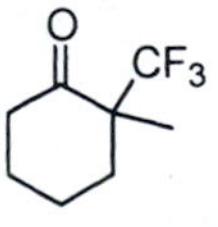 **1g**	58
3[d]	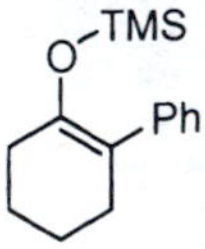 **4h**	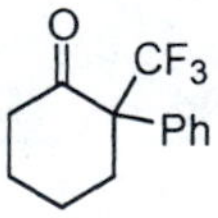 **1h**	45 (45)

[a] Et_3B was added in flat 15 s.
[b] Determined by ^{19}F NMR using BTF as an inernal standard. The values in ( ) refer to the yields of isolated products.
[c] Silyl enol ether of α -Me cyclohexanone consists of thermodynamic and kinetic enol ethers (87 : 13).
[d] Silyl enol ether of α -Ph cyclohexanone contains only thermodynamic enol ether.

Radical Reaction Mechanism

In investigating the effect of the amount of radical initiator (Et_3B) on yields of **1b**, an interesting result was obtained. The relationship between the amount of Et_3B and the product yield is shown in Figure 2. In the case of titanium ate enolates, the yields gradually decreased as the amount of Et_3B decreased (ketone:LDA:$Ti(O^iPr)_4$=1.0:1.0:1.0 and =1.0:1.6:1.6). In sharp contrast, this is not the case with lithium enolates; α-CF_3 product could be obtained in good yield (71%) even when the amount of Et_3B was reduced to 5 mol%. This indicates that the radical cycle does not work well for titanium ate enolate but does works well for lithium enolate.

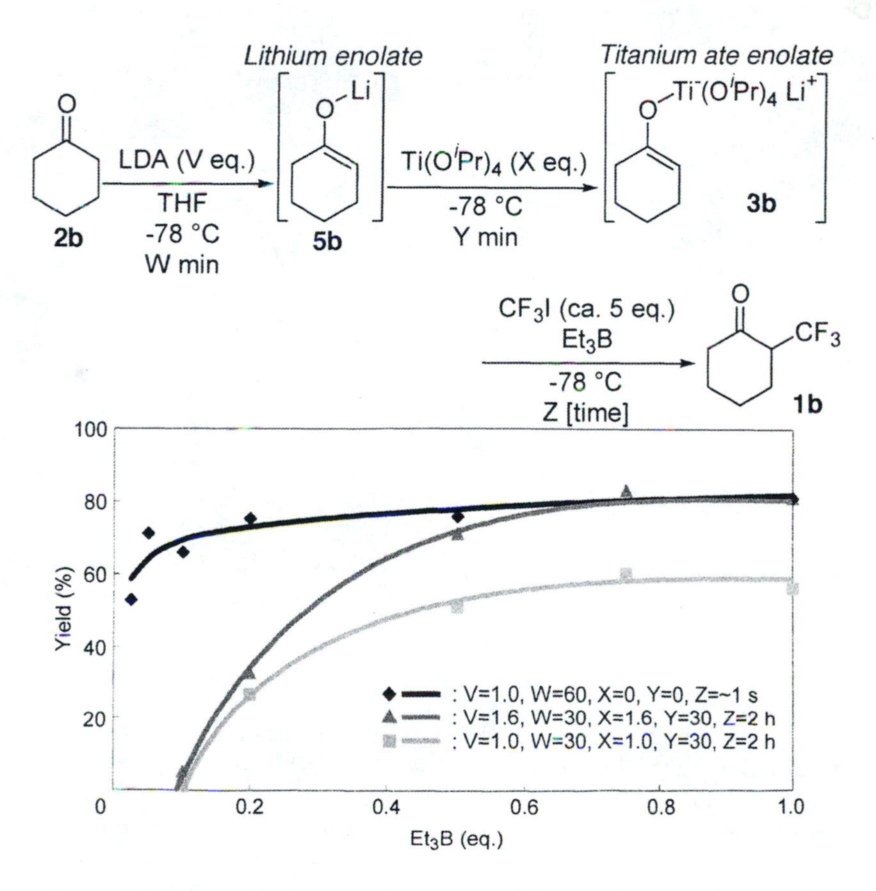

Figure 2. Relationship between the amount of Et_3B and the yields of α-CF_3 ketones.

Considering the fact that titanium has two stable oxidation states (III and IV), two reaction mechanisms could be proposed (Figure 3). First, CF_3 radical reacts with enolate (step (1)). From the radical adduct (**6**), one possibility is Path A; elimination of metal from ketyl radical intermediate (**6**) by reducing its oxidation number by one along with the formation of the α-CF_3 product (step (2)). Regeneration of CF_3 radical can be achieved by the reaction of CF_3I with metal (step (3)). Another possibility is Path B; ketyl radical intermediate (**6**) directly reacts with CF_3I to regenerate CF_3 radical along with α-CF_3 product.

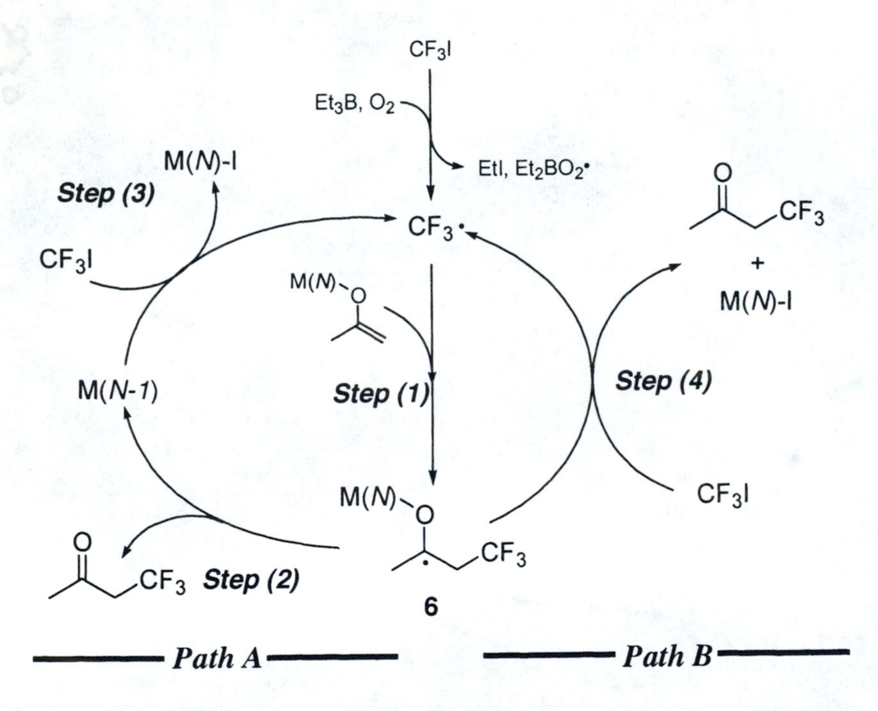

Figure 3. Proposed CF_3 radical addition mechanism.

A computational study was carried out to evaluate the reaction mechanism and to uncover the origin of the difference between titanium ate and lithium enolates (*24*). All the calculations were performed with the GAUSSIAN 03 program package (*10*). All the structures were optimized at UB3LYP/631+LAN (LANL2DZ for titanium, I, 6-31+G* for others) level (*11*) Then the energies were recalculated in the presence of polarizable dielectric (THF, ε=7.58) as described by the CPCM model (*25*) at

UB3LYP/6311+LAN (LANL2DZ for titanium, I, 6-311+G* for others) level. The energies shown in this report are Gibbs free energies and thus contain zero-point, thermal and entropy effects at 195.15 K (experimental reaction temperature) and 1 atm pressure.

Energy diagram of the reaction mechanism is shown in Figures 4, and 5 and schematic reaction path is shown in Figure 6. In the case of titanium ate enolate (Figure 4), CF_3 radical addition step (step (1)) does not have significant reaction barrier and thus readily takes place. Ti(IV) ketyl radical intermediate could proceed in either step (2) or step (4). For step (2), the reaction barrier is very low (3.8 kcal/mol). Ti(III) species, which was generated after step (2), could react as an one electron reductant to CF_3I but can not abstract I^- from the radical anion of CF_3I to regenerate CF_3 radical. Even when Ti(IV) ketyl radical proceeds to step (4) to provide α-CF_3 product, CF_3 radical could not be regenerated in step (3).

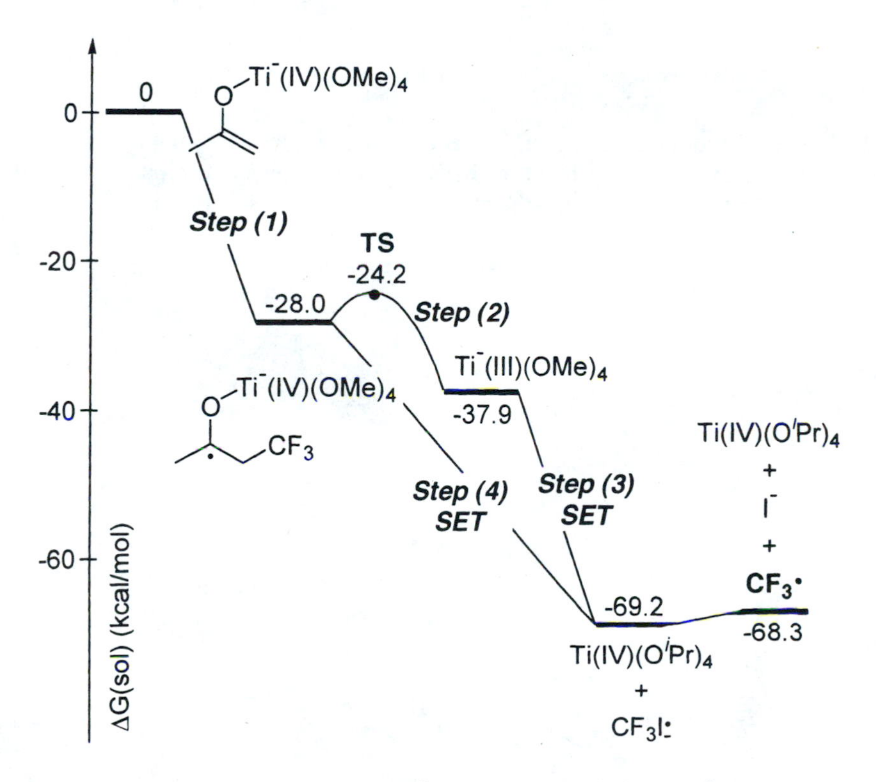

Figure 4. Energy diagram of the radical trifluoromethylation of titanium ate enolate.

In the case of lithium enolate (Figure 5), CF_3 radical addition step (step (1)) is not of significant reaction barrier. The difference between the reactions of titanium ate enolate and lithium enolate could be found at step (2). Although the reaction barrier of step (2) of Ti(IV) ketyl radical is very low (3.8 kcal/mol), lithium ketyl radical is of high reaction barrier (at least 27.2 kcal/mol) and thus lithium ketyl radical would not proceed to step (2) but to (4). Another difference between titanium and lithium was observed at step (4). Lithium ketyl radical could react with CF_3I to produce α-CF_3 product and Li^+ species. Li^+ could abstract I^- from radical anion of CF_3I to regenerate CF_3 radical (SET: single electron transfer). Another possibility is AT (atom transfer) which could regenerate CF_3 radical in one step. Effective regeneration of CF_3 radical is based on exothermic bond formation of Li-I in either (SET or AT) case. The difference between titanium ate enolate and lithium enolate in Figure 6 is attributed to the bond forming ability of the metal species with I^- to regenerate CF_3 radical.

Conclusion

We have developed the direct generation and synthetic application of α-CF_3 metal enolates. The present methods can be recognized as a paradigm shift to the direct radical trifluoromethylation of metal enolates in fluorine chemistry.

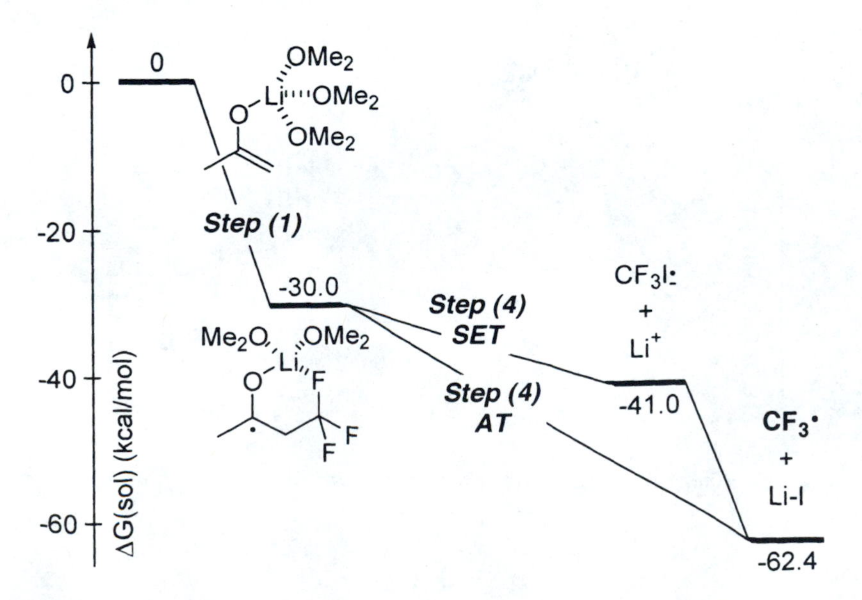

Figure 5. Energy diagram of the radical trifluoromethylation of lithium enolate.

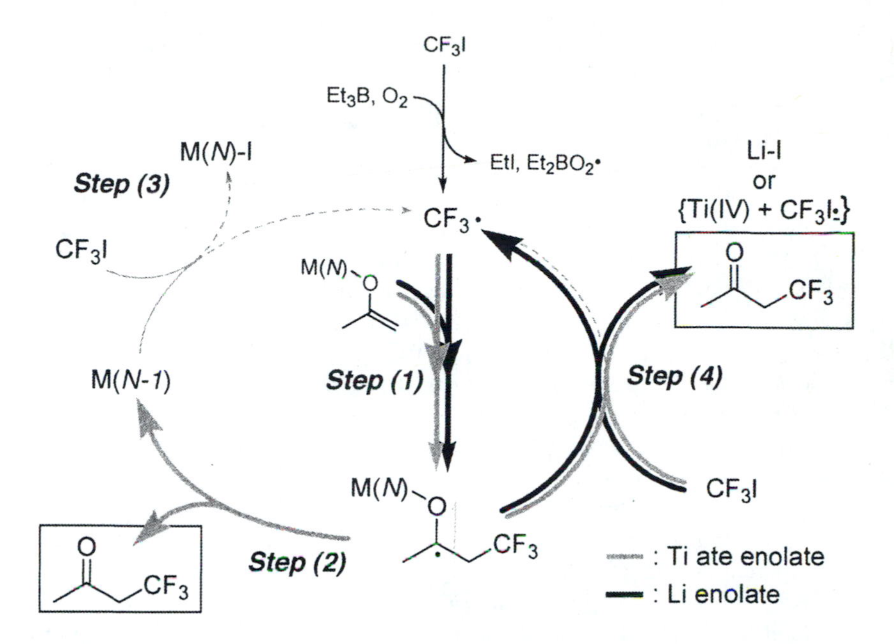

Figure 6. Reaction mechanism of radical trifluoromethylation of titanium ate and lithium enolates.

α-CF_3 carbonyl compounds can be widely employed to synthesize organofluorine compounds because of their important applications in biological and material sciences. Further application to the "enantioselective" synthesis of CF_3 compounds (*1 i*) of specific absolute configurations should emerge to provide highly enantio-enriched organofluorine compounds with particularly high physiological activity and remarkable physical properties.

References

1 a) Shimizu, M.; Hiyama, T. *Angew. Chem. Int. Ed.* **2005**, *44*, 214. b) Ma, J.-A.; Cahard, D. *Chem. Rev.* **2004**, *104*, 6119. c) Mikami, K.; Itoh, Y.; Yamanaka, M. *Chem. Rev.* **2004**, *104*, 1. d) Mikami, K.; Itoh, Y.; Yamanaka, M. *Fine Chemical* **2003**, *32(1)*, 35. e) Mikami, K.; Itoh, Y.; Yamanaka, M. *Fine Chemical* **2003**, *32(2)*, 11. f) Hiyama, T.; Kanie, K.; Kusumoto, T.; Morizawa, Y.; Shimizu, M. *Organofluorine Compounds*; Springer-Verlag: Berlin Heidelberg, 2000. g) Soloshonok, V. A. Ed. *Enantiocontrolled Synthesis of Fluoro-Organic Compounds*; Wiley: Chichester, 1999. h) Chambers, R. D. Ed. *Organofluorine Chemistry*, Springer, Berlin 1997. i) Iseki, K. *Tetrahedron* **1998**, *54*, 13887. j) Ojima, I.; McCarthy, J. R.; Welch, J. T. Eds. *Biomedical Frontiers of Fluorine*

Chemistry; American Chemical Society: Washington DC, 1996. k) Smart, B. E., Ed. *Chem. Rev.* **1996**, *96*, No. 5 (Thematic issue of fluorine chemistry). l) Banks, R. E.; Smart, B. E.; Tatlow, J. C. Eds. *Organofluorine Chemistry: Principles and Commercial Applications*; Plenum Press: New York. 1994. m) Kitazume, T.; Ishihara, T.; Taguchi, T. *Chemistry of Fluorine*; Koudansha: Tokyo, 1993. n) Ishikawa, N. Ed. *Synthesis and Reactivity of Fluorocompounds*; CMC: Tokyo, Vol. 3, 1987. o) Ishikawa, N.; Kobayashi, Y. *Fluorine Compounds*; Koudansha: Tokyo, 1979. p) Hudlicky, M. *Chemistry of Organic Fluorine Compounds, 2nd edn*; Ellis Horwood: Chichester, 1976.

2 a) Smart, B. E. *J. Fluorine Chem.* **2001**, *109*, 3. b) Schlosser, M. *Angew. Chem. Int. Ed.* **1998**, *37*, 1496. c) O'Hagan, D.; Rzepa, H. S. *Chem. Commun.* **1997**, 645.

3 a) Huheey, J. E. *J. Phys. Chem.* **1965**, *69*, 3284. b) Banus, J.; Emeléus, H. J.; Haszeldine, R. N. *J. Chem. Soc.* **1951**, 60. c) Kobayashi, Y.; Yoshida, T.; Kumadaki, I. *Tetrahedron Lett.* **1979**, 3865.

4 Only limited example of covalently bonded α-CF_3 metal enolate are reported. Al enolate of ketone: Ishihara, T.; Kuroboshi, M.; Yamaguchi, K.; Okada, Y. *J. Org. Chem.* **1990**, *55*, 3107. B enolate of amide: a) Ishihara, T.; Kuroboshi, M.; Yamaguchi, K. *Chem. Lett.* **1990**, 211. b) Kuroboshi, M.; Ishihara, T. *Bull. Chem. Soc. Jpn.* **1990**, *63*, 1191.

5 Itoh, Y.; Yamanaka, M.; Mikami, K. *J. Am. Chem. Soc.* **2004**, *126*, 13174.

6 a) Harrison, C. R. *Tetrahedron Lett.* **1987**, *28*, 4135. b) Evans, D. A.; Clark, J. S.; Metternich, Novack, V. J.; Sheppard, G. S. *J. Am. Chem. Soc.* **1990**, *112*, 866. c) Evans, D. A.; Urpí, F.; Somers, T. C.; Clark, J. S.; Bilodeau, M. T. *J. Am. Chem. Soc.* **1990**, *112*, 8215. d) Evans, D. A.; Rieger, D. L.; Bilodeau, M. T.; Urpí, F. *J. Am. Chem. Soc.* **1991**, *113*, 1047. e) Esteve, C.; Ferreró, M.; Romea, P.; Urpí, F.; Vilarrasa, J. *Tetrahedron Lett.* **1999**, *40*, 5079. f) Tanabe, Y.; Matsumoto, N.; Higashi, T.; Misaki, T.; Itoh, T.; Yamamoto, M.; Mitarai, K.; Nishii, Y. *Tetrahedron* **2002**, *58*, 8269. g) Solsona, J. G.; Romea P.; Urpí, F.; Vilarrasa, J. *Org. Lett.* **2003**, *5*, 519.

7 When the enolate was quenched with D_2O, the yield of **2a** was only 81%. This is presumably due to the protonation by HCl salt of Et_3N (formed as a byproduct of enolate formation).

8 Defluorination took place when the $TiCl_3$-enolate of α-CF_3 ketone was warmed above -78 °C. Therefore, low temperature reaction is essential.

9 a) Plenio, H. *Chem. Rev.* **1997**, *97*, 3363. b) Murphy, E. F.; Murugavel, R.; Roesky, H. W. *Chem. Rev.* **1997**, *97*, 3425. c) Murray-Rust, P.; Stallings, W. C.; Monti, C. T.; Preston, R. K.; Glusker, J. P. *J. Am. Chem. Soc.* **1983**, *105*, 3206. d) Shimoni, L.; Glusker, J. P.; Bock, C. W. *J. Phys. Chem.* **1995**, *99*, 1194. e) Shibakami, M.; Sekiya, A. *Bull. Chem. Soc. Jpn.* **1993**, *66*, 315. f) Fernandez-G., J. M.; Rodriguez-Romero, A.; Panneerselvam, K.; Soriano-Garcia, M. *Acta. Crystallogr. Sect. C* **1995**, *51*, 1643. g) Murphy, E. F.; Lubben, T.; Herzog, A.; Roesky, H. W.; Demsar, A.; Noltemeyer, M.; Schmidt, H.-G. *Inorg. Chem.* **1996**, *35*, 23. h) Murphy, E. F.; Yu, P.–H.; Dietrich, S.; Roesky, H. W.; Parisini, E.; Noltemeyer, M. *J.*

Chem. Soc., Dalton Trans. **1996**, 1983. i) Herzog, A.; Roesky, H. W.; Jäger, F.; Steiner, A. *Chem. Commun.* **1996**, 29. j) Shah, S. A. A.; Dorn, H.; Voigt, A.; Roesky, H. W.; Parisini, E.; Schmidt, H.–G.; Noltemeyer, M. *Organometallics* **1996**, *15*, 3176.

10 All calculations were performed using the Gaussian 98 and 03 program package. Frisch, M. J.; Trucks, G. W.; Schlegel, H. B.; Scuseria, G. E.; Robb, M. A.; Cheeseman, J. R.; Montgomery, Jr., J. A.; Vreven, T.; Kudin, K. N.; Burant, J. C.; Millam, J. M.; Iyengar, S. S.; Tomasi, J.; Barone, V.; Mennucci, B.; Cossi, M.; Scalmani, G.; Rega, N.; Petersson, G. A.; Nakatsuji, H.; Hada, M.; Ehara, M.; Toyota, K.; Fukuda, R.; Hasegawa, J.; Ishida, M.; Nakajima, T.; Honda, Y.; Kitao, O.; Nakai, H.; Klene, M.; Li, X.; Knox, J. E.; Hratchian, H. P.; Cross, J. B.; Bakken, V.; Adamo, C.; Jaramillo, J.; Gomperts, R.; Stratmann, R. E.; Yazyev, O.; Austin, A. J.; Cammi, R.; Pomelli, C.; Ochterski, J. W.; Ayala, P. Y.; Morokuma, K.; Voth, G. A.; Salvador, P.; Dannenberg, J. J.; Zakrzewski, V. G.; Dapprich, S.; Daniels, A. D.; Strain, M. C.; Farkas, O.; Malick, D. K.; Rabuck, A. D.; Raghavachari, K.; Foresman, J. B.; Ortiz, J. V.; Cui, Q.; Baboul, A. G.; Clifford, S.; Cioslowski, J.; Stefanov, B. B.; Liu, G.; Liashenko, A.; Piskorz, P.; Komaromi, I.; Martin, R. L.; Fox, D. J.; Keith, T.; Al-Laham, M. A.; Peng, C. Y.; Nanayakkara, A.; Challacombe, M.; Gill, P. M. W.; Johnson, B.; Chen, W.; Wong, M. W.; Gonzalez, C.; and Pople, J. A.; Gaussian, Inc., Wallingford CT, 2004.

11 a) Hay, P. J.; Wadt, W. R. *J. Chem. Phys.* **1985**, *82*, 270. b) Wadt, W. R.; Hay, P. J. *J. Chem. Phys.* **1985**, *82*, 284. c) Hay, P. J.; Wadt, W. R. *J. Chem. Phys.* **1985**, *82*, 299. d) Hehre, W. J.; Radom, L.; von Ragué Schleyer, P.; Pople, J. A. *Ab initio Molecular Orbital Theory*, Wiley: New York, **1986**, and references cited therein.

12 Although optimization was set to start from the structure close to lithium enolate, stationary point of the Ti-O-C bond was straight.

13 a) Nielson, A. J.; Schwerdtfeger, P.; Waters, J. M. *J. Chem. Soc., Dalton Trans.* **2000**, 529. b) Dobado, J. A.; Molina, J. M.; Uggla, R.; Sundberg, M. R. *Inorg. Chem.* **2000**, *39*, 2831.

14 There are some reports of trifluoromethylation using CF_3^+.: a) Yagupol'skii, L. M.; Kondratenko, N. V.; Timofeeva, G. N. *J. Org. Chem. USSR* **1984**, *20*, 115. b) Umemoto, T.; Ishihara, S. *J. Am. Chem. Soc.* **1993**, *115*, 2156. c) Umemoto, T.; Adachi, K. *J. Org. Chem.* **1994**, *59*, 5692.

15 Perfluoroalkylation of silyl and germyl enolates of esters and ketones: a) Miura, K.; Taniguchi, M.; Nozaki, K.; Oshima, K.; Utimoto, K. *Tetrahedron Lett.* **1990**, *31*, 6391. b) Miura, K.; Takeyama, Y.; Oshima, K.; Utimoto, K. *Bull. Chem. Soc. Jpn.* **1991**, *64*, 1542. Perfluoroalkylation of silyl enol ethers provided the products in good yields except for trifluoromethylation. Trifluoromethylation of ketone germyl enolates proceeds in good yield.

16 Trifluoromethylation of lithium enolate of imides: a) Iseki, K.; Nagai, T.; Kobayashi, Y. *Tetrahedron Lett.* **1993**, *34*, 2169. b) Iseki, K.; Nagai, T.;

Kobayashi, Y. *Tetrahedron: Asymm.* **1994**, *5*, 961. They have succeeded in trifluoromethylation by adopting Evans oxazolidinones with bulky substitutent at α position to suppress defluorination.

17 Trifluoromethylation of enamines: a) Cantacuzène, D.; Wakselman, C.: Dorme, R. *J. Chem. Soc., Perkin Trans. 1* **1977**, 1365. b) Kitazume, T.; Ishikawa, N. *J. Am. Chem. Soc.* **1985**, *107*, 5186.

18 Itoh, Y.; Mikami, K. *Org. Lett.* **2005**, *7*, 649.

19 Nozaki, K.; Oshima, K.; Utimoto, K. *J. Am. Chem. Soc.* **1987**, *109*, 2547.

20 Some reactions involving titanium ate enolates: a) Siegel, C.; Thornton, E. R. *J. Am. Chem. Soc.* **1989**, *111*, 5722. b) Bernardi, A.; Cavicchioli, M.; Marchionni, C.; Potenza, D.; Scolastico, C. *J. Org. Chem.* **1994**, *59*, 3690. c) Yachi, K.; Shinokubo, H.; Oshima, K. *J. Am. Chem. Soc.* **1999**, *121*, 9465. d) Han, Z.; Yorimitsu, H.; Shinokubo, H.; Oshima, K. *Tetrahedron Lett.* **2000**, *41*, 4415.

21 Some reactions involving other titanium ate comlexes: a) Reetz, M. T.; Wenderoth, B. *Tetrahedron Lett.* **1982**, *23*, 5259. b) Reetz, M. T.; Westermann, J.; Steinbach, R.; Wenderoth, B.; Peter, R.; Ostarek, R.; Maus, S. *Chem. Ber.* **1985**, *118*, 1421. c) Reetz, M. T.; Steinbach, R.; Westermann, J.; Peter, R.; Wenderoth, B. *Chem. Ber.* **1985**, *118*, 1441. d) Takahashi, H.; Kawabata, A.; Niwa, H.; Higashiyama, K. *Chem. Pharm. Bull.* **1988**, *36*, 803. e) Takahashi, H.; Tsubuki, T.; Higashiyama, K. *Synthesis* **1988**, 238. f) Takahashi, H.; Tsubuki, T.; Higashiyama, K. *Chem. Pharm. Bull.* **1991**, *39*, 260. g) Bernardi, A.; Cavicchioli, M.; Scolastico, C. *Tetrahedron* **1993**, *49*, 10913. h) Bernardi, A.; Marchionni, C.; Pilati, T.; Scolastico, C. *Tetrahedron Lett.* **1994**, *35*, 6357. i) Mahrwald, R. *Tetrahedron* **1995**, *51*, 9015.

22 a) Stork, G.; Hudrlik, P. F. *J. Am. Chem. Soc.* **1968**, *90*, 4462. b) Stork, G.; Hudrlik, P. F. *J. Am. Chem. Soc.* **1968**, *90*, 4464.

23 NMR study of a titanium ate enolate (ref. 20b) showed that the ketone was formed in the generation of the titanium ate enolate (although this is only mentioned in the footnote of Figure 2 of the article). This fact also supports the proposed mechanism.

24 Submitted for publication.

25 (a) Barone, V.; Cossi, M. *J. Phys. Chem. A* **1998**, *102*, 1995. (b) Cossi, M.; Rega, N.; Scalamani, G.; Barone, V. *J. Comput. Chem.* **2003**, *24*, 669. (c) Klamt, A.; Schüürmann, G. *J. Chem. Soc., Perkin Trans.2* **1993**, 799.

Chapter 2

Syntheses of Organofluorine Derivatives by a New Radical Process

Samir Z. Zard

Laboratoire de Synthèse Organique Associé au CNRS, Ecole Polytechnique, 91128 Palaiseau, France

New routes to organofluorine derivatives based on the powerful xanthate radical transfer technology are described. A special emphasis is placed on the synthesis of trifluoromethyl-substituted structures, including trifluoromethylketones and fluorinated aromatic and heteroaromatic substances of interest to the pharmaceutical and agrochemical industries.

Introduction

The enormous importance of organofluorine derivatives for the pharmaceutical and agrochemical industries and for material sciences has generated a steep demand for cheap, efficient, and flexible methods for the introduction of fluorine or fluorinated groups into various families of compounds *(1)*. Countless, and in many cases ingenious, approaches have been devised to cover the needs of academic and industrial chemists interested in organofluorine derivatives. Nevertheless, there is still a niche for new reactions allowing the direct introduction of fluorinated groups into the substrate or the synthesis of fluorinated building-blocks that can then be incorporated in a given synthetic scheme.

We have developed in recent years a new radical chemistry of xanthates **1** and related derivatives of general formula Z-C(=S)S-R which allows the generation and capture of various types of radicals *(2)*. We thus found that xanthates are capable of undergoing additions to various olefins by way of the simplified radical chain mechanism outlined in Scheme 1. The overall result is the addition of the elements of the xanthate across the olefinic bond to give adduct **4**. As the examples in the following pages will show, it turns out that this process is also a uniquely powerful tool for the synthesis of numerous fluorinated synthons and building blocks, many of which would not be easily accessible otherwise.

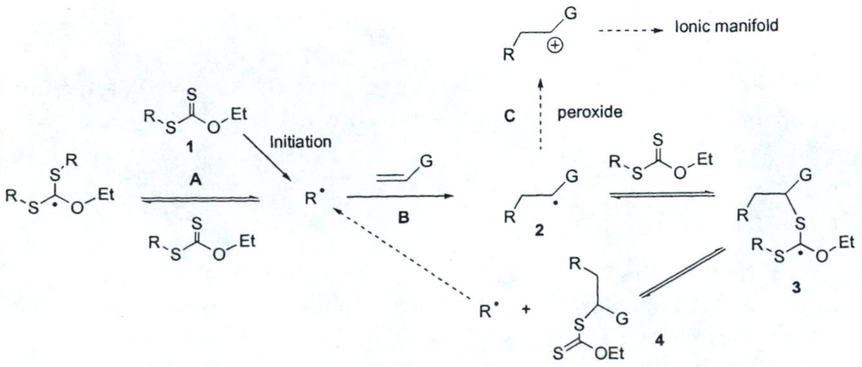

Scheme 1. *Simplified mechanism for the degenerative xanthate transfer.*

The subtleties of the mechanistic manifold will not be discussed in detail, but three general properties are especially important:

(a) The reaction of radical R• with its xanthate precursor (path **A**) is degenerate and does not consume the radical. This effective absence of competition forces the radical into reacting with the olefinic trap either in an intra- or intermolecular mode (path **B**). This translates into a unique ability to accomplish *intermolecular* additions to *un-activated* olefins and sets this method apart from essentially all other radical processes.

(b) The adduct is itself a xanthate. This allows the implementation of another radical sequence (which can in turn lead to yet another xanthate), or the xanthate group can be used as an entry into the extremely rich "ionic" chemistry of sulfur. Hence, a large array of synthetic transformations can now be used to introduce further diversity and complexity into the structure.

(c) It is important, for the chain process to be efficient, to select the reacting partners in such a way that the adduct radical **2** is less stable than the initial radical R•, in order to favour the fragmentation of intermediate **3** in the desired direction. This consideration is crucial, especially when intermolecular additions are concerned, as it provides a handle for stopping at the mono-adduct in preference to telomerisation. It is nevertheless possible, when the adduct radical **2** has a low oxidation potential and therefore easily oxidised into the corresponding cation, to use the peroxide both as an initiator and an oxidant (path **C**). Oxidising the radical opens a passageway from a radical to a polar reaction manifold and expands considerably the synthetic scope. This aspect will have its importance in the context of a tin-free reductive removal of the xanthate group and in cyclisations involving aromatic rings (vide infra).

The utility of this process is that it allows the inter- or intra-molecular creation of C—C bonds, even in the case of non-activated alkenes. As far as fluorinated groups are concerned, they can be present in the olefin, in the xanthate, or in both. The convergent synthesis of a di-fluoro nucleoside analogue **9** detailed in Scheme 2 illustrates the case where the fluorine atoms adorn the olefinic partner *(3)*. The efficient radical addition is followed by aminolysis of the adduct xanthate **6**, ring-closure to thiolactone **7**, and mild reduction to give a thiolactol intermediate. Finally, acetylation provides a substrate, **8**, capable of undergoing a Vorbrüggen type introduction of the thymine base.

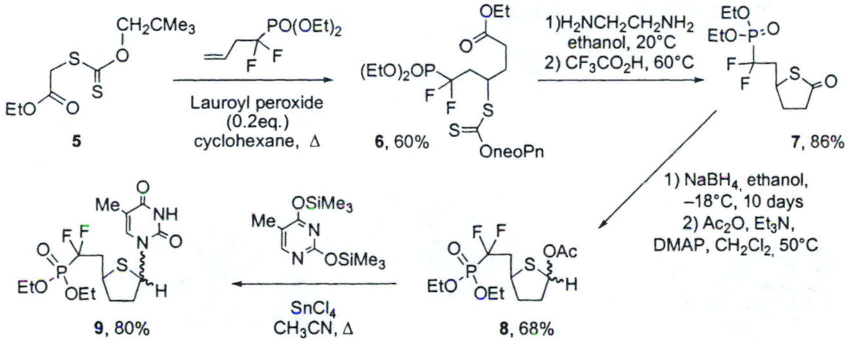

Scheme 2. *Synthesis of a nucleoside analogue.*

Alternatively, the fluorinated group can be attached to the xanthate. A number of reagents can thus be readily prepared and used to access to a large array of fluorinated synthons. For instance, starting from the hemiacetal of trifluoroacetaldehyde, crystalline xanthate **10** can be prepared on a multigram scale. Its addition to variously substituted olefins proceeds efficiently and cleanly to afford a multitude of structures possessing a geminally disposed trifluoromethyl and amido groups *(4)*. The examples pictured in Scheme 3 underscore the remarkable compatibility of this synthetic method with a broad diversity of functional groups present on the olefinic partner. The specific reactivity of the various substituents can be harnessed to further increase the complexity, for example by inducing ring-closures to form nitrogen or sulfur heterocycles.

Furthermore, the presence of the xanthate group in the addition products can be exploited in countless ways. In some cases, another radical addition can be accomplished, as shown by the transformation on the right in Scheme 4 leading to compound **11**, where almost every carbon is functionalised. Densely functionalised, complex fluorinated structures can thus be swiftly assembled *(4)*. Addition to vinyl acetate provides an adduct containing a masked aldehyde function, since the xanthate is now geminal to the acetoxy group. Heating this adduct with methanol and acid gives crystalline acetal **12**, a compound bearing both a protected amine and a masked aldehyde; it constitutes thus a very interesting synthon for the synthesis of trifluoromethylated heterocycles. The third example in Scheme 4 concerns the addition to N-vinyl pyrrolidone, which gives a product that can be thermolysed into enamide **13** in essentially quantitative yield *(4)*.

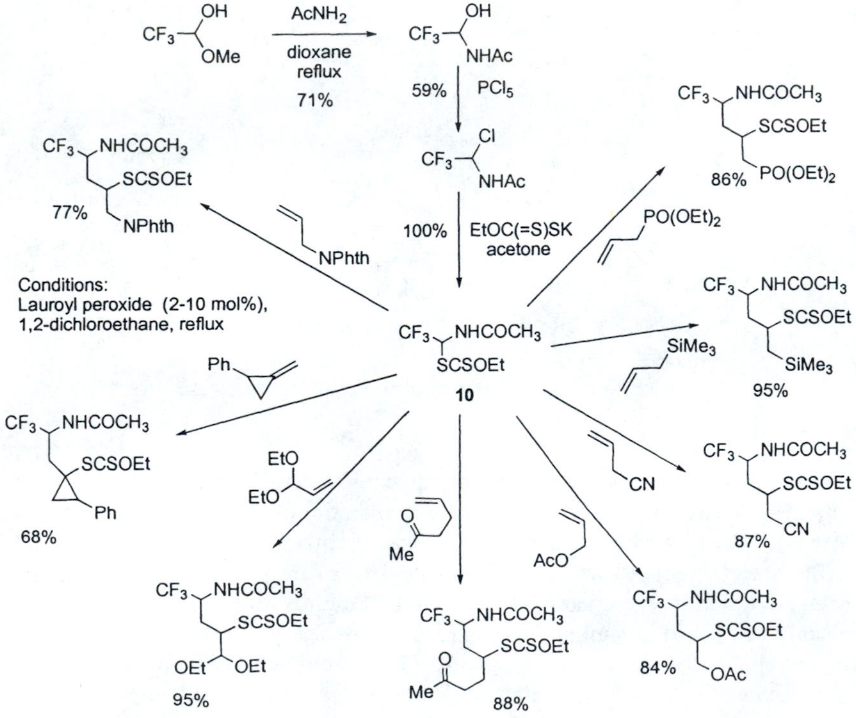

Scheme 3. *Examples of xanthate transfer radical additions.*

Enamides such as **13** are useful precursors for the synthesis of trifluoroalanine, trifluoroanalinal, and trifluoroalaninol, through ozonolysis of the olefinic bond followed by an appropriate oxidative or reductive work up of the ozonolysis medium. One synthetic application is depicted at the bottom of Scheme 4, where in situ condensation of the aldehyde with acetylacetone leads to trifluoromethylated pyrrole **14** through the classical Knorr condensation *(5)*.

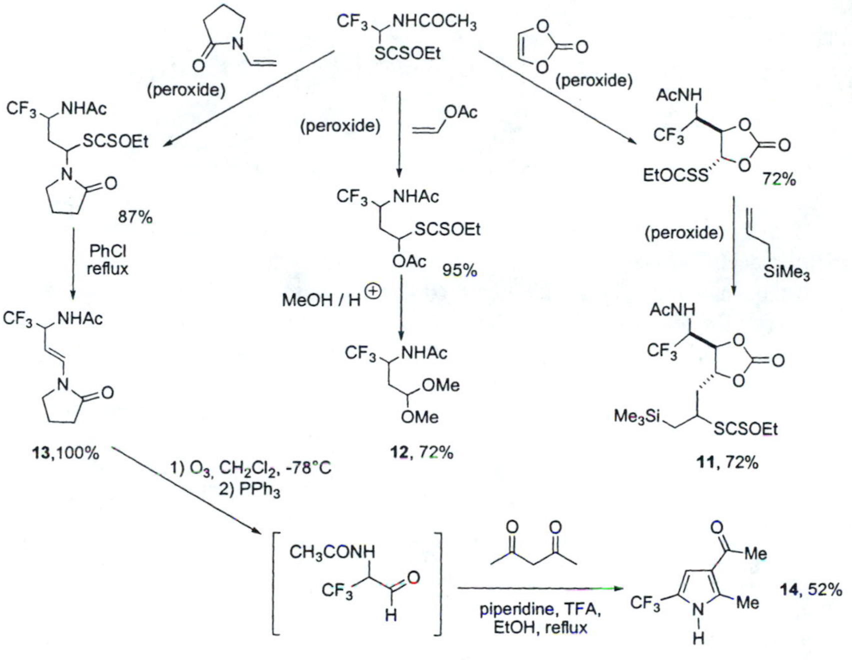

Scheme 4. *Further transformations of the adducts.*

It is interesting to note that in the absence of a trap, xanthate **15**, the benzoyl analogue of **10**, reacts with a stoichiometric quantity of lauroyl peroxide to give homodimer **16**, as 1:1 mixture of the *meso* and *d,l* diastereomers (Scheme 5). The benzoyl analogue was used to simplify isolation *(4)*. The formation of dimers by coupling of radicals is not surprising when dealing with radical intermediates; what is astonishing is the remarkably high yield *of the homodimer*, since one would have expected a complex mixture arising from anarchic and statistical recombinations of the various radicals in the medium. The mechanistic implications of this observation are exceedingly important and reveal some subtle aspects of the process *(6)*, but these will not be discussed here. Suffice it to say that this approach represents a unique, practical route to protected 2,3-diamino-1,1,1,4,4,4-hexafluorobutane, a hitherto unknown family of fluorinated vicinal diamines of potential use for the synthesis of unusual ligands for transition metal chemistry or for the construction of fluorinated nitrogen heterocycles.

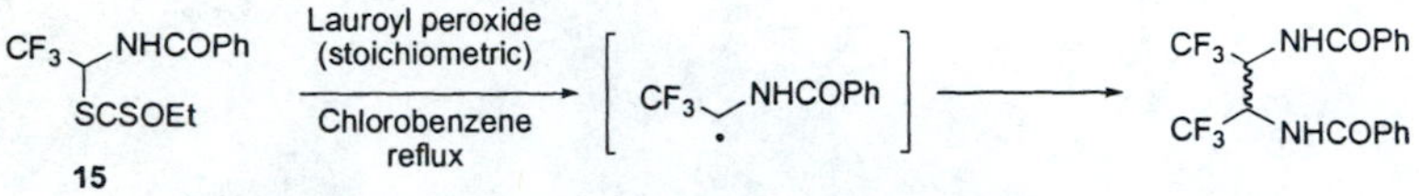

Scheme 5. *Formation of homodimers.*

Other useful xanthates can be obtained from the hemiacetal of trifluoroacetaldehyde. The acetoxy and chloro derivatives were prepared in a similar manner to the amide analogue **10** and found to react in the same way with a variety of olefinic traps, as indicated by the examples collected in Scheme 6 *(6)*. Again, notice the richness and density of the functional groups.

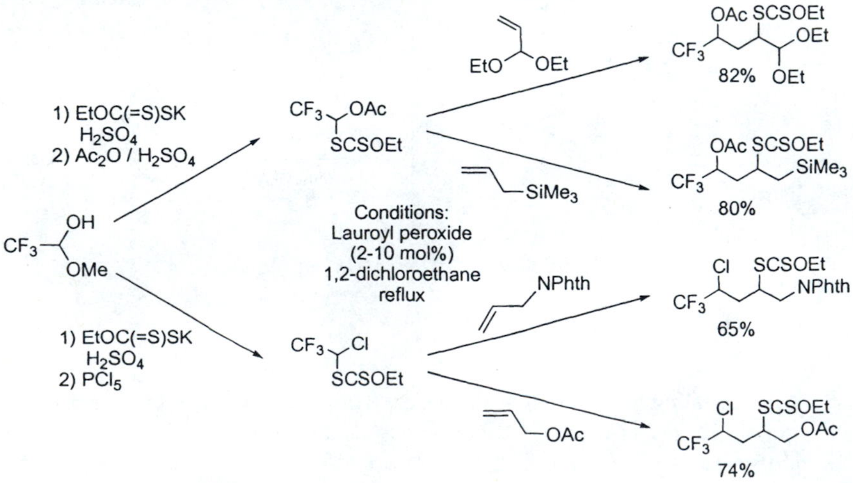

Scheme 6. *Synthesis and addition of xanthates derived from trifluoroacetaldehyde hemiacetal.*

It is possible to use the chemistry of xanthate to access trifluomethyl ketones by a radical pathway. The precursor is easily made by reacting a xanthate salt with commercially available 1-bromo-3,3,3-trifluoropropan-2-one *(7)*. The *O*-neopentyl xanthate was used in this case to increase lipophilicity and limit complications due to the reversible formation of the hydrate. Again, numerous olefins can be used as traps, and trifluoromethyl ketones that would otherwise be very difficult to obtain can be made in essentially one step. As far as we are aware, this is the only instance that trifluoroacetyl radicals have been applied in synthesis. The examples in Scheme 7 are representative of the tremendous potential of this approach. The first example corresponds to the synthesis of a masked aminoacid, whereas the second is a protected (Phth = phthalimido) amino ketone which should close to a trifluoromethyl piperidine derivative upon deprotection of the amino group.

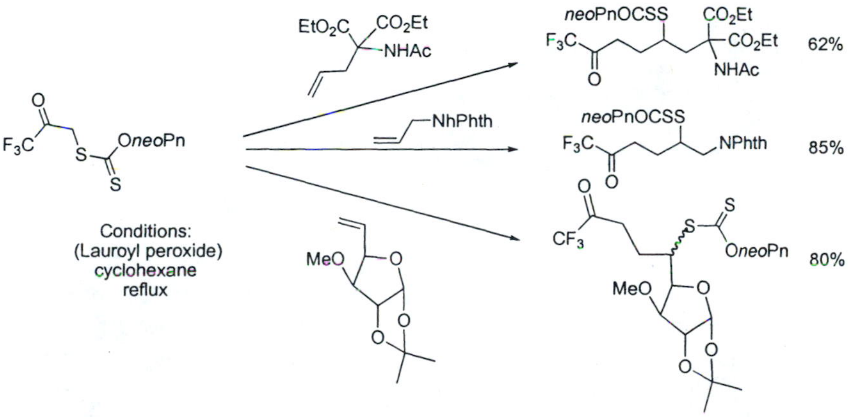

***Scheme** 7. Synthesis of trifluoromethyl ketones.*

A simple trifluoromethyl group can be introduced via the corresponding xanthate. Such xanthates, however, and not surprisingly, cannot be made by a simple nucleophilic substitution on a trifluoromethyl halide, as in the case of the previous derivatives. An indirect route via the decarbonylation of the *S*-trifluoroacyl xanthate **17** was devised to overcome this problem *(8)*. A "heavier" xanthate was employed to aid in handling the intermediates and to avoid having to manipulate a potentially volatile reagent. Xanthate **18** turns out to be a highly effective agent for the introduction of the trifluoromethyl group, as shown by the example displayed in Scheme 8. The olefinic partner can be varied in the same way as in the preceding schemes. A large variety of trifluoromethylated building blocks thus become accessible, starting from cheap, readily available starting materials.

The xanthate group in the product provides an entry into the exceptionally rich chemistry of sulfur and a plethora of subsequent transformations can be envisaged following the radical addition step. Nevertheless, reductive removal of the xanthate group is often also a desired transformation. Many traditional reagents (tributylstannane, Raney nickel, nickel boride) are capable of accomplishing this task cleanly. However, these reagents are not very convenient for large scale use because of toxicity, cost, or problems with purification.

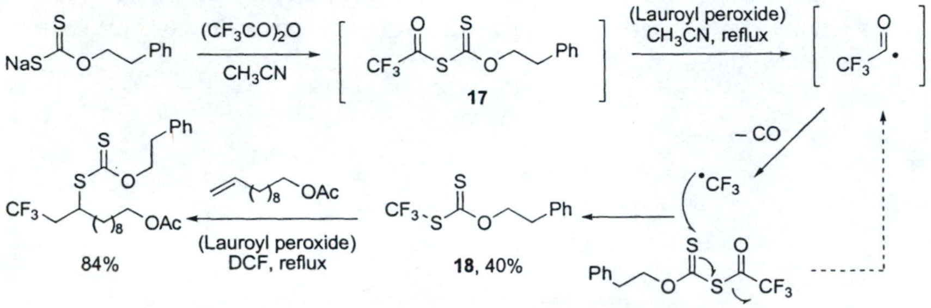

Scheme 8. Generation and capture of trifluoromethyl radicals.

We found that a simple, yet highly efficient reductive method existed, which took advantage of the electrophilic character of a radical centre adjacent to inductively acting (– I) electron withdrawing groups, and especially fluorine atoms. Reduction of compound **21**, obtained by radical addition of acetonitrile derived xanthate **19** to heptadecafluorodecene, is a case in point *(9)*. The xanthate is simply dissolved in refluxing cyclohexane and treated with a small amount of lauroyl peroxide. A chain reaction sets in, whereby the cyclohexane acts as the hydrogen atom source, as outlined in Scheme 9. Intermediate radical **22** is highly electrophilic in character, because of the powerful electron withdrawing effect exerted by the nearby fluorine atoms, yet unstabilised. It therefore rapidly abstracts a hydrogen atom from the solvent to give a cyclohexyl radical to propagate the chain. The importance of the fluorine atoms can be visualised, by comparing in a competition experiment, the behaviour of adduct **21** with that of the non-fluorinated analogue **20**. The latter is hardly affected under these conditions and recovered mostly intact.

In order to be able to reduce ordinary secondary xanthates that are not substituted by reactivity enhancing electron-withdrawing groups, it is necessary to use a better hydrogen atom donor than cyclohexane. A convenient candidate turns out to be 2-propanol *(10)*. Its hydrogen atom donor ability is much greater than cyclohexane, but this is counterbalanced by the fact that the resulting ketyl radical is too stabilised and therefore incapable of propagating the chain. This radical can, however, be oxidised by electron transfer to the peroxide into the corresponding cation, which then sheds a proton to give a molecule of acetone.

The peroxide acts therefore as an initiator and as a stoichiometric oxidant. The first application pictured in Scheme 10 illustrates the synthesis of a differentially protected 1,4-butanediamine *(4)*. The second transformation concerns the appending of a trifluoroacetonyl group on native pleuromutilin, a terpene antimicrobial *(11)*. This transformation is remarkable in that there is no need to protect any of the functional groups present in the substrate. It is perhaps worth stressing that a traditional approach to all these products would certainly be lengthy and tedious.

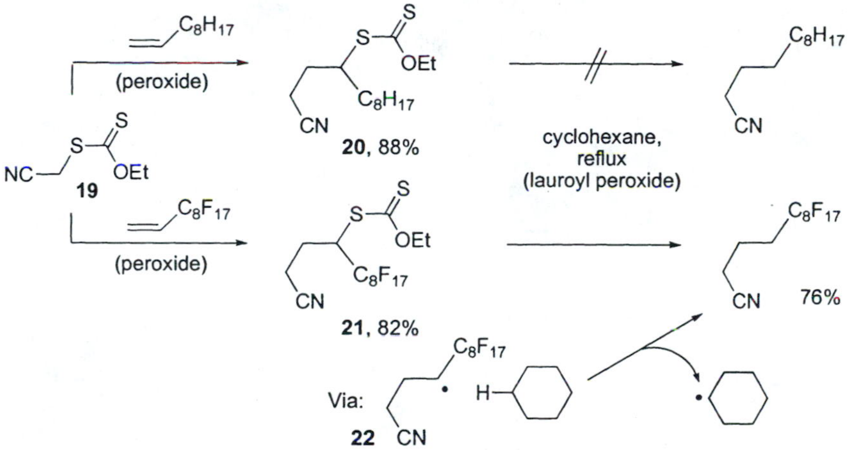

Scheme 9. *Reductions using cyclohexane.*

The fact that the peroxide can be used both as an initiator and as an oxidant opens numerous opportunities for the synthesis of fluorinated aromatic derivatives. These are highly valuable precursors for pharmaceuticals and agrochemicals, as well for material sciences. The underlying principle of this approach is to exploit the relatively long life of the intermediate radicals generated from the xanthate to force cyclisation on an aromatic or heteroaromatic ring and to use the peroxide to oxidise the adduct radical (a cyclohexadienyl radical in the case of benzene type aromatics) into the cation, which then rapidly aromatises by loss of a proton. The fluorine-bearing motif can be part of the aromatic structure or incorporated into the xanthate. An enormous advantage of this strategy is the fact that strongly electron withdrawing substituents such as trifluoromethyl groups, which severely retard Friedel-Crafts and other mechanistically related electrophilic substitutions commonly used in the synthesis of aromatic derivatives, do not generally affect much the radical cyclisation step. This expands considerably the range of substrates can thus be used in this approach, as illustrated by the examples discussed in the following paragraphs.

The synthesis of naphthalenes displayed in Scheme 11 relies on the intermolecular addition of a *p*-fluorophenacyl radical to vinyl pivalate, followed by ring closure onto the aromatic ring to give tetralone **23** *(12)*. Both of these radical steps are very easy to execute using the xanthate technology but would be almost impossible to accomplish as efficiently with any of the other radical methods currently available.

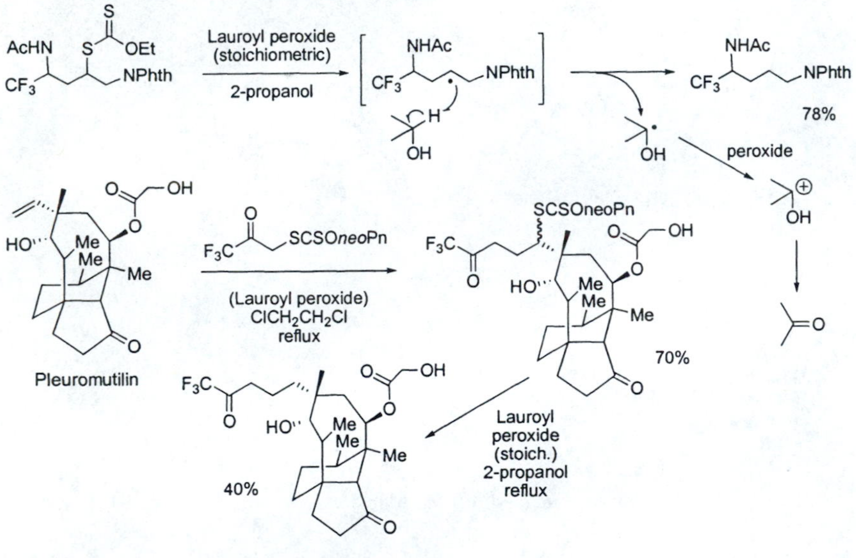

Scheme 10. *Reductive removal of the xanthate using isopropanol.*

Tetralone **23** represents a springboard to access a number of monofluoronaphthalenes through classical ionic modifications *(12)*. For instance, heating with acid causes aromatisation into naphthol **24**, whereas prior bromination α- to the ketone using pyridinium bromide perbromide gives ultimately bromofluoronaphthol **25**. It is also possible to perform a Wittig-Horner condensation without noticeable complications from the hindered pivalate ester; acid catalysed elimination then affords fluoronaphthaleneacetate **26**. By constructing the second ring in the napththalene, one avoids the formation of regioisomers that often plagues classical approaches.

Even more difficult cyclisations on aromatic rings can be effected, as shown by the two transformations in Scheme 12. The construction of δ-lactams by radical cyclisation starting with unsubstituted amides is notoriously difficult, even in the aliphatic series. Yet, we have found it possible to prepare trifluoromethyl substituted tetrahydroquinolinone **27** *(13)* and homophthalimide **28** *(14)*, albeit in only moderate yield. Nevertheless, it is worth stressing that the trifluoromethyl substituents are perfectly tolerated and that products with such substitution patterns, despite their apparent simplicity, would require significantly more circuitous conventional routes to attain.

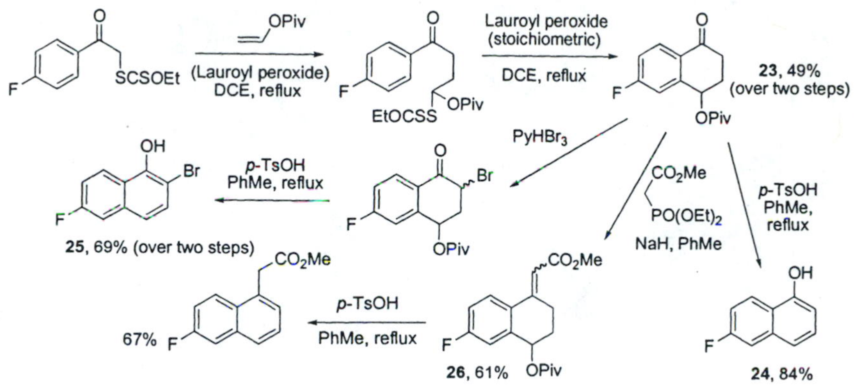

Scheme 11. *Synthesis of naphthalenes.*

Addition of fluorine bearing xanthates to *N*-allyl anilines represents an alternative pathway to fluorinated indolines. Three examples in Scheme 13 illuminate this strategy and involve an assortment of xanthates.

The first two imply the use of reagents **10** *(4)* and **17** *(8)*. In the last example, representing a trifluoromethyl ketone, an intramolecular Friedel-Crafts reaction mediated by cold sulphuric acid leads to new tricylic structure **29** in very good yield *(15)*. The mesyl on the nitrogen can be replaced by more easily removed protecting groups, such as esters or carbamates (e. g. Boc), and the indolines can be converted by oxidation into the more interesting indoles if needed.

There are other ways to use a xanthate group to mediate the synthesis of fluorinated compounds through radical intermediates. One contrivance pictured in Scheme 14 hinges on the addition to trifluoromethylvinylsulfone **30**, followed by β-elimination of an ethylsulfonyl radical *(16)*. The latter looses sulfur dioxide to generate a reactive ethyl radical that readily propagates the chain. The product in this transformation is an enol carbonate, the hydrolysis of which would produce in principle the corresponding trifluoromethyl ketone.

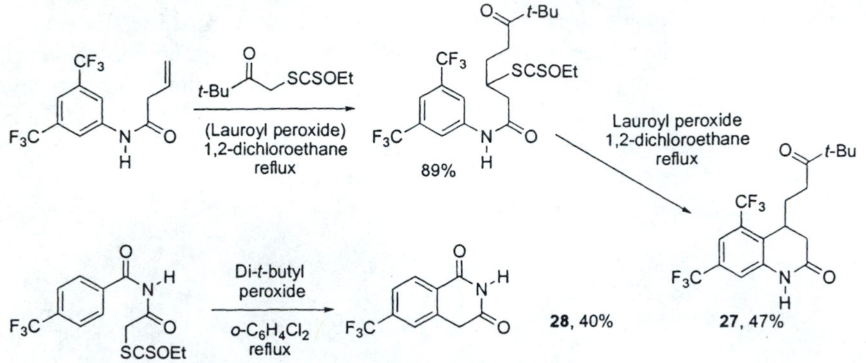

CF_3
F_3C
N
H
O
t-Bu
O
SCSOEt
(Lauroyl peroxide)
1,2-dichloroethane
reflux
CF_3
F_3C
N
H
O
O
t-Bu
SCSOEt
89%
Lauroyl peroxide
1,2-dichloroethane
reflux
O
t-Bu
CF_3
F_3C
N
H
O
27, 47%

F_3C
O
N
H
O
SCSOEt
Di-t-butyl
peroxide
o-$C_6H_4Cl_2$
reflux
F_3C
O
N
H
O
28, 40%

***Scheme 12.** Synthesis of dihydroquinolones and homophthalimides.*

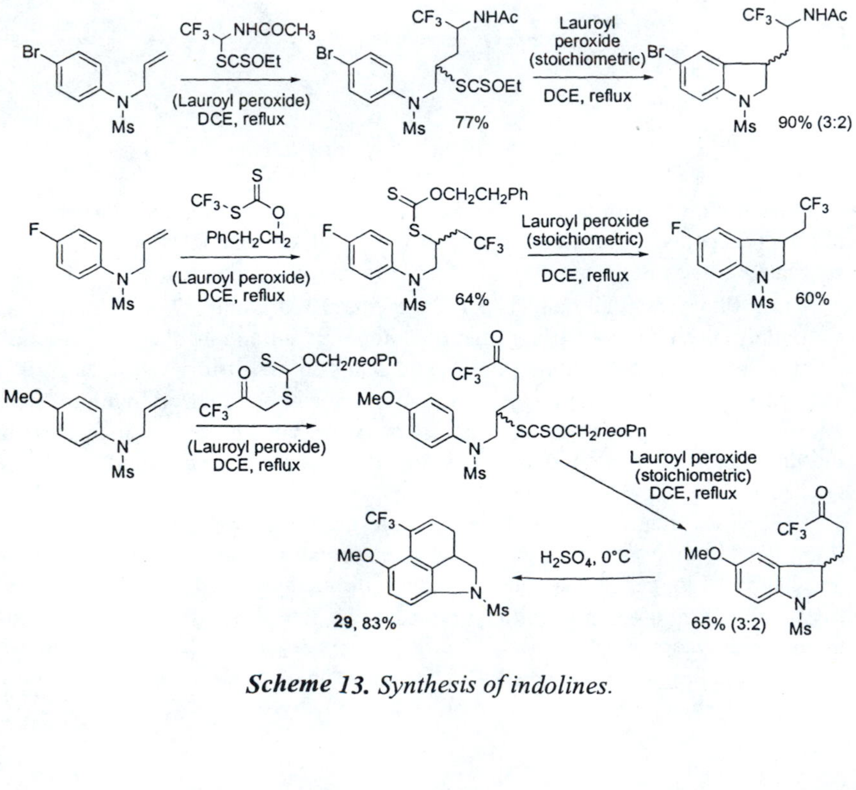

Br
N
Ms
CF_3 $NHCOCH_3$
SCSOEt
(Lauroyl peroxide)
DCE, reflux
Br
CF_3 NHAc
SCSOEt
N
Ms
77%
Lauroyl
peroxide
(stoichiometric)
DCE, reflux
Br
CF_3 NHAc
N
Ms
90% (3:2)

F
N
Ms
S
CF_3 S O
$PhCH_2CH_2$
(Lauroyl peroxide)
DCE, reflux
F
S OCH_2CH_2Ph
S
CF_3
N
Ms
64%
Lauroyl peroxide
(stoichiometric)
DCE, reflux
F
CF_3
N
Ms
60%

MeO
N
Ms
S OCH_2neoPn
O
CF_3
S
(Lauroyl peroxide)
DCE, reflux
MeO
O
CF_3
$SCSOCH_2neoPn$
N
Ms
Lauroyl peroxide
(stoichiometric)
DCE, reflux
O
CF_3
MeO
N
Ms
65% (3:2)
H_2SO_4, 0°C
CF_3
MeO
N
Ms
29, 83%

***Scheme 13.** Synthesis of indolines.*

The powerful radical exchange of xanthates has a broad scope for the inter- or intra-molecular creation of C—C bonds that extends beyond the realm of organofluorine compounds. The transformations discussed in the preceding paragraphs give a glimpse the numerous possibilities when this technology is applied to constructing novel fluorine containing structures and building blocks that may be useful for chemists interested in such compounds. The experimental procedure is quite simple and the reagents are cheap and readily accessible. Interestingly, the process is essentially self-regulating and the reactions, in contrast to the prevailing opinion conditioned chiefly by the better known stannane chemistry, often work best in very concentrated media and are easily scaled up.

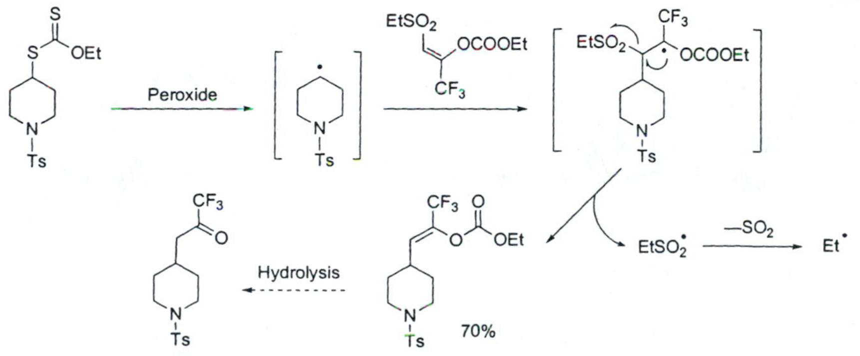

Scheme 14. *An alternative synthesis of trifluoromethyl ketones.*

Acknowledgements: I wish to express my heartfelt thanks to my collaborators, whose names appear in references and whose dedication and skills made this work possible. I also wish to gratefully acknowledge generous financial support from Ecole Polytechnique, the CNRS, Rhodia, Sanofi-Aventis, CONACyT (Mexico), and the Ministerio de Educación, Cultura y Deporte (Spain).

References

1. (a) Shimizu, M.; Hiyama, T. *Angew. Chem. Int. Ed. Engl.* **2005**, *44*, 214-231. (b) Chambers, R. D. *Fluorine in Organic Chemistry;* Blackwell Publishing: Oxford, 2004. (c) Kirsch, P. *Modern Fluoroorganic Chemistry;* Wiley-VCH: Weinheim, 2004. (d) *Topics in Current Chemistry - Organofluorine Chemistry: Techniques and Synthons*; Chambers, R. D., Ed.; Springer: Berlin, 1997, Volume 193. (e) *Biomedical Frontiers of*

Fluorine Chemistry; Ojima, I.; McCarthy, J. R.; Welch, J. T., Eds.; American Chemical Society: Washington, D.C., 1996. (f) *Fluoroorganic Chemistry: Synthetic Challenges and Biomedical Rewards*; Resnati, G.; Soloshonok, V. A., Eds. Tetrahedron Symposium-in-Print No. 58; *Tetrahedron* **1996**, *52*, 1-330. (g) *Fluorine-Containing Amino Acids*; Kukhar, V. P.; Soloshonok, V. A., Eds.; Wiley: Chichester, 1994. (h) *Organofluorine in Medicinal Chemistry and Biochemical Applications*; Filler, R.; Kobayashi, Y.; Yagupolski, L. M., Eds.; Elsevier: Amsterdam, 1993.

2. Zard, S. Z. *Angew. Chem. Int. Ed. Engl.* **1997**, *36*, 672-685. (b) Quiclet-Sire, B.; Zard, S. Z. *Phosphorus, Sulfur, and Silicon* **1999**, *153-154*, 137-154. (c) Zard, S. Z. in *Radicals in Organic Synthesis*; P. Renaud, P.; M. P. Sibi, M. P., Eds.; Wiley-VCH: Weinheim, 2001; Vol. 1, pp 90-108.
3. Boivin, J.; Ramos, L.; Zard, S. Z. *Tetrahedron Lett.* **1998**, *39*, 6877-6880.
4. Gagosz, F.; Zard, S. Z. *Org. Lett.* **2003**, *5*, 2655-2657.
5. Tournier, L. *PhD Thesis;* Ecole Polytechnique: Palaiseau, 2005.
6. Tournier, L.; Zard, S. Z. *Tetrahedron Lett.* **2005**, *46*, 455-459.
7. Denieul, M.-P.; Quiclet-Sire, B.; Zard, S. Z. *Chem. Commun.* **1996**, 2511-2512.
8. Bertrand, F.; Pevere, V.; Quiclet-Sire, B.; Zard, S. Z. *Org. Lett.* **2001**, *3*, 1069-1071.
9. Quiclet-Sire, B.; Zard, S. Z. *J. Am. Chem. Soc.* **1996**, *118*, 9190-9191.
10. Liard, A.; Quiclet-Sire, B.; Zard, S. Z. *Tetrahedron Lett.* **1996**, *37*, 5877-5880.
11. Bacqué, E. ; Pautrat, F. ; Zard, S. Z. *Chem. Commun.* **2002**, 2312-2313.
12. Cordero-Vargas, A. ; Pérez-Martin, I. ; Quiclet-Sire, B.; Zard, S. Z. *Org. Biomol. Chem.* **2004**, 2, 3018-3025.
13. Binot, G.; Zard, S. Z. *Tetrahedron Lett.* **2005**, *46*, 7503-7506.
14. Quiclet-Sire, B.; Zard, S. Z. *Chem. Commun.* **2002**, 2306-2307
15. Sortais, B. *PhD Thesis;* Univeristé Paris-Sud: Orsay: 2002.
16. Bertrand, F.; Leguyader, F.; Liguori, L.; Ouvry, G.; Quiclet-Sire, B.; Seguin, S.; Zard, S. Z. *C. R. Acad. Sci. Paris* **2001**, *II4*, 547-555.

Chapter 3

An Issue of Synthesis Strategy: To Make or Rather to Transform Organofluorine Compounds

Manfred Schlosser, Qian Wang, and Fabrice Cottet

Institute of Chemical and Engineering Sciences, Ecole Polytechnique Fédérale, CH–1015 Lausanne, Switzerland

Virtually all organofluorine compounds are man-made. This fact immediately brings up the critical question of how and at what moment to integrate fluorine into the carbon backbone. One may import the featured halogen from an inorganic source which can be either a nucleophilic (*e.g.*, hydrogen fluoride) or an electrophilic reagent (*e.g.*, perchloryl fluoride). Alternatively, one may deliver the required number of fluorine atoms attached to a C_1 species (*e.g.*, as a trifluoromethyl group). Finally, one may start from a commercial compound which carries the right number of fluorine atoms at the right places and which just needs to be fitted with the lacking functionality. The latter approach is particularly appealing if diversity-oriented, as opposed to target-oriented, synthesis is the objective.

Fluorine is undeniably *overrepresented in the life science arena*. According to recent estimates (*1*), up to, respectively, 20 and 40 % of new low molecular-weight active principles in the pharmaceutical and agricultural (phytosanitary) sectors harbor one or several atoms of fluorine. Subtle changes in shape and size, acidity, dipole moments, polarizability, lipophilicity, transport behavior and chemical and metabolic stability (or fragility) have been evoked to explain why the "dope" of the lightest halogen is so effective in fine-tuning biological properties and, as a corollary, in improving the chances of scoring a hit by almost tenfold (*2*).

Paradoxically, only a handful of organofluorine compounds occurs in nature and even those only in insignificant amounts. Thus, all commercial fluorine-containing substances are man-made. This raises a principal question in relation with synthesis and manufacture planning. Should the fluorine label be introduced at the very beginning or at the very end or somewhere in the middle of the product assembly chain ? Often people opt for one of the two extremes. They either attach the halogen to a prefabricated carbon backbone as late as possible or they elaborate a fluorine-adorned starting material in a multistep sequence to the desired final product. Whatever the practical approach, its merits and shortcomings have to be evaluated on a from case to case basis.

The Direct Fluorination

The ultimate precursor to any fluorinating reagent is the fluorspar (apatite, calcium fluoride) from which, among other things, hydrofluoric acid, sulfur tetrafluoride and elemental fluorine are derived. Any laboratory or pilot plant that disposes of the appropriate equipment and know-how to handle such corrosive liquids or gases feels tempted to make their own organofluorine chemicals. Whereas hydrogen fluoride (*3*) and potassium fluoride, for "halex" reactions (*4*), can be most advantageously applied on an industrial scale, academic researchers are more inclined to work with small quantities and to use more sophisticated reagents.

Electrophilic Fluorinating Reagents

The fluorination of electron-rich carbon entities can be readily achieved with perchloryl fluoride (*5*). This versatile reagent was first employed to convert diethyl malonate into diethyl fluoromalonate (**1**) and diethyl difluoromalonate (**2**). As the intermediate enolates were set free under acid-base equilibrating conditions, inevitably mixtures of zero-, mono- and difluorinated products were obtained (*6*).

$$CH_2(COOR)_2 \underset{\longleftarrow}{\xrightarrow{\text{Na-OR}}} Na^{\oplus}\,{}^{\ominus}CH(COOR)_2 \xrightarrow{FClO_3} CHF(COOR)_2 \underset{\longleftarrow}{\xrightarrow{\text{Na-OR}}} Na^{\oplus}\,{}^{\ominus}CF(COOR)_2 \xrightarrow{FClO_3} CF_2(COOR)_2$$

[$OR = OC_2H_5$] **1** **2**

Such "fluorine scrambling" can be prevented if one produces an organometallic compound irreversibly in an ethereal solvent. In this way, fluorocyclooctatetraene (**3**), a model compound for dynamic nmr studies, was accessed from bromocyclooctatetraene *via* the lithium species, if in just 10 % yield (*7*).

Br $\xrightarrow{LiC_4H_9}$ Li $\xrightarrow{FClO_3}$ F **3**

Prior to this landmark work, 2-thienyllithium, 6-methyl-2-thienyllithium and 2-benzothienyllithium, all generated from the halogen-free heterocycles by permutational hydrogen-metal rather than halogen-metal interconversion, provided the 2-fluoro derivatives **4a** – **4c** in 44 – 70 % yield when treated with perchloryl fluoride (*8*). It remains obscure why the authors failed to convert phenyllithium into fluorobenzene under the same conditions (*8*).

$\xrightarrow{LiC_4H_9}$ $\xrightarrow{FClO_3}$ **4**

[**a** : R = R' = H; **b** : R = H_3C, R' = H; **c** : R + R' = H_4C_6]

Actually, the method proved generally applicable (*9*). This was demonstrated by the preparation of open-chain or cyclic fluoroalkanes (**5** – **7**; 39 – 83 %), a fluoroalkene (**8**; 45 %), several fluoroarenes (**9** – **11**; 42 – 94 %) and, always in aprotic medium, α-fluoroacetophenone (**12**; 44 %). In the case of ω-fluorophenylacetylene, too reactive to survive in the presence of strong nucleophiles, only its selfcondensation product 1,4-diphenyl-1,3-butadiyne (**13**; 32 %) was isolated in moderate yield (*9*).

$C_{12}H_{25}$- F **5** **6** **7**

8 **9** **10** **11**

12 **13**

Perchloryl fluoride has the bad reputation of being explosive. This is not quite exact. When left in contact with ethereal solvents or amines, it triggers a radical chain process producing chloric acid. The latter behaves like a time bomb. At any moment it may disproportionate exothermally, under detonation, to hydrochloric acid and perchloric acid. To preclude any hazards one has to monitor continuously the consumption of the reagent (most simply by using a wet, potassium-iodide impregnated paper strip) and to avoid rigorously its accumulation in the presence of aprotic organic solvents.

Safety considerations fostered the development of *N*-fluorinated surrogates, in particular *N*-fluorosulfonamides (*10, 11*), *N*-fluoroquinuclidinium salts (*12*) and *N*-chloromethyl-*N*-fluoro-1,4-diazoniabicyclo[2.2.2]octane bis(tetrafluoroborate) (*13*). However, all these reagents suffer from drawbacks. Except those which are commercial, they have to be prepared using elemental fluorine. Most of them are insufficiently soluble in ethereal media and many other solvents and, all in all, they are only moderately reactive. The fluorinated camphenes ***Z*-14**, ***E*-14** and **15** were required for an olfactive comparison with the halogen-free parent substance (*14, 15*). Whereas 10,10-difluorocamphene (**15**) was easily obtained by Horner-Wittig reaction from camphenilone, the fluorination of (*Z*)- and (*E*)-10-camphenyllithium, generated by halogen/metal permutation with butyllithium from (*Z*)-10-bromocamphene or (*E*)-10-iodocamphene, succeeded solely when *N-tert*-butyl-*N*-fluorobenzenesulfonamide (*11b*) was employed (*15).*

(1.) X = Br → H
(2.) LiC_4H_9
(1.) LiC_4H_9 + $KOC(CH_3)_3$
(2.) I_2
LiC_4H_9
F-N SO$_2$Ph
$F_2CHP(O)(OC_2H_5)_2$ ⟶ $F_2C^{\ominus}$-$P(OLi)(OC_2H_5)_2$
Z-14
E-14
15

Nucleophilic Fluorination Reagents

A most convenient laboratory-scale reaction is the vicinal bromofluorination using *N*-bromosuccinimide in the presence of an excess of hydrofluoric acid, either neat or complexed with pyridine or an aliphatic amine. In this way,

(2-bromo-1-fluoroethyl)arenes and, after base-mediated dehydrobromination, (2-bromo-1,1-difluoroethyl)arenes (**16**) became readily available from styrenes (*16*). Alternatively, they could be prepared by heating α–bromomethyl aryl ketones with diethylaminosulfur trifluoride [DAST] (*17*).

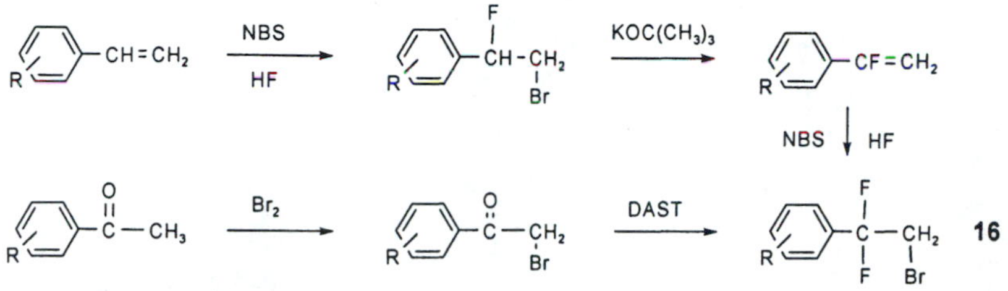

Although they reacted quite reluctantly, the (2-bromo-1,1-difluoroethyl)-arenes (**16**) were amenable to nucleophilic substitutions. The thioethers or primary alcohols thus obtained were oxidized by means of the Pummerer method or the Swern reagent to the corresponding α-aryl-α,α-difluoroacetaldehydes (**17**) (*18, 19*).

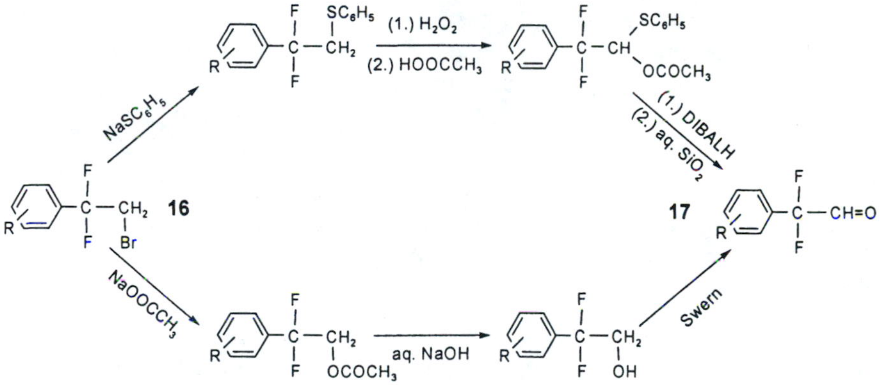

One of these β,β-difluoroacetaldeydes was elaborated to α-amino-β-(4-chloro-phenyl)-β,β-difluoropropionic acid (**18**; also obtained in enantiomerically enriched form). β-(4-Chlorophenyl)-β,β-difluoro-α-hydroxypropionic acid was formed as a by-product in the critical reductive amination of the key intermediate β-(4-chlorophenyl)-β,β-difluoro-α-oxo-propionic acid. The α-oxo acid proved to be a potent and selective inhibitor of arogenate dehydrogenase, the enzyme that catalyzes the last step of the phenylalanine biosynthesis (*19*).

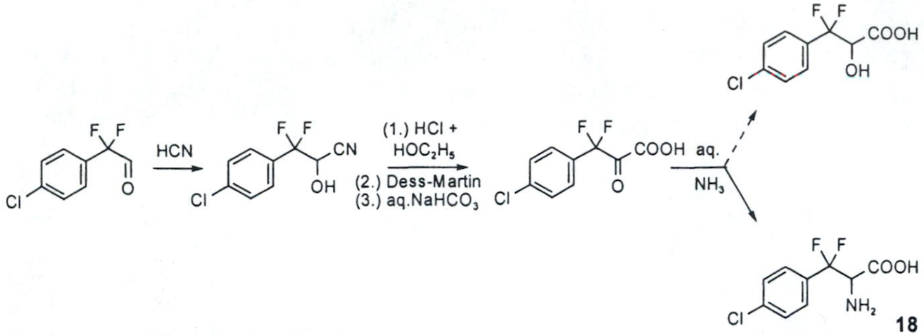

Package Deal

A trifluoromethyl substituent incorporated in an aliphatic, aromatic or heterocyclic structural motif can be ordinarily created by treating a carboxy function with sulfur tetrafluoride in the presence of anhydrous hydrogen fluoride (*20*). However, the entire CF_3 group may instead be delivered as a package from a suitable precursor, in particular the so-called Prakash-Ruppert reagent (*21*) trimethyl(trifluoromethyl)silane, to the target compound where it displaces a bromine or iodine atom. This attractive alternative option is based on the pioneering work of mainly Japanese researchers (*22* – *26*) and has been optimized recently (*27*).

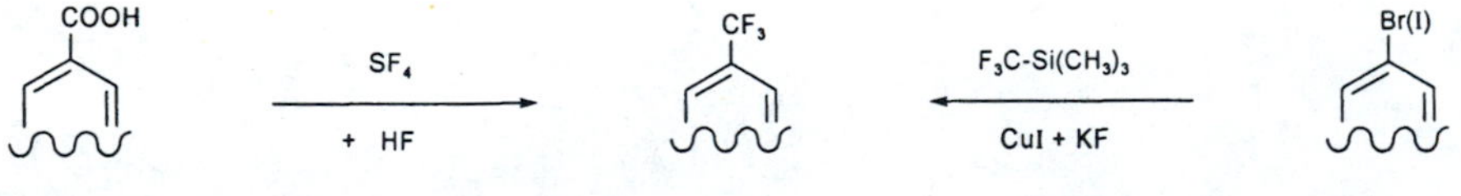

Our main objective was to extend the scope of the method toward sterically hindered arenes, a topic being featured elsewhere (*2*), and, on the other hand, to heterocyclic substrates. Thus, for example, 3-chloro-2-(trifluoromethyl)pyridine (46 %), 5-chloro-2-(trifluoromethyl)pyridine (72 %), 2-chloro-6-(trifluoromethyl)pyridine (89 %), 2-chloro-3-(trifluoromethyl)pyridine (91 %), 2-chloro-4-(trifluoromethyl)pyridine (90 %), 3-chloro-4-(trifluoromethyl)pyridine (69 %), 3-bromo-5-(trifluoromethyl)pyridine (72 %), 2-bromo-5-(trifluoromethyl) pyridine (42 %), 2-bromo-6-(trifluoromethyl)pyridine (93 %), 3-bromo-5-chloro-2-(trifluoromethyl)pyridine (69 %), 3,5-dibromo-2-(trifluoromethyl)pyridine (87 %) and 5-bromo-2-chloro-4-(trifluoromethyl)pyridine (64 %) were prepared (*27*, *28*). As a systematic comparison had revealed (*27*), the I/CF_3 displacement occurs more readily than the Br/CF_3 displacement and thus provides higher yields. Moreover, bromoiodoarenes can be selectively converted into bromo(trifluoromethyl)arenes which subsequently may be subjected to a heavy halogen/metal permutation followed by functionalization (*28*).

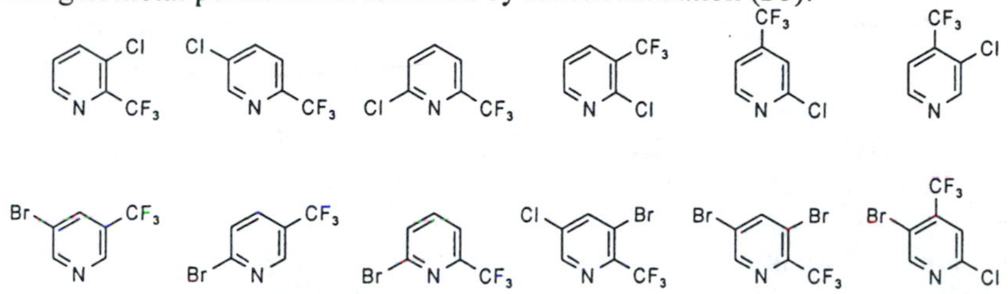

The C_1F_n "package delivery" option is not restricted to trifluoromethylation (n = 3). A famous industrial process is based on chlorodifluoromethane (n = 2) as a fluorine carrier. Upon deprotonation with sodium hydroxide and ensuing elimination of sodium chloride, it sets free difluorocarbene which readily combines with phenolates to afford aryl difluoromethyl ethers (*29*) which often exhibit intriguing biological activities (*1b, 30*). Dibromofluoromethyllithium can be readily produced from tribromofluoromethane (n = 1) by butyllithium-promoted halogen/metal permutation (*31*). The addition or condensation products obtained after electrophilic trapping should be amenable to selective debromination using, for example, tributyltin hydride as the reducing agent. *In situ* generated fluoromethoxide enables the straightforward preparation of fluoromethyl carboxylates (*32*). The structure of tris(dimethylamino)-sulfonium trifluoromethoxide (*33*) and reactions of trifluorothiolate (*34*) have to be also mentioned in this context.

Functionalization of Fluorinated Core Compounds

It is amazing how many fluoroarenes and fluoroheterocycles can be ordered from specialized suppliers at affordable prices. However, most of such core compounds are devoid of functional groups and need to be modified first before they qualify as *building blocks* in pharmaceutical or agricultural research. Organometallic intermediates are ideal for this purpose as they offer an almost illimited choice of possible transformations. The standard electrophile used to trap such species is of course carbon dioxide. A carboxylic acid, usually well crystalline, is isolated after neutralization. But in the same way one can employ *N,N*-dimethylformamide, aldehydes or ketones, oxiranes, sulfur dioxide, thionyl or sulfuryl chloride and a variety of phosphorus halides to access aldehydes, alcohols, sulfinic or sulfonic acids, phosphinic or phosphonic acids and numerous other classes of useful derivatives.

Product versatility is one big advantage of the organometallic approach to functionalized fluoro compounds, its *positional flexibility* another. Let us assume we have identified a lead structure and wish to couple it with a fluoroindolecarboxylic acid as a complementary subunit. Even if we confine the halogen and carboxy group to the benzo ring, there are still 12 possible regioisomers (for examples, see the end of this Chapter). Unless we know (how ?) in advance which substitution pattern satisfies the requirements best, we should make all possible combinations and test the resulting samples empirically. Such a *regiochemically exhaustive functionalization* is *a priori* a challenge to organic synthesis. To facilitate the task we have developed the "toolbox methods" (*35*). They will be illustrated below by representative examples.

The most acidic site of 2-fluoropyridine is the 3-position. When treated consecutively with lithium diisopropylamide in tetrahydrofuran and dry ice, the acid **19a** was isolated in 84 % yield after neutralization (*36*). The 3-position was blocked by the introduction there of a chlorine atom using 1,1,2-trichloro-1,2,2-trifluoroethane as the reagent (85 % yield) which at the same time activated the 4-position. Another metalation, this time performed with butyllithium at -75 °C, and trapping with dry ice afforded 3-chloro-2-fluoropyridine-4-carboxylic acid and, after reductive dechlorination 2-fluoropyridine-4-carboxylic acid (**19b**; 62 % yield over all). Whereas chlorine activates adjacent positions, trialkylsilyl groups screen them by steric hindrance. Thus, metalation of 3,4-dichloro-2-fluoropyridine and 3-chloro-2-fluoro-4-(trimethylsilyl)pyridine (formed by reaction of the 4-lithiated intermediate with trichlorotrifluoroethane and chlorotrimethylsilane in 80 and 90 % yield, respectively) followed by carboxylation and deprotection gave 6-fluoropyridine-3-carboxylic acid (**19c**; 57 % over all) and 6-fluoropyridine-2-carboxylic acid (**19d**; 22 % over all) (*36*).

The skilful deployment of protective groups was also crucial for the regioexhaustive functionalization of 3-fluorophenol, the methoxymethyl acetal of which was rapidly attacked by *sec*-butyllithium at the 2-position, as expected, leading to 2-fluoro-6-hydroxybenzoic acid (**20a**; 93 %). The two acids **20b** and **20c** became available *via* the same intermediate 2-chloro-1-fluoro-3-(methoxymethyl)benzene, a substrate blessed with "optional site selectivity" (*37*). Whereas butyllithium deprotonated the oxygen-adjacent 4-position (after deprotection 66 % of acid **20b** over all), lithium 2,2,6,6-tetramethylpiperidide (LITMP) attacked exclusively the fluorine-adjacent 6-position (61 % of acid **20c** over all). The access to the last acid **20d** (26 % over all) featured [(2-fluoro-6-(methoxymethoxy)phenyl]trimethylsilane as the key intermediate. This was subjected to metalation with the LIC-KOR superbase, chlorination, new metalation (with LITMP), carboxylation and deprotection (*38*).

[OR = OCH_2OCH_3]

α-Alkoxyalkyl entities such as methoxymethyl or 2-tetrahydropyranyl are excellent protective groups for phenolic hydroxyls as they can be easily attached and removed. Moreover, compared to simple alkyl groups, they accelerate *ortho*-metalation. It is nevertheless fortunate to dispose of a complementary option. The triisopropyl group sterically and electronically shields the vicinity of the phenolic oxygen atom and thus directs the attack of the base (in the given context always *sec*-butyllithium complexed with *N,N,N',N",N"*-pentamethyldiethylenetriamine) to a remote position. Thus, depending on the protective group employed, 2-, 3- and 4-(trifluoromethoxy)phenols (**21a** – **21c**; scheme overleaf) were functionalized either at an OH- or an OCF_3-adjacent site (*40*).

This peripheral control of site selectivity became the cornerstone of the regioexhaustive functionalization of 3-chlorophenol. Chlorine, which had played

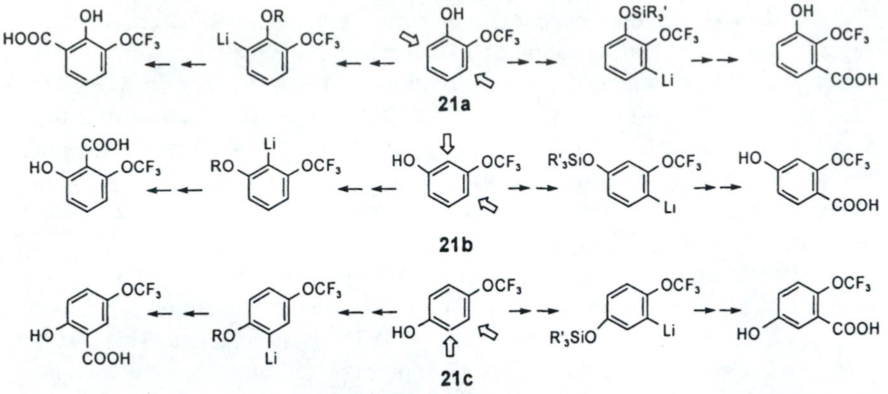

[RO = H_3COCH_2; R'_3SiO = $(^iH_7C_3)_3SiO$]

a decisive role in the regioselective derivatization of 3-fluorophenol (see above), was, for obvious reasons, not eligible as a protective group. The 2-chloro-6-hydroxybenzoic acid (**22a**; 84 %) and the 4-chloro-2-hydroxybenzoic acid (**22b**; 23 % over all) were prepared by metalation and subsequent carboxylation of 1-chloro-3-(methoxymethoxy)benzene either directly or after prior blockage of the 2-position with a trimethylsilyl group (*40*). The 2-chloro-4-hydroxybenzoic acid (**22c**; 50 %) was made from 1-chloro-3-(triisopropylsilyl)oxybenzene by metalation and carboxylation whereas the 3-chloro-5-hydroxybenzoic acid (**22d**; 21 % over all) was obtained after a multistep sequence consisting of metalation, bromination, heavy halogen migration (*23*), halogen/metal permutation and carboxylation (*38*).

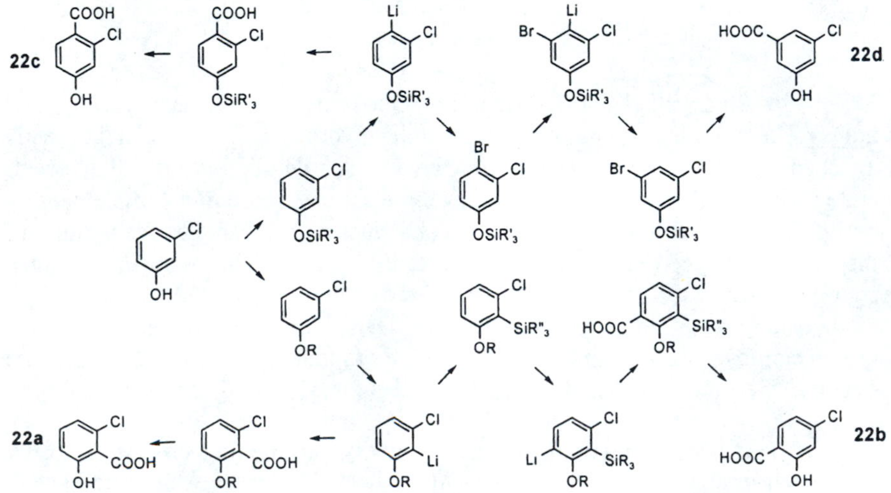

As the last examples show convincingly, the "toolbox methods" (*23*) are not only applicable to fluoro and trifluoromethyl bearing aromatic and heterocyclic substrates but also to chloro and oxygen substituted analogs. The site selective transformation of 4,7-dichloroquinoline, the precursor to the antimalarial chloroquine, sustains this claim. The substrate offering threefold "optional site selectivity", the three acids **23a** (74 %), **23b** (17 %) and **23c** (54 %) were selectively produced by merely varying the metalating reagent (LITMP in diethyl ether, butyllithium in the presence of lithium 2-dimethylaminoethoxide in diethyl ether and lithiumdiisopropylamide in tetrahydrofuran, respectively). In order to synthesize acid **23d** (12 % over all), one had to "silence" consecutively the 3- and 8-position by attaching trimethylsilyl groups. Thereafter deprotonation occurred exclusively at the 6-position (*41*). In order to access the remaining acid **23e**, one would have to introduce a 5-bromine atom and to promote its dislocation to the 4-position by deprotonation-triggered heavy halogen migration (*23*) before terminating in halogen/metal permutation, carboxylation and protodesilylation.

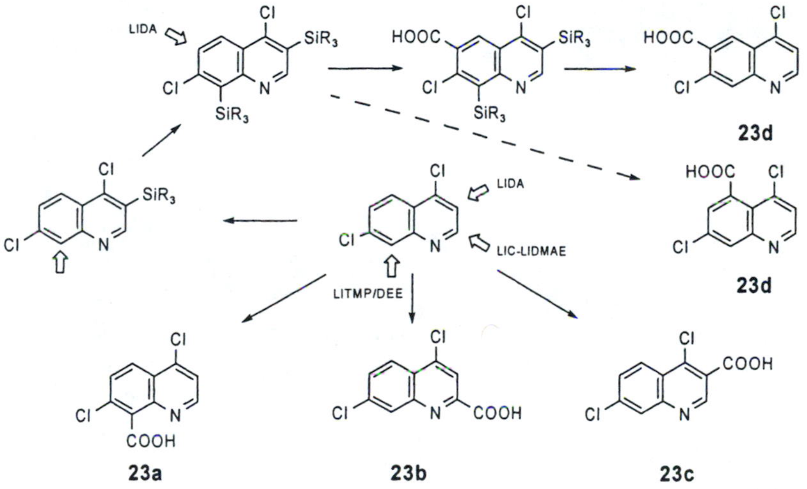

Until recently not a single functionalized derivative of 4-, 5-, 6- or 7-fluoroindole had been reported. We have elaborated a convenient entry to all twelve regioisomers wherein both the functional group and the halogen are accommodated at the benzo ring (*42*).

Only 6-fluoroindole underwent metalation directly, whereas the three other isomers, unless *N*-silyl protected, proved inert toward even strongest bases. Thus, the 6-fluoroindole-7-carboxylic acid (**24a**, 60 %) was obtained in a most straightforward manner (*42*). After attachment of a triisopropylsilyl group metalation took place instead at the sterically unhindered 5-position thus paving the way to the acid **24b** (71 % over all). Treatment of the 5-lithiated intermediate with iodine gave a 5-iodo derivative (59 %) which was converted into its 4-iodo isomer (88 %) by deprotonation-triggered heavy halogen migration. Halogen/metal permutation followed by carboxylation and deprotection eventually provided 6-fluoroindole-4-carboxylic acid (**24c**; 90 %) (*42*).

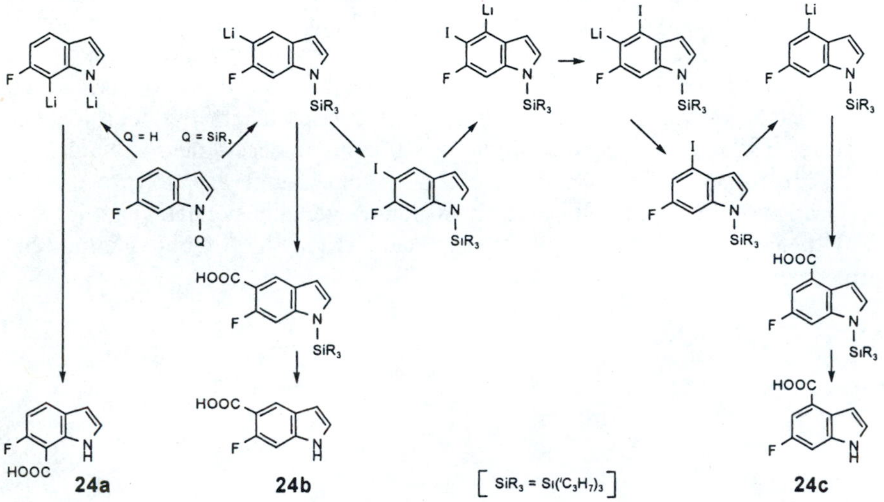

24a **24b** [$SiR_3 = Si(^iC_3H_7)_3$] **24c**

A Synopsis

"Methodology", which means the *knowledge* and critical *evaluation* of methods, is the basis for synthesis planning in any field, organofluorine chemistry included. What makes the latter area special are coincidences such as the notorious lability of fluorine labels in most aliphatic and many heterocyclic structures and inconveniences in the handling of high-performance reagents. Therefore, practical progress is especially welcome.

One frequently recommended strategy is to minimize fluorine-related trouble by transferring the halogen to an organic backbone as late as possible, in other words toward the end of a reaction sequence. This advice is particularly apt when a single fluorine atom has to be introduced into a prefabricated skeleton or when a C_1F_n entity is to be attached packagewise to a substrate.

Other rules apply when the fluorine label is already incorporated into a commercially available starting material. The elaboration of such a core compound to the targeted final product by a sequence of reaction steps can be accomplished in various ways. Being simple, safe and cheap, organometallic methods, as exemplified in the preceding Chapter, offer unique advantages in this respect, in particular functional group versatility (*43*, *44*) and regiochemical flexibility (*44*).

Acknowledgment

This work was supported by the Schweizerische Nationalfonds zur Förderung der wissenschaftlichen Forschung, Bern (grants 20-55'303-98, 20-63'584-00 and 20-100'336-02).

References

1. (a) Schlosser, M.; Michel, D.; *Tetrahedron* **1986**, *52,* 99 – 108; (b) Dingwall; J. G.; *Pestic. Sci.* **1994**, *41*, 259 – 267 [*Chem. Abstr.* **1994**, *212*, 127765]; (c) Maienfisch, P.; Hall, R. G.; *Chimia* **2004**, *58,* 93 – 99.
2. Schlosser, M.; manuscript submitted.
3. Mietchen, R.; Peters, D.; in *Organo-Fluorine Compounds* (eds.:. Baasner, B.; Hagemann, H.; Tatlow; J. C.), *Houben-Weyl : Methods of Organic Chemistry*, *Vol. E10a*, Thieme, Stuttgart, **2000**, pp. 95 – 158.
4. Subramanian, L. R.; Siegemund, G.; in *Organo-Fluorine Compounds* (eds.: Baasner, B.; Hagemann, H.; Tatlow; J. C.), *Houben-Weyl : Methods of Organic Chemistry*, *Vol. E10a*, Thieme, Stuttgart, **2000**, pp. 548 – 596.
5. Zupan, M.; in *Organo-Fluorine Compounds* (eds.: Baasner, B.; Hagemann, H.; Tatlow, J. C.), *Houben-Weyl : Methods of Organic Chemistry*, *Vol. E 10a*, Thieme, Stuttgart, **2000**, pp. 265 – 269.
6. Inman, C. E.; Oesterling, R. E.; Tyczkowski, E. A; *J. Am. Chem. Soc.* **1958**, *80,* 6533 – 6535.
7. Gwynn, D. E.; Whitesides, G. M.; Roberts, J. D.; *J. Am. Chem. Soc.* **1965**, *87,* 2862 – 2864.
8. Schuetz, R. D.; Taft, D. D.; O'Brien, J. P.; Shea, J. L.; Mork, H. M.; *J. Org. Chem.* **1963**, *28,* 1420 – 1422.
9. Schlosser, M.; Heinz, G.; *Chem. Ber.* **1969**, *102,* 1944 – 1953.
10. Furin, G. G.; in *Organo-Fluorine Compounds* (eds.: Baasner, B.; Hagemann, H.; Tatlow; J. C), *Houben-Weyl : Methods of Organic Chemistry*, *Vol. E10a*, Thieme, Stuttgart, **2000**, pp. 470 – 499.

11. (a) Barnette, W.; *J. Am. Chem. Soc.* **1984**, *106,* 452 – 454; (b) Lee, S. H.; Schwartz, J.; *J. Am. Chem. Soc.* **1986**, *108,* 2445 – 2447.
12. (a) Banks, R. E.; Sharif, I.; *J. Fluorine Chem.* **1988**, *41,* 297 – 300; (b) Banks, R. E.; Besheesh, M. K.; Khaffaf, S. N.; Sharif, I.; *J. Fluorine Chem.* **1997**, *54,* 207 – 207.
13. (a) Lal, S. G.; *J. Org. Chem.* **1993**, *58,* 2791 - 2796; (b) Matthews, D. P.; Miller, S. C.; Jarvi, E. T.; Sabol, J. S.; McCarthy, J. R.; *Tetrahedron Lett.* **1993**, *34,* 3057 – 3060; (c) Banks, R. E.; Sharif, I.; Pritchard, R. G.; *Acta Crystallogr., Sect. C* **1993**, *49,* 492 – 495 [*Chem. Abstr.* **1993**, *118*, 223259] (d) Banks, R. E.; Lawrence, N. J.; Popplewell, A. L.; *J. Chem. Soc., Chem. Commun.* **1994**, 343 - 344.
14. Garamszegi, L.; Schlosser, M.; *Chem. Ber.* **1997**, *102,* 77 – 82.
15. Wang, Q.; Schlosser, M.; unpublished results (**1998**).
16. Suga, H.; Hamatani, T.; Guggisberg, Y.; Schlosser, M.; *Tetrahedron* **1990**, *46,* 4255 – 4260.
17. Guggisberg, Y.; *Doctoral Dissertation*, Ecole Polytechnique Fédérale, Lausanne, **1992**, pp. 40 and 76.
18. Suga, H.; Schlosser, M.; *Tetrahedron* **1990**, *46,* 4261 – 4264.
19. Schlosser, M.; Brügger, N.; Schmidt, W.; Amrhein, N.; *Tetrahedron* **2004**, *60,* 7731 – 7742.
20. Dmowski, W.; in *Organo-Fluorine Compounds* (eds.: Baasner, B.; Hagemann, H.; Tatlow; J. C.), *Houben-Weyl : Methods of Organic Chemistry*, *Vol. E10a*, Thieme, Stuttgart, **2000**, pp. 321 – 394.
21. (a) Ruppert, I.; Schlich, K.; Volbach, W.; *Tetrahedron Lett.* **1984**, *25,* 2195 – 2198; (b) Krishnamurti, R.; Bellew, D. R.; Prakash, G. K. S.; *J. Org. Chem.* **1991**, *56,* 984 – 989.
22. (a) Kobayashi, Y.; Kumadaki, I.; *Tetrahedron Lett.* **1969**, *10,* 4095 – 4096; (b) Kobayashi, Y. ; Kumadaki, I.; Sato, S.; Hara, N.; Chikami, E.; *Chem. Pharm. Bull.* **1970**, *18,* 2334 – 2339; (c) Kobayashi, Y.; Kumadaki, I.; Ohsawa, A.; Yamada, T.; *Chem. Pharm. Bull.* **1972**, *20,* 1839 – 1839.
23. McLoughlin, V. C. R. ; Thrower, I.; *Tetrahedron* **1969**, *25,* 5921 – 5940.
24. Konderatenko, N. V.; Vechirko, E. P.; Yagupolskii, L. M.; *Synthesis* **1980**, 932 – 933.
25. (a) Burton, D. J.; Wiemers, D. M.; *J. Am. Chem. Soc.* **1985**, *107,* 5014 – 5015; (b) Wiemers, D. M.; Burton, D. J.; *J. Am. Chem. Soc.* **1986**, *108,* 832 – 834.
26. Urata, H.; Fuchikami, T.; *Tetrahedron Lett.* **1991**, *32,* 91 – 94.
27. Cottet, F.; Schlosser, M.; *Eur. J. Org. Chem.* **2002**, 327 – 330.
28. (a) Cottet, F.; Marull, M.; Mongin, F.; Espinosa, D.; Schlosser, M.; *Synthesis* **2004**, 1619 – 1624; (b) Cottet, F.; Schlosser, M.; *Eur. J. Org.*

Chem. **2004**, 3793 – 3798; (c) Cottet, F.; Schlosser, M.; *Tetrahedron* **2004**, *60,* 3793 – 3798.

29. (a) Clark, R. F.; Simons, J. H.; *J. Am. Chem. Soc.* **1995**, *77,* 6618 – 6618; (b) Miller, T. G.; Thanassi, J. W.; *J. Org. Chem.* **1960**, *25,* 2009 – 2012; (c) Langlois, B.; *J. Fluorine Chem.* **1988**, *41,* 247 – 261.
30. Leroux, F.; Jeschke, P.; Schlosser, M.; *Chem. Rev.* **2005**, *105,* 827 – 856.
31. (a) Kuroboshi, M.; Yamada, N.; Tokebe, Y.; Hiyama, T.; *Synlett* **1995**, 987 – 988; (b) Shimizu, M.; Yamada, N.; Takebe, Y.; Hata, T.; Kuroboshi, M.; Hiyama, T.; *Bull. Soc. Chim. Jpn.* **1998**, *71,* 2903 – 2921.
32. Schlosser, M.; Limat, D.; *Tetrahedron* **1995**, *51,* 5807 – 5812.
33. Farnham, W. B.; Smart, B. E.; Middleton, W. J.; Calabrese, J. C.; Dixon, D. A.; *J. Am. Chem. Soc.* **1985**, *107,* 4565 – 4567.
34. Adams, D. J.; Clark, J. H.; Heath, P. A.. Hausen, L. B.; Sanders, V. C.; Tavener, S. J.; *J. Fluorine Chem.* **2000**, *101,* 187 – 191.
35. Schlosser, M.; *Angew. Chem.* **2005**, *116,* 380 – 398; *Angew. Chem. Int. Ed. Engl.* **2005**, *44,* 376 – 393.
36. Bobbio, C.; Schlosser, M.; *J. Org. Chem.* **2005**, *70,* 3039 – 3045.
37. (a) Katsoulos, G.; Takagishi, S.; Schlosser, M.; *Synlett* **1991**, 731 – 732; (b) Mongin, F.; Maggi, R.; Schlosser, M.; *Chimia* **1996**, *50,* 650 – 652.
38. Marzi, E.; Bobbio, C.; Cottet, F.; Schlosser, M.; *Eur. J. Org. Chem.* **2005**, 2116 – 2123.
39. (a) Schlosser, M.; Strunk, S.; *Tetrahedron Lett.* **1984**, *25,* 741 – 744; (b) Schlosser, M.; *Pure Appl. Chem.* **1988**, *60,* 1627 – 1634; (c) Schlosser, M.; *Modern Synth. Methods* **1992**, *6*, 227 - 271.
40. Marzi, E.; Schlosser, M.; *Tetrahedron* **2005**, *61,* 3393 – 3401.
41. Schlosser, M.; Marull, M.; unpublished results (**2003**).
42. (a) Schlosser, M.; Ginanneschi, A.; Leroux, F.; *Eur. J. Org. Chem.,* in press (**2006**); (b) Ginanneschi, A.; *Doctoral Dissertation*, Ecole Polytechnique Fédérale, Lausanne (**2005**).
43. Schlosser, M.; Gorecka, J.; Castagnetti, E.; *Eur. J. Org. Chem.* **2003**, 452 – 462.
44. Schlosser, M. (ed.); *Organometallics in Synthesis : A Manual,* 2nd edition, Wiley, Chichester, **2004**.

Chapter 4

Nitrogen-Containing Organofluorine Compounds through Metathesis Reactions

Santos Fustero[1,2,*], Juan F. Sanz-Cervera[1,2], Carlos del Pozo[1], and José Luis Aceña[2]

[1]**Departamento de Química Orgánica, Universidad de Valencia, E–46100 Burjassot, Spain**
[2]**Laboratorio de Moléculas Orgánicas, Centro de Investigación Príncipe Felipe, E–46013 Valencia, Spain**

The metathesis reaction is a convenient method for the preparation of cyclic and heterocyclic compounds, of which fluorinated nitrogen-containing derivatives are of increasing interest in the search for new biologically active molecules. Several examples have been studied, including fluorinated α- and β-amino acids, uracils, piperidines, lactams, and lactones.

Introduction

Organofluorine compounds, although practically unknown in Nature, are nowadays essential for the development of new biologically active molecules in both medicinal chemistry and crop protection (*1-4*). The small size and high electronegativity of fluorine atoms endow them with unique properties that make their introduction into organic molecules a standard procedure for modulating the molecules' physical, chemical, and biological characteristics. Particularly interesting are the fluorinated analogues of proteins, nucleic acids, and their building blocks (*5-6*) because these analogues can frequently act as mimetics of their parent compounds, but with improved biological activities. For this reason, new methods for the synthesis of nitrogen-containing organofluorinated derivatives are in great demand.

At the same time, the metathesis reaction is currently recognized as one of the most potent tools for the creation of carbon-carbon bonds (*7*), as is clearly reflected by its widespread use in the syntheses of different classes of natural products and compounds of biological interest (*8*). Specifically, the ring-closing metathesis (RCM) facilitates the preparation of medium- and large-size cyclic compounds, which just a decade ago required both longer and much more complicated syntheses. By the same token, cross metathesis (CM) reactions allow for the preparation of acyclic structures by exchanging alkylidene moieties. The efficiency of this method has been made possible by the development of stable, yet highly reactive ruthenium-based catalysts such as **1-3** (Figure 1), which are compatible with the presence of a variety of functional groups.

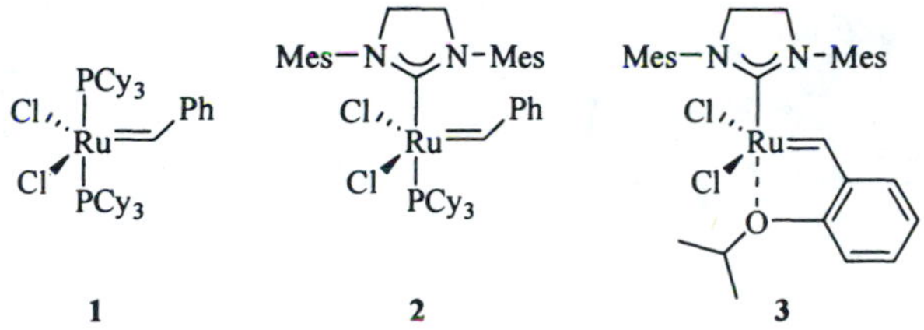

Figure 1. Structures of Grubbs' first (1) and second generation (2) and Grubbs-Hoveyda (3) ruthenium catalysts.

Metathesis reactions thus constitute a key step in many syntheses of both cyclic and acyclic nitrogenous compounds (*9-10*). The aim of this review is to show the most recent applications of metathesis processes for preparing organofluorine nitrogen-containing derivatives, with a special focus on the synthesis of fluorinated amino acids, uracils, and other nitrogen heterocycles.

Synthesis of Cyclic Fluorinated Amino Acids

Fluorinated α-Amino Acids

The insertion of unnatural amino acids is a common strategy for synthesizing modified peptides with more stable secondary and tertiary structures, as well as for improving their bioavailability. For this purpose, fluorinated amino acids, the syntheses of which have been recently reviewed (*11-12*), are widely employed. While several approaches for the preparation of cyclic fluorinated α-amino acids have been reported, only a few have involved RCM processes. One pioneering example is the synthesis of racemic five- to seven-membered fluorinated α-amino acid derivatives with the nitrogen atom included in the ring, which was reported by Osipov, Dixneuf *et al.* This synthesis starts from fluorinated imines **4** derived from methyl 3,3,3-trifluoropyruvate or 3-chloro-3,3-difluoropyruvate (Figure 2). Regioselective addition of alkenyl Grignard reagents affords amines **5**, in which a second unsaturated substituent is introduced on the nitrogen atom to yield dienes **6**. The subsequent RCM reaction of **6** in the presence of catalyst **1** affords racemic cyclic α-amino esters **7**, which bear a quaternary trifluoromethyl or chlorodifluoromethyl group as a substituent (*13-14*).

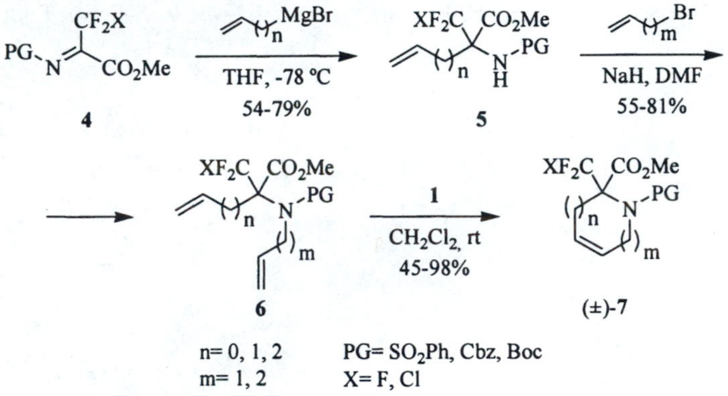

n= 0, 1, 2 PG= SO_2Ph, Cbz, Boc
m= 1, 2 X= F, Cl

Figure 2. Synthesis of five- to seven-membered fluorinated α-amino esters.

Our research group has described the preparation of enantiomerically pure fluorinated β-amino alcohols starting from 2,2-difluoro-4-pentenoic acid **8** (*15*) (Figure 3). Conversion of this acid into imidoyl chloride **9** with Uneyama's method (*16*) was followed by condensation with the lithium carbanion derived from optically pure sulfoxide **10**; the subsequent reduction of the imino function gave β-amino sulfoxide **11** in a 99:1 diastereomeric ratio. After changing the amino protecting group, transformation of sulfoxide **12** into β-amino alcohol **13** was accomplished by means of a non-oxidative Pummerer reaction, and subsequent *N*-allylation or homoallylation was accompanied by oxazolidinone ring formation. Finally, an RCM of compounds **14** provided bicyclic β-amino alcohol derivatives **15** in high yields. These, in turn, served as direct precursors to the corresponding seven- and eight-membered cyclic α-amino acids **16**, which incorporate a *gem*-difluoro unit within the ring (*17*).

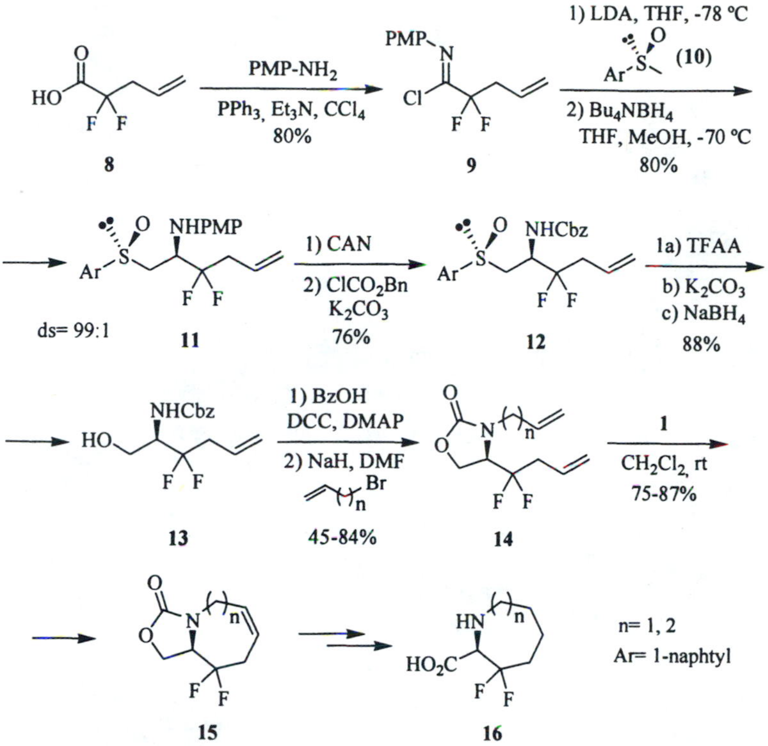

Figure 3. *Synthesis of bicyclic fluorinated β-amino alcohol derivatives.*

More recently, we have developed a different synthetic route to cyclic α-amino acids that also starts from fluorinated acid **8**, which is first converted into chiral imidoyl chloride **17** derived from *O*-(methyl)-α-phenylglycinol (Figure 4). Palladium-catalyzed alkoxycarbonylation of the derived imidoyl iodide in the presence of 2-(trimethylsilyl)ethanol (TMSE-OH) gives imino ester **18**, and addition of allylzinc bromide affords amino ester **19** in a fully regio- and diastereoselective manner. A subsequent RCM reaction furnishes compound **20**, and then removal of the chiral auxiliary and double bond reduction through hydrogenation yields **21**. Finally, TBAF-assisted ester hydrolysis produces cyclic amino acid **22**, the absolute configuration of which was determined by subjecting its acetamide derivative **23** to X-ray analysis (*18*).

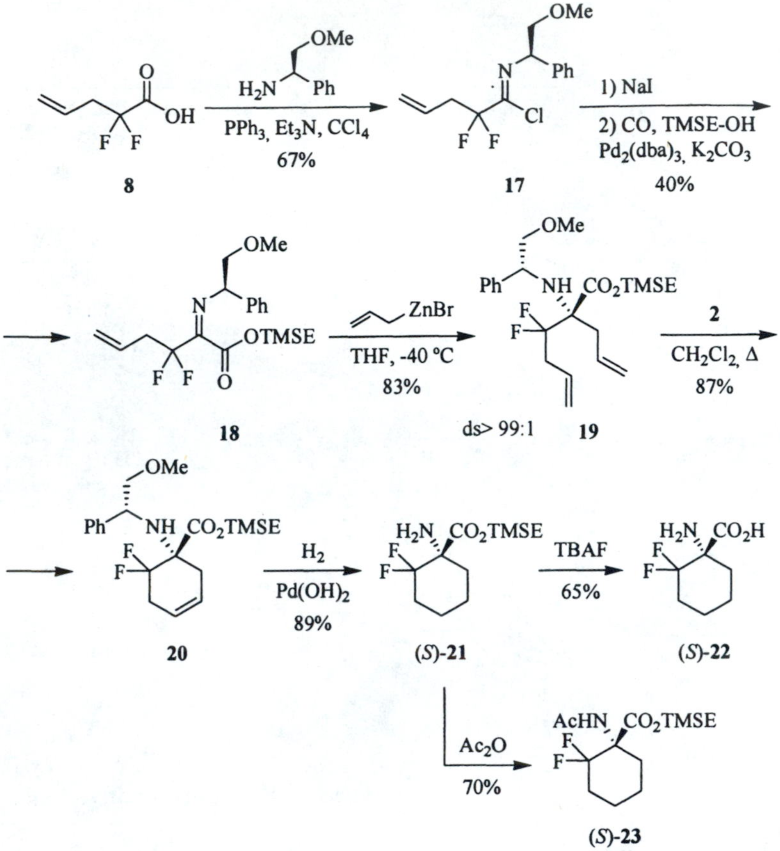

Figure 4. Synthesis of enantiomerically pure fluorinated cyclic α-amino acids.

Fluorinated β-Amino Acids

The β-analogues of proteinogenic amino acids are interesting target compounds that can be found in the structure of biologically active natural products (*19-21*). In addition, they can form oligomers (β-peptides) with defined folded structures (*22*). While syntheses of fluorinated acyclic β-amino acids are relatively common in the literature (*23-26*), those of their cyclic derivatives are much rarer; in fact, only our group has reported their synthesis with the aid of metathesis reactions. To begin, we prepared racemic seven-membered β-amino esters containing a *gem*-difluoro unit, again starting from imidoyl chloride **9** (Figure 5). Reaction with the lithium enolate derived from ethyl 4-pentenoate afforded the imino ester **24**. Next, an RCM reaction in the presence of catalyst **2** afforded the cyclic imino ester **25**, which in turn was stereoselectively reduced with $NaCNBH_3$ to yield racemic *cis*-β-amino ester **26**. Reversing the order of steps (imine reduction *before* the metathesis reaction) produced a 1:1 mixture of *cis* and *trans* β-amino esters and also caused a decrease in the yield (27%) of the metathesis process. Compound **26** could be transformed into saturated derivative **27** either through hydrogenation or by removing the *p*-(methoxy)phenyl (PMP) protecting group with ceric ammonium nitrate (CAN) to yield **28** (*27*).

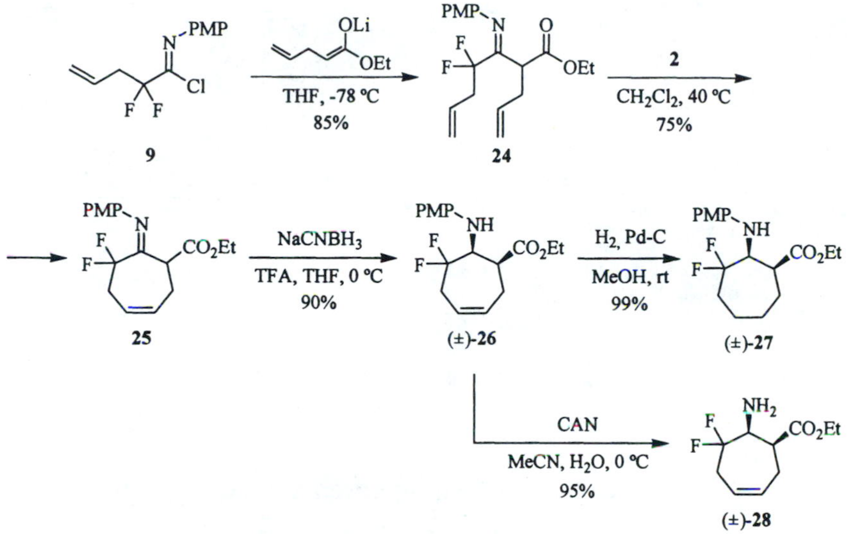

Figure 5. Synthesis of seven-membered β-amino esters by means of ring-closing metathesis.

Despite its effectiveness in the aforementioned synthesis, this synthetic strategy was not suitable for the preparation of five- and six-membered β-amino acid derivatives. For this reason, we developed an alternative route, employing a cross-metathesis (CM) reaction instead of an RCM (Figure 6). In this manner, imidoyl chlorides **9**, **29**, and **30**, along with varying lengths of their unsaturated chains, were reacted with ethyl acrylate in the presence of catalyst **2** to give compounds **31-33**. To the best of our knowledge, this is the first reported example of a metathesis reaction in the presence of an imidoyl halide functionality. After the double bond was hydrogenated, the resulting imidoyl chlorides were reacted with lithium diisopropylamide (LDA) to produce an intramolecular Dieckmann-type condensation that led to enamino esters **34-36**. These compounds were efficiently reduced under several different conditions in sequence to obtain selectively *cis*-β-amino esters **37**, a protected difluorinated analogue of the fungicide agent *cis*-pentacin, and **38**, along with the previously obtained **27**. In addition, an enantioselective version of this synthesis has also been carried out with a chiral auxiliary in the ester group in order to prepare enantiomerically pure β-amino acid derivatives (*28*).

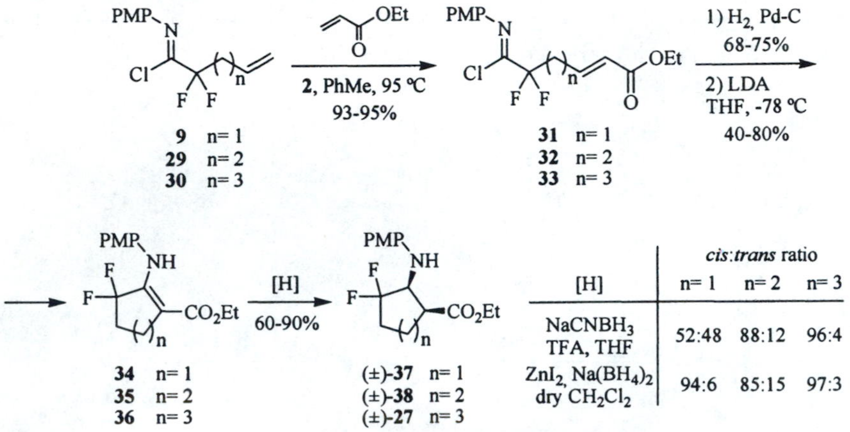

[H]	*cis*:*trans* ratio n= 1	n= 2	n= 3
$NaCNBH_3$, TFA, THF	52:48	88:12	96:4
ZnI_2, $Na(BH_4)_2$, dry CH_2Cl_2	94:6	85:15	97:3

Figure 6. Synthesis of five- to seven-membered β-amino esters by means of cross metathesis.

Synthesis of Fluorinated Uracils

The metathesis reaction has been widely employed for the synthesis of nucleoside derivatives, usually to build the sugar ring. Only occasionally has it been used to prepare fluorinated analogues (*29*). We became interested in the synthesis of uracils with a fluorinated group at C-6 due to their potential as agrochemicals (*30-31*). In this context, we prepared two new families of bicyclic difluorinated uracils starting from nitrile **39**, which was obtained in three steps from acid **8** (Figure 7). Addition of the lithium enolates of acetate derivatives to **39** afforded enamino esters **40**, which were further reacted with several different isocyanates to yield uracils **41**. When a suitable olefin substituent was present in the molecule (R^1= vinyl or allyl), direct RCM cyclization produced six- and seven-membered bicyclic uracils **42**. In contrast, palladium-mediated *N*-allylation of **41** (R^1= H) furnished diene **43**, which was then cyclized to give **44** (*32*).

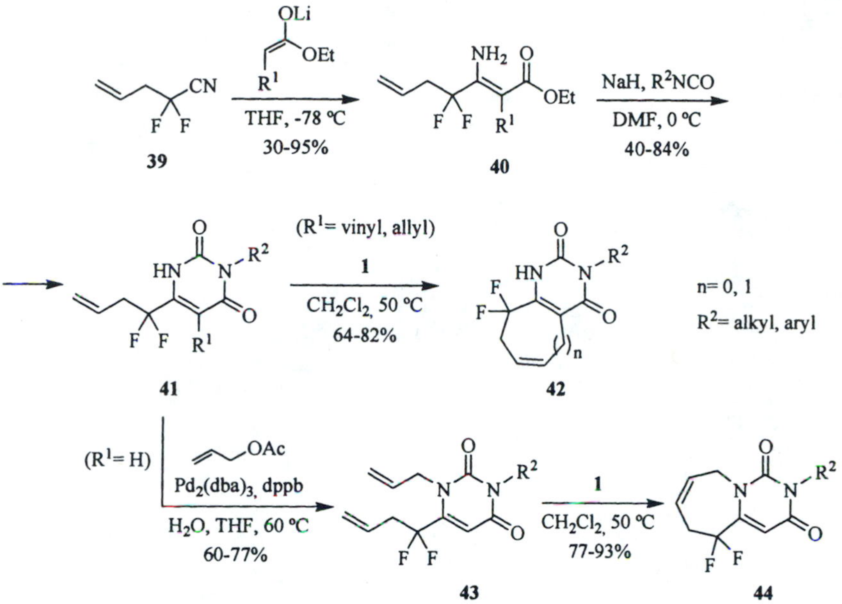

Figure 7. *Synthesis of bicyclic fluorinated uracils.*

Synthesis of Fluorinated Piperidines

Nitrogen heterocycles, including pyrrolidines and piperidines, constitute structural features of many natural products and drug candidates, and are now readily accessible through RCM reactions (*33*). Consequently, several fluorinated nitrogen heterocycles have been prepared by means of RCM processes, although with several limitations. Thus, when Brown *et al.* tried to prepare both pyrrolidines and piperidines starting from fluoroolefins **45** (Figure 8), the approach was effective only in the latter case (n= 2), yielding fluorinated piperidine **46**. In contrast, the allyl amine (n= 1) underwent a double bond isomerization followed by enamine hydrolysis, which led to the formation of compound **47** (*34*). This isomerization is probably mediated by a ruthenium hydride species formed as a result of the decomposition of catalyst **2** (*35*).

Billard *et al.* employed a fluoral hemiketal as starting material for the preparation of an RCM precursor (Figure 9). Thus, allylation of compound **48** followed by imine hydrolysis and protection afforded trifluoromethylated homoallylamine **49**. *N*-Allylation gave diene **50**, and its RCM cyclization furnished **51**, which was hydrogenated and deprotected to yield racemic trifluoropipecoline hydrochloride **52** (*36*).

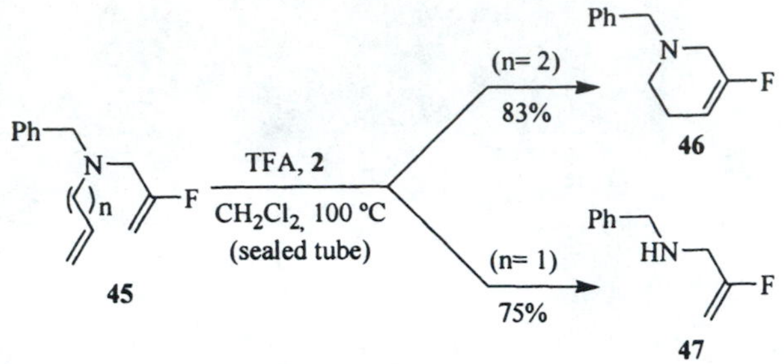

Figure 8. Synthesis of fluorinated piperidines.

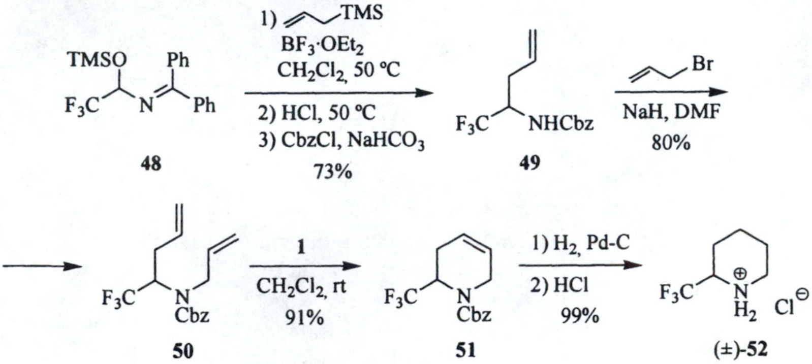

Figure 9. Synthesis of trifluoropipecoline.

More recently, Bonnet-Delpon *et al.* also made use of imines derived from fluoral in order to prepare an RCM precursor (Figure 10). To this end, chiral imine **53** was bis-allylated in a one-pot process to obtain diene **54** with a high degree of diastereoselectivity, although the configuration of the new stereocenter was not determined. A subsequent RCM reaction produced trifluoromethyl piperidine **55** with no loss of optical purity (*37*).

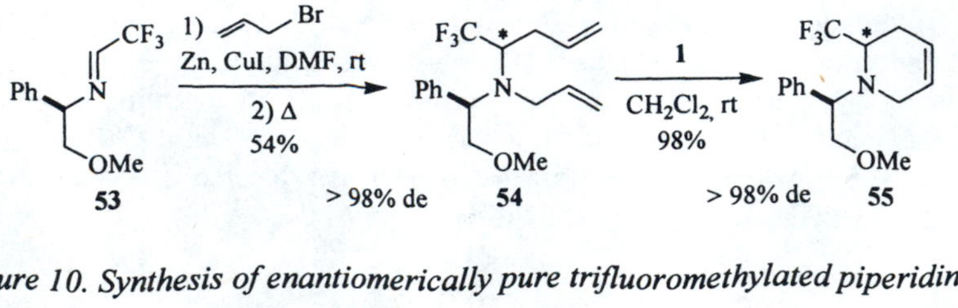

Figure 10. Synthesis of enantiomerically pure trifluoromethylated piperidines.

Synthesis of Fluorinated Lactams, Lactones and Aza-Lactones

The synthesis of lactam rings by means of RCM reactions is quite straightforward since the acyclic precursors are readily available through acylation of secondary alkenyl amines. In this way, Haufe *et al.* prepared fluorinated five- and six-membered lactams starting from amines **56** (Figure 11). Acylation with α-fluoroacrylic acid produced **57**, and subsequent RCM with catalyst **2** afforded five- and six-membered lactams **58**. The higher homologues (n= 3 and 4) failed to yield the corresponding seven- and eight-membered lactams under the same conditions (*38*). Similar results were obtained by Rutjes *et al.* (*i.e.* **59** → **60**), but the synthesis of a five-membered trifluoromethylated lactam was unsuccessful (*39-40*).

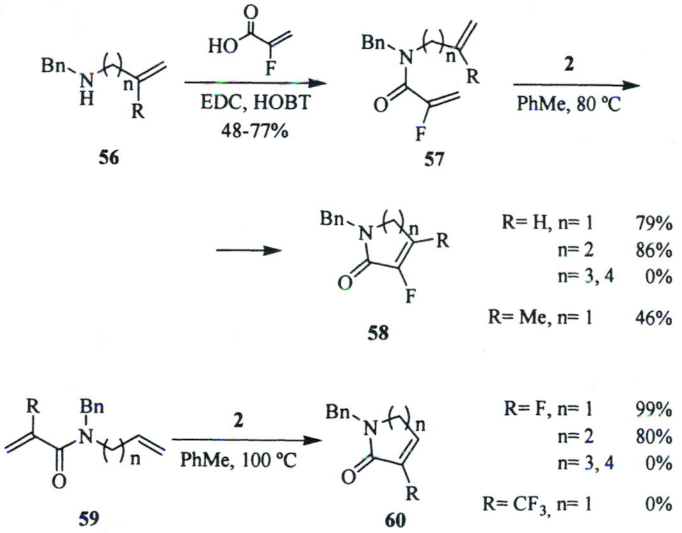

Figure 11. Synthesis of fluorinated lactams.

We have recently shown that regioselective preparation of unsaturated fluorinated lactams containing a *gem*-difluoro unit in the ring system is possible by RCM and carefully choosing of the appropriate ruthenium catalyst, solvent, and reaction conditions (Figure 12). For this purpose we used dienyl amide **61**, which was prepared in two steps from acid **8** (n=1,2). Treatment of **61** with catalyst **1** in refluxing CH_2Cl_2 afforded exclusively the normal RCM product **62** whereas the use of catalyst **2** in toluene at reflux gave the isomeric lactams **63**. It was also possible to isomerize the double bond starting from **62** under the same reaction conditions. As in the case described above, this double bond isomerization was probably promoted by a ruthenium hydride species formed as a result of the decomposition of catalyst **2** at high temperatures (*35*). Interestingly, the other possible regioisomer was not detected due to a stereoelectronic effect exerted by the difluoromethylene group that compels the double bond to migrate exclusively towards the nitrogen atom. This observation was supported by the fact that non-fluorinated analogues produced mixtures of isomers towards both sides under the same reaction conditions. For the synthesis of a six-membered lactam, we chose a different strategy: **61** was first isomerized to **64** in the presence of a ruthenium hydride, with subsequent RCM cyclization furnishing lactam **65** (*41*).

Fluorinated analogues of naturally occurring macrolactones can be a source of derivatives with improved biological activities, as was demonstrated with the preparation of a trifluoromethylated analogue of the antitumoral compounds epothilones. Moreover, their synthesis is usually performed by means of RCM cyclizations (*i.e.* **66** → **67**) (Figure 13) (*42*). Accordingly, we decided to prepare different classes of fluorinated macrolactones (Figure 14).

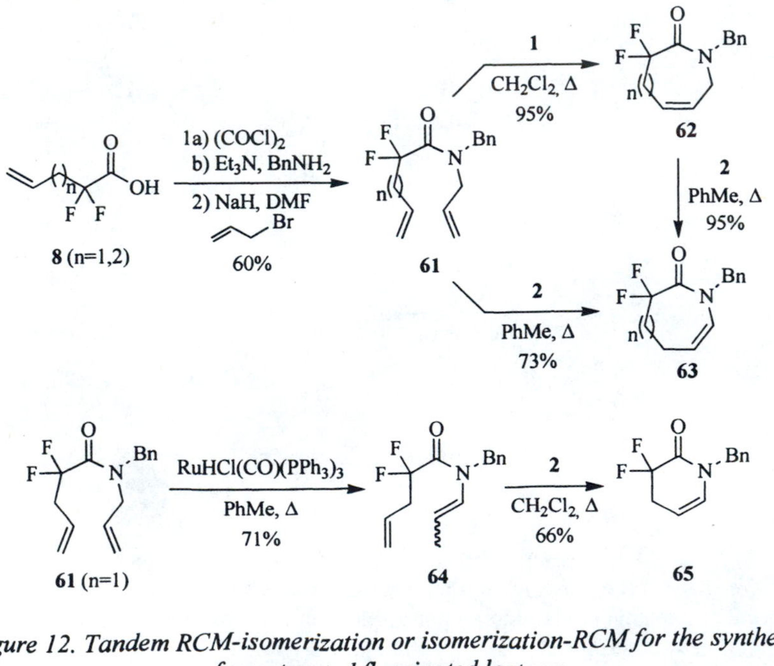

Figure 12. Tandem RCM-isomerization or isomerization-RCM for the synthesis of unsaturated fluorinated lactams.

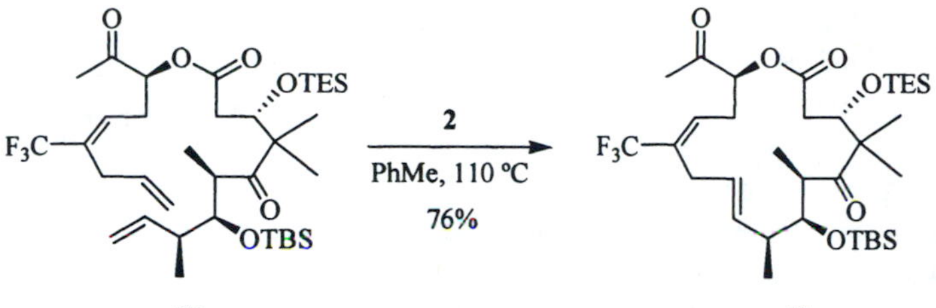

O O O OTES F_3C O OTBS

66

2

PhMe, 110 °C

76%

O O O OTES F_3C O OTBS

67

Figure 13. Synthesis of a precursor of a trifluoromethylated analogue of the epothilones by ring-closing metathesis.

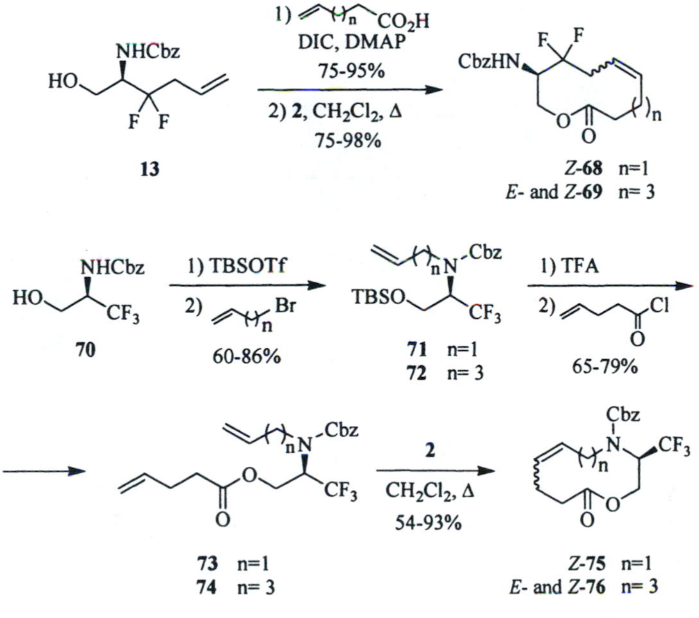

NHCbz HO F F

13

1) n CO_2H
DIC, DMAP
75-95%

2) **2**, CH_2Cl_2, Δ
75-98%

F F CbzHN O O n

Z-**68** n=1
E- and *Z*-**69** n= 3

NHCbz HO CF_3

70

1) TBSOTf
2) n Br
60-86%

n N Cbz TBSO CF_3

71 n=1
72 n= 3

1) TFA
2) Cl O
65-79%

n N Cbz O O CF_3

73 n=1
74 n= 3

2
CH_2Cl_2, Δ
54-93%

Cbz N CF_3 n O O

Z-**75** n=1
E- and *Z*-**76** n= 3

Figure 14. Synthesis of 10- and 12-membered fluorinated lactones and aza-lactones.

Thus, acylation of enantiomerically pure fluorinated amino alcohol **13** followed by an RCM reaction with catalyst **2** afforded 10- and 12-membered lactones **68** and **69**, respectively, both with a difluoromethylene unit. While **68** was obtained as the *Z* isomer only, the higher homologue **69** proved to be an equimolecular mixture of *E* and *Z* olefins. In contrast, the synthesis of the corresponding 11-membered lactone was much less effective since ring-contraction products were predominant in the reaction mixture. A different class of aza-macrolactones was prepared from trifluoromethylated β-amino alcohol **70**, which was obtained in a similar manner to compound **13**. Thus, hydroxyl protection and *N*-alkylation with allyl bromide or 4-pentenyl bromide led to **71** and **72**, respectively, and further deprotection and acylation with 4-pentenoyl chloride afforded **73** and **74**. Finally, an RCM reaction in the presence of catalyst **2** produced 10- and 12-membered trifluorometylated aza-lactones **75** and **76**, the latter as a 3:1 mixture of *E* and *Z* isomers (*43*).

Conclusions

In summary, metathesis reactions have been used quite intensively in the last few years for the preparation of organofluorinated nitrogen-containg compounds. These methods have made possible the preparation of numerous systems that otherwise would have been much more difficult to obtain. While the vast majority of examples of the application of metathesis reactions correspond to RCM, we also believe that CM reactions will follow in their development in the immediate future.

Acknowledgements

The authors wish to thank the Ministerio de Educación y Ciencia (BQU2003-01610) and the Generalitat Valenciana of Spain (GR03-193 and GV05/079) for financial support. C. P. thanks the MEC for a Ramón y Cajal contract and J. L. A. thanks the CIPF for a postdoctoral fellowship.

References

1. *Organo-Fluorine Compounds, Houben-Weyl vol. E10 a-c*; Baasner, B.; Hagemann, H.; Tatlow, J. C., Eds.; Georg Thieme Verlag KG: Stuttgart-New York, 2000.

2. *Enantiocontrolled Synthesis of Fluoroorganic Compounds: Stereochemical Challenges and Biomedical Targets;* Soloshonok, V. A., Ed.; John Wiley and Sons: Chichester, 1999.
3. Böhm, H.-J.; Banner, D.; Bendels, S.; Kansy, M.; Kuhn, B.; Müller, K.; Obst-Sander, U.; Stahl, M. *ChemBioChem* **2004**, *5*, 637-643.
4. Jeschke, P. *ChemBioChem* **2004**, *5*, 570-589.
5. Jäckel, C.; Koksch, B. *Eur. J. Org. Chem.* **2005**, 4483-4503.
6. Pankiewicz, K. W. *Carbohydr. Res.* **2000**, *327*, 87-105.
7. *Handbook of Metathesis, vols. 1-3*; Grubbs, R. H., Ed.; Wiley-VCH Verlag GmbH: Weinheim, 2003.
8. Nicolaou, K. C.; Bulger, P. G.; Sarlah, D. *Angew. Chem. Int. Ed.* **2005**, *44*, 4490-4527.
9. Philips, A. J.; Abell, A. D. *Aldrichimica Acta* **1999**, *32*, 75-89.
10. Vernall, A. J.; Abell, A. D. *Aldrichimica Acta* **2003**, *36*, 93-105.
11. *Fluorine Containing Amino Acids: Synthesis and Properties*; Kukhar, V. P.; Soloshonok, V. A., Eds.; John Wiley and Sons: Chichester, 1995.
12. Qiu, X.-L.; Meng, W.-D.; Qing, F.-L. *Tetrahedron* **2004**, *60*, 6711-6745.
13. Osipov, S. N.; Bruneau, C.; Picquet, M.; Kolomiets, A. F.; Dixneuf, P. H. *Chem. Commun.* **1998**, 2053-2054.
14. Osipov, S. N.; Artyushin, O. I.; Kolomiets, A. F.; Bruneau, C.; Picquet, M.; Dixneuf, P. H. *Eur. J. Org. Chem.* **2001**, 3891-3897.
15. Lang, R. W.; Greuter, H.; Romann, A. J. *Tetrahedron Lett.* **1988**, *29*, 3291-3294.
16. Uneyama, K. *J. Fluorine Chem.* **1999**, *97*, 11-25.
17. Fustero, S.; Navarro, A.; Pina, B.; García Soler, J.; Bartolomé, A.; Asensio, A.; Simón, A.; Bravo, P.; Fronza, G.; Volonterio, A.; Zanda, M. *Org. Lett.* **2001**, *3*, 2621-2624.
18. Fustero, S.; Sánchez-Roselló, M.; Rodrigo, V.; del Pozo, C.; Sanz-Cervera, J. F.; *submitted to publication.*
19. Fülöp, F. *Chem. Rev.* **2001**, *101*, 2181-2204.
20. Liu, M.; Sibi, M. P. *Tetrahedron* **2002**, *58*, 7991-8035.
21. Lelais, G.; Seebach, D. *Biopolymers (Peptide Science)* **2004**, *76*, 206-243.
22. Cheng, R. P.; Gellman, S. H.; DeGrado, W. F. *Chem. Rev.* **2001**, *101*, 3219-3232.
23. Fustero, S.; Sanz-Cervera, J. F.; Soloshonok, V. A. In *Enantioselective Synthesis of β-Amino Acids, 2nd Edition*; Juaristi, E.; Soloshonok, V. A., Eds.; John Wiley and Sons, Inc.: Hoboken, NJ, 2005; pp 319-350.
24. Fustero, S.; Pina, B.; Salavert, E.; Navarro, A.; Ramírez de Arellano, M. C.; Simón Fuentes, A. *J. Org. Chem.* **2002**, *67*, 4667-4679.

25. Hook, D. F.; Gessier, F.; Noti, C.; Kast, P.; Seebach, D. *ChemBioChem* **2004**, *5*, 691-706.
26. Chung, W. J.; Omote, M.; Welch, J. T. *J. Org. Chem.* **2005**, *70*, 7784-7787.
27. Fustero, S.; Bartolomé, A.; Sanz-Cervera, J. F.; Sánchez-Roselló, M.; García Soler, J.; Ramírez de Arellamo, C.; Simón Fuentes, A. *Org. Lett.* **2003**, *5*, 2523-2526.
28. Fustero, S.; Sánchez-Roselló, M.; Sanz-Cervera, J. F.; Aceña, J. L.; del Pozo, C.; Fernández, B.; Bartolomé, A.; Asensio, A.; *submitted to publication*.
29. Amblard, F.; Nolan, S. P.; Agrofoglio, L. A. *Tetrahedron* **2005**, *61*, 7067-7080.
30. Fustero, S.; Salavert, E.; Sanz-Cervera, J. F.; Piera, J.; Asensio, A. *Chem. Commun.* **2003**, 844-845.
31. Fustero, S.; Piera, J.; Sanz-Cervera, J. F.; Catalán, S.; Ramírez de Arellano, C. *Org. Lett.* **2004**, *6*, 1417-1420.
32. Fustero, S.; Catalán, S.; Piera, J.; Sanz-Cervera, J. F.; Fernández, B.; Aceña, J. L.; *submitted to publication*.
33. Deiters, A.; Martín, S. F. *Chem. Rev.* **2004**, *104*, 2199-2238.
34. Salim, S. S.; Bellingham, R. K.; Satcharoen, V.; Brown, R. C. D. *Org. Lett.* **2003**, *5*, 3403-3406.
35. Hong, S. H.; Day, M. W.; Grubbs, R. H. *J. Am. Chem. Soc.* **2004**, *126*, 7414-7415.
36. Gille, S.; Ferry, A.; Billard, T.; Langlois, B. R. *J. Org. Chem.* **2003**, *68*, 8932-8935.
37. Magueur, G.; Legros, J.; Meyer, F.; Ourévitch, M.; Crousse, B.; Bonnet-Delpon, D. *Eur. J. Org. Chem.* **2005**, 1258-1265.
38. Marhold, M.; Buer, A.; Hiemstra, H.; van Maarseveen, J. H.; Haufe, G. *Tetrahedron Lett.* **2004**, *45*, 57-60.
39. De Matteis, V.; Van Delft, F. L.; de Gelder, R.; Tiebes, J.; Rutjes, F. P. J. T. *Tetrahedron Lett.* **2004**, *45*, 959-963.
40. De Matteis, V.; Van Delft, F. L.; Tiebes, J.; Rutjes, F. P. J. T. *Eur. J. Org. Chem.* **2006**, 1166-1176.
41. Fustero, S.; Sánchez-Roselló, M.; Jiménez, D.; Sanz-Cervera, J. F.; del Pozo, C.; Aceña, J. L. *J. Org. Chem.* **2006** (in press).
42. Rivkin, A.; Chou, T.-C.; Danishefsky, S. J. *Angew. Chem. Int. Ed.* **2005**, *44*, 2838-2850.
43. Fernández, B., Ph. D. Dissertation in progress, Universidad de Valencia, Spain.

Chapter 5

Highly Diastereoselective Anodic Fluorination

Toshio Fuchigami and Toshiki Tajima

Development of Electronic Chemistry, Tokyo Institute of Technology, Nagatsuta, Midori-ku, Yokohama 226–8502, Japan

This chapter summarizes diastereoselective electrochemical fluorination developed in our laboratory. This includes diastereoselective anodic fluorination of open-chain and cyclic compounds containing heterocyclic compounds mainly via intramolecular 1,2- and 1,3-asymmetric induction. The stereocontrol mechanisms are illustrated. The related work by others are also covered in this chapter.

Introduction

Synthesis of chiral compounds having a fluorine at a chiral center has been becoming of much importance in medicinal and analytical chemistry together with material science. Among various synthetic methods, asymmetric direct fluorination is the most ideal method from an aspect of atom economy. Asymmetric direct fluorination has been successfully carried out using chiral fluorinating reagents, however, asymmetric anodic fluorination is quite difficult in general, due to the small size of a fluoride ion and necessity of use of polar solvents for electric conductivity (*1-4*). Therefore, so far only a few studies have been reported on asymmetric anodic fluorination. In this chapter, successful examples of asymmetric anodic fluorination are briefly illustrated.

Diastereoslective Anodic Fluorination of Open-Chain Compounds

Laurent and his co-workers investigated diastereoselective anodic benzylic fluorination of α-phenylacetates via intramolecular asymmetric induction using various chiral auxiliaries as an ester group (*5*). They obtained the fluorinated product with up to 60% de when 8-phenylmenthyl group was used as a chiral auxiliary as shown in Scheme 1.
The moderate to low diastereoselectivity of above examples seems to be attributable to the long distance between the reaction site and the chiral auxiliary moiety as well as non rigid conformation of the cationic intermediate. Therefore, we next investigated diastereoselective anodic fluorination via 1,2-asymmetric induction as follows.

At first, anodic fluorination of 2-methoxypropyl phenyl sulfide was investigated in Et_3N-3HF/MeCN (*7*), however, the diastereoselectivity of the product was as low as 30% de. In sharp contrast, the diastereoselectivity of the anodic fluorination of sulfide having two methoxy groups at β- and γ-positions increased to 60% de.

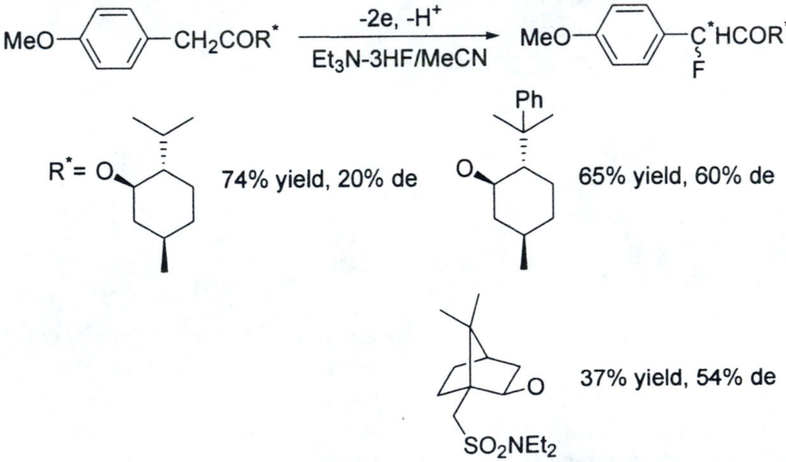

MeO–C6H4–CH_2COR^* —(-2e, $-H^+$; Et_3N-3HF/MeCN)→ MeO–C6H4–C^*HCOR^* (F)

R^* = O-menthyl: 74% yield, 20% de

O-8-phenylmenthyl: 65% yield, 60% de

O-bornyl (SO_2NEt_2): 37% yield, 54% de

Scheme 1. Diastereselective Anodic Fluorination of α-phenylacetates having chiral auxiliaries

We studied similar diastereoselective anodic fluorination of α-phenylsulfenyl esters having various chiral auxiliaries (*6*). 8-Phenylmenthyl group also gave the highest diastereoselectivity, however the de of the product was very low. The diastereoselectivity was also affected by the supporting electrolyte and it was found that that Et_4NF-3HF showed a better diastereoselectivity than Et_3N-3HF or Et_3N-2HF as shown in Scheme 2.

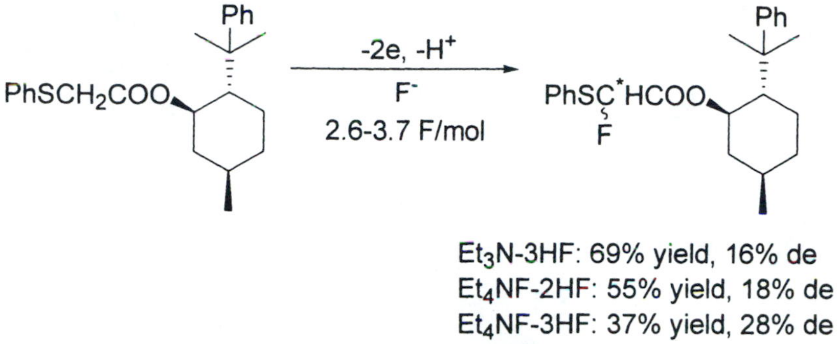

Ph

$PhSCH_2COO$ → ($-2e, -H^+$; F^-; 2.6-3.7 F/mol) → $PhSC^*HCOO$ (F)

Ph

Et_3N-3HF: 69% yield, 16% de
Et_4NF-2HF: 55% yield, 18% de
Et_4NF-3HF: 37% yield, 28% de

Scheme 2. Effect of supporting fluoride salts on diastereoselectivity of anodic fluorination

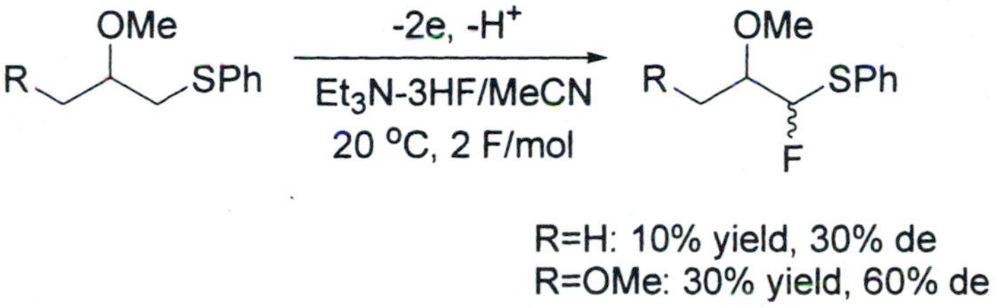

OMe

R SPh → ($-2e, -H^+$; Et_3N-3HF/MeCN; 20 °C, 2 F/mol) → R SPh (OMe, F)

R=H: 10% yield, 30% de
R=OMe: 30% yield, 60% de

Scheme 3. Diastereoselective anodic fluorination of 2-methoxypropyl phenyl sulfide

Then, protecting group effect of 1,2-diol moieties on the diastereoselective anodic fluorination was investigated (*7*). Among the protecting groups, the 2-spirocyclohexyl-1,3-dioxolan-4-yl group gave the highest diastereoselectivity (**2a**) as shown in Scheme 4. 1,3-Dioxolanyl and 2,2-dimethyl-1,3-dioxolanyl groups gave ca. 70% de (**2b**, **2c**), while 1,3-dioxolanonyl group decreased the de drastically to 13% (**2d**).
Interestingly, the diastereoselectivity, de is not affected appreciably by the reaction temperature although de is changed significantly depending on the solvents as shown in Scheme 5.
Thus, the diastereoselectivity of the anodic fluorination was found to be greatly affected by the protecting groups, supporting fluoride salts, and electrolytic solvents.

Based on the X-ray analysis of the major diastereomer of the product **2a**, the configuration was found to be (1*R*,2*S*). On the basis of the result, we considered the detailed mechanism of the stereochemical control in nucleophilic reactions as shown in Scheme 6 using the Felkin-Anh model generally employed for this type of reaction via a cationic intermediate. For the two possible conformations **C** and **D**, fluoride ion attack at the activated carbon atom is usually explained to occur from the other side of the

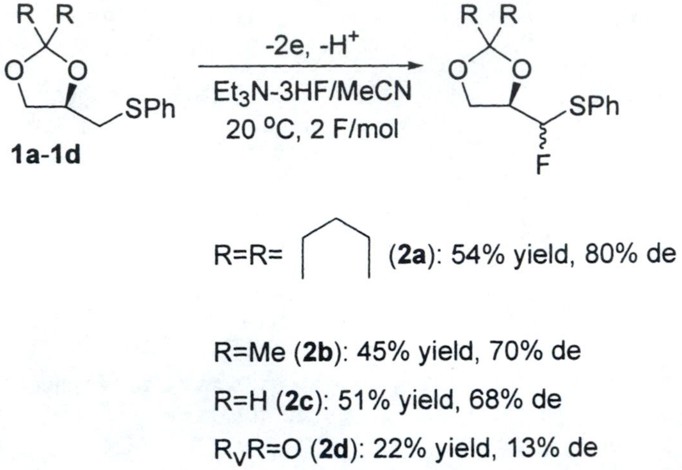

R R, O O, SPh — 1a-1d

$-2e, -H^+$

Et_3N-3HF/MeCN

20 °C, 2 F/mol

R R, O O, SPh, F

R=R= (cyclohexylidene) (**2a**): 54% yield, 80% de

R=Me (**2b**): 45% yield, 70% de

R=H (**2c**): 51% yield, 68% de

R,R=O (**2d**): 22% yield, 13% de

Scheme 4. Protecting group effect of 1,2-diol moieties on the diastereoselective anodic fluorination

1a: O O, SPh

$-2e, -H^+$

Et_3N-3HF/Solvent

3-6 F/mol

2a: O O, SPh, F

at 20 °C in MeCN (78% de)
DME (51% de)
DME/MeCN (61% de)
at 40 °C in MeCN (78% de)
DME (52% de)
DME/MeCN (59% de)

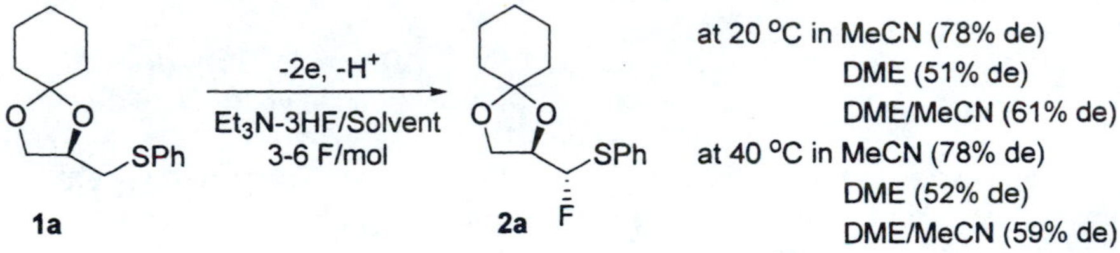

Scheme 5. Effect of electrolytic temperature on diastereoselective anodic fluorination

perpendicular C-O bond, and conformation **D** should experience less steric repulsion with the incoming fluoride ion with the possible electrostatic attraction of the β-oxygen atom and the sulfenium ion. Thus, the reaction of a fluoride ion was realized from the *si* face of conformation **D**, leading to the formation of the (1*R*, 2*S*) isomer, which is totally consistent with the X-ray analysis data.

In contrast, in the case of sulfide **1d**, the electron density of the ring oxygen atoms should decrease due to the electron-withdrawing carbonyl group, which decreases the electrostatic attraction of **1d** (**D'** conformation), as shown in Scheme 7. Therefore, the enegy difference between **C'** and **D'** would become much less than that between **C** and **D**. Thus, a fluoride ion may also attack the activated carbon of **C'** in spite of steric repulsion between the incoming fluoride ion and the perpendicular C-O bond. Consequently, the diastereoselectivity of the fluorination of **1d** decreases.

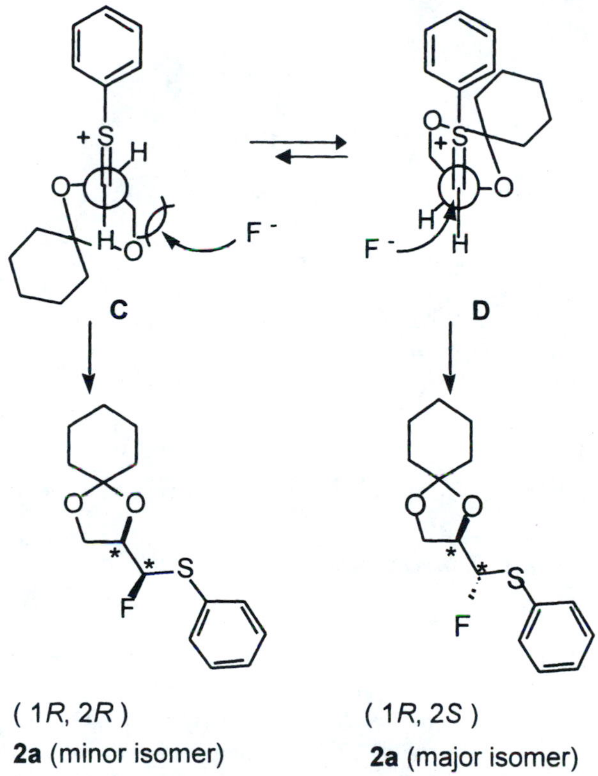

C

D

(1*R*, 2*R*)
2a (minor isomer)

(1*R*, 2*S*)
2a (major isomer)

Scheme 6. Mechanism of stereochemical control

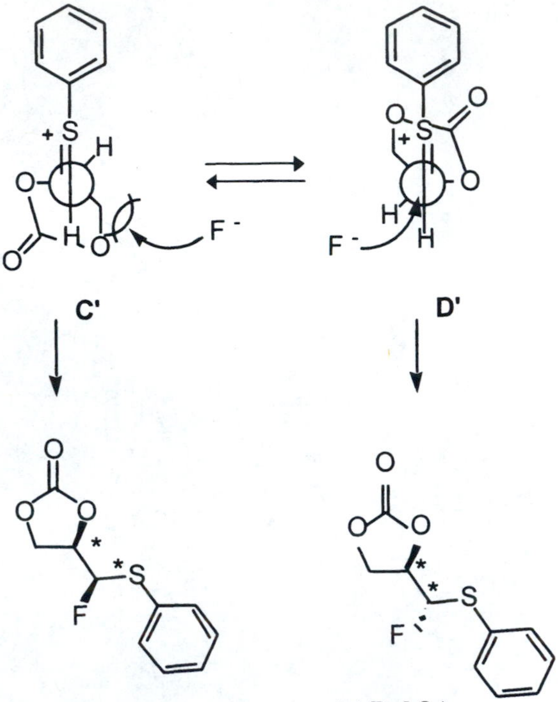

C' D'

(1*R*, 2*R*)
2d (minor isomer)

(1*R*, 2*S*)
2d (major isomer)

Scheme 7. Mechanism of stereochemical control

Chemical fluorination of sulfide **1a** as a model compound using *N*-fluoropyridinium salts did not proceed at all, while the use of selectfluor provided the fluorinated product **2a** in extremely poor yield (5%) and with much lower diastereoselectivity (32% de) compared with anodic fluorination.

Next, we investigated diastereoselective anodic fluorination of 3,3,3-trifluoro-2-methoxypropyl aryl sulfides **3** (*8*). As shown in Table 1, highly diastereoselective fluorination proceeded to provide the desired α-fluorinated products. This is in sharp contrast to the low diastereoselectivity (30% de) in the case of anodic fluorination of the corresponding non-fluorinated prototype 2-methoxypropyl phenyl sulfide. It is noted that the diastereoselectivity was always very high regardless of aryl groups and electrolytic conditions. Even at a higher temperature, the de increased to 74% as shown in Table 1 (Run 7). As shown in Runs 1~4, the diastereoselectivity was not significantly affected by the electrolytic temperature. Although the diastereoselectivity of anodic fluorination of α-phenylthioacetates having chiral auxiliaries was affected significantly by supporting fluoride salts, particularly the number of fluoride ions in the salts (Scheme 2) (*6*), such effect was not observed in these cases.

In consideration to these results and X-ray analysis of a major diastereomer, such high diastereoselectivity can be explained by the participation of the neighboring methoxy group as shown in Scheme 8.

Table 1 Diastereoselective Anodic Fluorination Aryl 3,3,3-Difluoro-2-methoxypropyl Sulfide

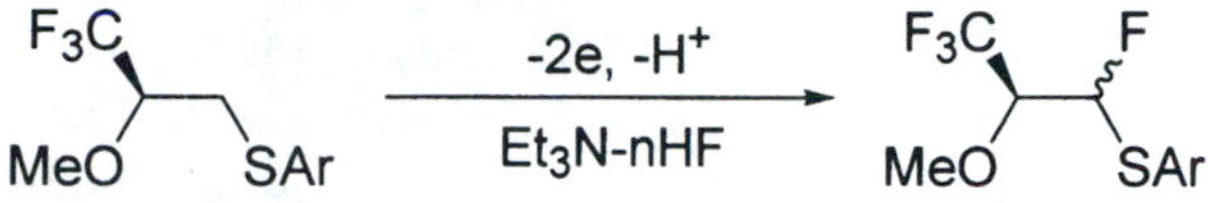

F_3C / MeO / SAr —(−2e, $-H^+$; Et_3N-nHF)→ F_3C / F / MeO / SAr

3a: Ar=Ph
3b: Ar=*p*-ClC_6H_4
3c: Ar=2-Naph

4a-4c

Run	3	n	Temp/°C	Electricity/F mol^{-1}	Yield/%	de/%[a]
1	**3a**	3	15	2.34	96	77
2	**3a**	3	20	2.40	91	76
3	**3a**	4	25	2.15	83	78
4	**3a**	5	35	1.95	74	80
5	**3b**	3	10	2.00	85	80
6	**3c**	3	25	2.50	33	61
7	**3c**	3	45	4.80	47	74

[a] The configuration of major diastereomer of **4** is (1*R*, 2*S*)

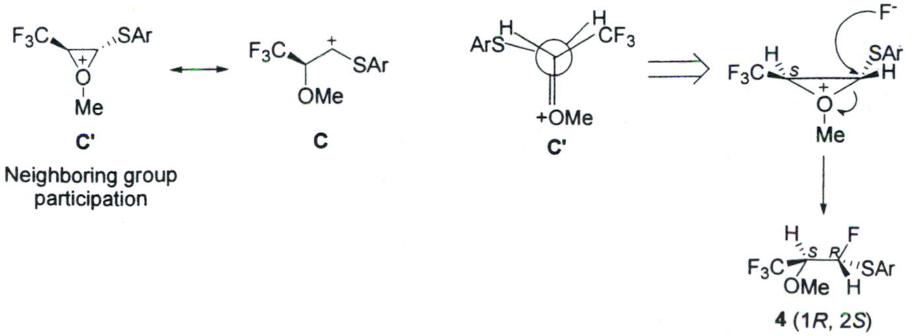

F3C
SAr
O
Me
C'
Neighboring group participation
F3C
SAr
OMe
C
ArS
H
H
CF3
+OMe
C'
F-
F3C
H
S
SAr
H
O
Me
H
F
S R
F3C
SAr
OMe H
4 (1R, 2S)

Scheme 8. Mechanism of stereochemical control

Diastereoslective Anodic Fluorination of Cyclic Compounds

Diastereoselectivity of anodic fluorination of 2-substituted 4-thiazolidinones was affected by both substitutents Ar group and R group (*9*). When the Ar group is more bulky and at the same time the R group is less bulky, the diastereselectivity is higher as shown in Scheme 9. Anodic fluorination of 1,3-oxathiolanones and 1,3-dithiolanones was comparatively studied using Et_3N-3HF and Et_4NF-4HF (*10,11*). The diastereoselectivity was found to depend on substituent R group as well as supporting fluoride salt. When the R group of 1,3-dithiolanone is highly bulky mesityl group, 60 % de was obtained using Et_4NF-4HF as shown in Scheme 10 (*11*). In this case, the use of Et_3N-3HF resulted in decrease of the de (46%). Intramolecular 1,3-asymmetric induction is thus generally low, and only when substituents Ar (Scheme 9) and R (Scheme 10) at the 2-position on the heterocyclic rings are bulky, the diastereoselectivity is moderate.

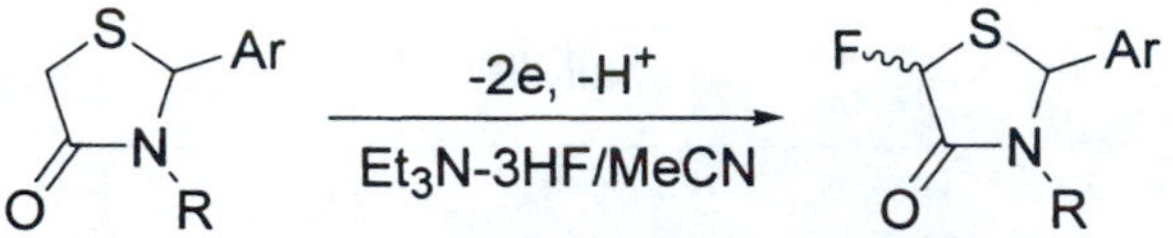

-2e, $-H^+$

Et_3N-3HF/MeCN

Ar=Ph, R=H: 30% de
Ar=Ph, R=Ph: 14% de
Ar=1-Naphthyl, R=Me: 42% de

Scheme 9. Diastereoselective anodic fluorination of 2-substituted 4-thiazolidinones

-2e, $-H^+$

F^-/MeCN

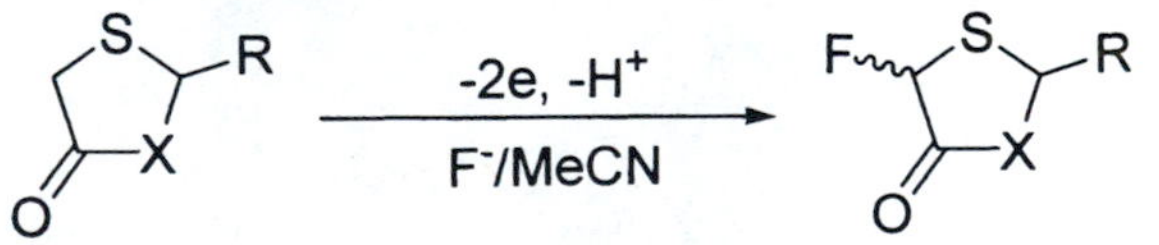

X=O, R=*i*-Pr: 14% de (Et_4NF-4HF)
X=O, R=*p*-CNC_6H_4: 22% de (Et_4NF-4HF)
X=S, R=*o*-Tol: 34% de (Et_4NF-4HF)
20% de (Et_3N-3HF)
X=S, R=Mesityl: 60% de (Et_4NF-4HF)
46% de (Et_3N-3HF)

Scheme 10. Diastereoselective anodic fluorination of 1,3-oxathiolanones and 1,3-dithiolanones

On the other hand, highly diastereoselective anodic fluorination of chiral 1,3-oxathiolanones derived from camphorsulfonamides was realized as shown in Scheme 11 (*12*). A single diastereomer was always obtained regardless of bulkiness of the *N*-alkyl groups on the sulfonamide and electrolytic conditions. Particularly when dimethoxyethane (DME) containing Et_4N-4HF was used, the corresponding monofluorinated products were obtained in good yields as a single diastereomer. The absolute configuration of starting materials are (1*S*,2*R*), while that of the fluorinated carbon of the products is *S*-form. Therefore, it is reasonable that a fluoride ion predominantly attacks the anodically generated cationic intermediate from the less hindered *re* face, because the sulfonamide group blocks the attack from the *si* face as shown in Scheme 12. In this case, camphor chiral auxiliaries keep a rigid, three-demensionally asymmetric environment in the reaction media during the electrolysis (Scheme 12). In sharp contrast, chemical fluorination using various *N*-fluoropyridinium salts provided ring opening product solely and fluorination did not proceed at all as shown in Scheme 13. Therefore, the electrochemical method is much superior to the conventional chemical method.

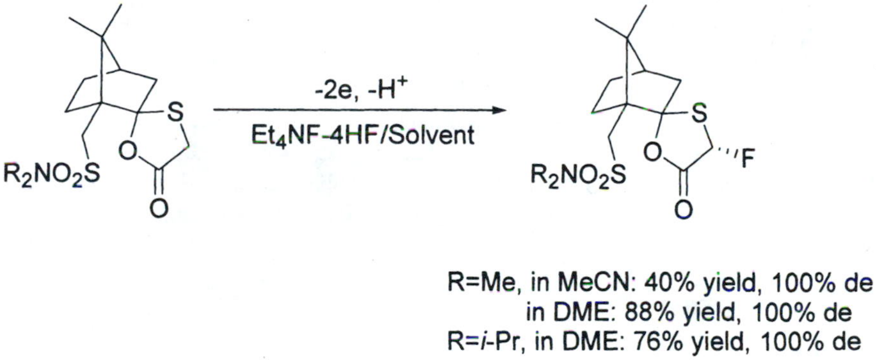

R_2NO_2S (camphor-derived 1,3-oxathiolanone) —[-2e, $-H^+$; Et_4NF-4HF/Solvent]→ R_2NO_2S (fluorinated product, ···F)

R=Me, in MeCN: 40% yield, 100% de
in DME: 88% yield, 100% de
R=*i*-Pr, in DME: 76% yield, 100% de

Scheme 11. Highly diastereoselective anodic fluorination of chiral 1,3-oxathiolanones derived from camphorsulfonamides

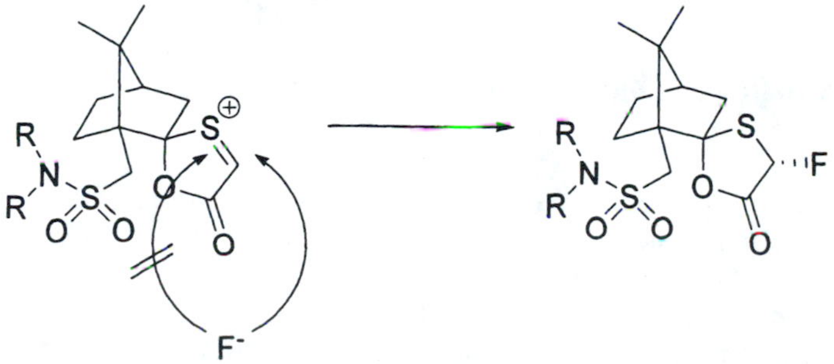

Scheme 12. Mechanism of stereochemical control

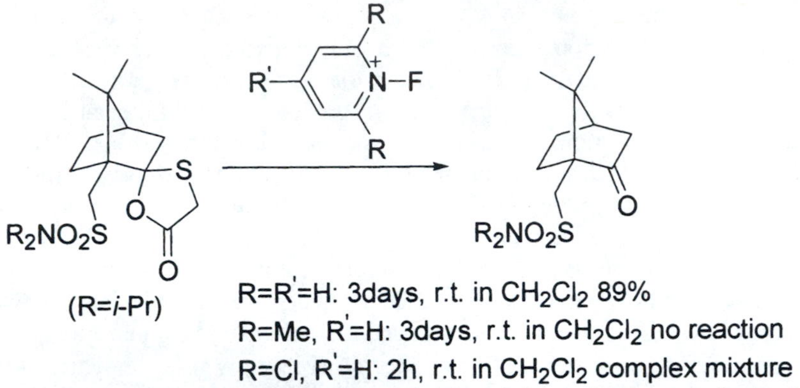
R
R'
N–F
R
S
O
R2NO2S
O
R2NO2S
O
(R=i-Pr)
R=R'=H: 3days, r.t. in CH2Cl2 89%
R=Me, R'=H: 3days, r.t. in CH2Cl2 no reaction
R=Cl, R'=H: 2h, r.t. in CH2Cl2 complex mixture

Scheme 13. Chemical fluorination of chiral 1,3-oxathiolanone derived from camphorsulfonamide

Next, intramolecular 1,2-asymmetric induction of heterocyclic compounds was studied as follows. Anodic fluorination of 3-alkyl-3-methoxycarbonyl-4-thianones was investigated and it was found that the diastereoselectivity was greatly affected by a R group as shown in Scheme 14 (*13*). When R is methyl group, only one diastereoisomer was formed although the yield was extremely low. On the other hand, when R is benzyl group, the yield was increased to ca. 50%, however, the de value was decreased to 44%. In sharp contrast, when R is isopropyl group, the diastereoselectivity was decreased drastically to only 8 %. Since a less bulky group gives higher diastereoselectivity, this diastereoselectivity seems to be highly controlled by the bulkiness of the substituents, R groups.

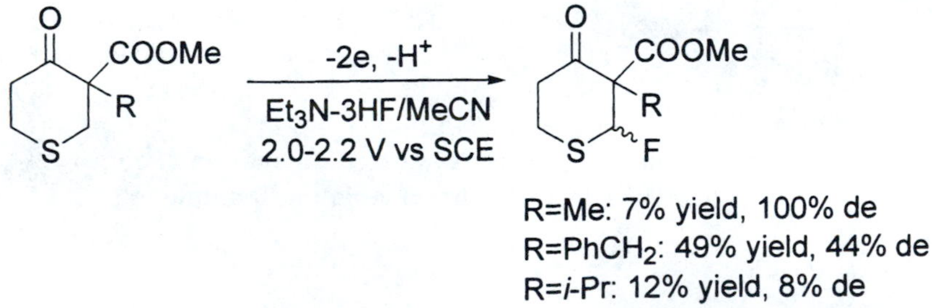
O
COOMe
R
S
-2e, -H+
Et3N-3HF/MeCN
2.0-2.2 V vs SCE
O
COOMe
R
S
F
R=Me: 7% yield, 100% de
R=PhCH2: 49% yield, 44% de
R=i-Pr: 12% yield, 8% de

Scheme 14. Anodic fluorination of 3-alkyl-3-methoxycarbonyl-4-thianones

We also investigated anodic fluorination of *N*-substituted thiazolidines derived from *L*-systeine as shown in Scheme 15 (*14*). The fluorination proceeded in DME with moderate to high diastereoselectivity. Interestingly, the diastereoselectivity increases with an increase of HF content in the supporting fluoride salts and almost 100% de is obtained using Et_4NF-5HF.

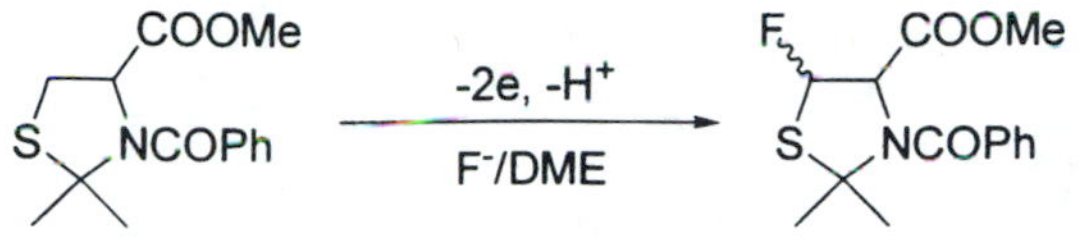

COOMe

S NCOPh

-2e, $-H^+$

F^-/DME

F COOMe

S NCOPh

Et_3N-3HF: 78% yield (78% de)
Et_3N-4HF: 78% yield (94% de)
Et_4NF-5HF: 55% yield (99% de)

Scheme 15. Effect of HF content in the supporting fluoride salts on diastereoselective anodic fluorination

The diastereoselectivity shown in Scheme 15 seems to be controlled by the steric repulsion between the incoming fluoride ion and the carbomethoxy group. However, as shown in Scheme 16, the acyl group on the nitrogen atom greatly affects the diastereoselectivity. The diastereselectivity increases with an increase of the bulkiness of the acyl group.

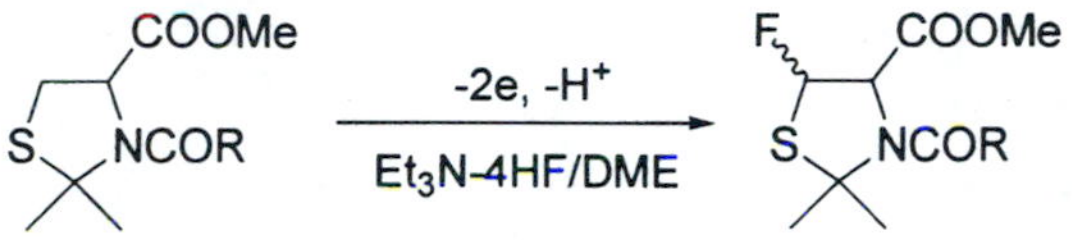

COOMe

S NCOR

5a-5d

-2e, $-H^+$

Et_3N-4HF/DME

F COOMe

S NCOR

6a: R=*p*-Tol: 91% yield, 95% de
6b: R=Ph: 78% yield, 94% de
6c: R=Me: 53% yield, 80% de
6d: R=H: 52% yield, 59% de

Scheme 16. Effect of acyl group on diastereoselective anodic fluorination

The sterecontrol can be explained on the basis of PM3 calculation of the minimum energy structures of the cation intermediates as follows. As shown in Fig. 1, the phenyl group may be perpendicular to the thiazolidine ring in the case of $\mathbf{5b}^+$. Thus, one side of the thiazolidine would be sterically blocked by both the carbomethoxy and phenyl groups. Therefore, a fluoride ion should attack $\mathbf{5b}^+$ predominantly from the less hindered side. In sharp contrast, in the case of the $\mathbf{5d}^+$, the much smaller formyl group may be coplanar to the heterocyclic ring. This seems to cause much lower diastereoselectivity. Therefore, the diastereoselectivity observed in the anodic fluorination of **5** is reasonable.

Diastereselective anodic fluorination of 1,3-oxazolidine derived from *L*-threonine 7 is also successful as shown in Scheme 18 (*15*). The stereoselectivity can be explained as follows. When a fluoride ion attacks the cationic carbon atom of 7^+ from the opposite side of the adjacent methyl group, strong steric repulsion between the methyl and carbomethoxy groups should be formed. Therefore, a fluoride ion attacks preferentially 7^+ from the same side as the methyl group as shown in Scheme 19.

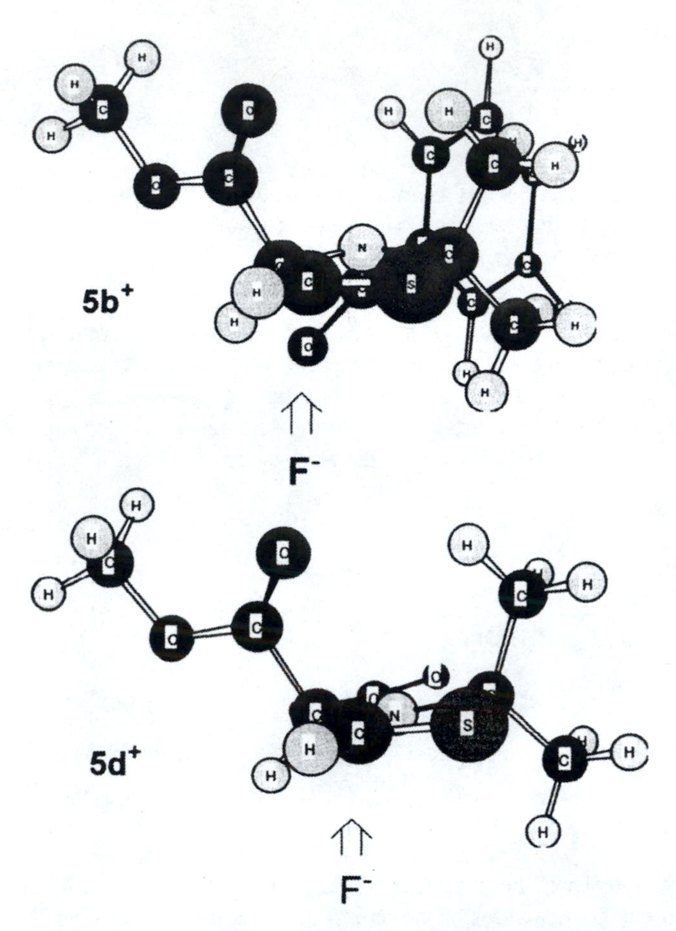

Scheme 17. Stereocontrol mechanism on the basis of PM3 calculation of the minimum energy structures of the cation intermediates of thiazolidines

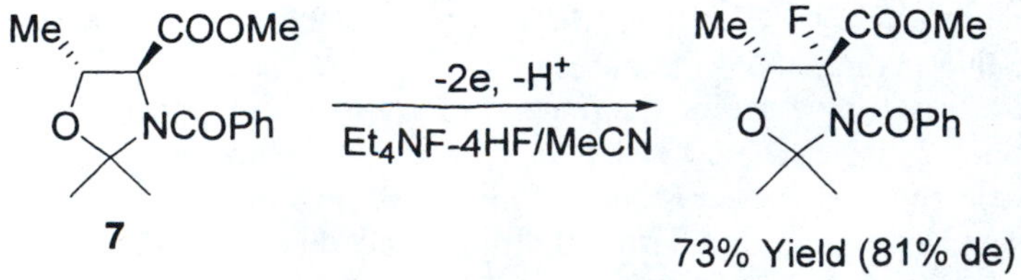

Me, COOMe, O, NCOPh, **7**

$\xrightarrow[\text{Et}_4\text{NF-4HF/MeCN}]{-2e,\ -H^+}$

Me, F, COOMe, O, NCOPh

73% Yield (81% de)

Scheme 18. Diastereoselective anodic fluorination of 1,3-oxazolidine derived from *L*-threonine

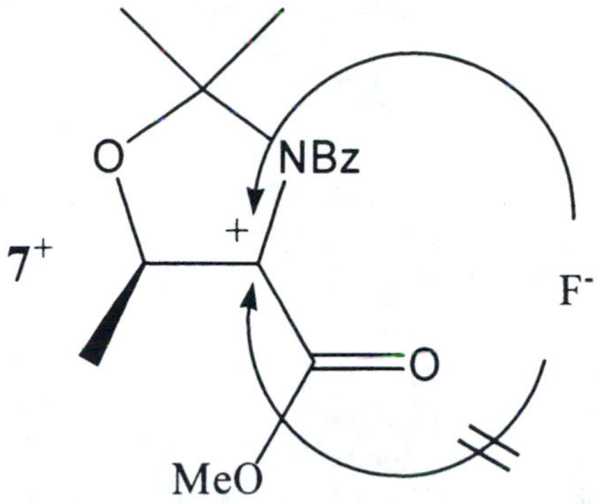

Scheme 19. Mechanism of stereochemical Control

Laurent and his co-workers carried out the anodic difluorination of camphanyl enol ester derivatives as shown in Scheme 20 (*16*). In these cases, the diastereoselectivity of the difluorination was not observed at all. However, diastereomers were readily separated by liquid chromatography, followed by hydrolysis to provide optically active 2-fluoro-1-tetralones (>95% ee). Their absolute configuration was also established.

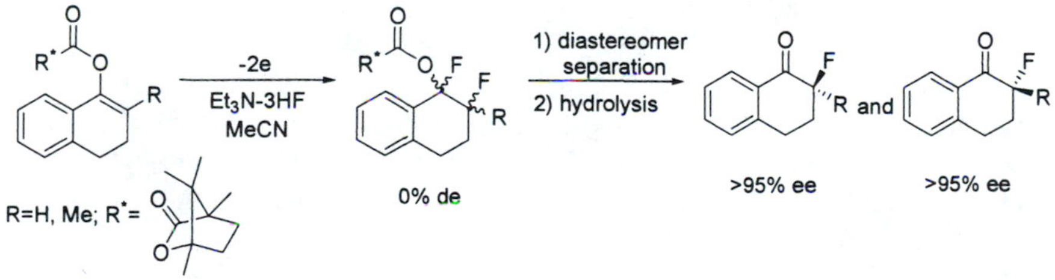

Scheme 20. Preparation of optically active 2-fluoro-1-tetralones by anodic difluorination of camphanyl enol ester derivatives

Conclusion

In this capter, we have described recent results on the diastereoselective anodic fluorination of various open-chain and cyclic compounds including heterocycles. It is hoped that the information in this chapter will be helpful for the development in modern organofluorine chemistry.

References

1. Hiyama, T. *Organofluorine Compounds*; Springer-Verlag: Berlin, 2000.
2. Fuchigami, T. In *Advances in Electron-Transfer Chemistry*; Mariano, P. S., Ed.; JAI Press: Connecticut, 1999; vol.6, p 41-130.
3. Fuchigami, T. In *Organic Electrochemistry*, 4th ed.; Lund, H.; Hammerich, O., Eds.; Marcel Dekker: New York, 2001; Chapter 25.
4. Fuchigami, T.; Nonaka, T.; Schafer, H.J. In *Encyclopedia of Electrochemistry, Vol 8. Organic Electrochemistry*; Schafer, H.J., Ed. Wiley-VCH: Weinheim, 2004; Chapter 13.
5. Kabore, L.; Chebli, S.; Faure, R.; Laurent, E.; Marquet, B. *Tetrahedron Lett.* **1990**, *31*, 3137-3140.
6. Narizuka, S.; Koshiyama, H.; Konno, A.; Fuchigami, T. *J. Fluorine Chem.* **1995**, *73*, 121-127.
7. Suzuki, K.; Fuchigami, T. *J. Org. Chem.* **2004**, *69*, 1276-1282.
8. Fuchigami, T.; Furuta, S. 193rd Electrochemical Society Meeting, Abstract No. 931 (1998).
9. Fuchigami, T,; Narizuka, S.; Konno, A. *J. Org. Chem.* **1992**, *57*, 3755-3757.
10. Higashiya, S.; Narizuka, S.; Kono, A.; Maeda, T.; Momota, K.; Fuchigami, T. *J. Org. Chem.* **1999**, *64*, 133-137.
11. Fuchigami, T.; Narizuka, S.; Konno, A.; Momota, K. *Electrochim. Acta*, **1998**, *43*, 1985-1998.
12. Baba, D.; Yang, Y.-J.; Uang, B.-J.; Fuchigami, T. *J. Fluorine Chem.* **2003**, *121*, 93-96.
13. Narizuka, S.; Konno, A.; Matsuyama, H.; Fuchigami, T. *Electrochemistry (formerly DENKI KAGAKU)*, **1993**, *61*, 868-869.
14. Baba, D.; Ishii, H.; Higashiya, S.; Fujusawa, K.; Fuchigami, T. *J. Org. Chem.* **2001**, *66*, 7020-7024.
15. Baba, D.; Fuchigami, T. *Tetrahedron Lett.* **2002**, *43*, 4805-4808.
16. Ventalon, F. M.; Faure, R.; Laurent, E. G.; Marquet, B. S. *Tetrahedron Asymm.* **1994**, *5*, 1909-1912.

Chapter 6

A Stereospecific Entry to Functionalized *cis*-1,2-Difluoroalkenes

Anilkumar Raghavanpillai[1,2] and Donald J. Burton[1,*]

[1]Department of Chemistry, The University of Iowa, Iowa City, IA 52242
[2]Current Address: DuPont Central Research and Development, Experimental Station, Wilmington, DE 19880

The stereospecific preparation of various 1,2-difluoroalkenyl synthons and their utilization for the synthesis of the corresponding functionalized 1,2-difluoroalkenes is discussed. The focus of this review is on the recent developments in our laboratory on the stereospecific introduction of several *cis*-1,2-difluoroalkenylorganometallic building blocks and their functionalization for the preparation of synthetically intricate *cis*-1,2-difluoroalkenes, such as *cis*-1,2-difluorostyrenes, *cis*-2-iodo-1,2-difluorostyrenes, *cis*-2-alkylsubstituted-1,2-difluorostyrenes, *cis*-aryl substituted 2,3-difluoroacrylic esters, *cis*-2,3-difluoro-4-oxo-substituted 2-butenoates, symmetrically and unsymmetrically substituted *cis*-1,2-difluorostilbenes, *cis*-1-arylperfluoroalkenes and *cis*-1-iodoperfluoroalkenes. The terminology '*cis*' or '*trans*'-1,2-difluoro refers to fluorines on the same or opposite side of the double bond of the 1,2-difluoroalkenyl species.

Introduction

1,2-Difluoroalkenes are versatile organofluorine building blocks because of their wide range of applications in the area of fluoropolymers, liquid crystalline materials and biologically active agents (*1, 2, 3, 4, 5, 6*). Generally, the physical and biological properties of 1,2-difluoroalkenes largely depend on their geometry, and the development of the corresponding synthons *via* a *cis* or *trans* selective route would allow studies on the stereochemical, synthetic, kinetic, and biological aspects of compounds involving such building blocks. However, practical methods for the stereospecific introduction of the 1,2-difluoroalkenyl unit remains as a considerable synthetic challenge. Although several methodologies for the introduction of *trans*-1,2-difluoroethenyl synthons and substituted *trans*-1,2-difluoroalkenes are well established, the corresponding sterospecific synthesis of *cis*-1,2-difluoroethenyl synthons and functionalized *cis*-1,2-difluoroalkenes is still relatively underdeveloped. Herein we briefly review some of the approaches for the stereospecific preparation of several *cis*-1,2-difluoroethenyl building blocks and their utilization for the synthesis of various substituted *cis*-1,2-difluoroalkenes.

1,2-difluoroalkenyl synthons: An overview

A large number of methods have been developed for the preparation of 1,2-difluoroalkenes which involve the utilization of Wittig type chemistry *(7, 8, 9*), addition-elimination reactions (*10, 11, 12, 13, 14, 15, 16, 17, 18, 19, 20*), dehydrohalogenation reactions (*21, 22, 23, 24, 25, 26*) and synthesis involving 1,2-difluoroalkenyl organometallic synthons (*27, 28, 29, 30, 31*). The classical route to RCF=CFX, where X = F, Cl or R_F and R is alkyl or aryl, involves reactions of the corresponding fluoroalkenes [CF_2=CFX] with organolithium compounds RLi. For example, reaction of phenyllithium with a number of perfluoroalkenes produced 1-phenylsubstituted fluoroalkenes in 20-60% yields together with products resulting from the successive displacement of fluorines by phenyl groups (*10*). Reaction of substituted aryl Grignard reagents with perfluoroalkenes have been reported to produce corresponding aryl substituted 1,2-difluoroalkenes as a mixture of *trans* and *cis* isomers with *trans* isomer as the favored product. (*12, 13*). Dehydrohalogentaion reactions have been widely used for the synthesis of 1,2-difluoroalkenes from the corresponding suitable precursors. Leroy synthesized various 1,2-difluoroalkenes by the dehydrofluorination of the corresponding 1,2,2-trifluoro precursors using *t*-BuOK base (*21*). In this report, the *cis*-1,2-difluoroalkene was obtained as the sole product and the *cis*-selectivity was attributed to the increased stability of the *cis*-isomer compared to the *trans* one. Similar to Leroy's work, fluoroalkenes of the general formula *cis*-RCF=CFCF$_3$ (R = cyclopentyl, cyclohexyl) were stereospecifically synthesized recently by the dehydofluorination of the corresponding alkanes using *t*-BuOK base (*24*). Hiyama and co-workers synthesized *cis*-2-alkylsubstituted-1-bromo-1,2-difluoroalkenes (RCF=CFBr) stereospecifically by dehydrobromination of the corresponding brominated precursor (*22*). These workers have also developed synthetic methods for both *cis* and *trans*-1-aryl-perfluoroalkenes where the *trans*-1-arylperfluoroalkenes prepared by the dehydrohalogenation of the corresponding precursor using DBU upon photoisomerization produced pure *cis*-1-and aryl-perfluoroalkenes (*23*). We have recently reported a similar method for the synthesis of 1-aryl-perfluoroalkenes by dehydrofluorination of the corresponding precursor using lithium hexamethyldisilazide (*25, 26*). 1,3-Dicarbonyl compounds have been shown to undergo reaction with nucleophilic fluorinating agents, such as DAST or Deoxofluor™ to produce a mixture of *cis* and *trans* isomers (1:1) of 2,3-difluoroeneones (*32, 33*).

The most important and widely used approach for the synthesis of 1,2-difluoroalkenes is *via* the corresponding 1,2-difluoroalkenylorganometallic synthons. Tarrant and co-workers reported an exclusive formation of *trans*-perfluoropropenyllithium (*trans*-CF$_3$CF=CFLi) during the low temperature metallation of a 1:1 mixture of *cis* and *trans*-CF$_3$CF=CFH (*34*). They proposed that the *cis*-lithium reagent [*cis*-CF$_3$CF=CFLi] readily isomerized to the *trans* isomer during the metallation process. However, reinvestigation of this preparation in our laboratory did not detect any isomerization of the *cis*-lithium reagent, instead metallation of a mixture of 80:20 trans and *cis*-CF$_3$CF=CFH, produced corresponding lithium reagents with the retention of stereochemical integrity of the starting alkenes (*tran/cis* 80:20) (*35*). Although the preparation and synthetic utility of 1,2-difluoroalkenyl synthons *via* the corresponding lithium reagent was largely exploited by Normant and co-workers, a significant disadvantage of this route was the highly unstable nature of the 1,2-difluoroalkeneyllithium reagents at ambient temperatures (*27, 36, 37, 38*). It was also reported that the *trans*-difluorovinyllithium reagents were more stable than their *cis* analogues (*36, 39*) Thus, the alkyl substituted *trans*-difluoroethenyllithium reagent could be prepared at -30 °C in THF and decomposed above -5 °C, whereas the corresponding *cis*-difluorovinyllithium was prepared in a mixture of ether-THF at -110 °C, but decomposed when warmed to -80 °C (Scheme 1) (*36*).

Trans-1,2-difluoroalkenylmagnesium reagents are well known and were prepared from the readily available *trans*-1,2-difluoroalkenylhalides either by direct reaction with magnesium or exchange reaction with organomagnesium halides at a relatively higher temperature (-20 to 5 °C) (*40, 41, 42*). For example, *trans*-perfluoroalkenylmagnesium halides were synthesized by the reaction with magnesium in THF at rt and utilized for the reaction with a variety of n-C_7H_{15}(F)C=C(F)(H) —[n-BuLi, THF, -30 °C]→ n-C_7H_{15}(F)C=C(F)(Li) —[HCHO]→ n-C_7H_{15}(F)C=C(F)(CH_2OH)

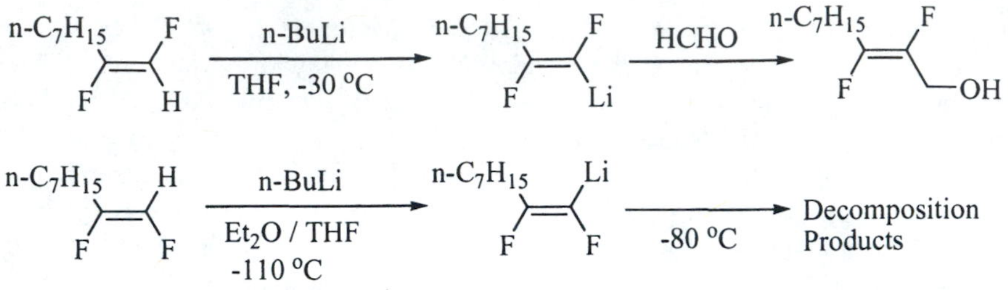

n-C_7H_{15}(F)C=C(H)(F) —[n-BuLi, Et_2O / THF, -110 °C]→ n-C_7H_{15}(F)C=C(Li)(F) —[-80 °C]→ Decomposition Products

Scheme 1. Preparation of alkyl substituted trans and cis-1,2-difluoroalkenyllithium reagents at low temperature.

electrophiles to produce corresponding *trans*-perfluoroalkenylated products (Scheme 2) (*40, 41, 42, 43, 44*).

Trans-perfluoroalkenylmagnesium reagents also react with phosphrous (III) chlorides to produce the corresponding phosphonites which upon reaction with hexafluoroacetone produced the corresponding dioxaphospholanes (*45, 46, 47, 48*). Similarly, *trans*-1,2-difluoro-2-(pentafluorosulfanyl)ethenyl magnesium iodide [*trans*-$SF_5CF{=}CFMgI$] [generated from *trans*-1,2-difluoro-1-iodo-2-(pentafluorosulfanyl)ethene] reacted with selected phosphorous (III) chlorides to produce the corresponding phosphonite derivatives which upon reaction with hexafluoroacetone produced the corresponding penatfluorosulfanyl dioxaphospholanes (Scheme 3) (*48*).

Since the utility of 1,2-difluoroalkenyllithium reagents have been limited by their poor thermal stability, attention has been focused on the development of more stable 1,2-difluoroalkenyl synthons. Transmetallation of *trans*-1,2-difluoroalkenyllithium or magnesium reagent with an organotin halide like tributyltin chloride provides the more stable *trans*-1,2-difluoroalkenyltin reagent. (*44, 49*). Under Barbier conditions, where the tributyltin halide was used *in situ* during the metallation process, the yield of *trans*-1,2-difluoroalkenyltin regent was significantly improved (Scheme 4). Normant and co-workers developed several 1,2-difluoroalkenylzinc reagents by trans-metallation of 1,2-difluoroalkenyllithium or magnesium reagents with zinc halide at low temperature (Scheme 4) (*50, 51, 52, 53, 54, 55, 56*). The 1,2-difluoroalkenylzinc reagents could also be conveniently prepared by a zinc insertion reaction with 1,2-difluoroalkenyl halides at ambient temperature (Scheme 4) (*31, 57, 58*). The stereochemical integrity of the 1,2-difluroalkenyl halide was retained during the preparation of the zinc reagent thus providing the corresponding *cis* or *trans*-1,2-difluoroalkenyl zinc reagents from stereochemically pure *cis* or *trans*-alkenyl iodides. The 1,2-difluoroalkenylzinc reagents synthesized by either route exhibited excellent thermal stability at room temperature (*57*).

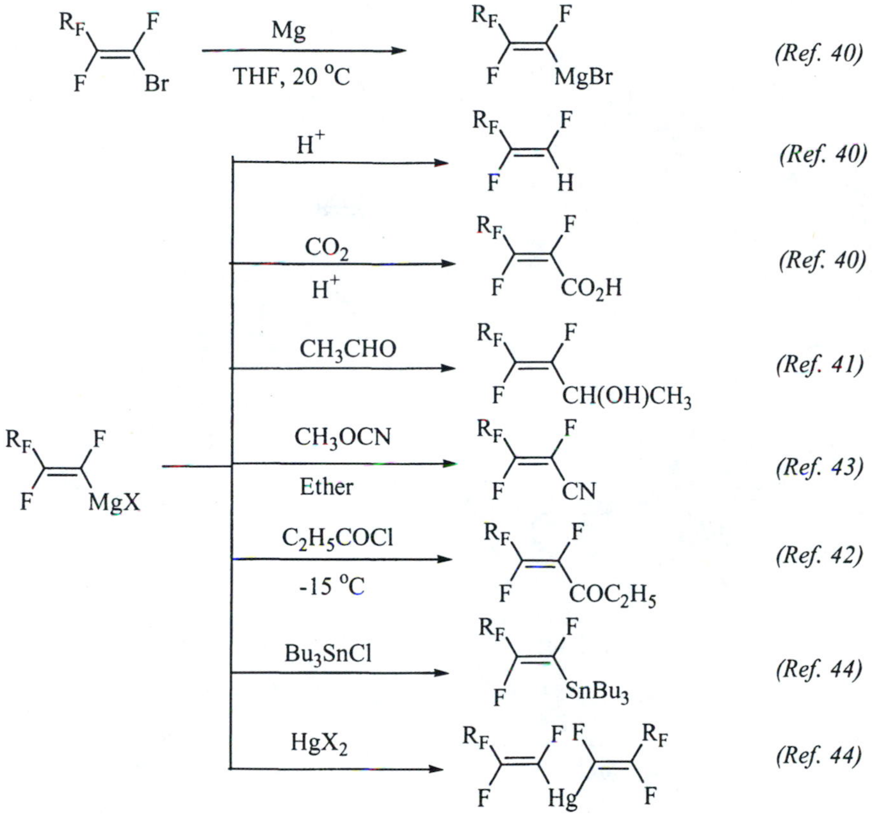

(Ref. 40)

(Ref. 40)

(Ref. 40)

(Ref. 41)

(Ref. 43)

(Ref. 42)

(Ref. 44)

(Ref. 44)

Scheme 2. Preparation of trans-perfluoroalkenylmagnesium reagents and their reaction with various electrophiles.

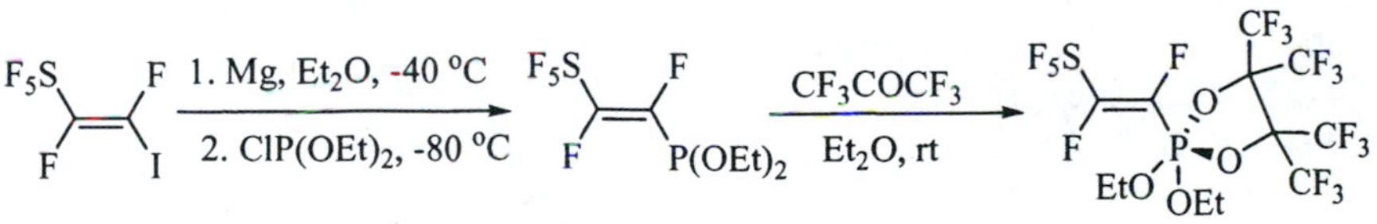

Scheme 3. Synthesis of trans-1,2-difluoro-2-(pentafluorosulfanyl)ethenyl phosphonites.

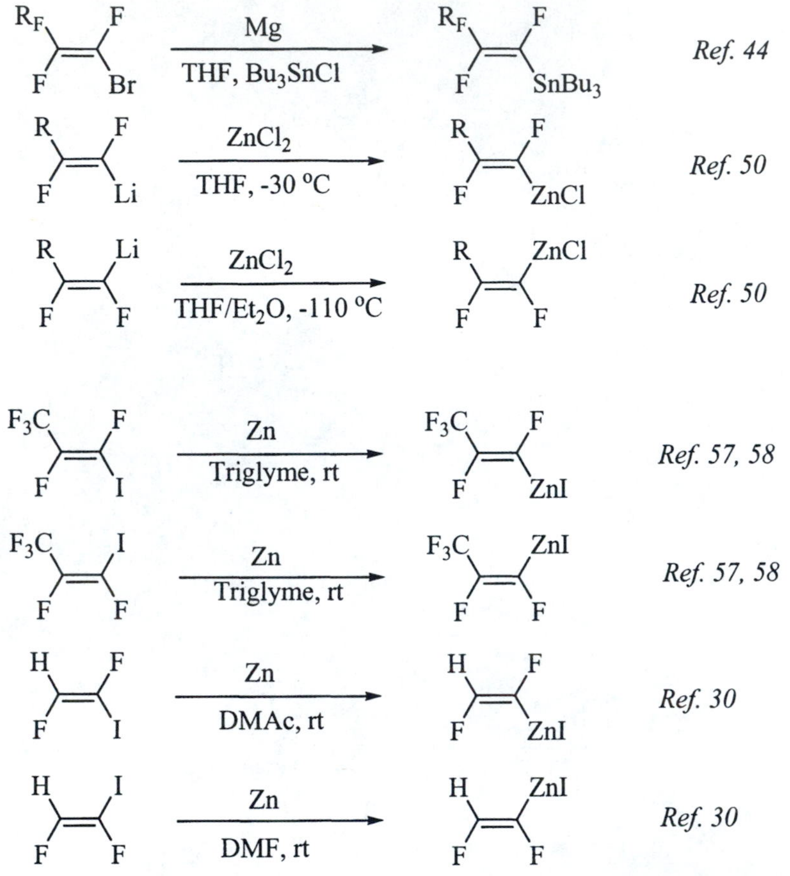

Reactant	Reagents	Product	Reference
R_F, F / F, Br	Mg; THF, Bu_3SnCl	R_F, F / F, $SnBu_3$	*Ref. 44*
R, F / F, Li	$ZnCl_2$; THF, -30 °C	R, F / F, ZnCl	*Ref. 50*
R, Li / F, F	$ZnCl_2$; THF/Et_2O, -110 °C	R, ZnCl / F, F	*Ref. 50*
F_3C, F / F, I	Zn; Triglyme, rt	F_3C, F / F, ZnI	*Ref. 57, 58*
F_3C, I / F, F	Zn; Triglyme, rt	F_3C, ZnI / F, F	*Ref. 57, 58*
H, F / F, I	Zn; DMAc, rt	H, F / F, ZnI	*Ref. 30*
H, I / F, F	Zn; DMF, rt	H, ZnI / F, F	*Ref. 30*

Scheme 4. Stereospecific preparation of 1,2-difluoroalkenyltin and zinc reagents

The 1,2-difluoroalkenylzinc reagents underwent palladium catalyzed cross-coupling reaction with aryl, or alkenyl iodides to produce the corresponding 1,2-difluoroalkenes or dienes in good isolated yield (Scheme 5). Similar to the 1,2-difluoroalkenylzinc reagents, 1,2-difluoroalkenylcadmium reagents were prepared either by transmetallation of the corresponding lithium reagent with cadmium halide or by cadmium insertion into the corresponding alkenyl bromides (*27, 59*). The stereochemistry of the alkenyl halide was retained during the cadmium insertion reaction. The 1,2-difluroalkenylcadmium reagents also showed impressive thermal stability and excellent reactivity towards various electrophiles (*27, 59*).

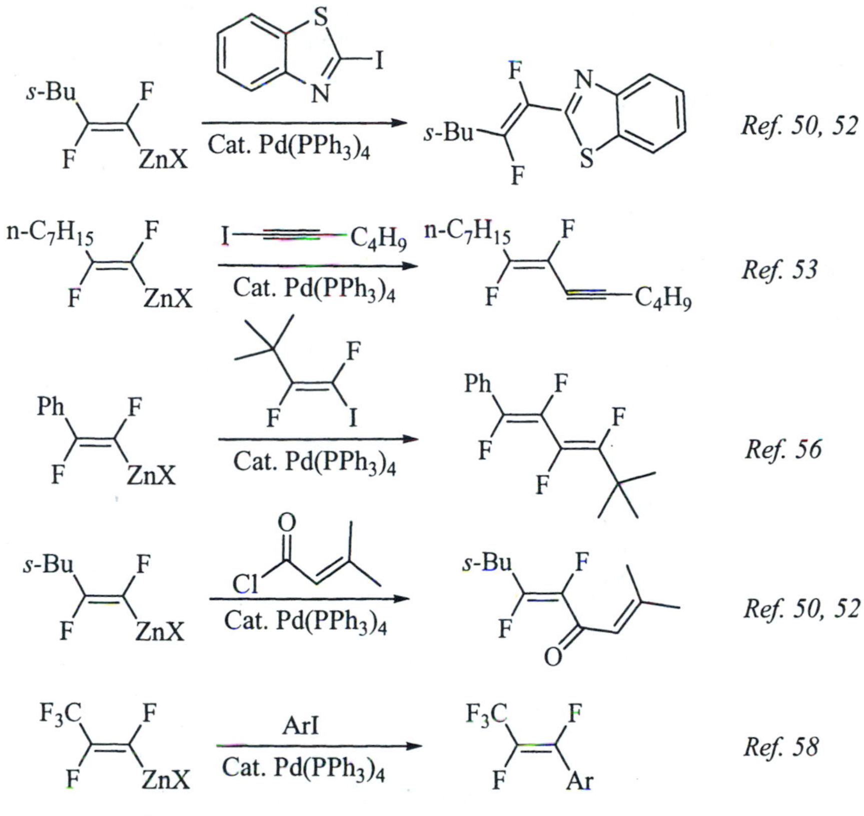

Scheme 5. Stereospecific synthesis of various 1,2-difluoroalkenes via the corresponding 1,2-difluoroalkenylzinc reagents.

Normant and co-workers also prepared several *trans*-1,2-difluoroalkenes from trifluoroalkenyltrialkylsilane intermediates prepared by the metallation of the trifluoroalkenyl halides with an alkyllithium in presence of the trialkylsilylchloride (*28, 54, 60, 61*). *Trans*-(2-alkyl or 2-aryl)-1,2-difluoroethenylsilanes were synthesized by addition-elimination reaction of the trifluoroethenylsilane with alkyl or aryl lithium reagents and the resulting lithium reagent was transformed to potential synthons for the introduction of *trans*-1,2-disubstituted-1,2-difluoroalkenes (*29, 30, 38, 54, 62, 63, 64*). The synthetic potential of these silanes was further investigated in our laboratory and in a stereospecific transformation, *trans*-(2-alkyl or 2-aryl)-1,2-difluoroethenylsilanes were converted to the *trans*-(2-alkyl or 2-aryl)-1,2-difluoroethenyl stannanes as illustrated in Scheme 6. The *trans*-(2-alkyl or 2-aryl)-1,2-difluoroethenyl stannanes were then utilized for the palladium catalyzed cross-coupling reaction with aryl or alkenyl iodides to obtain the corresponding *trans*-1,2-disubstituted-1,2-difluoroalkenes (*65*). Recent work in our laboratory has demonstrated that 1,2-difluoroalkenyl stannanes could be obtained in a stereospecific fashion *via* the correposnding thermally stable zinc or cadmium reagents (*66, 67*).

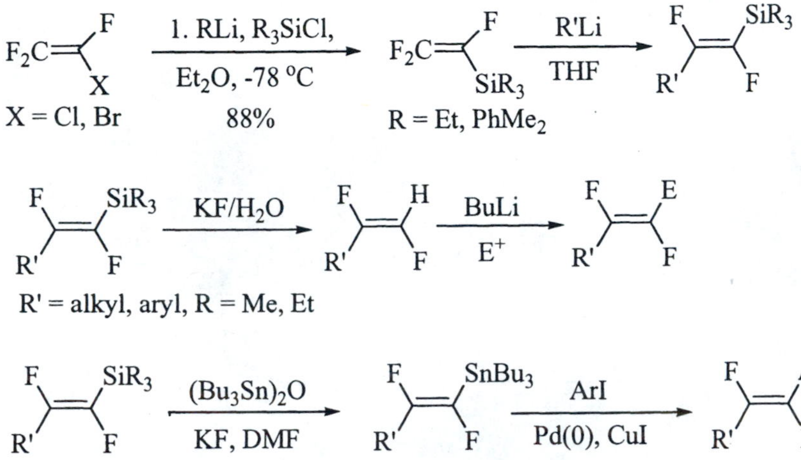

$F_2C{=}C(F)X$; X = Cl, Br — 1. RLi, R_3SiCl, Et_2O, -78 °C, 88% → $F_2C{=}C(F)SiR_3$; R = Et, $PhMe_2$ — R'Li, THF → F, SiR_3, R', F

F, SiR_3, R', F ; R' = alkyl, aryl, R = Me, Et — KF/H_2O → F, H, R', F — BuLi, E^+ → F, E, R', F

F, SiR_3, R', F — $(Bu_3Sn)_2O$, KF, DMF → F, $SnBu_3$, R', F — ArI, Pd(0), CuI → F, Ar, R', F

Scheme 6. Synthesis of trans-2-substituted-1,2-difluoroethenylsilanes and stannanes and their transformation to trans-1,2-disubstituted-1,2-difluoroalkenes.

Trifluoroethenyltrialkylsilane upon reduction with $LiAlH_4$ produced a 95: 5 mixture of *trans* and *cis*-1,2-difluoroethenyltrialkylsilanes, which were utilized for the introduction of various other useful 1,2-difluoroethenylsynthons (Scheme 7) (*29, 30*) For example, *trans*-(1,2-difluoro-2-iodoethenyl)triethylsilane was obtained in quantitative yield by the reaction of a 95: 5 mixture of *trans* and *cis*-1,2-difluoroethenyltrialkylsilane with *n*-BuLi followed by treatment with iodine. The *trans*-(1,2-difluoro-2-iodoethenyl)triethylsilane was further transformed to *trans*-1,2-difluoroethenyl iodide which was used as a synthon for the generation of the *trans*-1,2-difluoroethenylzinc reagent. The *trans*-1,2-difluoroethenylzinc reagent upon palladium catalyzed cross-coupling reaction with aryl iodides produced the corresponding *trans*-1,2-difluorostyrenes (*68*).In another transformation, *trans*-1,2-difluoroethenyltrialkylsilanes were converted to the trans-1,2-difluoroethenyltributylstannane by the reaction with KF in the presence of the tributyltin chloride or tributyltin oxide (30).

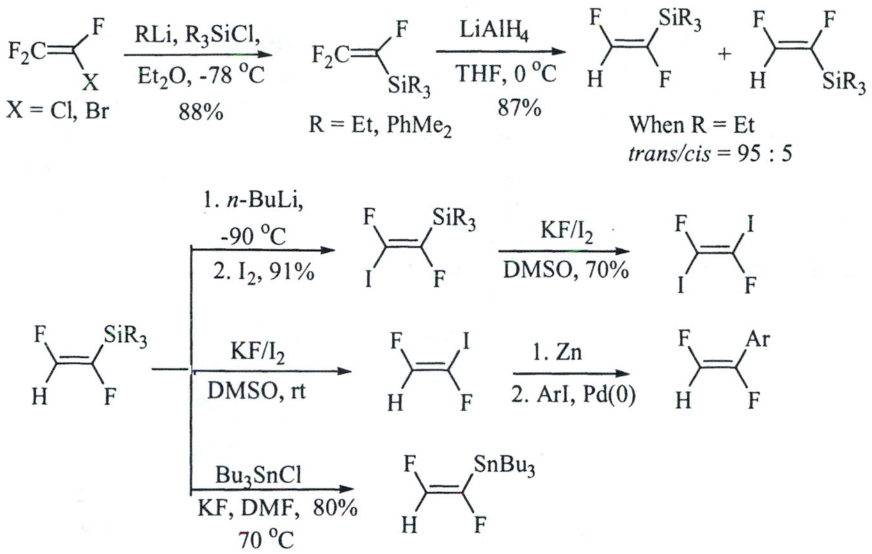

Scheme 7. Synthesis of trans-1,2-difluoroethenyltrialkylsilanes and its utilization for the preparation of various 1,2-difluoroethenyl synthons.

In addition to the above discussed 1,2-difluoroalkenylorganometallic synthons, several other useful 1,2-difluoroalkenyl organometallic or metalloid compounds have been reported, which involve 1,2-difluoroalkenyl mercury, copper, silver, germanium, boron, xenon compounds. The 1,2-difluoroalkenylcopper reagents are prepared by transmetallation of the corresponding alkenyllithium with cuprous salts at very low temperature (*27*). A more convenient approach was developed in our laboratory where, the metathesis of the corresponding *trans*-1,2-difluoroalkenylzinc or cadmium reagents with cuprous halides produced the corresponding copper reagent stereospecifically. The *trans*-1,2-difluoroalkenylcopper reagents exhibit excellent thermal stability and readily undergo functionalization reactions. Metathesis of the *trans*-pentafluoropropenyl cadmium reagent with silver trifluoroacetate produced the corresponding silver reagent (*27*). The 1,2-difluoroalkenylmercury organometallics have also been reported and were prepared by a Barbier reaction of the corresponding alkenyl bromide with magnesium in the presence of mercuric halides (*27, 49*). Brisdon and co-workers recently synthesized *trans*-1,2-difluoroalkenylgermanium compounds, where trifluoroethenyltriphenylgermanium ($Ph_3GeCF{=}CF_2$), prepared by the metallation of HFC-134a with *n*-BuLi and triphenylgermanium bromide, upon reaction with $LiAlH_4$ or a range of organolithium reagents (RLi, R = H, Me, *n*-Bu, *t*-Bu, Ph) produced the corresponding *trans*-1,2-difluoroethenylgermanes (*trans*-$Ph_3GeCF=CFR$) (*69*). Frohn and co-workers recently reported the synthesis of 1,2-difluoroalkenylboron and xenon compounds (*70, 71, 72*). In their preparation, *cis* and *trans*-polyfluoroalk-1-enyltrifluoroborates were generated by the nucleophilic addition of corresponding *cis* or *trans*-RCF=CFLi to $B(OCH_3)_3$ followed by hydrolysis and fluoride substitution with $K[HF_2]$ in aqueous HF (Scheme 8) (*70, 71*). Photoisomerization of the mixture of *cis* and *trans*-polyfluoroalk-1-enyltrifluoroborates to the corresponding *cis*- borates was attempted, but was only moderately successful (*73*). The *cis*-1,2-difluoroalk-1-eneylboranes upon reaction with XeF_2 produced the corresponding *cis*-1,2-difluoroalk-1-eneylxenon (II) salts, whereas the corresponding *trans*-isomers under similar conditions failed to give the corresponding xenon salt (*72*).

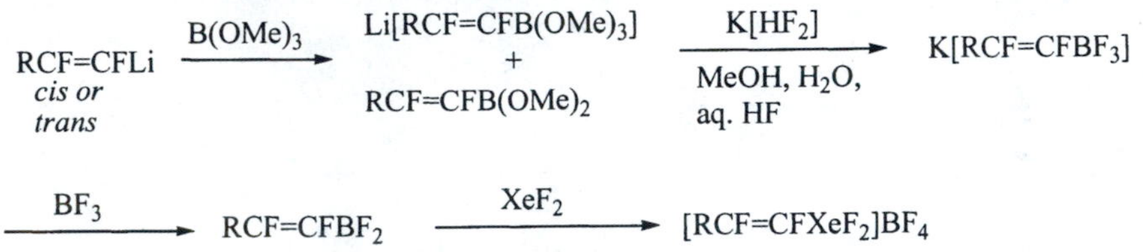

$$\underset{\textit{cis or trans}}{RCF{=}CFLi} \xrightarrow{B(OMe)_3} \begin{matrix} Li[RCF{=}CFB(OMe)_3] \\ + \\ RCF{=}CFB(OMe)_2 \end{matrix} \xrightarrow[\text{MeOH, } H_2O\text{, aq. HF}]{K[HF_2]} K[RCF{=}CFBF_3]$$

$$\xrightarrow{BF_3} RCF{=}CFBF_2 \xrightarrow{XeF_2} [RCF{=}CFXeF_2]BF_4$$

Scheme 8. Preparation of 1,2-difluoroalk-1-eneyltrifluoroborates and the cis-1,2-difluoroalk-1-enylxenon salts via corresponding lithium reagents.

Stereospecific synthesis of *cis*-1,2-difluoroalkenylsynthons

Except for a few literature reports on dehydrohalogenation reactions (*21, 24, 74*) the synthesis of *cis*-1,2-difluoroethenyl compounds have been mainly limited to examples in which the *cis*-isomer has been observed as the minor component of a *trans/cis* mixture. Literature reports of isomerization of *trans*-1,2-difluoroalkenes are rare. Exceptions are the isomerization-bromination of $(CH_3)CF=CFCO_2Et$ (*cis/trans* 1:1) noted earlier (*32*); the isomerization of *trans*-α,β-difluoro-α,β-unsaturated ketones with Me_3SiI or HCl (*19*) and the bromodesilylation of *trans*-(n-C_7H_{15})FC=$CFSiMe_3$ to afford predominately *cis*-(n-C_7H_{15})FC=CFBr (*cis/trans* 90:10) (*75*).

Our goals in this area of research were: (i) to develop a useful, practical and general route to a *cis*-1,2-difluoroethenyl building block that could be utilized for conversion into functionalized *cis*-1,2-difluoroethenyl derivatives; and (ii) to determine if the thermodynamic stability of *cis*-1,2-difluoroethenes was maintained in substituted *cis*-1,2-difluoroethenyl derivatives, such as RCF=CFH. Our initial work focused on the utilization of 1,1,2-trifluoro-2-trialkylsilylethene.

This trisubstituted alkene had been previously reported by Hiyama (62, 63) and Normant (54) and more importantly can be readily prepared from the commercially available $F_2C{=}CFCl$ or $F_2C{=}CFBr$ as discussed earlier in this report (Scheme 6 and 7). To facilitate isolation of $F_2C{=}CFSiR_3$ from solvent and by-products, the R_3SiCl employed was either Et_3SiCl or $PhMe_2SiCl$. Subsequent reduction of the trifluorotrialkylsilyethene with $LiAlH_4$ provided predominately the *trans*-1,2-difluorotrialkylethene as outlined in Scheme 7 (29, 54, 63). Subsequently, in a key transformation, we discovered that this *trans/cis* mixture of 1,2-difluoro-1-trialkylsilylethenes could be readily isomerized to a *cis/trans* mixture (enriched in the *cis*-isomer) by irradiation with ultraviolet light and a catalytic amount of phenyl disulfide, (29), as illustrated in Scheme 9.

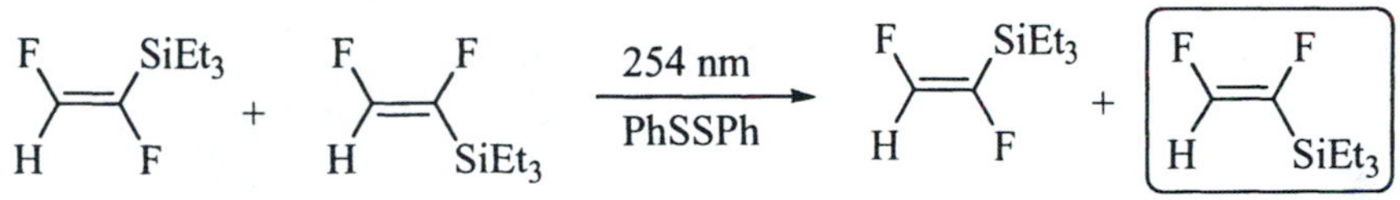

254 nm, PhSSPh

trans/cis = 95 : 5 → *trans/cis* = 5 : 95

Scheme 9. Photoisomerization of trans-1,2-difluoro-1-trialkylsilylethenes to cis-1,2-difluoro-1-trialkylsilylethenes

Since these silylethene derivatives can potentially be further functionalized *via* the silyl site or the 1-hydro site, this key discovery provided a unique opportunity for the development of pure *cis*-1,2-difluoroethenyl functionalized building blocks. For example *cis*-1,2-difluorotriethylsilylethene was converted to *cis*-1,2-difluorotriethylsilylethenyl stannane by the treatment with LTMP and tributyltin chloride, which underwent iododestannation or bromodestannation to produce the corresponding iodo or bromo substituted *cis*-1,2-difluorotriethylsilylethene (Scheme 10).

The *cis*-1,2-difluorotrialkylethenylsilane was converted to the corresponding *cis*-1,2-difluoro-2-iodoethene *via* two routes (Scheme 11), either by the reaction with KF/I_2 or by the treatment with $Bu_3SnOSnBu_3$ and catalytic KF in DMF (*30, 31, 76*).

Sterospecific synthesis of cis-2-alkylsubstituted-1,2-difluoroalkenes

Several alkyl or cycloalkyl substituted 1,2-difluoro-alkenes were prepared *via* methodology reported by Normant (*38, 51, 77*) as outlined in Scheme 12. These alkenes were then irradicated with ultraviolet light and catalytic PhSSPh to provide a *cis/trans* mixture enriched in the *cis*-isomer. The *cis/trans* mixture

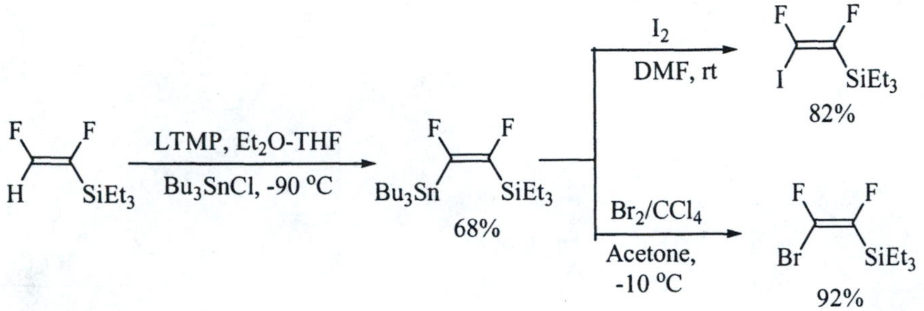

$$\text{cis-}F(H)C{=}C(F)SiEt_3 \xrightarrow[Bu_3SnCl,\ -90\ ^{\circ}C]{LTMP,\ Et_2O\text{-}THF} \text{cis-}F(Bu_3Sn)C{=}C(F)SiEt_3\ (68\%)$$

$$\xrightarrow[DMF,\ rt]{I_2} \text{cis-}F(I)C{=}C(F)SiEt_3\ (82\%)$$

$$\xrightarrow[Acetone,\ -10\ ^{\circ}C]{Br_2/CCl_4} \text{cis-}F(Br)C{=}C(F)SiEt_3\ (92\%)$$

Scheme 10. Transformation of cis-1,2-difluoro-1-trialkylsilylethenes to cis-1,2-difluorotriethylsilylethenyl stannane and cis-1,2-difluoro-2-iodo (bromo)-triethylsilylethenes

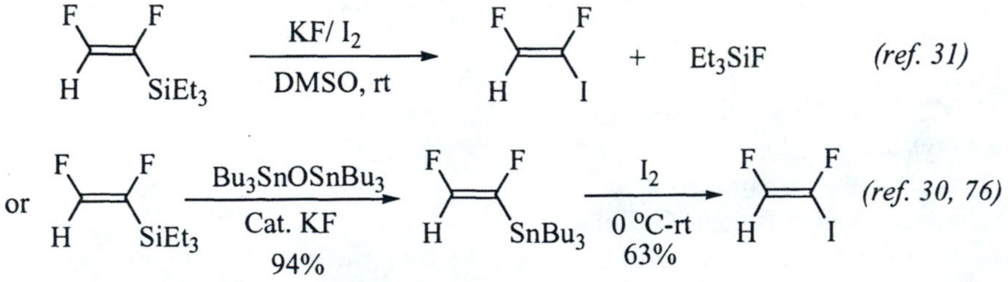

$$\text{cis-}F(H)C{=}C(F)SiEt_3 \xrightarrow[DMSO,\ rt]{KF/\ I_2} \text{cis-}F(H)C{=}C(F)I + Et_3SiF \quad \textit{(ref. 31)}$$

or

$$\text{cis-}F(H)C{=}C(F)SiEt_3 \xrightarrow[Cat.\ KF,\ 94\%]{Bu_3SnOSnBu_3} \text{cis-}F(H)C{=}C(F)SnBu_3 \xrightarrow[0\ ^{\circ}C\text{-}rt,\ 63\%]{I_2} \text{cis-}F(H)C{=}C(F)I \quad \textit{(ref. 30, 76)}$$

Scheme 11. Transformation of cis-1,2-difluoro-1-trialkylsilylethenes to cis-1,2-difluoro-2-iodoethenes

obtained after irradiation could be fractionally distilled to give the *cis*-isomer in >95% isomeric purity. When the *cis*-alkene was subjected to irradiation with PhSSPh, the *cis/trans* isomeric ratios obtained were identical to the isomer ratios obtained on irradiation of the *trans*-isomer, suggesting that equilibrium had been attained. The capability to obtain the *cis*-isomer in high purity *via* distillation provided the opportunity to readily prepare *cis*-functionalized 1,2-difluorosubstituted alkenes. Typical examples are cited in Table 1.

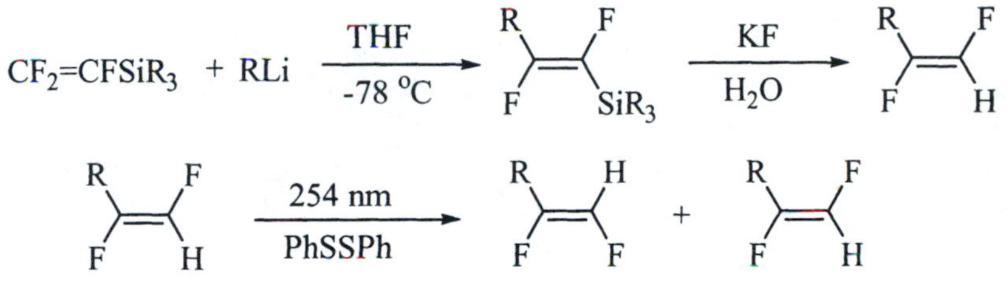

$CF_2{=}CFSiR_3$ + RLi —(THF, -78 °C)→ R(F)C=C(F)SiR_3 —(KF, H_2O)→ R(F)C=C(F)H

R(F)C=C(F)H —(254 nm, PhSSPh)→ R(F)C=C(H)F + R(F)C=C(F)H

R		
R = *n*-Bu	80	: 20
= *t*-Bu	98	: 2
= *s*-Bu	85	: 15
= C_6H_{11}	91	: 9

Scheme 12. Synthesis of cis-2-alkylsubstituted-1,2-difluoroalkenes

Table 1. Stereospecific preparation of *cis*-1-alkyl-2-arylsubstituted-1,2-difluoroalkenes

F(R)C=C(F)H —(t-BuLi, Bu_3SnCl, -90 °C)→ F(R)C=C(F)$SnBu_3$ —(ArI, $Pd(PPh_3)_4$)→ F(R)C=C(F)Ar

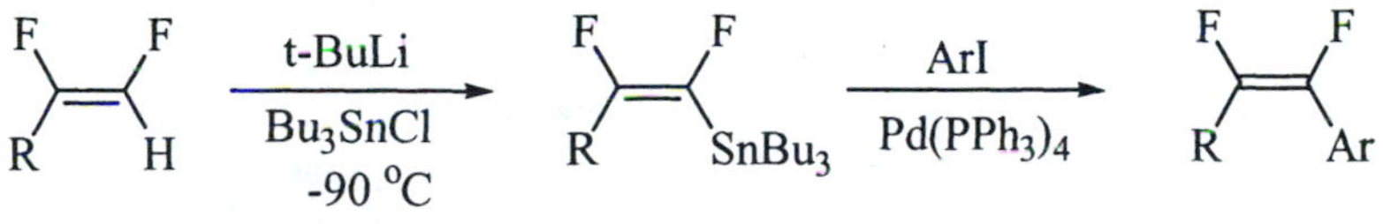

Entry	*R*	*Ar*	*Yield (%)*[a]
1	*n*-Bu	$3\text{-}NO_2C_6H_4\text{-}$	91
2	*s*-Bu	$4\text{-}MeOC_6H_4\text{-}$	82
3	*n*-C_6H_{13}	$2\text{-}NO_2C_6H_4\text{-}$	86

[a] Configuration assigned on the basis of $^3J_{FF\,(cis)}$ = 10-15 Hz.

Wesolowski (*78*) utilized this methodology to prepare the corresponding *cis*-1,2-difluoro-2-iodo substituted alkenes, which on carboalkoxylation stereospecifically afforded the *cis*-3-alkyl-2,3-difluoroacrylates, as illustrated in Scheme 13.

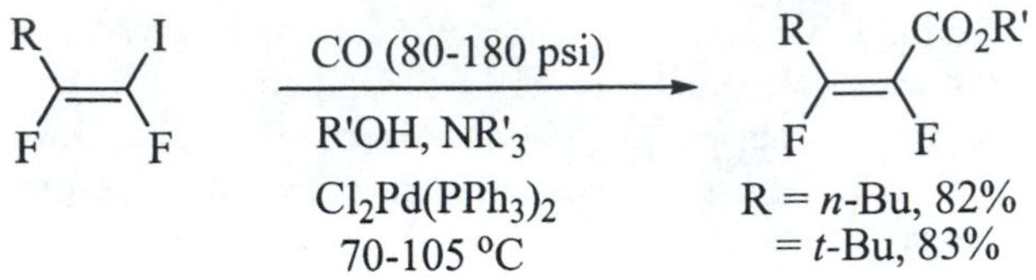

$$\mathrm{(F)(R)C{=}C(F)(I)} \xrightarrow[\substack{\mathrm{R'OH,\ NR'_3} \\ \mathrm{Cl_2Pd(PPh_3)_2} \\ 70\text{-}105\ ^\circ\mathrm{C}}]{\mathrm{CO\ (80\text{-}180\ psi)}} \mathrm{(F)(R)C{=}C(F)(CO_2R')}$$

R = *n*-Bu, 82%
= *t*-Bu, 83%

Scheme 13. Synthesis of cis-3-alkyl-2,3-difluoroacrylates

The above examples demonstrate the utility of the *cis*-1,2-difluoroethenyl-trialkylsilane to be converted into other useful *cis*-1,2-difluoroethenyl building blocks. The major obstacle at this junction of our research activities was the problem of scale-up of these procedures. The cost of Et_3SiCl or $PhMe_2SiCl$ was a major problem for the large scale synthesis of the key intermediate, namely the 1,2-difluoroethenyltrialkylsilane, on a practical scale. Fortunately, a co-worker, Vinod Jairaj, in our laboratory provided an answer to this dilemma. He found that the readily available zinc reagent, [$F_2C{=}CFZnBr$], could be easily functionalized with chlorotrimethylsilane and catalytic Cu(I)Br to give (cleanly) 1,1,2-trifluoro-2-trimethylsilylethene, as shown in Scheme 14 (*79*).

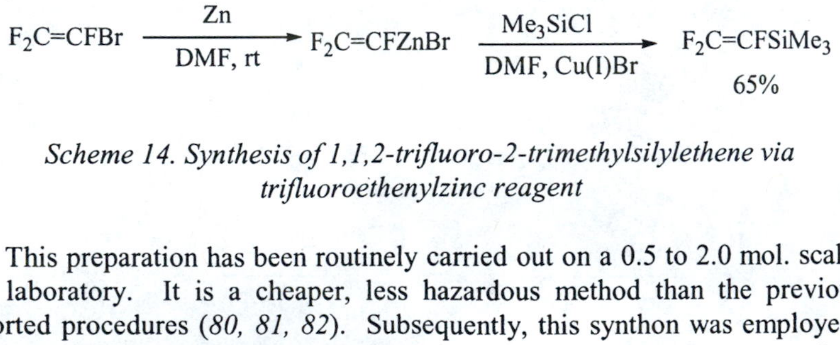

$$\mathrm{F_2C{=}CFBr} \xrightarrow[\mathrm{DMF,\ rt}]{\mathrm{Zn}} \mathrm{F_2C{=}CFZnBr} \xrightarrow[\mathrm{DMF,\ Cu(I)Br}]{\mathrm{Me_3SiCl}} \mathrm{F_2C{=}CFSiMe_3}$$

65%

Scheme 14. Synthesis of 1,1,2-trifluoro-2-trimethylsilylethene via trifluoroethenylzinc reagent

This preparation has been routinely carried out on a 0.5 to 2.0 mol. scale in our laboratory. It is a cheaper, less hazardous method than the previously reported procedures (*80, 81, 82*). Subsequently, this synthon was employed in place of the Et_3Si-group noted above in our further development of new functionalized building blocks. In one of those transformations, the *cis*-1,2-difluorotrimethylsilylethene was transformed to *cis*-1,2-difluorotrimethylsilylethenyl stannane by an *in situ* metallation using 4-methoxy-LTMP in the presence of Bu_3SnCl. Palladium catalyzed cross-coupling reaction of the *cis*-1,2-difluorotrimethylsilylethenyl stannane with trifluoroethenyl iodide under Stille-Liebeskind conditions [Pd(0)/CuI] produced the *cis*-trimethylsilyl-1,2,3,4-pentafluoro-1,3-butadiene. The *cis*-trimethylsilyl-1,2,3,4-pentafluoro-1,3-butadiene was further transformed to *cis*-tributylstannyl-1,2,3,4-pentafluoro-1,3-butadiene by the treatment with $Bu_3SnOSnBu_3$ and catalytic KF (Scheme 15) in DMF (*83*).

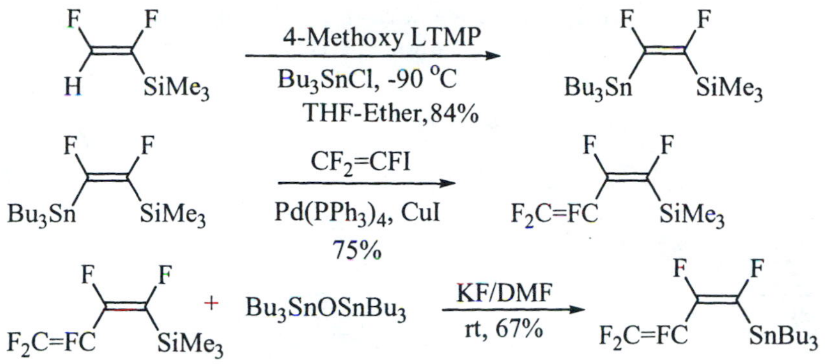

$(F)HC=C(F)SiMe_3$ → 4-Methoxy LTMP; Bu_3SnCl, -90 °C; THF-Ether, 84% → $Bu_3Sn(F)C=C(F)SiMe_3$

$Bu_3Sn(F)C=C(F)SiMe_3$ → CF_2=CFI; $Pd(PPh_3)_4$, CuI; 75% → F_2C=FC(F)C=C(F)$SiMe_3$

F_2C=FC(F)C=C(F)$SiMe_3$ + $Bu_3SnOSnBu_3$ → KF/DMF; rt, 67% → F_2C=FC(F)C=C(F)$SnBu_3$

Scheme 15. Synthesis of cis-trimethylsilyl-1,2,3,4-pentafluoro-1,3-butadiene and and cis-tributylstannyl-1,2,3,4-pentafluoro-1,3-butadiene.

Stereospecific synthesis of cis-1,2-difluorostyrenes, cis-1,2-difluoro-1-idodostyerenes and cis-1,2-difluorostilbenes

The *cis*-1,2-difluoroiodoethene synthon (*cis*-CHF=CFI) synthesized by the previously discussed methodology was used for a convenient preparation of *cis*-1,2-difluorostyrenes. The *cis*-1,2-difluoroiodoethene was transformed to *cis*-1,2-difluoroalkenylzinc reagent (*cis*-CHF=CFZnI) by a zinc insertion reaction. The zinc reagent, thus generated, was then utilized for the palladium catalyzed cross-coupling reaction with aryl iodides to obtain *cis*-1,2-difluorostyrenes (*31, 84*), as illustrated in Table 2.

The *cis*-1,2-difluorostyrenes prepared by this methodology were then converted to *cis*-1,2-difluoro-1-iodostyrenes *via* two approaches (Table 3). In Method A, metallation of the *cis*-1,2-difluorostyrene with lithium tetramethylpiperidide (LTMP) at low temperature, followed by *in situ* trapping with Bu_3SnCl, stereospecifically afforded the corresponding *cis*-1,2-difluoroethenylstannanes, which on reaction with iodine produced the *cis*-1,2-difluoro-1-iodostyrenes. In Method B, *n*-BuLi was utilized to metallate the *cis*-1,2-difluorostyrene at low temperatures; quenching the corresponding ethenyllithium with iodine gave the *cis*-1,2-difluoro-1-iodostyrenes (*31*). These results are summarized in Table 3.

Table 2. Stereospecific preparation of *cis*-1,2-difluoroalkenylzinc reagent and *cis*-1,2-difluorostyrenes

$$\text{FHC=CFI} \xrightarrow[\text{DMF}]{\text{Zn}} \text{FHC=CF(ZnI)} \xrightarrow[\text{DMF, rt, 1-3 h}]{\text{ArI, Pd(PPh}_3)_4} \text{FHC=CFAr}$$

cis/trans = 95:5

Entry	*ArI*	*Cis-1,2-difluorostyrenes*	*Yield (%)*[a]
1	C_6H_5	C_6H_5 CF=CFH	65
2	4-$CH_3OC_6H_4I$	4-$CH_3OC_6H_4$CF=CFH	85
3	4-$NO_2C_6H_4I$	4-$NO_2OC_6H_4$CF=CFH	66 [b]
4	2-$NO_2C_6H_4I$	2-$NO_2OC_6H_4$CF=CFH	78
5	4-$CH_3C_6H_4I$	4-$CH_3OC_6H_4$CF=CFH	72
6	4-$EtO_2CC_6H_4I$	4-$EtO_2CC_6H_4$CF=CFH	93
7	4-$CH_3C(O)C_6H_4I$	4-$CH_3C(O)C_6H_4$CF=CFH	71
8	3-ClC_6H_4I	3-ClC_6H_4CF=CFH	60
9	1,4-$C_6H_4I_2$	4-HFC=CFC_6H_4CF=CFH	80
10	4-$CF_3C_6H_4I$	4-$CF_3C_6H_4$CF=CFH	70
11	2-$(CH_3)_2CHC_6H_4I$	2-$(CH_3)_2CHC_6H_4$CF=CFH	55 [c]

[a] Isolated yield of *cis*-isomer. [b] Isolated as 95:5 *cis/trans* mixture. [c] Reaction conditions, 60 °C, 8 h.

These *cis*-1,2-difluorovinylstannanes/iodides can be subsequently employed in Pd(0) coupling reactions to prepare symmetrical or unsymmetrical *cis*-1,2-difluorostilbenes; a class of compounds that has been relatively inaccessible by most current methodologies. A typical example is illustrated in Scheme 16 (*85*).

Sterospecific preparation of cis and trans-aryl substituted 2,3-difluoroacrylic esters

Aryl substituted 2,3-difluoroacrylic esters are interesting and useful synthons and several methods have been established for the synthesis of the *trans*-ester (*50, 52, 78, 86*). The first stereospecific preparation of the *cis*-substituted 2,2-difluoroacrylic esters was described earlier in this report *via* carboalkoxylation of the corresponding 1,2-difluoro-1-iodoalkenes (where R = alkyl). When R = aryl or substituted aryl, the carboalkoxylation also (*78*) worked well. However, a major limitation of this methodology was the requirement of pure alkyl or aryl substituted vinyl iodides, since each precursor

Table 3. Stereospecific preparation of *cis*-1,2-difluoro-1-idodostyerenes

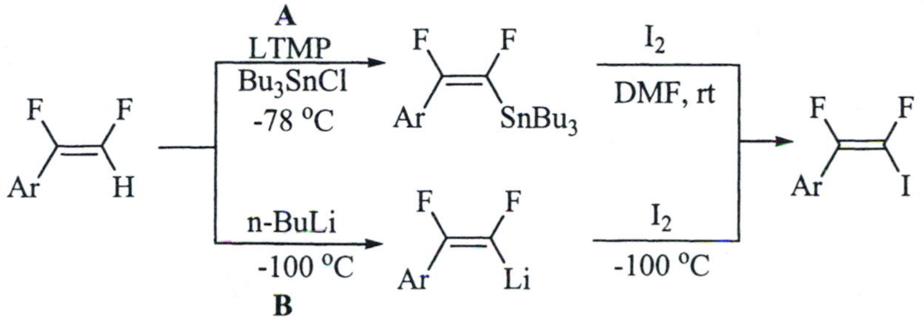

A: LTMP, Bu_3SnCl, -78 °C → ArCF=CF$SnBu_3$ → I_2, DMF, rt → ArCF=CFI

B: n-BuLi, -100 °C → ArCF=CFLi → I_2, -100 °C → ArCF=CFI

Entry	*Product*	*Method*	*Yield (%)*[a]
1	$4\text{-}CH_3OC_6H_4CF{=}CFI$	A	87
2	$4\text{-}CH_3OC_6H_4CF{=}CFI$	B	85
3	$C_6H_5CF{=}CFI$	B	83
4	$4\text{-}CH_3C_6H_4CF{=}CFI$	B	65
5	$2\text{-}(CH_3)_2CHC_6H_4CF{=}CFI$	B	54
6	$4\text{-}CF_3C_6H_4CF{=}CFI$	A	92
7	$4\text{-}CF_3C_6H_4CF{=}CFI$	B	81
8	$4\text{-}EtO_2CC_6H_4CF{=}CFI$	B	39 [b]
9	$4\text{-}EtO_2CC_6H_4CF{=}CFI$	A	77 [c]
10	$3\text{-}ClC_6H_4CF{=}CFI$	A	85 [c]

[a] Isolated yield based on ArCF=CFH. [b] LTMP was used instead *n*-BuLi. [c] One-pot procedure; vinylstannane intermediate was treated directly with I_2.

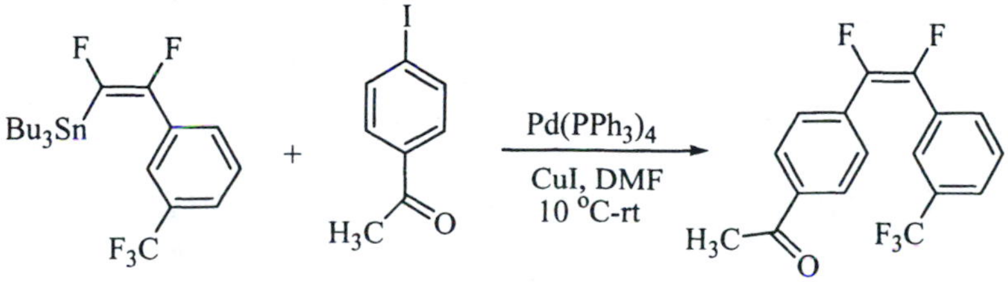

$Pd(PPh_3)_4$, CuI, DMF, 10 °C-rt

Scheme 16. Stereospecific synthesis of unsymmetrical cis-1,2-difluorostilbenes

needed to be prepared independently. To circumvent this problem, a more general approach to the aryl analogues was designed using a common synthon, where the introduction of the aryl group occurs in the final stage of the synthetic sequence. Thus, both the *cis* and *trans*-2,3-difluoro-3-stannylacrylate synthons were developed. The *trans*-2,3-difluoro-3-stannylacrylic ester was readily prepared from the corresponding *trans*-1,2-difluorotrialkylsilylethene *via* stereospecific stannyl/silyl exchange (*30*) (Scheme 17). Subsequent coupling with aryl iodides under Stille-Liebeskind conditions (Pd(0)/CuI) afforded the corresponding aryl substituted *trans*-2,3-difluoroacrylic esters (*87*). With this synthon a stereospecific synthesis of fluorinated dienes could also be achieved *via* coupling of the *trans*-2,3-difluoro-3-stannylacrylic ester synthons with ethenyl halides under Stille-Liebeskind conditions (*87, 88*) (Scheme 17). For example, reaction of *trans*-2,3-difluoro-3-(tributylstannyl)acylate with *cis*-1-iodostyrenes under $Pd(PPh_3)_4$/CuI catalysis afforded the corresponding ethyl (2*E*,4*Z*)-2,3-difluoro-5-phenyl-2,4-pentadienoate in 80% yield (*88*).

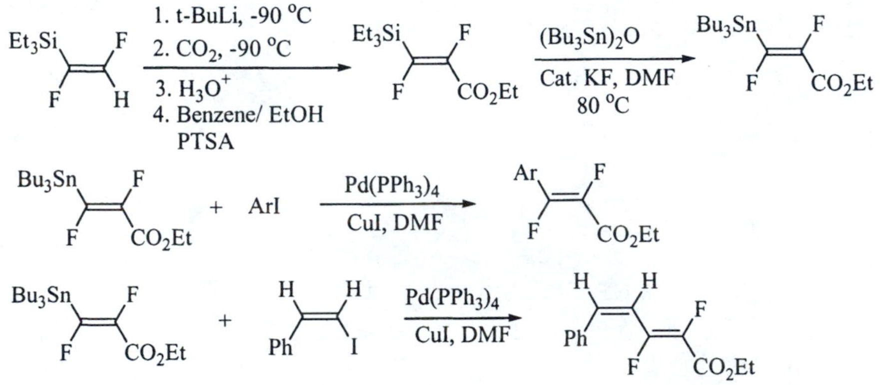

Et_3Si, F / F, H — 1. t-BuLi, -90 °C; 2. CO_2, -90 °C; 3. H_3O^+; 4. Benzene/ EtOH PTSA → Et_3Si, F / F, CO_2Et — $(Bu_3Sn)_2O$; Cat. KF, DMF 80 °C → Bu_3Sn, F / F, CO_2Et

Bu_3Sn, F / F, CO_2Et + ArI — $Pd(PPh_3)_4$; CuI, DMF → Ar, F / F, CO_2Et

Bu_3Sn, F / F, CO_2Et + H, H / Ph, I — $Pd(PPh_3)_4$; CuI, DMF → H, H / Ph / F / F, CO_2Et

Scheme 17. Synthesis of trans-2,3-difluoro-3-stannylacrylic ester synthon and their Pd(0) catalyzed coupling reaction with aryl or alkenyl iodides.

Due to the instability of the ethenyllithium reagent produced *via* metallation of *cis*-1,2-difluoro-3-triethylsilylethene, the corresponding ethyl-(*cis*)-2,3-difluoropropenoate could not be prepared *via* the route utilized for the trans-analogue (cf. above). Therefore, a new route was designed to prepare this key intermediate (*87*). First, a stereospecific palladium catalyzed carboalkoxylation of *cis*-1,2-difluoro-1-iodoethenes in ethanol produced the corresponding *cis*-2,3-difluoroacylate. In the second step, in situ metallation with LDA in the presence of Bu_3SnCl at low temperature afforded the *cis*-2,3-difluoro-3-(tributylstannyl)acrylate stereospecifically (Scheme 18). Coupling the *cis*-acrylate synthons with aryl iodides or vinyl halides under Stille-Liebeskind conditions produced the isomerically pure *cis*-aryl substituted acrylic esters or conjugated dienes in good yields (*87*). This methodology and typical examples is outlined in Table 4.

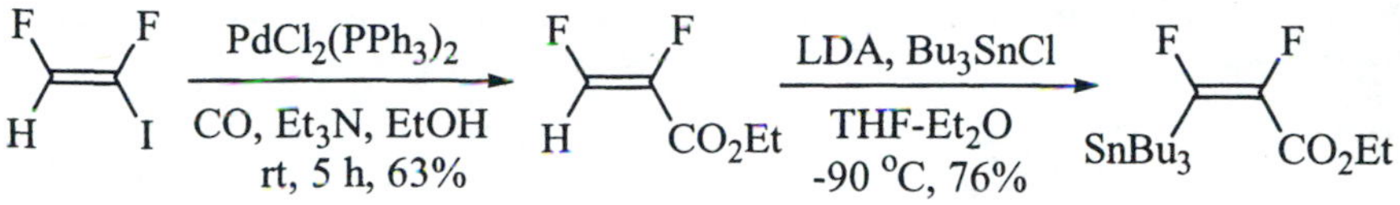

F F, H I → [$PdCl_2(PPh_3)_2$; CO, Et_3N, EtOH; rt, 5 h, 63%] → F F, H CO_2Et → [LDA, Bu_3SnCl; $THF\text{-}Et_2O$; -90 °C, 76%] → F F, $SnBu_3$ CO_2Et

Scheme 18. Synthesis of cis-2,3-difluoro-3-stannylacrylic ester synthon.

Table 4. Pd(0) catalyzed coupling reaction of *cis*-2,3-difluoro-3-stannylacrylic ester synthon with aryl or alkenyl iodides.

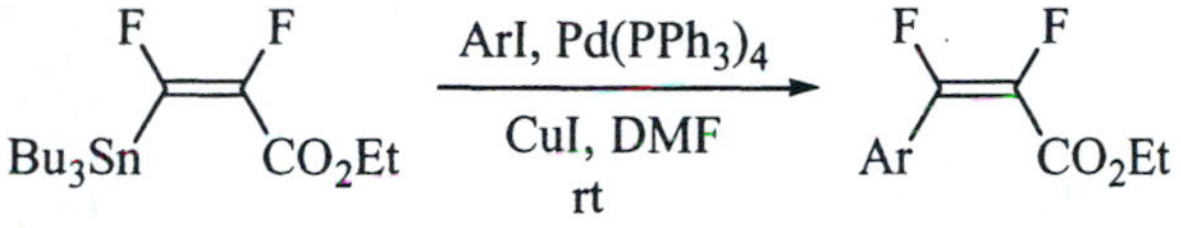

F F, Bu_3Sn CO_2Et → [ArI, $Pd(PPh_3)_4$; CuI, DMF; rt] → F F, Ar CO_2Et

ArI	*Product*	*Isolated Yiled (%)*[a]
Iodobenzene	(*cis*)-$C_6H_5CF{=}CFCO_2Et$	75
4-Iodoanisole	(*cis*)-4-$CH_3OC_6H_4CF{=}CFCO_2Et$	83
1-Iodo-4-nitrobenzene	(*cis*)-4-$O_2NC_6H_4CF{=}CFCO_2Et$	78
1-Iodo-3-nitrobenzene	(*cis*)-3-$O_2NC_6H_4CF{=}CFCO_2Et$	86
2-Iodothiophene	(*cis*)-$C_4H_3SCF{=}CFCO_2Et$	73
3-Iodoanisole	(*cis*)-3-$CH_3OC_6H_4CF{=}CFCO_2Et$	74
1-Bromo-4-iodobenzene	(*cis*)-4-$BrC_6H_4CF{=}CFCO_2Et$	69
(*E*)-$CHBr{=}CHCO_2Et$	(2*Z*,4*E*)-$EtO_2CCH{=}CHCF{=}CFCO_2Et$	94
(Z)-$CHBr{=}CHCO_2Et$	(2*Z*,4*Z*)-$EtO_2CCH{=}CHCF{=}CFCO_2Et$	86

[a] Configuration assigned on the basis of $^3J_{FF\,(cis)}$ = 0-22 Hz.

Sterospecific preparation of cis-2,3-difluoro-4-oxo-substituted 2-butenoates

The *cis*-2,3-difluoro-3-(tributylstannyl) acrylate synthons could also be utilized for an efficient copper (I) iodide catalyzed coupling reaction with acyl halides to obtain the *cis*-2,3-difluoro-4-oxo-substituted 2-butenoates stereospecifically (Table 5) (*89*) Similar reaction under Stille conditions did not produce *cis*-2,3-difluoro-4-oxo-substituted 2-butenoates whereas under Stille-Liebeskind conditions [$Pd(PPh_3)_4/CuI$] , the coupled product was formed only in moderate yield (*89*).

Table 5. Stereospecific synthesis of *cis*-2,3-difluoro-4-oxo-substituted 2-butenoates

F F

Bu_3Sn CO_2Et + RCOCl —CuI (10 mol%), DMF, rt→ F F, R, O, CO_2Et

Entry	*RCOCl*	*Product*	*Isolated Yiled (%)*
1	C_6H_5COCl	*cis*-$C_6H_5COCF{=}CFCO_2Et$	75
2	*p*-$O_2NC_6H_4COCl$	*cis-p*-$O_2NC_6H_4COCF{=}CFCO_2Et$	70
3	*n*-$C_9H_{19}COCl$	*cis-n*-$C_9H_{19}COCF{=}CFCO_2Et$	83
4	*p*-$MeOC_6H_4COCl$	*cis-p*-$MeOC_6H_4COCF{=}CFCO_2Et$	67
5	Me_3CCOCl	*cis*-$Me_3CCOCF{=}CFCO_2Et$	59
6	*p*-IC_6H_4COCl	*cis-p*-$IC_6H_4COCF{=}CFCO_2Et$	60

The active intermediate in this conversion is the stable *cis*-2,3-difluoroethenylcopper reagent, produced by tin/Cu(I) exchange (Scheme 19). This intermediate could be observed spectroscopically *via* ^{19}F NMR and could be generated independently from the corresponding *cis*-2,3-difluoroethenylzinc reagent (Scheme 19) This result represents the first direct observation of a vinylcopper intermediate in the Stille coupling of vinylstannanes (*89*).

Synthesis of 3,4-difluoro-6-substituted-2-pyrones from cis-2,3-difluoro-3-iodoacrylic acid synthons.

The *cis*-2,3-difluoro-3-iodoacrylic acid, noted above, readily reacts with terminal alkynes under Pd(0) catalysis (Sonogashira reaction) to stereospecifically afford the *cis*-2,3-difluoro-5-substituted-pent-2-en-4-ynoic acids, which cyclize (under the Pd(0) catalysis) to give the 3,4-difluoro-6-substituted-2-pyrones (*90*). This simple one-step route provides the first, facile,

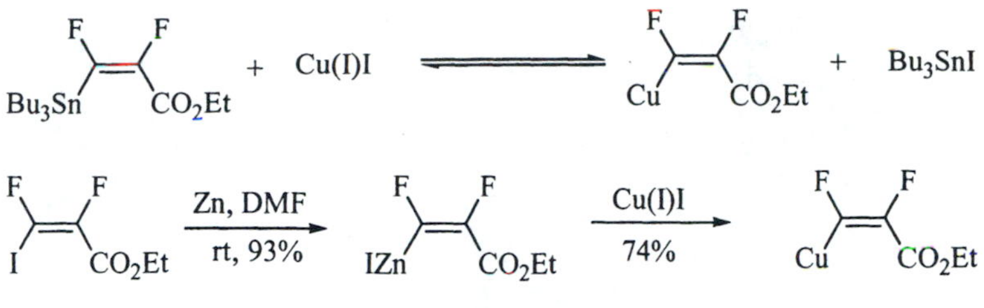

Scheme 19. Formation of cis-2,3-difluoroethenylcopper via tin/Cu(I) excahange as well as from cis-2,3-difluoroethenylzinc reagent

Table 6. Synthesis of 3,4-difluoro-6-substituted-2-pyrones

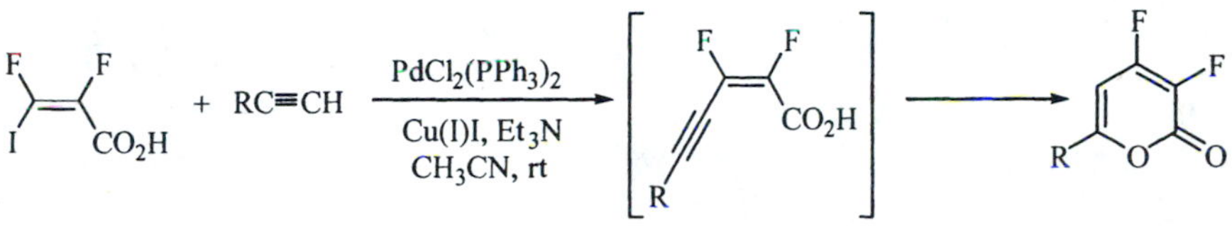

Entry	R	Time	Isolated Yiled (%)
1	C_6H_5	24	62
2	n-C_5H_{11}	24	59
3	$C_6H_5CH_2CH_2$	24	64
4	p-$CF_3C_6H_4$	12	69
5	p-$MeOC_6H_4$	16	71
6	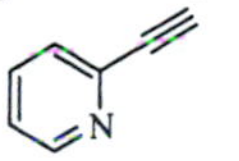 [2-ethynylpyridine structure]	24	43

general synthesis of this class of 2-pyrones. Table 6 summarizes some typical 2-pyrones prepared by this one-pot methodology.

A slight modification of the above route also provides a key entry to synthesize 3,4-difluoro-5-iodo-6-substituted-2-pyrones (*90*) *via* an electrocyclization reaction illustrated in Scheme 20. The iodo-substituted 2-pyrones thus generated, readily undergo Negishi or Sonogashira coupling reactions to give the 5,6-disubstituted-3,4-difluoro-2-pyrones.

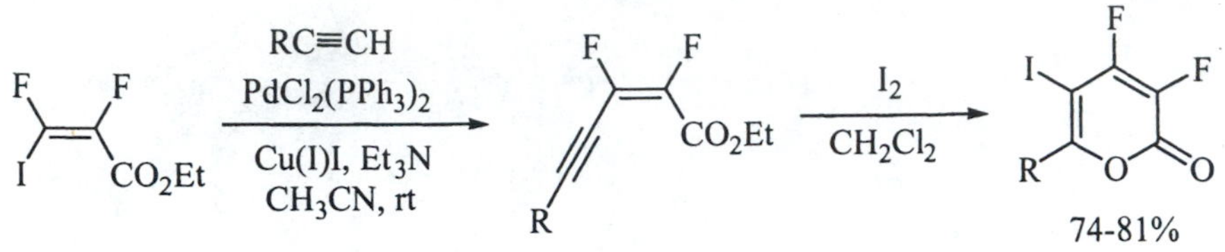

RC≡CH, $PdCl_2(PPh_3)_2$, Cu(I)I, Et_3N, CH_3CN, rt; I_2, CH_2Cl_2; 74-81%

Scheme 20. Synthesis of 3,4-difluoro-5-iodo-6-substituted-2-pyrones

"Bis"-cis-1,2-difluoroalkenyl synthons

Most of the 1,2-difluorosynthons described above were "mono" synthons capable of a single coupling reaction. The first "bis"-synthon developed in our laboratory was *trans*-(1,2-ethenediyl)bis[tributylstannane] (*91*), which was prepared by the metallation of *trans*-1,2-difluorotributylstannane with LTMP followed by transmeallation with Bu_3SnCl (Scheme 21). The stable bis-stannane thus generated could be coupled with aryl iodides under Stille-Liebeskind conditions to afford a clean, efficient route to symmetrically substituted *trans*-1,2-difluorostilbenes.

91 % *trans*; $Bu_3SnOSnBu_3$, TBAF (cat), THF; 97% *trans* after column chromatography

1. LiTMP, THF, -90 °C; 2. Bu_3SnCl

1. LiTMP, THF, -90 °C; 2. Bu_3SnCl; Decomposition of the intermediate lithium reagent

2ArI, $Pd(PPh_3)_4$, Cu(I)I, DMF/THF, rt or 60 °C; Ar = symmetrical aromatic groups, 51-89 %

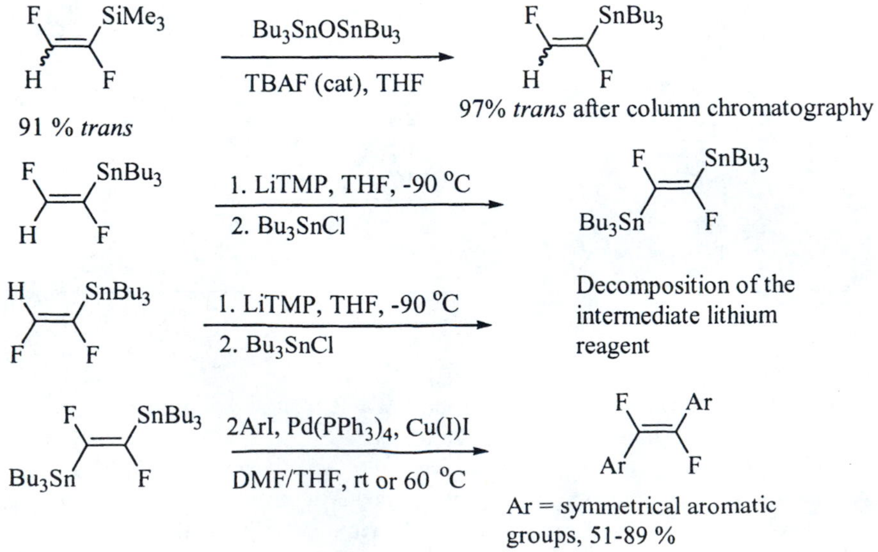

Scheme 21. Synthesis of trans-(1,2-ethenediyl)bis[tributylstannane] and symmetrically substituted trans-1,2-difluorostilbenes.

Other "bis" synthons under development in our laboratory are the *cis* and *trans*-2-halo-1,2-difluoro-1-iodoalkenes (where X = Br, Cl) (*92*).These synthons should be capable of sequential functionalization by metal-catalyzed coupling reactions and thus capable of the synthesis of both *cis* and *trans* – symmetrical or unsymmetrical functionalized derivatives (for eg. *cis* and *trans* 1,2-difluorostilbenes) (Scheme 22).

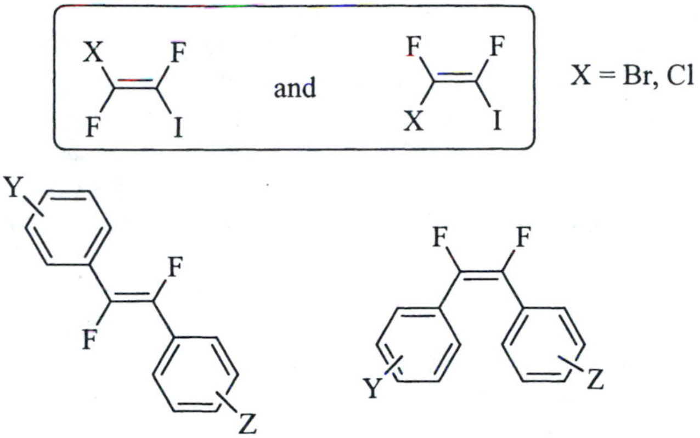

X = Br, Cl

Scheme 22. Cis and trans-2-halo-1,2-difluoro-1-iodoalkene, possible bis-synthon for the sequential introduction of unsymmetrical aryl groups.

Stereospecific synthesis of cis-1-Arylperfluoroalkenes.

The addition-elimination or dehydrohalogenetaion methods reported for the preparation of 1-arylperfluoroalkenes generally produce exclusively the *trans*-isomer or a mixture of isomers favoring *trans* as the major product (*10, 11, 12, 13*). The first synthesis of stereospecifically pure *cis*-1-arylperfluoropropenes were achieved in our laboratory *via* the Pd(0) coupling of *cis*-perfluoropropenylzinc reagent (*cis*-$CF_3CF{=}CFZnI$) with aryl iodides. The *cis*-perfluoralkenylzinc reagents were generated from *cis*-1-iodoperfluoroalkenes ($R_FCF{=}CFI$) (R_F = CF_3, C_2F_5) (*57, 58*) which in turn was prepared from *cis*-$R_FCF{=}FH$. The *cis*-$R_FCF{=}FH$ was not readily available and was prepared by a stereospecific phosphodefluoridation of perfluoroalkenes ($R_FCF{=}CF_2$, R_F = CF_3, C_2F_5-) (*8, 93, 94, 95*) to obtain corresponding *trans*-$R_FCF{=}FH$, which was then isomerized using SbF_5 to obtain desired *cis*-$R_FCF{=}FH$ (*93, 96*). But this methodology for the preparation of *cis*-$R_FCF{=}FH$ was not feasible for perfluoroalkenes with longer chains due to the formation of internal isomers in addition to the desired 1-hydroperfluoroalkene (*97*) Recent work of Frohn and co-workers demonstrated that metallation of $C_6F_{13}CF_2CH_2F$ under conditions similar to that of the metallation of HFC-134a at low temperature produced the corresponding perfluoroalkenyllithium ($C_6F_{13}CF{=}CFLi$) in a *cis*-selective fashion (>95%) (*72, 98*). This observation prompted us to investigate a similar process for the preparation of the corresponding *cis*-perfluoroalkenylzinc reagents (*cis*-$R_FCF{=}CFZnCl$) *via* our *in situ* metallation strategy developed for the generation of trifluoroalkenyl and halodifluoroalkenylzinc reagents (*99, 100, 101, 102, 103, 104*). The *cis*-perfluoroalkenylzinc reagents thus generated could be effectively used in a coupling reaction with aryl iodides to produce *cis*-1-arylperfluoroalkenes. Thus, commercially available $CF_3CF_2CH_2F$ was metallated using LDA in the presence of $ZnCl_2$ at 15 °C, to produce the corresponding perfluoropropenylzinc reagent in 67% yield with selectivity in favor of the *cis*-isomer [(*cis/trans*)-$CF_3CF{=}CFZnCl$ = ~82:18] (*74*). When the metallation was performed at -78 °C, an improved yield (81%) and selectivity [(*cis/trans*)-$CF_3CF{=}CFZnCl$ = ~89:11] of the perfluoropropenylzinc reagent was observed (Scheme 23). The zinc reagent was then hydrolyzed using acetic acid to obtain corresponding 1-hydroperfluoropropene with similar selectivity [(*cis/trans*)-$CF_3CF{=}CFH$ = 87:13]. The perfluoropropenylzinc reagent underwent Pd(0) catalyzed cross-coupling reaction with aryl iodides to produce the corresponding 1-arylperfluoropropenes in excellent isolated yield (*74*). The preferred *cis*-geomerty of the zinc reagent was retained during the coupling process and no isomerization was observed.

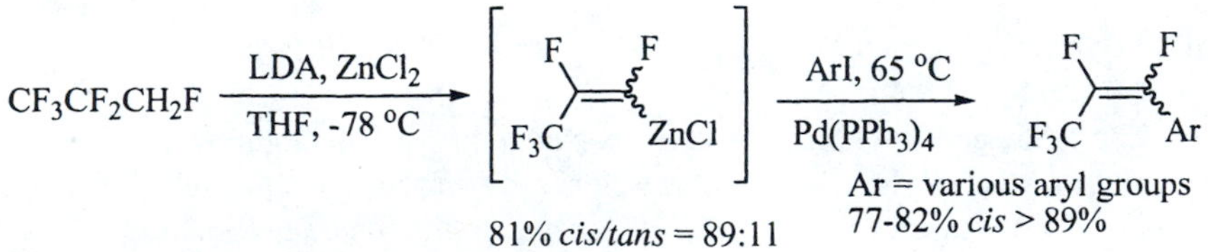

$CF_3CF_2CH_2F$ —(LDA, $ZnCl_2$ / THF, -78 °C)→ [F, F, F_3C, ZnCl] —(ArI, 65 °C / $Pd(PPh_3)_4$)→ F, F, F_3C, Ar

81% *cis/tans* = 89:11

Ar = various aryl groups
77-82% *cis* > 89%

Scheme 23. Synthesis of cis-perfluoropropenylzinc reagent and cis-1-arylperfluoroprop-1-enes

After the successful synthesis of the *cis*-pentafluoropropenylzinc reagent and *cis*-1-arylperfluoroprop-1-enes, the *in situ* metallation of higher homologues of *1H,1H*-perfluoroalkenes ($C_4F_9CF_2CH_2F$, $C_5F_{11}CF_2CH_2F$, $C_6F_{13}CF_2CH_2F$ and $C_{10}F_{21}CF_2CH_2F$) were performed and the results are summarized in Table 7 (*74*). It was observed that the size of the R_F group influenced the selectivity in the elimination process; excellent *cis*-selectivity, up to 99%, was observed for longer sized R_F groups (C_2F_5, C_4F_9, C_5F_{11}, C_6F_{13}). Due to the poor solubility of *1H,1H*-perfluorododecane in THF, the metallation of $C_{10}F_{21}CF_2CH_2F$ was performed in a THF-diethyl ether solution at a higher temperature (-10 °C) to produce the perfluorododec-1-enylzinc reagent in 61% yield with a slightly diminished selectivity (~93% *cis*). Palladium catalyzed cross-coupling reaction of each of the perfluoroaleknylzinc reagents with a variety of aryl iodides were then performed to obtain the corresponding 1-arylperfluoroalkenes in excellent isolated yield with the retention of *cis*-stereochemistry (*74*). Representative examples for the coupling reaction of perfluoroalkenylzinc reagents with iodobenzene are summarized in Table 7. The excellent *cis*-selectivity observed during the metallation of $R_FCF_2CH_2F$ could be due to the preferred syn-clinal conformation of $R_FCF_2CH_2F$ over the less favored anti-periplanar conformation. The syn-clinal effect is attributed to the relatively large electron flow from the $\sigma_{C\text{-}H}$ bond towards to the $\sigma^*_{C\text{-}F}$ relative to the electron flow from $\sigma_{C\text{-}F}$ bond to the $\sigma^*_{C\text{-}F}$ bond in the anti-periplanar conformation (*74, 105*).

Table 7. Synthesis of various *cis*-perfluoroalkenyl zinc reagents and *cis*-1-phenylperfluoroalkenes

$R_FCF_2CH_2F$ —(LDA, $ZnCl_2$ / THF, -78 °C)→ [R_F(F)C=C(F)ZnCl] —(C_6H_5I, 65 °C / $Pd(PPh_3)_4$)→ R_F(F)C=C(F)C_6H_5

Entry	R_F	Yield (%)[a] of $R_FCF{=}CFZnCl$	*cis/trans* ratio [a] $R_FCF{=}CFZnCl$	Yield (%) [b] of $R_FCF{=}CFC_6H_5$	*cis/trans* ratio [a]
1	CF_3	80	89:11	80	87:13
2	C_2F_5	76	94:6	89	93:7
3	C_4F_9	75	96:4	81	96:4
4	C_5F_{11}	83	99:1	89	99:1
5	C_6F_{13}	79	98:2	81	98:2
6	$C_{10}F_{21}$	61	93:7	78	94:6

[a] The *Z/E* ratio determined by ^{19}F NMR ($^3J_{FF\ (trans)}$ = ~132 Hz, $^3J_{FF\ (cis)}$ = ~12 Hz). [b] Isolated yield based on iodobenzene.

Stereospecific synthesis of cis-1-iodoperfluoroalkenes

The *cis*-1-iodo-perfluoroalkenes are less conveniently available and difficult to synthesize, especially the higher homologues [the higher homologues other than $R_F = CF_3, C_2F_5$ could not be prepared by the SbF_5 catalyzed isomerization of trans-$R_FCF{=}CFH$ (*8, 93, 96, 97*). *Cis*-perfluoroalkenylzinc reagents generated by the metallation of *1H,1H*-perfluoroheptene and *1H,1H*-perfluorooctene, upon iodinolysis, produced the corresponding 1-iodoperfluoroalkenes in good yield, with >98% *cis*-selectivity (Scheme 24) (*74*).

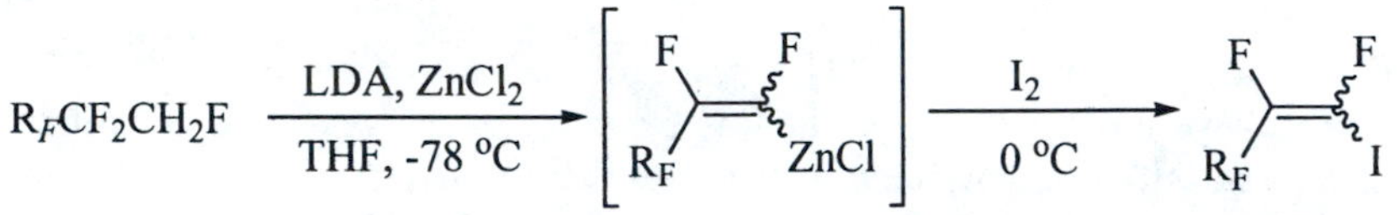

$$R_FCF_2CH_2F \xrightarrow[\text{THF, -78 °C}]{\text{LDA, ZnCl}_2} \left[R_F(F)C{=}C(F)ZnCl \right] \xrightarrow[\text{0 °C}]{I_2} R_F(F)C{=}C(F)I$$

$R_F = n\text{-}C_5F_{11}$, 83%, 98% *cis* $R_F = n\text{-}C_5F_{11}$, 68%, 98% *cis*
$R_F = n\text{-}C_6F_{13}$, 79%, 99% *cis* $R_F = n\text{-}C_6F_{13}$, 70%, 99% *cis*

Scheme 24. Synthesis of cis-1-iodo-perfluoroalkenes

Conclusion

In summary, we have presented the synthesis of several 1,2-difluoroalkenyl synthons and their utilization for the stereospecific preparation of the corresponding 1,2-difluoroalkenes. By the utilization of a variety of *cis*-1,2-difluoroalkenyl organometallic synthons, it was demonstrated that several synthetically intricate *cis*-1,2-difluoroalkenes such as *cis*-1,2-difluorostyrenes, *cis*-2-iodo-1,2-difluorostyrenes, *cis*-2-alkylsubstituted-1,2-difluorostyrenes, *cis*-aryl substituted 2,3-difluoroacrylic esters, *cis*-2,3-difluoro-4-oxo-substituted 2-butenoates, symmetrically and unsymmetrically substituted *cis*-1,2-difluorostilbenes, *cis*-1-arylperfluoroalkenes and *cis*-1-iodoperfluoroalkenes could be readily prepared. *Cis*-2,3-difluoroacrylic acid synthons could be used as an efficient synthon for a one-pot synthesis of 5 (4, 5 di) substituted 2,3-difluoro-2-pyrones.

References

1. Organofluorine Compounds: Chemistry and Applications, Hiyama, T. Ed, Springer-Verlag, Berlin Heidelberg, 2000, chapter 5, pp 137-177. chapter 6, pp 183-233.

2. Organofluorine Chemistry: Principles and Commercial Applications, Banks, R.E.; Smart, B. E.; Tatlow, J. C., Eds.; Plenum Press: New York, 1994.
3. Schmiegel, W. W. in Chemistry of Organic Fluorine Compounds II: A critical Review, Hudlicky, M.; Pavlath, A.E., Eds.; ACS Monograph, 1995, pp 1101-1125.
4. Selective Fluorination in Organic and Bioorganic Chemistry, ACS Symp. Ser. No. 456; Welch, J. T., Ed.; American Chemical Society: Washington, DC, 1991.
5. Fluorine in Biological Chemistry, Welch, J. T.; Eswarakrishnan, S. Eds.Wiley; New York, 1991.
6. Biomedical Frontiers of Fluorine Chemistry, ACS Symp. Ser. No. 639; Ojima, I.; McCarthy, J. R.; Welch, J. T., Eds.; American Chemical Society: Washington, DC, 1996.
7. Burton, D. J.; Yang, Z. Y.; Qiu, W. *Chem. Rev.* **1996**, 96, 1641-1716.
8. Burton, D. J.; Cox, D.G. *J. Am. Chem. Soc.* **1983**, 105, 650-651.
9. Okada, Y.; Kuroboshi, M.; Ishihara, T. *J. Fluorine. Chem.* **1998**, 41, 435-438.
10. Dixon, S. *J. Org. Chem.* **1956**, 21, 400-403.
11. Dmowski, W. *J. Fluorine Chem.* **1982**, 21, 201-219.
12. Dmowski, W. *J. Fluorine Chem.* **1981**, 18, 25-30.
13. Albadri, R.; Moreau, P.; Commeyras, A. *Nouveu Journal De Chimie.* **1982**, 6, 581-587.
14. Khodkevich, O. M.; Rybakova, L. F.;Stepanov, M. V.; Panov, E. M.; Chernoplakova, V. A.; Starostina, T. A.; Kocheshkov, K. A.; *Doklady Akademii Nauk SSSR*, **1976**, 229, 645-646.
15. Chuit, C.; Sauvetre, R.; Masure, D.; Baudry, M.; Normant, J. F. *J. Chem. Res. Synopses*, **1977**, 4, 104.
16. Timperley, C. M.; Waters, M. J.; Greenall. J. A. *J. Fluorine Chem.* **2006**, 127, 249-256.
17. Chen, L. S.; Tamborski, C. *J. Fluorine. Chem.* **1982**, 20, 341-348.
18. Nguyen, T.; Rubienstein, M.; Wakselman, C. *J. Fluorine Chem.* **1978**, 11, 573-589.
19. Martinet, P.; Sauvetre, R.; Normant, J. F. *Bull. Soc. Chim. Fr.* **1990**, 27, 86-92.
20. Tozer, M. J; Herpin, T.F. *Tetrahedron* **1996**, 52, 8619-8683.
21. Leroy. J. *J. Org. Chem.* **1981**, 46, 206-209.
22. Kuroboshi, M.; Yamada, N.; Takabe, Y.; Hiyama, T. *Tetrahedron Lett.* **1995**, 36, 6271-6274.
23. Kuroboshi, M.; Hiyama, T. *Chem. Lett.* **1990**, 1607-1610.
24. Chambers, R. D.; Fuss, R. W.; Spink, R. C. H.; Greenhall, M. P.; Kenwright, A. M.; Batsanov, A. S.; Howard, J. A. K. *J. Chem. Soc., Perkin Trans. 1*, **2000**, 1623-1638.

25. Anilkumar, R.; Burton, D. J.; *Tetrahedron Lett.* **2003**, 44, 6661-6664.
26. Anilkumar, R.; Burton, D.J.; *J. Fluorine Chem.* **2005**, 126, 1174-1184.
27. Burton, D. J.; Yang, Z.Y.; Morkan, P. *Tetrahedron*, **1994**, 50, 2993-3063 and references cited therein
28. Normant, J. F.; *J. Organomet. Chem.* **1990**, 400, 19-34.
29. Fontana, A. S.; Davis, C. R; He Y-B.; Burton, D. J. *Tetrahedron*, **1996**, 52, 37-44.
30. Xue, L.; Lu, L.; Pederson, D. S; Liu, Q.; Narske, M. R.; Burton, D. J. *J. Org. Chem.* **1997**, 62, 1064-1071.
31. Davis, C. R.; Burton, D. J. *J. Org. Chem.* **1997**, 62, 9217-9222.
32. Asato A. E.; Liu, R. S. H. *Tetrahedron Lett.* **1986**, 3337-3340.
33. Singh, R. P.; Majumder, U.; Shreeve, J. M. *J. Org. Chem.* **2001**, 66, 6263-6267.
34. Tarrant, P.; Whitfield Jr, R.W.; Summerville, R.H. *J. Fluorine Chem.* **1971/72**, 1, 31–40.
35. Hahnfeld, J. L.; Burton, D.J. *Tetrahedron Lett.* **1975**, 773–776.
36. Gillet, J. P.; Sauvetre, R.; Normant, J.F. *Synthesis* **1986**, 355–360.
37. Martin, S. Sauvetre, R.; Normant, J. F. *Tetrahedron Lett.* **1982**, 23, 4329–4332.
38. Martin, S.; Sauvetre, R.; Normant, J.F. *Tetrahedron Lett.* **1983**, 24, 5615–5618.
39. Kvicala, J.; Hrabal, R.; Czerenek, J.; Bartosova, I.; Paleta, O.; Welter, A. *J. Fluorine Chem.* **1992**, 113, 211–218.
40. Moreau, P.; Dalverny, G.; Commeyras, A. *J. Chem. Soc. Chem. Commun.* **1976**, 174-175.
41. Moreau, P.; Albadri, R; Redwane, N.; Commeyras, A. *J. Fluorine Chem.* **1980**, 15, 103–116.
42. Moreau, P.; Redwane, N.; Commeyras, A. *Bull. Soc. Chim. Fr.* **1984**, 117–123.
43. Thoai, N.; Rubinstein, M.; Wakselman, C. *J. Fluorine Chem.* **1982**, 20, 271–276.
44. Redwane, N.; Moreau, P.; Commeyras, A. *J. Fluorine Chem.* **1982**, 20, 699–713.
45. Wessolowski, H.; Roeschenthaler, G-V.; Winter, R.; Gard, G. L. *Z. Naturforsch, B*: **1991**, 46, 126-8.
46. A. Gentzsch, G.-V.; Röschenthaler. *Chemiker Ztg.* **1990**, 114, 6-7.
47. Chernega, A. N.; Kolomeitsev, A. A.; Yagupolskij, Y. L. *J. Fluorine Chem.* **1993**, 60, 263-273.
48. Wessolowski, H.; Gentzsch, A.; Röschenthaler, G.-V.; Gard, G. L. *Heteroatom. Chem.* **1997**, 8, 467-471.
49. Moreau, P.; Redwane, N.; Zissis, J. P.; *J. Fluorine Chem.* **1982**, 20, 715–726.

50. Gillet, J. P. Sauvetre, R.; Normant, J. F. *Synthesis* **1986**, 538–543.
51. Tellier, F.; Sauvetre, R.; Normant, J.F. *J. Organomet. Chem.* **1986**, 303, 309–315
52. Gillet, J. P.; Sauvetre, R.; Normant, J. F. *Tetrahedron Lett.* **1985**, *26*, 3999-4002.
53. Tellier, F.; Sauvetre, R.; Normant, J. F. *J. Organomet. Chem.* **1987**, 328, 1-13
54. Martinet, P.; Sauvetre, R.; Normant, J. F. *J. Organomet. Chem.* **1989**, 367, 1-10.
55. Tellier, F. ; Sauvetre, R.; Normant, J. F. *J. Organomet. Chem.* **1989**, 364, 17–28.
56. Tellier, F.; Sauvetre, R.; Normant, J. F. ; *J. Organomet. Chem.* **1985**, 292, 19-28.
57. Hansen, S. W.; Spawn, T. D.; Burton. D. J. *J. Fluorine Chem.* **1987**, 35, 415-420.
58. Heinze, P. L.; Burton, D. J. *J. Org. Chem.* **1988**, 53, 2714-2720.
59. Burton, D.J.; Hansen, S. W. ; *J. Fluorine Chem*, **1986**, 31, 461-465.
60. Martin, S.; Sauvetre, R.; Normant, J. F. ; *Bull. Soc. Chim Fr.* **1986**, 6, 900-905.
61. Martin, S.; Sauvetre, R.; Normant, J. F. *Tetrahedron Lett.* **1986**, 27, 1027-1030.
62. Hiyama, T.; Nishide, K.; Obayashi, M. *Chem. Lett.* **1984**, *10*, 1765-1768.
63. Fujita, M.; Obayashi, M.; Hiyama, T. *Tetrahedron* **1988**, *44*, 4135-4145.
64. Bainbridge, J. M.; Corr, S.; Kanai, M.; Percy, J. M. *Tetrahedron Lett.* **2000**, 41, 971-974.
65. Lu, L.; Burton, D.J.; *Tetrahedron. Lett.* **1997**, 38, 7673-7676.
66. Burton, D. J.; Jairaj, V. *J. Fluorine. Chem.* **2005**, 126, 797-801.
67. Burton, D. J.; Jairaj, V. *J. Fluorine. Chem.* **2004**, 125, 673-680.
68. Liu, Q.; Burton, D.J. *Tetrahedron Lett.* **2000**, 41, 8045-8048.
69. Brisdon, A. K.; Crossley, I. R.; Pritchard, R. G.; Warren, J. E. *Inorg Chem.* **2002**, 41, 4748-4755.
70. Frohn, H.-J.; Bardin, V. V. *Z. Anorg. Allg. Chem.* **2001**, 627, 2499-2505.
71. Frohn, H.-J.; Bardin, V. V. *J. Organomet. Chem.* **2001**, 631, 54-58.
72. Frohn, H-J; Adonin, N. Y.; Bardin, V. V. *Z. Anorg. Allg. Chem.* **2003**, 629, 2499-2508.
73. Bardin, V. V.; Frohn, H.-J. *Z. Anorg. Allg. Chem.* **2002**, 628, 721-722.
74. Raghavanpillai, A.; Burton, D. J. *J. Fluorine. Chem.* **2006**, 127, xxxx.
75. Gouyon, T.; Sauvetre, R.; Normant, J. F. *J. Organomet. Chem.* **1990**, 394, 37-44.
76. Wesolowski, C. A. University of Iowa, unpublished results.
77. Martin, S.; Sauvetre, R.; Normant, J. F. *J. Organometal. Chem.* **1984**, 264, 155-161.

78. Wesolowski, C. A.; Burton, D. J. *Tetrahedron Lett.* **1999**, *40*, 2243-2246.
79. Jairaj, V.; Burton, D.J. *J. Fluorine Chem.* **2003**, 121, 75-77.
80. Stepanov, A. Martynov, B. L.; Tomilov, A. P. *Russ. J. Electrochem.* **2000**, 36, 190-192.
81. Petrov, V. A.; Mlsna, T. E.; Des Marteau, D. D. *Mendeleev Commun.* **1993**, 6, 240.
82. Bardin, V. V.; Pressman, L. S.; Cherstkov,V. F. *Synth. Commun.* **1995**, 25, 2425-2433.
83. Lim C.; Burton, D. J.; Wesolowski, C. A. *J. Fluorine. Chem.* **2003**, 119, 21-26.
84. Davis, C.R.; Burton, D. J. *Tetrahedron Lett.* **1996**, 37, 7237-7240.
85. Wesolowski, C. A. University of Iowa, unpublished results.
86. Zhang, Q.; Lu, L. *Tetrahedron Lett.* **2000**, 41, 8545-8548.
87. Wang, Y.; Lu, L.; Burton, D. J. *J. Org. Chem.* **2005**, 70, 10743-10746.
88. Zhang, X.; Lu, L.; Burton, D. J. *Coll. Czech. Chem. Commun.* **2002**, 67, 1247-1261.
89. Wang, Y.; Burton, D. J. *Org. Lett.* **2006**, 8, 1109-1111.
90. Wang, Y.; Burton, D. J. Submitted for publication.
91. Liu, Q.; Burton, D. J. *Org. Lett.* **2002**, 4, 1483-1485.
92. Unpublished work of Chongsoo Lim, University of Iowa.
93. Burton, D. J.; Spawn, T. D.; Heinze, P. L.; Bailey, A. R.; Shin-ya, S. *J. Fluorine. Chem.* **1989**, 44, 167-174.
94. Burton, D. J.; Shin-ya, S.; Howells, R. D. *J. Fluorine. Chem.* **1980**,15, 543-546.
95. Burton, D. J. Shin-ya, S.; Howells, R. D. *J. Am. Chem. Soc.* **1979**, 101, 3689-3690.
96. Filyakova, T. I.; Belen'kii, G. G.; Lur'e, E. P.; Zapevalov, A. Y.; Kolenko, I.. P.; German, L. S. *Izv. Akad. Nauk. SSSR. Ser. Khim.* **1979**, 681-683.
97. Filyakova, T. I.; Zapevalov, A. Y.; Kodess, M. I.; Kurykin, M. A.; German, L. S. *Izv. Akad. Nauk. Ser. Khim.* **1994**, 1614-1619.
98. Frohn, H.-J.; Bardin, V. V. *J. Fluorine. Chem.* **2004**, 123, 43-49.
99. Anilkumar, R. Burton, D. J. *Tetrahedron Lett.* **2002**, 43, 2731-2733.
100. Raghavanpillai, A.; Burton, D. J.; *J. Org. Chem.* **2004**, 69, 7083-7091.
101. Anilkumar, R.; Burton, D. J. *Tetrahedron Lett.* **2002**, 43, 6979-6982.
102. Anilkumar, R ; Burton, D. J. *J. Fluorine. Chem.* **2005**, 126, 833-841.
103. Anilkumar, R ; Burton, D. J. *J. Fluorine. Chem.* **2004**, 125, 561-566.
104. Anilkumar, R ; Burton, D. J. *J. Fluorine. Chem.* **2005**, 126, 455-461.
105. Mikami, K.; Itoh, Y.; Yamanaka, M. *Chem. Rev.* **2004**, 104, 1-16.

Chapter 7

Stereoselective Synthesis of Polyfluorinated *Exo*-Tricyclononenes and Norbornenes

Viacheslav A. Petrov

DuPont Central Research and Development, Experimental Station, P.O. Box 80328, Wilmington, DE 19880–0328

This chapter summarizes the recent progress in the area of *exo*-selective synthesis of polyfluorinated tricyclo[4.2.1.$0^{2,5}$]non-7-enes (tricyclononenes) and bicyclo[2.2.1]hept-5-enes (norbornenes). The overview of the synthesis polyfluorinated *exo*- tricyclononenes by reactions of quadricyclane and to lesser extend, norbornadiene is given in the first part of the review. The second part contains information on stereoselective synthesis of polyfluorinated *exo*- norbornenes and transformations of *exo*- tricyclononenes.

Publication No. 8709 from DuPont Central Research and Development.

1. Introduction

Rapid development of microlithography using 193 and 157 nm wavelength light revealed a pressing need for new functional materials having array of properties, such as high transparency in vacuum UV region of spectrum and etch resistance, in combination with solubility of polymer in organic solvents and basic developer. That resulted in creation of completely new group of polyfluorinated amorphous polymers and simultaneously, new synthetic methods for production of fluorinated polycyclic monomers. Norbornene- and tricyclononene- based materials, containing hexafluoro-*iso*-propoxy- or carbalkoxy- functional groups [1-14] quickly became popular, due to unique combination of properties, but most importantly, the ability to polymerize under either radical or metal catalyzed conditions. It was rapidly discovered however, that the geometry of functional polycyclic monomer is very important and the orientation of the substituent in norbornene fragment significantly affects the polymerization process. An *exo*- substituted norbornenes and tricyclononenes were found to give polymers with substantially higher molecular weights, and at notably better conversion of relatively expensive monomers. [1-4] This practically important discovery triggered rapid development of new strategies for *exo* - selective synthesis of functional polyfluorinated norbornenes and tricyclononenes.

This chapter is an attempt to provide an overview of this area synthetic fluoroorganic chemistry and it summarizes recent progress in the area of *exo*-selective synthesis of polyfluorinated tricyclo[4.2.1.0$^{2.5}$]non-7-enes (tricyclononenes) and bicyclo[2.2.1]hept-5-enes (norbornenes).

2. Synthesis of Polyfluorinated *exo*-Tricyclononenes Using Cycloaddition Reactions of Quadricyclane (1) and Norbornadiene (7)

Tetracyclo[3.2.0.0^{2,7}0$^{4.6}$]heptane (quadricyclane, **1**) is an attractive hydrocarbon synthon. It readily reacts with a wide variety of electron deficient olefins and acetylenes forming *exo*- tricyclo[4.2.1.0^{2,5}]non-7-enes as the result of highly stereoselective [$2_\sigma+2_\sigma+2_\pi$] cycloaddition process. [15,16] The chemistry of **1** was well developed in past 40 years and was reviewed recently. [17] The chemical behavior of **1** is defined by high strain energy (78.7 kcal/mol), [18] relatively high thermal stability ($t_{1/2}$>14 h, 140°C), [19] low ionization potential (7.40 eV, adiabatic) [20] and one of the lowest oxidation potential known for saturated hydrocarbons [$E_{1/2}$=0.91 V (SCE)]. [21] Compound **1** readily combines with a wide variety of electrondeficient polyfluorinated addends including fluoroolefins, [22-25] acetylenes, [26] nitrogen- [27-29] and sulfur- [30-32] containing substrates, giving the corresponding *exo*- tricyclononenes. Although isomeric norbornadiene (**7**) has very different reactivity (see below), sometimes, it also can be used as starting material for the synthesis of polyfluorinated *exo*-tricyclononenes.

2.1. Synthesis of Polyfluorinated *exo*- Tricyclo[4.2.1.0$^{2.5}$]non-7-enes

In his pioneering work, Smith demonstrated that electrondeficient olefins react with **1** forming *exo*-tricyclo[4.2.1.0$^{2.5}$]non-7-enes. For example, tetracyanoethylene, exothermally reacts with **1** producing *exo*-3,3,4,4-tetracyanotricyclo[4.2.1.0$^{2.5}$]non-7-ene (**2**) in high yield. [15]

1 + $(CN)_2C{=}C(CN)_2$ —exotherm., CH_2Cl_2→ **2**, 78%

The ability of **1** to react with fluorinated olefins was recognized by Sargent, who reported the reaction of 1,2-bis(trifluoromethyl)-3,3-difluorocyclopropene (**3**) and perfluorocyclopropene (**4**) and **1**. [22] Compound **3** reacts at low temperature selectively forming tetracyclodecene **5** as the result of $[2_\pi+2_\sigma+2_\sigma]$ cycloaddition process.

1 + **3** —-45 to 25°C, neat→ **5**, 76%

Perfluorocyclopropene **4** has lower reactivity - the formation of the cycloadduct was not observed after 24 h at ambient temperature, although, at elevated temperature, surprisingly, tricyclononene **6** was formed probably, as the result of thermal cyclopropyl-allyl rearrangement of the intermediate **6a**.

1 + F F C F2 **4** —100°C, 21h, neat→ [F F CF2 **6a**] → F F F F **6**

Norbornadiene (**7**) has distinctly different reactivity. Fast reaction with **3** at ambient temperature leads to the adduct **8**, as the result of selective homo-Diels-Alder reaction.[22]

7 + **3** —25°C, CH_2Cl_2→ F_3C CF_3 F F **8**, quant.

Both perfluoromethylenecyclopropane (**9**) and $(CF_3)_2C{=}C(CN)_2$ (**10**) react with **1** producing cycloadducts **11** and **12**, respectively[23,24]

1 —25°C→ $F_2C{=}$(CF_2, CF_2) **9** → F_2C, CF_2, CF_2 **11**

1 —25°C, $(CF_3)_2C{=}C(CN)_2$ **10**→ F_3C CF_3 NC CN **12**, 85%

In general, nonfluorinated olefins do not interact with **1**, unless the double bond is activated by electron withdrawing group. For example, ethylene does not interact with **1** even at elevated temperature (110°C, 72h),[25] but acrylonitrile or t-butylacrylate both form the corresponding adducts upon heating.[33,34] Polyfluorinated acrylates behave similarly. Methyl ester of both αtrifluorometyl- (**13**) and perfluoro- (**14**) acrylic acids udergo cycloaddition with **1** forming **15a,b** and **16a,b**, respectively. [4,12]

$$1 + CX_2{=}C(CF_3)C(O)OCH_3 \xrightarrow[24\text{-}72h]{96\text{-}100^{\circ}C}$$

13, X=H;
14, X=F

CF_3, $C(O)OCH_3$, CX_2

15a or **16a**

+ $C(O)OCH_3$, CF_3, CX_2

15b or **16b**

X=H; **15a,b** yield 94%;
ratio **15a : 15b** = 68 : 32;
X=F; **16a,b** yield 22% after 72h at 100°C; 73% after 72h at 100°C and 14 d at 25°C;
ratio **16a : 16b** = 51 : 49

The formation of two isomers in both cases is the result of different orientation of $-CF_3$ and $-C(O)OCH_3$ substituents relative to one-carbon bridge of the norbornene fragment. The structure of the isomer with *syn*- orientation of –C(O)OH group (isolated by fractional crystallization of the hydrolysis products of esters **16a,b**) was confirmed by single crystal X-ray diffraction. [4] Compared to esters, acryloyl halides have significantly higher reactivity. For example, α-trifluoromethyl acryloyl chloride (**17**) reacts with **1** exothermally, forming cycloadduct **18a,b** in a high yield. Perfluoroacryloyl fluoride (**19**) has similar reactivity, producing tricyclononene **20a,b** under mild conditions; [35,36] less active α-fluoroacryloyl fluoride (**21**) gives the cycloadduct **23a,b** only at elevated temperature (100°C, 16h, ether). [35,36]

Despite the fact that the Baylis-Hillman adduct of hexafluoroacetone and t-butylacrylate (**23**) [37] contains hexafluoro-*iso*-propanol group (see Sec. 2.5), the cycloaddition to **1** surprisingly results in the formation of the corresponding cycloadduct **24a,b** with high preference for isomer **24a** [35,36] having *syn*- orientation of ester group according to single crystal X-ray diffraction.

$$1 + CX_2{=}CYC(O)Z \xrightarrow[12\text{-}36h]{t}$$

17, X=H;Y=CF_3; Z=Cl
19, X=Y= Z=F;
21, X=H, Y,Z=F

18a,b, X=H;Y=CF_3; Z=Cl;
ratio 3:1, yield 74.1%;
20a,b X=Y= Z=F; ratio 63 : 37;
yield 87%;
22a,b X=H, Y= Z=F; ratio 63 : 37;
yield 51%;

$$1 + CH_2{=}C[C(CF_3)_2OH]C(O)OC(CH_3)_3 \xrightarrow{100^{\circ}C,10h}$$

23

24a + **24b**

24a,b, 65%;
ratio**24a:24b** - 97:3

Despite the fact that the reaction of 1,1,3,3,3-pentafluoro-2-propenyl benzoate (**25**) with quadricyclane is slow even at elevated temperature, it results in moderate yield of the cycloadduct **26a,b**. [35,36]

In contrast to ethylene, tetrafluoroethylene (**27**) or hexafluoropropene (**28**) slowly react with **1** at elevated temperature.[25]

Polarization of the double bond significantly affects the reactivity of the fluoroolefin and yields of cycloadducts **37** and **38** obtained in the reaction of two isomeric butenes **30** and **31** are indicative of its importance. Jugged based on yields of **34-39a,b**, the reactivity of fluoroolefins increases proportionally to the number of fluoroalkyl groups attached to the double bond. This conclusion was confirmed by relative rates of cycloaddition of fluoroolefins to **1** measured by high temperature kinetic NMR. Indeed, relative reactivities of fluoroolefins **28-30, 32, 33** correlate well with the number of perfluoroalkyl groups attached to double bond. Counter-intuitively however, the reactivity reaches maximum for

$1 + CF_2{=}C[OC(O)C_6H_5]CF_3$ **25** $\xrightarrow[\text{glass}]{125^{\circ}C, 144h}$ **26a,b**, 61% (bicyclic product with $OC(O)C_6H_5$, CF_3, CF_2)

1+ **27-33** $\xrightarrow{t}$ (bicyclic product with X, Y, CX'Y)

Olefin:	Temp., °C, (Time, h)	Product (yield %)
27, $CF_2{=}CF_2$	110(72)	**34**, 10
28, $CF_3CF{=}CF_2$	110(48)	**35a,b**, 22
29a,b, $CF_3CF{=}CFCF_3$	110(48)	**36a,b**, 30
30, $CH_2{=}C(CF_3)_2$	100(16)	**37**, 63
31, trans-$CF_3CH{=}CHCF_3$	110(72)	**38**, trace
32, $(CF_3)_2C{=}CFC_2F_5$	90(12)	**39a,b**, 77[a]
33, $(CF_3)_2C{=}C(CF_3)_2$	109	-

[a] ratio **39a:39b** = 93:7

trisubstituted- [$(CF_3)_2C{=}CFC_2F_5$ (**32**)] (rather than tetrasubstituted- olefin [$(CF_3)_2C{=}C(CF_3)_2$ (**33**)]).[25] Unexpectedly low reactivity of **33** is the result of steric repulsion between two bulky CF_3- substituents of the olefin and one of the protons of the methylene bridge of **1** in the transition state, assuming that the addition proceeds selectively from *exo-* side of quadricyclane. This explanation also agrees with significant preference for the formation of one isomer in cycloadducts **24a,b** (ratio 97:3), **39a,b** (ratio 93:7) in case of olefins (**23** and **32**), containing sterically demanding substituents. The cycloaddition of olefin **40** is another example. It results in preferential formation of isomer **40a** with *anti-* orientation of five-membered ring.[25]

The mechanism of the reaction of fluorinated olefins with **1** is not clear, although some experimental data suggest that it may be concerted.[25]

Although compound **7** reacts with fluorinated ethylenes (**27**,[1,38] $CCl_2{=}CF_2$),[38] perfluorinated vinyl ethers,[36] hexafluoropropene (**28**)[38] at significantly higher temperature (180-200°C), surprisingly, the reaction leads to preferential formation of the corresponding tricyclononenes. Compared to

1 + 40 → (90°C, 6h) 40a + 40b

40a,b, 44%,
ratio **40a** : **40b** - 98:2

cycloadditions involving **1**, this process is less stereoselective, but *exo*- isomer **34** is the major product in reaction of **27** and **7** (86% yield).[1] Since thermal [2+2] cycloaddition is forbidden as concerted process by Woodward-Hoffman rule, the mechanism of this transformation probably, involves the formation biradicial intermediate, typical for thermal [2+2] reactions of polyfluorinated olefins.[39]

Fluorinated dienes also readily undergo cycloaddition to **1**. For example, hexafluoro-butadiene-1,3 (**41**) gives mixture of isomers **42a,b**. Interestingly, the cycloaddition of 1,1-bis(trifluoromethyl)butadiene-1,3 (**43**) proceeds exclusively through CH_2=CH- group, leading to **44a,b**. It is worth mentioning, that based on relative rates of cycloaddition, diene **43** is almost three times more reactive towards **1**, compared to $(CF_3)_2C=CH_2$ (**30**). This is an indication that $(CF_3)_2C$=CH- group in **43** acting as very strong electron withdrawing substituent.[40]

Fluorinated acetylenes **45** and **46** react **1** under relatively mild conditions forming cyclic dienes **47** and **48**, respectively.[26]

2.2 Cycloaddition of Fluorinated Carbonyl Compounds to 1 and 7

Nonactivated ketones, such as cyclopentanone or acetophenone do not react with **1** even at elevated temperature and high pressure.[41,17] Usually cycloaddition involving C=O group require photochemical activation.[17]

$$1 + Z_2C{=}CY'{-}CY{=}CX_2 \xrightarrow{110^\circ C,\ 24h}$$ Y, $CY'{=}CZ_2$, CX_2

41,	X,Y,Y',Z=F	**42a,b**, 55%
43,	X,Y,Y'=H, Z=CF_3	**44a,b**, 81%

$$1 + R{-}{\equiv}{-}CF_3 \xrightarrow[9h\text{-}3d]{40\text{-}70^\circ C}$$ R, CF_3

45,	R=C(O)OCH_3	**47**, 96%
46,	R=CF_3,	**48**, 98%

However, the introduction of strong electron withdrawing substituents to C=O results in significant activation of carbonyl group. For example, dimethylglyoxalate (**49**) reacts with **1** upon heating, forming the corresponding oxetane - *exo*- 3-oxa-4,4-bis-(carbmethoxy)tricyclo[4.2.1.$0^{2,5}$]non-7-ene (**50**). [41] Introduction of fluorine into molecule of *p*-quinone alters the reactivity significantly and tetrafluoro-*p*-quinone slowly reacts with **1** at ambient temperature, giving the predominantly fluorinated norbornenoxetane.[42] As it was shown recently, [43] the introduction of perfluoroalkyl substituents into carbonyl compounds has pronounced effect on the reactivity of carbonyl group towards **1**. In contrast to inert acetone (often used as a solvent for reactions of **1**), hexafluoroacetone (**51**) combines with **1** *exothermally* producing *exo*- 3-oxa-4,4-bis(trifluoromethyl)tricyclo[4.2.1.$0^{2,5}$]non-7-ene (**52**) in high yield. [43] It should be pointed out, that this process (*as many other reactions of activated polyfluorinated materials and 1*) is *highly* exothermic [strain energy of quadricyclane 78.7 kcal/mol (!) [18]] and should be carried out in inert solvent and at low temperature to control the heat evolved during the reaction.

Bis(trifluoromethyl)ketene **53** reacts with **1** selectively forming oxetane **54**. It should be pointed out, that, the reaction of norbornadiene (**7**) and **53** proceeds differently, resulting in the formation of isomeric cyclobutane **55a** and homo-Dies-Alder adduct **55b**. [44]
Similarly, interaction between **7** [44,45] and **51** or 1,3-dichlorotetrafluoroacetone [44] at elevated temperature (100-150°C) leads to the corresponding homo Diels-Alder adducts exclusively.

Fluorinated α-dicarbonyl compounds $CF_3C(O)C(O)OCH_3$ (**56**) and $CF_3C(O)C(O)CF_3$ (**57**) combine with **1** selectively producing monoadducts **58a,b** [46] and **59a,b**, respectively.

$O=C(CF_3)_2$
51

1 —— 2-4h, 0-20^{0}C, ether ——

$C(CF_3)_2$, O

52, 95%

$O=C=C(CF_3)_2$
53

$C=C(CF_3)_2$, O

54, 92%

7 + 53 —— 25°C, 2.5 d, C_6H_6 ——

O, $C(CF_3)_2$ + O, $C(CF_3)_2$

55a, yield 31%   **55b**, yield 42%

1 + $CF_3C(O)C(O)R$ —— 10-30°C, 2-6h, ether ——

C(O)R, CF_3, O

56   R=OCH_3 **58a,b** 73%, ratio 70:30

57   R=CF_3 **59a,b**, 94.5%, ratio 55:45

The introduction of even one CF_3-substituent has a pronounced effect on the reactivity of C=O group. In sharp contrast to acetophenone or acetone, [41,46] 1,1,1-trifluoroacetophenone (**60**), 1,1,1-trifluoro-3-bromo- (**61**) and 1,1,1-trifluoro-(**62**) acetones react all with **1** at elevated temperature giving the oxetanes **63-66** in good yield. [46]

The reaction is quite general and polyfluorinated acetones **66-68**, acyl fluorides **72,73** and carbonyl fluoride (**74**) all combine with **1** giving the corresponding *exo*-3-oxa-4,4'-fluoroalkyltricyclo[4.2.1.$0^{2,5}$]non-7-enes **69-71**, **75-77** in moderate to high yield. [46]

CF_3, C_6H_5, O

63a,b, 57%, ratio 74:26

100°C, 18h ← **1** → $CF_3C(O)CH_3$ **62**, 90°C, 36h

$C_6H_5C(O)CF_3$ **60**

25-90°C, 3h ↓ $CF_3C(O)CH_2Br$ **61**

CH_3, CF_3, O

65a,b, 22%, ratio 63:37

CH_2Br, CF_3, O

64a,b, 87%, ratio 58:42

CF_2H, CF_3, O

69a,b, 88%, ratio 55:45

$CF_3(O)CF_2H$ **66**, 0-10°C, 3h ← **1** → $CF_3C(O)CF_2Br$ **67**, 0-20°C, 5h

10-30°C, 12h ↓ $CF_3C(O)CCl_2H$ **68**

CF_2Br, CF_3, O

70a,b, 83%, ratio 57:43

CCl_2H, CF_3, O

71a,b, 77%, ratio 60:40

The observed reactivity of fluorinated carbonyl compounds towards **1** correlates well with the reduction of electrophilicity of the carbonyl group, decreasing in the following order: [46]

$CF_3C(O)CF_2X$ (X=F,H,Br) ≈ $FSO_2CF_2C(O)F$ ≈ $CF_3C(O)C(O)R$ (R=CF_3, OCH_3) > $CF_3C(O)CH_2Br$ > $CF_3C(O)F$ > $CF_3C(O)CH_3$ > $C(O)F_2$.

The mechanism of cycloaddition polyfluorinated carbonyl compounds to **1** is not clear. The fact that most of these reactions develop transient color (usually faint yellow) may suggests formation of charge transfer complexes [47] or

75a,b, ratio 75:25; yield 60% ← 25°C, 18h, $CF_3C(O)F$ 72 — 1 — 15-30°C, 6h, $FSO_2CF_2C(O)F$ 73 → 76a,b, ratio 63:37; yield 92%

60°C, 16h, $C(O)F_2$ 74 ↓

77, 48%

radical-ion pairs, [48] however, more work need to establish true mechanism of this process. [46]

The formation of two isomers in the reaction of nonsymmetrical carbonyl compounds and **1** is the result of *syn- vs. anti-* orientation of substituents attached to the carbon of *exo*-oxetane ring. In major isomer, smaller substituent occupies *syn-* position, in order to minimize steric interaction with hydrogen of methylene bridge. The assignment of structure of isomers **75a,b** was carried using 2D ^1H-^{19}F HOESY NMR and NOE values observed for major isomer **75a** are consistent with

a) *syn-* orientation of fluorine substituent;
b) *anti-* orientation of CF_3-;
c) *exo-* orientation of oxetane ring.[46]

It should be pointed out, that fluorinated acyl chlorides have very different reactivity compared to acyl fluorides. For example, the reaction of $CF_3C(O)Cl$ (**78**) and **1**, leading to the mixture of norbornene **79a** (major component), along with oxetanes **79a,b** and small amount (~3%) of other byproducts. [46]

1 + $CF_3C(O)Cl$ **78** —(0-25°C, 14 h)→ **79a** + **79b,c**

Ratio **79a :79b:79c** = 50:30:17
yield of mixture - 90%;

On the other hand, the cycloadduct **81a,b** can be conveniently prepared by direct reaction of ethylhemiacetal of trifluoracealdehyde (**80**) and **1** at elevated temperature. [46]

1 + $CF_3CH(OC_2H_5)OH$ **80** —85-90°C, 18h→ **81a,b**

Ratio **81a : 81b** - 88:12;
yield 44%;

2.3. Cycloaddition of Fluorinated Nitrogen-Containing Compounds to 1

Imine of hexafluoroacetone has significantly lower reactivity towards **1** compared to hexafluoroacetone. Even prolonged heating (110°C, 16h) of the mixture **1** and $(CF_3)_2C{=}NH$ results in formation of trace amount of the corresponding cycloadduct. [29] However, more electron deficient fluorinated azines **82, 83** combine with **1** at elevated temperature producing *exo*-3-azatricyclo[4.2.1.0^{2,5}]non-7-enes **84a,b** and **85a,b**, respectively. [49,28,27]

1 + $[CF_3CX{=}N]_2$ —50-70°C, 2d, CH_2Cl_2→ product (X, CF_3, N, $N{=}CXCF_3$)

82, X=F
83, X=CF_3

84a,b, X=F, 80%
85a,b, X=CF_3, 54%

Anti- Perfluoro-3-azapentene-2 (**86**) has similar reactivity, giving the cycloadduct **87a,b** in high yield [29]

1 + $C_2F_5N{=}C(F)CF_3$ **86** —90 °C,12 h→ **87a** (F, CF_3, N, C_2F_5) + **87b** (CF_3, F, N, C_2F_5)

yield **87a,b,** 80%;
ratio **87a:87b** - 75:25

Syn- orientation of fluorine substituent in major isomer **87a** was confirmed by 2D ^{1}H-^{19}F HOESY NMR. [29]

Perfluorinated nitriles **88-90** react with **1** at elevated temperature forming *exo*-3-aza-4-perfluoroalkyltricyclo[4.2.1.0^{2,5}]non-3,7-dienes (**91-93**) [29] - a new, interesting class of polycyclic materials having two double bonds with significant difference in reactivity (for reactions of see Sec. 4.0).

$$\mathbf{1} + R_f\text{-CN} \xrightarrow[\text{ether}]{100\ ^{\circ}\text{C}, 12\ \text{h}}$$

R$_f$, N

88	$R_f = CF_3$	**91**, 78%
89	$R_f = C_2F_5$	**92**, 56%
90	$R_f = n\text{-}C_3F_7$	**93**, 81%

The activity of nitriles towards **1** strongly depends on the nature of fluorinated substituent. For example, the replacement of perfluoroalkyl group by less electronwithdrawing pentafluorophenyl results in total loss of reactivity of C_6F_5CN towards **1**. [29] Contrary to reaction of **1**, nitrile **88** combines with norbornadiene (**7**) giving exclusively homo Diels-Alder adduct. [44]

2.4 Cycloaddition of Fluorinated Sulfur-Containing Compounds to 1

Thiocarbonyl compounds have exceptionally highly reactivity towards **1**.[17] Thiophosgene (**94**) for example, combines with **1**, forming cycloadduct **95** in high yield, [50]

$$\mathbf{1} + Cl_2C{=}S \xrightarrow[CH_2Cl_2]{-78\text{ to }20^{\circ}\text{C}}$$

94

CCl$_2$, S

95, quant.

in addition, bis(trifluoromethyl)thioketene (**96**) selectively gives thiethane **97** under mild conditions. [31]

$1 + (CF_3)_2C{=}C{=}S$ (**96**) $\xrightarrow[CH_2Cl_2]{20\text{-}30^{\circ}C}$ **97**, 91%

Although hexafluorothioacetone (**98**) is know to be an excellent dienophile, [51] limited stability of this material at ambient temperature restricts its use in the synthesis. Fortunately, **98** can be conveniently generated "*in situ*" bv treatment of readily available dimer **99** [52] with alkali metal fluoride catalyst. [53-63] Recently, this methodology was extended for the preparation of cycloadduct **100**, by the reaction of **99** and **1**. [32]

1 + 1/2 **99** $\xrightarrow[\text{diglyme, CsF}]{20\text{-}30^{\circ}C}$ **100**, 74-95%

The cycloaddition also can be carried out in the absence of the catalyst, at elevated temperature (100-110°C) however, it results in lower yield of **100** (60%), due to side reactions of **99**.[52] Similarly, the reaction of difluorothiophosgene dimer (**101**) and **1** proceeds at elevated temperature without the catalyst, resulting in products **102a-c**.

1 + **101** $\xrightarrow[24h]{120^{\circ}C}$ **102a**, 13% + **102b**, 4% + **102c**, 17%

Reactions of **99** or **101** with **1** in the absence of catalyst are thought to be initiated by electron transfer from quadricyclane to electrondeficient cyclic sulfide, leading to the formation of biradical intermediate. [32]

In conclusion, high reactivity of **1** towards wide variety of polyfluorinated olefins, carbonyl-, nitrogen- and sulfur- containing compounds, and exceptionally high selectivity of the cycloaddition makes this methodology simple, flexible and valuable synthetic tool for the stereoselective preparation of polyfluorinated *exo*- tricyclo[4.2.1.0^{2,5}]non-7-enes.

2.5 Miscellaneous Reactions Involving Quadricyclane and Fluorinated Substrates

Quadricyclane is well known to be sensitive to the action of acids.[17] Recently it was found that even relatively weak acids, such as tertiary fluorinated alcohols $RC(CF_3)_2OH$, react with **1** forming an isomeric mixture of norbornene and nortricyclane, with significant predominance of the former.[3]

$1 + RC(CF_3)_2OH$ —70-80°C→ $OC(CF_3)_2R$ (norbornene) + $OC(CF_3)_2R$ (nortricyclane)

105a, 106a **105b, 106b**

103, $R{=}CH_2C(CF_3)_2OH$;
104, $R{=}C(CF_3)_2OH$

105a,b, $R{=}CH_2C(CF_3)_2OH$, 87%, ratio 70:30

106a,b, $R{=}C(CF_3)_2OH$, 59%, ratio 80:20

The reactivity of fluorinated alcohols towards **1** is clearly governed by the acidity. Hexafluoro-*iso*-propanol (**107**, pK_a=9.3[64]) reacts with **1** exothermally upon the addition to the cold solution of **1** in CH_2Cl_2, giving the mixture of norbornene **108a** and nortricyclane **108b**.

$1 + (CF_3)_2CHOH$ (**107**) —CH_2Cl_2, exothermic→ $OCH(CF_3)_2$ (**108a**) + $OCH(CF_3)_2$ (**108b**)

108a,b, 65%, ratio 40:60

However, less acidic CF_3CH_2OH (**109**, pK_a=12.4[64]) does not interact with **1** at ambient temperature and the addition of the alcohol requires acid catalyst, resulting in the mixture of two isomeric products **110a,b**, with significant prevalence of nortricyclane **110b**.

Hexafluoropentandione-1,3 (**111**) in reaction with **1** behaves as an acid, rather than carbonyl compound and adds to **1** through hydroxyl group of the enol form giving selectively the mixture of *trans*- **112a,b**, with slight preference for norbornene **112a**.

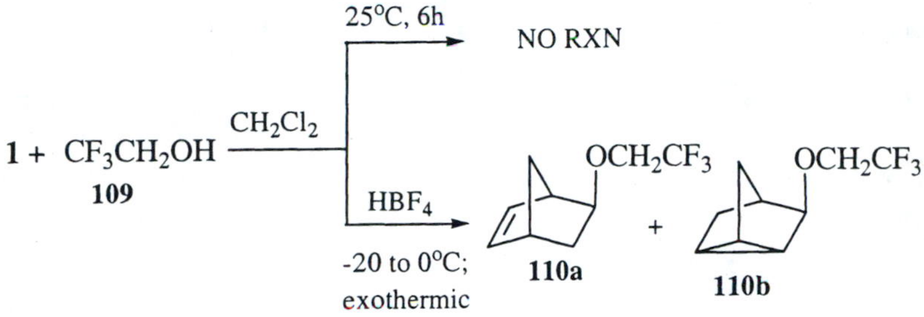

$$\mathbf{1} + CF_3CH_2OH\ (\mathbf{109}) \xrightarrow{CH_2Cl_2}$$

25°C, 6h → NO RXN

HBF_4, -20 to 0°C; exothermic → OCH_2CF_3 **110a** + OCH_2CF_3 **110b**

110a,b, yield 55%,
ratio **110a:110b** - 2:98

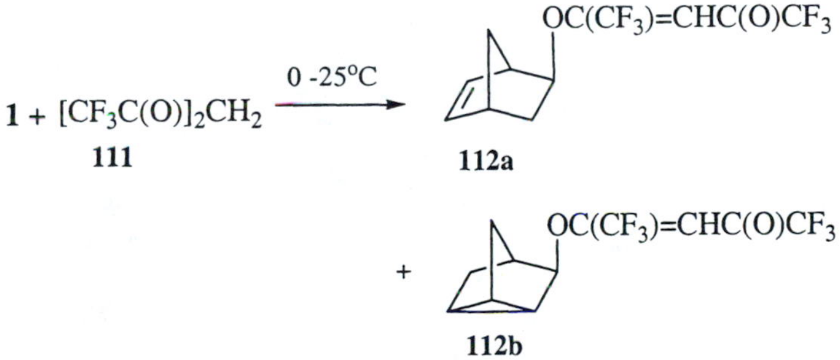

$$\mathbf{1} + [CF_3C(O)]_2CH_2\ (\mathbf{111}) \xrightarrow{0\text{ -}25^{\circ}C}$$

$OC(CF_3)=CHC(O)CF_3$ **112a**

+ $OC(CF_3)=CHC(O)CF_3$ **112b**

112a,b, yield 79%,
ratio **112a:112b** - 65:35

3.0. Synthesis of exo- Substituted Norbornenes.

3.1. Selective Synthesis of *exo*- 5(6)-Fluoroalkylnorborn-2-enes

There are two main strategies for the synthesis of substituted *exo*-5(6)-fluoroalkylnorborn-2-enes. They are based on either the attachment of fluorinated fragment to hydrocarbon starting material having desired geometry, such as *exo*- norborn-2-ene-5-ol (**113**) or *exo*- selective introduction of fluorinated group into hydrocarbon having norbornane skeleton.

The preparation of variety highly transparent polymer binders for 157 nm microlithography was enabled by successful development of efficient synthesises of functional *exo*- norbornenes, such as compounds **115** and **117**. This monomers were prepared using regioselective ring opening reaction of epoxide **114** [65] by the corresponding salts of *exo*- norborn-2-ene-5-ol (**113**) or *exo, exo*- norborn-2-ene-5,6-diol (**116**). [1,66]

X OH + CF_3 / O / CF_3 —1) NaH, 2)H^+, DMF→ X $OCH_2C(CF_3)_2OH$

113, X=H
116, X=OH

114

115, X=H, 90%, *exo* - isomer > 98%
117, X=$OCH_2C(CF_3)_2OH$, 87%; *exo,exo* - isomer > 96 %

The structure of diol **117** having *exo-, exo-* orientation of both fluoroalcohols groups was confirmed by single crystal X-ray diffraction. *Exo*-5-Norbornene-2-methanol in reaction with **114** gives the corresponding adduct in 88 % yield. [67]

Exo- Norbornene **120** containing hexafluoro-*iso*-propanol group direc-

C(O)Cl + 2 $CF_3Si(CH_3)_3$ —CsF, CH_3CN→ $C(CF_3)_2OH$

118 **119** **120**, 36.6%

tly attached to the ring was prepared by the reaction of *exo*- acyl chloride **118** with excess of Ruppert-Prakash reagent [$CF_3Si(CH_3)_3$,[68,69] **119**]
Despite of the presence in the reaction mixture fluoride anion, epimerization of either starting materials or products was not observed in this reaction. The synthesis of fluorinated norbornene monomer **122a,b** also is based of the reaction of acyl fluoride **121a,b** with **119**.[14,13]

C(O)F + 2 $CF_3Si(CH_3)_3$ → (KF/CsF, THF) $C(CF_3)_2OH$, F

121a,b **119** **122a,b**, 90%; *exo /endo* - 70:30

In this case the ratio of **122a,b** is defined by ratio of *exo/endo*- isomers in starting material **121a,b**. The synthesis of **121a,b**, relies on surprisingly high preference for the formation of *exo*- isomer in cycloaddition of acryloyl fluoride **21** and cyclopentadiene.[70, 71] It should be pointed out, that commonly used for the synthesis of norbornenes Diels-Alder cycloaddition of olefins to cyclopentadiene, can not be applied for the preparation of *exo*- substituted polyfluorinated norbornenes, due to low *exo*- selectivity in the majority of these processes. For example, in the reaction of cyclopentadiene with either $CH_2=CHCH_2C(CF_3)_2OH$ or $CH_2=CHC_4F_9$ the amount of the corresponding *exo*- isomers does not exceed 20%.[72,12]

Recently developed new, stereoselective method for the preparation of 5-fluoro-alkylnorborn-2-enes is an illustration of the second synthetic strategy[2] The synthesis is based on the ability of perfluorinated radicals add to the double bond exclusively from *exo*- side of norbornenes.[73,74] The three-step synthesis includes:

a) radical addition of fluoroalkyl iodide **124** or **125** to 5-chloronorborn-2-ene (**123**), resulting in the formation of isomeric mixture of products **126a,b** or **127a,b**;

b) reduction of iodides **126a,b** or **127a,b** by hydrogen to form **128a,b** or **129a,b**;

c) regeneration of double bond by elimination of HCl using $(CH_3)_3COK$.

Due to high selectivity of all three steps, the process can be run without isolation and purification of intermediates and overall yields are high, reaching 65% in case of **130** and resulting in > 96% selectivity for *exo*- isomer.

Cl + R_fI —70-90°C, AIBN→ Cl, R_f, I —H_2, Pd/C, 80 °C, $NaHCO_3$, C_2H_5OH→ Cl, R_f

123

124 R_f= $CH_2C(CF_3)_2OH$ 126a,b

125 R_f= n-C_4F_9 127a,b

128a,b

129a,b

45-70 °C, THF, $KOC(CH_3)_3$ (excess)

R_f

130, R_f=$CH_2C(CF_3)_2OH$, *exo* > 96%, total yield 60 - 65%

131, R_f=n-C_4F_9, *exo* > 96%, total yield 30 - 40%

3.2. Ring Opening Reactions of Polyfluorinated *exo* - 3-Oxa-4,4-fluoroalkyltricyclo[4.2.1.0^{2,5}]non-7-enes

Oxetane ring in *exo* - 3-oxa-4,4-bis(trifluoromethyl)tricyclo[4.2.1.0^{2,5}]non-7-ene (**52**) has low reactivity and the double bond can be hydrogenated without cleavage of oxetane ring.[46] On the other hand, **52** slowly react with butyl lithium at ambient temperature. The formation of two isomeric adducts **132a,b** results from the addition n-C_4H_9Li across polarized double bond of **52**.[46] Stereoselectivity of this process is surprisingly high and both **132a** and **132b** formed as single isomer, with *exo*- orientation of substituents.

The replacement of CF_3 group by $-CH_2Br$, significantly increases the reactivity of oxetane **64a,b** towards alkyl lithium reagents and results in interesting ring opening process. The treatment of **64a,b** with two equivalents of butyl lithium leads to the selective formation of **134**,[46] as the result of intermediate **133** reaction with second mole of n-C_4H_9Li, followed by shift of the double bond and elimination of LiF. Strong evidence in the favor of this mechanism was obtained in the reaction of **64a,b** with excess of less nucleophilic CH_3Li. The reaction carried out and quenched at -70°C, indeed resulted in high yield formation of alcohol **135**.

132a

132 b

Yield 65%; ratio **132a:132b** - 87:13

133

134, 65%

135, 81%

3.3 Ring Opening Reactions of *exo-* 3-Thia-4,4-bis(trifluoromethyl)tricyclo[$4.2.1.0^{2,5}$]non-7-ene (100)

The presence of two powerful electron-withdrawing substituents in $-C(CF_3)_2$-S- group leads to generation of positive charge on sulfur and therefore soft nucleophiles are able to attack sulfur atom of this moiety. Transformations of this type were reported for hexafluorothioacetone [75] and 4-alkoxy-2,2-bis(trifluoromethyl)thiethanes. [54,55] Compound **100** has similar reactivity, forming norbornenes **136-142** in reaction with alkyl- lithium and magnesium reagents, as the result of selective ring opening of thiacyclobutane ring. [76]

RMgX; THF; 0-5°C, 0.5- 1 h ← 100 → RLi; THF/hexane; -70°C, 1 h

140, R=C_6H_5, 77%
141, R=$CH(CH_3)_2$, 84%
142, R= $CH=CH_2$, 80%

136, R=CH_3, 88%
137, R= n-C_4H_9, 86%
138, R= $N[CH(CH_3)_2)]_2$ 68%
139, R=C_2F_5, 85%

The reaction proceeds with the preservation of geometry starting material, leading to exclusive formation of *exo, exo-* norbornenes **136-142**. Although compound **100** does not interact with fluoride anion (KF, DMF, 70°C, 8h), the reaction with $R_fSi(CH_3)_3$/MF mixture (R_f=CF_3 and C_2F_5; MF=KF or CsF) results in moderate yield formation of the corresponding ring opening product **143** (R=CF_3) or **139**, respectively. [76]

4.0. Chemical Transformations of Polyfluorinated *exo-* 4,4'-Fluoroalkyltricyclo[4.2.1.0^{2,5}]non-7-enes and *exo-* 3-Aza-4-Perfluoroalkyltricyclo[4.2.1.0^{2,5}]non-3,7-dienes

Due to rigid structure of tricyclononenes, functional groups can be chemically modified without altering the geometry of starting material, as long as transformation does not involve tricyclononene framework. This approach was employed for the preparation of variety functional monomers. For example, -C(O)Z (Z=Cl or F) in tricyclononenes **18a,b**, **20a,b, 22a,b** was converted into hexafluoro-*iso*-propanol group by the action of $CF_3Si(CH_3)_3$ (**119**). [36]

$(CH_3)_3COK$, DMF ← **18a,b, 20a,b, 22a,b** → **119**, KF or CsF

147a,b, X=H;Y=CF_3; 60%
148a,b, X=Y=F; 87%
149a,b, X=H, Y=F; 71%

18a,b, 20a,b, 22a,b
Z=Cl or F;
X=H, F

144a,b, X=H;Y=CF_3; 24%
145a,b, X=Y=F; 60%
146a,b, X=H, Y=F; 60%

The reaction with $(CH_3)_3COK$ resulted in esters **147-149**, containing acid-labile *t*-butyl group. [36] The synthesis of interesting monomer **150**, containing relatively acidic hydroxyl group directly attached to cyclobutane ring was achieved by basic hydrolysis of ester **26a,b**

26a,b $\xrightarrow[KOH]{CH_3OH}$

150a,b, 100%

Diene **44a,b** or oxetane **54**, containing one electrondeficient double bond were converted into the corresponding epoxides **151a,b** or **152**, by action of sodium hypochlorite [36,40]

44a,b or **54** $\xrightarrow[cat.]{NaOCl}$

151a,b 72%

152, 94%

Cyclic azadienes **91-93** rapidly react with silane **119** in the presence of CsF catalyst selectively forming stable silanes **153-155** and less electrophilic C=C bond remains unchanged in this process. It should be pointed out, that the addition of silane is also highly stereoselective, proceeding exclusively from less hindered *anti-* face of azabutane ring and resulting in formation of only one isomer, having R_f group in *syn-* position.[29] Hydrolytically stable silanes **153-155** were converted into the corresponding hydrochlorides **156,157** by treatment with HCl. [29] The reaction of **91-93** with *m*-chloroperoxibenzoic acid (MCPBA) proceeds with oxidation of both double bonds leading to oxazriridines **158-160**. The structure of **160** having *exo-* orientation of epoxide and *anti-* orientation of oxaziridine rings was firmly established by single crystal X-ray diffraction.

R_f

MCPBA 91-93 119

CH_3CN, 2d, 25°C THF, CsF

N CF$_3$ Si(CH_3)$_3$

158, R_f=CF_3,835%
159, R_f=C_2F_5, 72%
160, R_f=*n*-C_3F_7, 24%

HCl

153, R_f=CF_3, 85%
154, R_f=C_2F_5, 73%
155, R_f=*n*-C_3F_7, 68%

R_f N CF$_3$ H·HCl

156, R_f=CF_3;
157, R_f=C_2F_5

5.0. Conclusion

Last ten years have been a period of intensive development in the area of *exo*- selective synthesis of polyfluorinated polycyclic materials. Several new methods for the preparation of *exo*- substituted norbornenes are available now, because of strong demand for new sophisticated functional polymers from semiconductor industry. In general, the stereoselective synthesis of new monomers is based either on the chemical transformations of *exo*-tricyclononenes (readily available through reactions of fluorinated substrates with quadricyclane or sometimes norbornadiene), or reactions of norbornenes having right geometry with fluorinated substrates. Both approaches were successfully applied for the synthesis of polyfluorinated norbornenes and tricyclononenes. The list of fluorinated functional monomers incorporated into variety of new fluoropolymers includes norbornenes **34, 115, 117, 122a,b,** [14,13,77] **130,**[1,66,36] tricyclononenes **24a,b, 144a,b-150**, and oxetanes **52, 76a,b** and **152** [1,66,36] and some polymers have truly outstanding properties. For example, partially protected copolymer of tetrafluoroethylene and bis(fluoroalcohol)-norbornene **117** was demonstrated to function as the binder in photoresist formulation, imaged using 157 nm wavelength UV light. [66] This was possible due to unique combination of low absorbance ($1\mu m^{-1}$) in deep UV region of spectrum, sufficient dissolution rate and high etch-resistance of the polymer. [66]

Acknowledgement: This review is respectfully dedicated in memory of Professor Ivan Ludvigovich Knunyants on the occasion of the 100th anniversary of his birthday.

References

1. Feiring, A. E.; Crawford, M. K.; Farnham, W. B.; Feldman, J.; French, R. H.; Leffew, K. W.; Petrov, V. A.; Schadt, F. L.; Wheland, R. C.; Zumsteg, F. C. *J. Fluorine Chem.* **2003,** *122*, 11-16.
2. Feiring, A. E.; Petrov, V. A.; Schadt, F. L. (to DuPont) US Pat. 6,875,555 (2005).
3. Marsella, J. A.; Afbdourazak, A. H.; Carr, R. V. C.; Markley, T. J.; Robertson III, E. A. *Proc. SPIE-Int. Soc. Opt. Eng.* **2004,** *5376*, 266-275.
4. Sanders, D. P.; Connor, E. F.; Grubbs, R. H.; Hung, R. J.; Osborn, B. P.; Chiba, T.; MacDonald, S. A.; Willson, C. G.; Conley, W. *Macromolecules* **2003,** *36*, 1534-1542.
5. Crawford, M. K.; Farnham, W. B.; Feiring, A. E.; Feldman, J.; French, R. H.; Leffew, K. W.; Petrov, V. A.; Qiu, W.; Schadt, F. L.; Tran, H. V.; Wheland, R. C.; Zumsteg, F. C., Jr. *Proc. SPIE-Int. Soc. Opt. Eng.* **2003,** *5039*, 80-92.
6. Crawford, M. K.; Farnham, W. B.; Feiring, A. E.; Feldman, J.; French, R. H.; Leffew, K. W.; Petrov, V. A.; Qui, W.; Shadt III, F. L.; Tran, H. V.; Wheland, R. C.; Zumsteg, F. C. *Proc. SPIE-Int. Soc. Opt. Eng.* **2003,** *5039*, 80.
7. Crawford, M. K.; Farnham, W. B.; Feiring, A. E.; Feldman, J.; French, R. H.; Leffew, K. W.; Petrov, V. A.; Schadt, F. L., III; Zumsteg, F. C. *J. Photopolym. Sci. Technol.* **2002,** *15*, 677-687.
8. Crawford, M. K.; Farnham, W. B.; Feiring, A. E.; Feldman, J.; French, R. H.; Leffew, K. W.; Petrov, V. A.; Shadt III, F. L.; Zumsteg, F. C. *J. Photopolym. Sci. Technol.* **2002,** *15*, 677.
9. Crawford, M. K.; Feiring, A. E.; Feldman, J.; French, R. H.; Periyasamy, M. P.; Schadt, F. L., III; Smalley, R. J.; Zumsteg, F. C., Jr.; Kunz, R. R.; Rao, V.; Liao, L.; Holl, S. M. *Proc. SPIE-Int. Soc. Opt. Eng.* **2000,** *3999*, 357-364.
10. Crawford, M. K.; Feiring, A. E.; Feldman, J.; French, R. H.; Petrov, V. A.; Schadt, F. L., III; Smalley, R. J.; Zumsteg, F. C., Jr. *Proc. SPIE-Int. Soc. Opt. Eng.* **2001,** *4345*, 428-438.
11. Zumsteg, F. C.; Leffew, K. W.; Feiring, A. E.; Crawford, M. K.; Farnham, W. B.; Petrov, V. A.; Schadt, F. L., III; Tran, H. V. *J. Photopolym. Sci. Technol.* **2005,** *18*, 467-469.
12. Tran, H. V.; Hung, R. J.; Chiba, T.; Yamada, S.; Mrozek, T.; Hsieh, Y.-T.; Chambers, C. R.; Osborn, B. P.; Trinque, B. C.; Pinnow, M. J.; MacDonald, S. A.; Willson, C. G.; Sanders, D. P.; Connor, E. F.; Grubbs, R. H.; Conley, W. *Macromolecules* **2002,** *35*, 6539-6549.
13. Ishikawa, T.; Kodani, T.; Koh, M.; Moriya, T.; Araki, T.; Aoyama, H.; Yamashita, T.; Toriumi, M.; Hagiwara, T.; Furukawa, T.; Itani, T.; Fujii, K. *Proc. SPIE-Int. Soc. Opt. Eng.* **2004,** *5376*, 169-177.

14. Ishikawa, T.; Kodani, T.; Yoshida, T.; Moriya, T.; Yamashita, T.; Toriumi, M.; Araki, T.; Aoyama, H.; Hagiwara, T.; Furukawa, T.; Itani, T.; Fujii, K. *J. Fluorine Chem.* **2004,** *125*, 1791-1799.
15. Smith, C. D. *J. Am. Chem. Soc.* **1966,** *88*, 4273-4274.
16. Tabushi, I.; Yamamura, K.; Yoshida, Z. *J. Am. Chem. Soc.* **1972,** *94*, 787-792.
17. Petrov, V. A.; Vasilev, N. V. *Curr. Org. Synthesis* **2006,** *3*, 175-213.
18. Hall, H. K., Jr.; Smith, C. D.; Baldt, J. H. *J. Am. Chem. Soc.* **1973,** *95*, 3197-3201.
19. Hammond, G. S.; Turro, N. J.; Fischer, A. *J. Am. Chem. Soc.* **1961,** *83*, 4674-4675.
20. Taylor, G. N. *Chem. Phys. Lett.* **1971,** *10*, 355-360.
21. Gassman, P. G.; Yamaguchi, R.; Koser, G. F. *J. Org. Chem.* **1978,** *43*, 4392-4393.
22. Sargeant, P. B. *J. Am. Chem. Soc.* **1969,** *91*, 3061-3068.
23. Smart, B. E. *J. Am. Chem. Soc.* **1974,** *96*, 929.
24. Holyoke, C. W., Jr. (to DuPont). PCT WO 1995006631.
25. Petrov, V. A.; Davidson, F.; Krusic, P. J.; Marchione, A. A.; Marshall, W. J. *J. Fluorine Chem.* **2005,** *126*, 601-610.
26. Barlow, M. G.; Suliman, N. N. E.; Tipping, A. E. *J. Fluorine Chem.* **1995,** *73*, 61-67.
27. Barlow, M. G.; Pritchard, R. G.; Suliman, N. S.; Tipping, A. E. *Acta Crystallogr., Sect. C: Cryst. Struct. Commun.* **1994,** *C50*, 553-556.
28. Barlow, M. G.; Suliman, N. N.; Tipping, A. E. *J. Chem. Soc., Chem. Commun.* **1993**, 9-11.
29. Petrov, V. A.; Davidson, F.; Marshall, W. *J. Fluorine Chem.* **2004,** *125*, 1621-1628.
30. Raasch, M. S. (to DuPont), US Pat. 3406184, 1968.
31. Raasch, M. S. *J. Org. Chem.* **1970,** *35*, 3470-3483.
32. Petrov, V. A.; Krespan, C. G.; Marshall, W. *J. Fluorine Chem.* **2005,** *126*, 1332-1341.
33. Noyori, R.; Umeda, I.; Kawauchi, H.; Takaya, H. *J. Am. Chem. Soc.* **1975,** *97*, 812-820.
34. Connor, E. F.; Younkin, T. R.; Henderson, J. I.; Hwang, S.; Grubbs, R. H.; Roberts, W. P.; Litzau, J. J. *J. Polym. Sci., Part A: Polym. Chem.* **2002,** *40*, 2842-2854.
35. Feiring, A. E.; Schadt, F. L., III; Petrov, V. A.; Smart, B. E.; Farnham, W. B. (to DuPont) PCT WO 2004/014964, 2004.
36. Feiring, A. E.; Crawford, M. K.; Farnham, W. B.; Feldman, J.; French, R. H.; Junk, C.; Leffew, K. W.; Petrov, V. A.; Qiu, W.; Schadt III, F. L.; Tran, H. V.; Zumsteg, F. C. *Macromolecules, in press,,* **2006**.
37. Petrov, V. A.; Marshall, W. J.; Krespan, C. G.; Cherstkov, V. F.; Avetisian, E. A. *J. Fluorine Chem.* **2004,** *125*, 99-105.
38. Brasen, W. R. (to DuPont), US Pat 2928865, 1960.

39. Smart, B. E. In Chemistry of Fluorine Compounds II. A Critical Review.Eds. Hudlicky, M., Pavlath, A., ACS Monograph 187, Washington DC, 1995, pp. 767-796.
40. Petrov, V. A. *J. Fluorine Chem. accepted for publication,* **2006**.
41. Papadopoulos, M.; Jost, R.; Jenner, G. *J. Chem. Soc., Chem. Commun.* **1983**, 221-222.
42. Fabris, F.; De Lucchi, O.; Valle, G.; Cossu, S. *Heterocycles* **1995,** *41*, 665-673.
43. Petrov, V. A.; Davidson, F.; Smart, B. E. *J. Fluorine Chem.* **2004,** *125*, 1543-1552.
44. Vasil'ev, N. V.; Truskanova, T. D.; Buzaev, A. V.; Romanov, D. V.; Zatonskii, G. V. *Russ. Chem. Bull.* **2005,** *54*, 1038-1040.
45. Komokia, H.; Sumida, S.; Kawamura, S.; Miazawa, S.; Maeda, K. (to Glass Glass) PCT WO2004/096786, 2004.
46. Petrov, V. A.; Davidson, F.; Smart, B. E. *J. Fluorine Chem.* **2004,** *125*, 1543-1552.
47. Tabushi, I.; Yamamura, K. *Tetrahedron* **1975,** *31*, 1827-1831.
48. Papadopoulos, M.; Jenner, G. *Nouv. J. Chim.* **1983,** *7*, 463-464.
49. Barlow, M. G.; Suliman, N. N. E.; Tipping, A. E. *J. Fluorine Chem.* **1995,** *72*, 147-156.
50. Jenner, G.; Papadopoulos, M. *Tetrahedron Lett.* **1985,** *26*, 725-726.
51. Middleton, W. J. *J. Org. Chem.* **1965**, 1390-1394.
52. England, D. C. *J. Org. Chem.* **1981,** *46*, 147-153.
53. Postovoi, S. A.; Vol'pin, I. M.; Delyagina, N. I.; Galakhov, M. V.; Zeifman, Y. V.; German, L. S. *Izv. Akad. Nauk, Ser. Khim.* **1989**, 1380-1383.
54. Kitazume, T.; Otaka, T.; Takei, R.; Ishikawa, N. *Bull. Chem. Soc. Jpn.* **1976,** *49*, 2491-2494.
55. Kitazume, T.; Ishikawa, N. *Bull. Chem. Soc. Jpn.* **1974,** *47*, 785-786.
56. Kitazume, T.; Ishikawa, N. *Chem. Lett.* **1973,** *No. 3*, 267-268.
57. Kitazume, T.; Ishikawa, N. *Bull. Chem. Soc. Jpn.* **1973,** *46*, 3285-3288.
58. Ishikawa, N.; Kitazume, T. *Bull. Chem. Soc. Jpn.* **1973,** *46*, 3260-3263.
59. Ishikawa, N.; Kitazume, T. *Chem. Lett.* **1972**, 947-948.
60. Burton, D. J.; Inouye, Y. *J. Fluorine Chem.* **1981,** *18*, 459-473.
61. Burton, D. J.; Inouye, Y. *Tetrahedron Lett.* **1979**, 3397-3400.
62. Zubovics, Z.; Ishikawa, N. *J. Fluorine Chem.* **1976,** *8*, 43-54.
63. Kitazume, T.; Ishikawa, N. *Bull. Chem. Soc. Jpn.* **1975,** *48*, 361-362.
64. Eberson, L.; Hatrtshorn, M. P.; Pearson, P. *J. Chem. Soc. Perkin II* **1995**, 1735-1744.
65. Petrov, V. A. *Synthesis* **2002**, 2225-2231.
66. Feiring, A. E.; Crawford, M. K.; Farnham, W. B.; French, R. H.; Leffew, K. W.; Petrov, V. A.; Schadt III, F. L.; Tran, H. V.; Zumsteg, F. C. *Macromolecules* **2006,** *39*, 1443-1448.

67. Petrov, V. A. Feiring, A. E.; Feldman, J. (to DuPont) US 6,653,419, 2003.
68. Prakash, G. K. S.; Mandal, M. *J. Fluorine Chem.* **2001,** *112*, 123-131.
69. Babadzhanova, L. A.; Kirij, N. V.; Yagupolskii, Y. L.; Tyrra, W.; Naumann, D. *Tetrahedron* **2005,** *61*, 1813-1819.
70. Ito, H.; Saito, A.; Kakuuchi, A.; Taguchi, T. *Tetrahedron* **1999,** *55*, 12741-12750.
71. Essers, M.; Mueck-Lichtenfeld, C.; Haufe, G. *J. Org. Chem.* **2002,** *67*, 4715-4721.
72. Perez, E.; Laval, J. P.; Bon, M.; Rico, I.; Lattes, A. *J. Fluorine Chem.* **1988,** *39*, 173-196.
73. Brace, N. O. *J. Org. Chem.* **1962,** *27*, 3027-3032.
74. Feiring, A. E. *J. Org. Chem.* **1985,** *50*, 3269-3274.
75. Middleton, W. J.; Sharkey, W. H. *J. Org. Chem.* **1965**, 1384-1390.
76. Petrov, V. A. *Mendeleev Commun., accepted for publication,* **2006**.
77. Hagiwara, T.; Furukawa, T.; Itani, T.; Fujii, K.; Ishikawa, T.; Koh, M.; Kodani, T.; Moriya, T.; Yamashita, T.; Araki, T.; Toriumi, M.; Aoyama, H. *Proc. SPIE-Int. Soc. Opt. Eng.* **2004,** *5376*, 159-168.

Chapter 8

Catalytic In-Situ Generation of Trifluoroacetaldehyde from Its Hemiacetal and Successive Direct Aldol Reaction with Ketones

Kazumasa Funabiki and Masaki Matsui

Department of Materials Science and Technology, Faculty of Engineering, Gifu University, 1-1, Yanagido, Gifu 501–1193, Japan

Trifluoroacetaldehyde ethyl hemiacetal reacted with unmodified ketones in the presence of 30 to 50 each mol% of amines and acids at ambient temperature, affording the corresponding β-hydroxy-β-trifluoromethylated ketones in good yields with good to excellent diastereoselectivities. Furthermore, *L*-proline-catalyzed asymmetric direct aldol reaction of trifluoroacetaldehyde ethyl hemiacetal with unmodified ketones also proceeded smoothly to produce the corresponding β-hydroxy-β-trifluoromethylated ketones with good to excellent diastereo- and enantio-selectivities.

Introduction

Trifluoroacetaldehyde (CF_3CHO) is one of the most important C2 building block for the synthesis of the α-trifluoromethlyated alcohols, which are important core moieties of liquid crystals and antidepressant. Therefore, it has been widely used in aldol, Mukaiyama aldol, ene, Friedel-Crafts, Morita-Bayllis-Hillman reaction, and so on. Despite significant progress in the area of efficient asymmetric synthesis of trifluorometylated molecules using CF_3CHO *(1)*, the method for the generation of CF_3CHO from its hemiacetal or hydrate is still dependent on the early protocol, which includes a serious indispensable conditions such as use of an excess amount of concentrated sulfuric acid under high reaction temperature *(2)*. Therefore, an environmentally-friendly, practical, and efficient method for the *in situ* generation of CF_3CHO attended by its simultaneous stereoselective carbon-carbon bond formation reaction is really required.

Recently, we have found that a stoichiometric amount of enamines or imines are capable to *in situ* generation of CF_3CHO as well as the successive carbon-carbon bond formation reaction under mild conditions, producing β-hydroxy-β-trifluoromethylated ketones in good to excellent yields *(3)* (Figure 1).

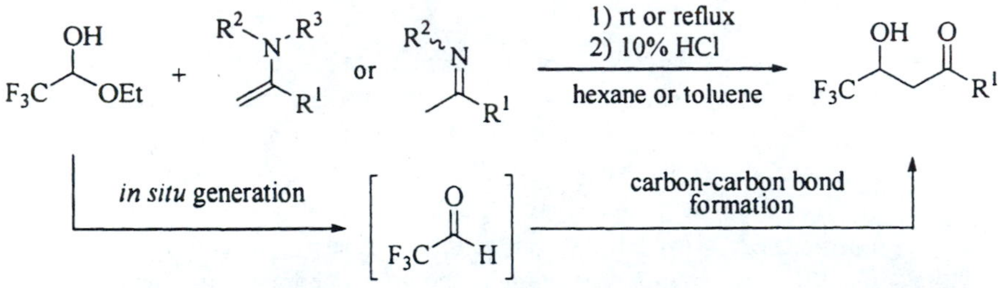

Figure 1. CF_3CHO Ethyl Hemiacetal with Enamines or Imines.

This reaction can serve a new expedient method for the general, practical, and regioselective synthesis of β-hydroxy-β-trifluoromethyled ketones.

As shown in Figure 2, this method can be applied to the asymmetric synthesis of chiral β-hydroxy-β-trifluoromethylated ketones by the use of chiral imines *(4)*. Furthermore, the other fluorinated aldehyde hemiacetal or hydrate as well as chloral hydrate can be used to produce chiral β-hydroxy-β-difluoromethylated, or pentafluoroethylated or trichloromethylated ketones *(5)* in good to excellent yields with high enentioselectivities.

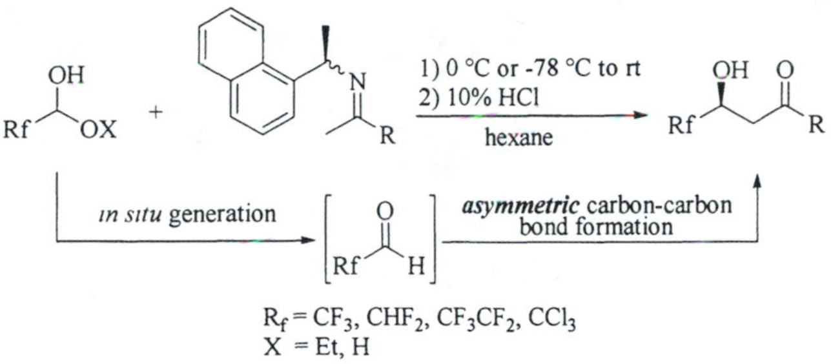

OH

Rf OX + N R

1) 0 °C or -78 °C to rt
2) 10% HCl
hexane

OH O
Rf R

in situ generation

O
Rf H

asymmetric carbon-carbon bond formation

$R_f = CF_3, CHF_2, CF_3CF_2, CCl_3$
X = Et, H

Figure 2. CF_3CHO Ethyl Hemiacetal with Chiral Imines.

However, in this method a couple of steps for the preparation of enamines and imines as well as for the hydrolysis of the intermediates producing β-hydroxy-β-trifluoromethylated ketones are absolutely necessary. For the further development of a new atom-economical method *(6)* for the synthesis of β-trifluormethylated aldol adducts, in this chapter we describe not only *in situ generation of CF_3CHO from its hemiacetal but simultaneous direct aldol reaction with unmodified ketones* by the use of the combination of a small (30-50 mol%) amount of amine and acid, affording the good yields of β-hydroxy-β-trifluoromethyl ketones with good to excellent diastereo- as well as enantio-selectivities.

The Direct Aldol Reaction of Trifluoroacetaldehyde Ethyl Hemiacetal with Ketones *(7)*

Optimization of the Reaction Conditions

CF_3CHO ethyl hemiacetal **1a** reacted smoothly with acetone **2a** in the presence of 50 each mol% of piperidine and acetic acid (CH_3CO_2H) at room temperature for 24 h to produce the β-hydroxy-β-trifluoromethylated ketone **3a** in 60% yield, together with a trace amount of 1,1,1,7,7,7-hexafluoro-2,6-dihydroxy-4-heptanone **4** (Table 1, entry 3). The direct aldol reactions of CF_3CHO ethyl hemiacetal **1a** with acetone **2a** in the presence of 50 mol % of piperidine under the various reaction conditions are summarized in Table 1 (Figure 3).

The use of only piperidine gave no product **3a** with a trace amount of bis-adduct **4** at various temperatures (entries 1 and 2). Other organic acids, such as trifluoroacetic acid and *p*-toluenesulfonic acid (*p*-TsOH), were not effective to give a trace amount of or no product (entries 4 and 5). Among the solid acids

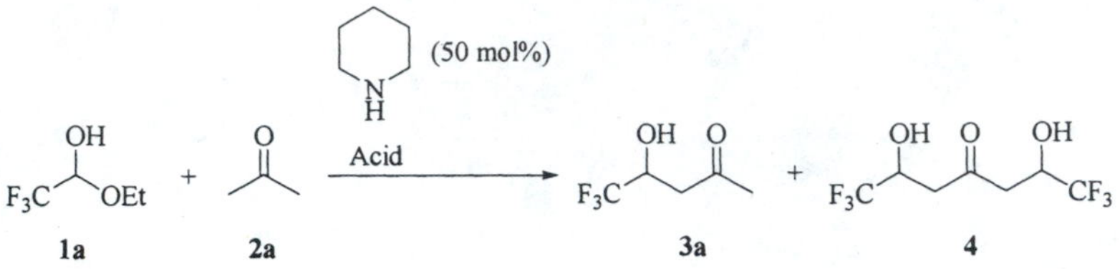

Figure 3. Reaction of CF_3CHO Ethyl Hemiacetal with Acetone.

Table 1. Direct Aldol Reaction of CF_3CHO Hemiacetal with Acetone in the Presence of 50 mol% of Piperidine with Various Acids

Entry[a]	*Acid*	*Conditions*	*Yield (%)*[b]
1	none	rt, 24 h	**3a** (0), **4** (10), **1a** (22), **1b** (16)
2	none	reflux, 3 h	**3a** (0), **4** (6), **1a** (17), **1b** (13)
3	CH_3CO_2H	rt, 24 h	**3a** (60), **4** (5), **1a** (0), **1b** (0)
4	CF_3CO_2H	rt, 24 h	**3a** (1), **4** (0), **1a** (32), **1b** (14)
5	*p*-TsOH	rt, 24 h	**3a** (0), **4** (0), **1a** (45), **1b** (11)
6	silica gel	rt, 24 h	**3a** (58), **4** (10), **1a** (0), **1b** (0)
7	MK10	rt, 24 h	**3a** (32), **4** (22), **1a** (0), **1b** (0)
8	$H_4SiW_{12}O_{49}$	rt, 24 h	**3a** (9), **4** (18), **1a** (18), **1b** (9)
9	Nafion R50	rt, 24 h	**3a** (0), **4** (8), **1a** (16), **1b** (19)

[a] All the reaction was carried out with trifluoroacetaldehyde ethyl hemiacetal **1a** (1 mmol) with piperidine (0.5 mmol) and organic acid (0.5 mmol) or solid acid (120 mg) in dry acetone **2a** (10 mL). [b] Yields were measured by ^{19}F NMR. Product **1b** is CF_3CHO hydrate.

examined, such as silica gel (Wakogel C200), Montmorillonite K10 (MK10), tungstosilicic acid ($H_4SiW_{12}O_{40}$), and Nafion R-50 (entries 6-9), silica gel was most effective for the direct aldol reaction with acetone to provide the aldol product **3a** in 58% yield, along with a 10% yield of **4** (entry 6).

Screening of a variety of amines are summarized in Figure 4. In both cases of CH_3CO_2H and silica gel, cyclic secondary amines, such as morpholine and 1-methylpiperazine as well as primary amines with *n*-propyl or *c*-hexyl group, were suitable for the reaction to give acceptable yields of the aldol product **3a**. The use of diethylamine as an acyclic secondary amine resulted in the increase of the formation of bis-adduct **4**, together with **3a** in 28-30% yields.

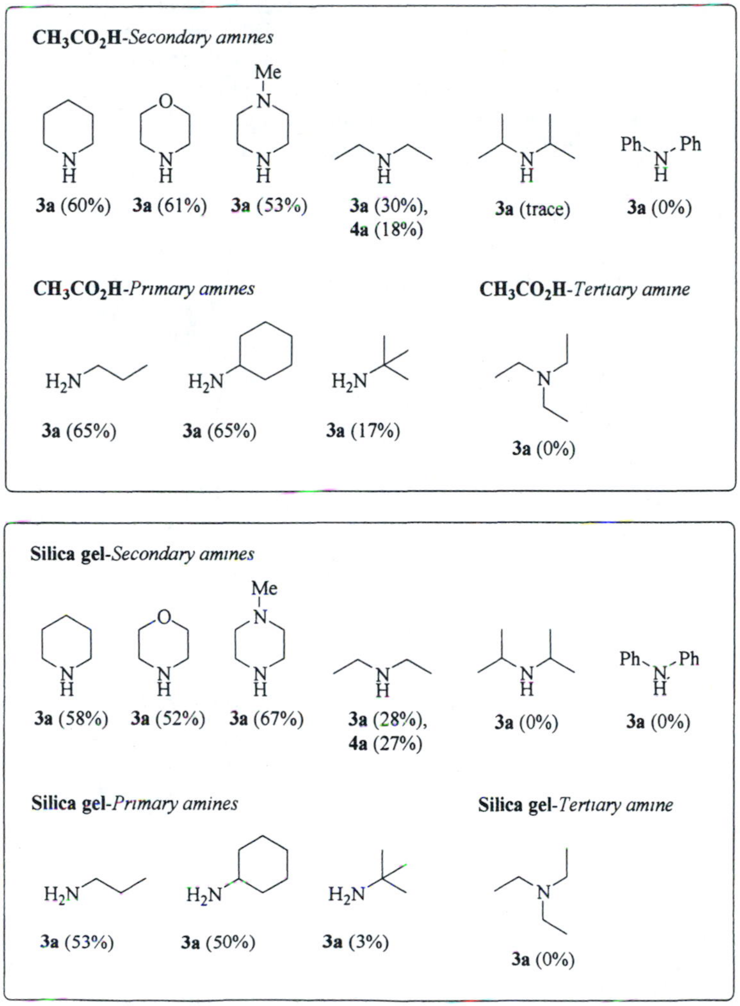

CH_3CO_2H-*Secondary amines*

Me
N
N H | O N H | N H | N H | N H | Ph N Ph H

3a (60%) **3a** (61%) **3a** (53%) **3a** (30%), **4a** (18%) **3a** (trace) **3a** (0%)

CH_3CO_2H-*Primary amines*

H_2N H_2N H_2N

3a (65%) **3a** (65%) **3a** (17%)

CH_3CO_2H-*Tertiary amine*

N

3a (0%)

Silica gel-*Secondary amines*

Me
N
N H | O N H | N H | N H | N H | Ph N Ph H

3a (58%) **3a** (52%) **3a** (67%) **3a** (28%), **4a** (27%) **3a** (0%) **3a** (0%)

Silica gel-*Primary amines*

H_2N H_2N H_2N

3a (53%) **3a** (50%) **3a** (3%)

Silica gel-*Tertiary amine*

N

3a (0%)

Figure 4. Screening of the Various Amines in the Presence of CH_3CO_2H or Silica Gel.

The reaction carrying *t*-butylamine or diisopropylamine having the bulky group produced only a small amount of the aldol product. Diphenylamine also was not effective for the reaction to afford no product, probably due to its low nucleophilicity. The use of triethylamine also gave no product.

These results may suggest that (1) the reaction proceeds *via* enamines or imines, which are produced from the corresponding ketone in the presence of the acid and the corresponding secondary or primary amines, and (2) the resulting enamines or imines react with CF_3CHO ethyl hemiacetal **1a** *via* not only *in situ* generation of CF_3CHO but also carbon-carbon bond formation reaction, followed by the hydrolysis, to give the corresponding β-hydroxy-β-trifluoromethyl ketones.

As shown in Figure 5, the reaction of hemiacetal **1a** with acetone **2a** in the presence of 30 each mol% of piperidine or *n*-propylamine and CH_3CO_2H afforded the corresponding aldol product **3a** in acceptable yields (53-62%). Reducing the amount of amines to 30 mol% and the amount of silica gel to 60% of the former amount gave rise to lowering the yield of **2a**, together with the formation of bis-adduct **4**, the formation of **1b** (0-6%), and the remaining **1a** (0-17%).

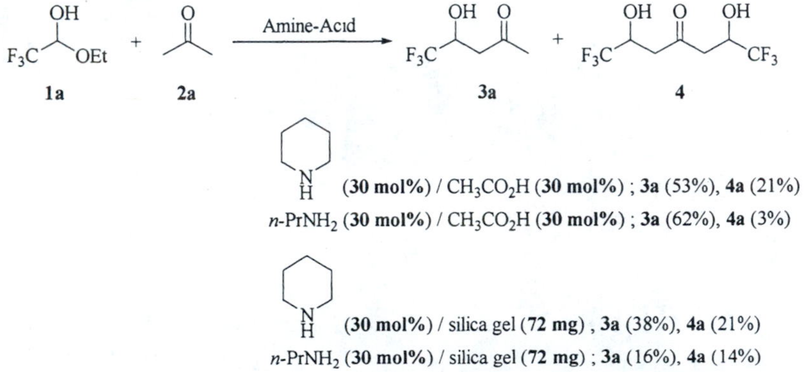

1a + **2a** —Amine-Acid→ **3a** + **4**

Piperidine (**30 mol%**) / CH_3CO_2H (**30 mol%**) ; **3a** (53%), **4a** (21%)
n-$PrNH_2$ (**30 mol%**) / CH_3CO_2H (**30 mol%**) ; **3a** (62%), **4a** (3%)

Piperidine (**30 mol%**) / silica gel (**72 mg**) , **3a** (38%), **4a** (21%)
n-$PrNH_2$ (**30 mol%**) / silica gel (**72 mg**) ; **3a** (16%), **4a** (14%)

Figure 5. Screening of the Amount of the Amines and Acids.

Irrespective of the amines as well as the acids, the use of DMSO as a polar co-solvent (DMSO-acetone = 4 : 1) resulted in decreasing the yield of **3a**, along with the increase of bis-adduct **4**, the formation of **1b** (0-19%), and the remaining **1a** (0-8%) (Figure 6).

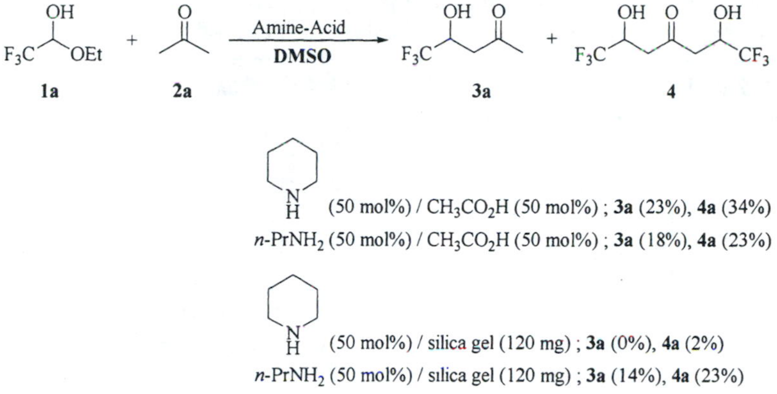

F_3C–CH(OH)–OEt (**1a**) + acetone (**2a**) → (Amine-Acid, **DMSO**) → **3a** + **4**

Piperidine (50 mol%) / CH_3CO_2H (50 mol%) ; **3a** (23%), **4a** (34%)
n-$PrNH_2$ (50 mol%) / CH_3CO_2H (50 mol%) ; **3a** (18%), **4a** (23%)

Piperidine (50 mol%) / silica gel (120 mg) ; **3a** (0%), **4a** (2%)
n-$PrNH_2$ (50 mol%) / silica gel (120 mg) ; **3a** (14%), **4a** (23%)

Figure 6. Effect of DMSO as a Co-solvent.

In all cases, mass balances in the yields are not good, probably due to these high volatility and self-polymerization of *in situ* generated CF_3CHO.

Importantly, this method could achieve a reduction of the steps for the synthesis of β-hydroxy-β-trifuoromethylated ketones. That is, a single manipulation includes multi steps in the reaction, such as (1) the formation of enamines or imines, (2) the enamines- or imines-assisted *in situ* generation of CF_3CHO, (3) the successive carbon-carbon bond formation reaction of CF_3CHO, and (4) the hydrolysis of the intermediates, producing β-hydroxy-β-trifluoromethylated ketones as well as reproduction of amines and acids.

Direct Aldol Reaction of Various Polyfluoroalkylaldehyde Hemiacetal or Hydrate with Cyclic Ketones

The results of the reaction of cyclic ketones with CF_3CHO ethyl hemiacetal **1a** or hydrate **1b**, difluoroacetaldehyde ethyl hemiacetal **1c**, and pentafluoropropionaldehyde hydrate **1d** are summarized in Table 2 (Figure 7).

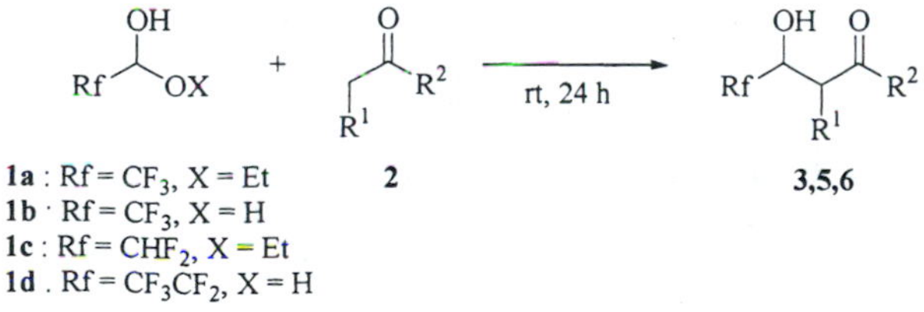

Rf–CH(OH)–OX + **2** → (rt, 24 h) → **3,5,6**

1a : Rf = CF_3, X = Et
1b : Rf = CF_3, X = H
1c : Rf = CHF_2, X = Et
1d : Rf = CF_3CF_2, X = H

Figure 7. Direct Aldol Reaction of Various Polyfluoroalkylaldehyde Hemiacetal or Hydrate with Cyclic Ketones.

Table 2. Direct Aldol Reaction of Polyfluoroalkylaldehyde Hemiacetal or Hydrate with Ketones using Amines and Acids

Entry[a]	*1*	*2*	R^1, R^2	*Method*	*3,5,6*	*Yield (%)*[b]	*Syn : Anti*[c]
1	**1a**	**2b**	$-(CH_2)_3-$	A	**3b**	82 (41)	93 : 7
2	**1a**	**2c**	$-(CH_2)_4-$	A	**3c**	45	40 : 60
3	**1b**	**2b**	$-(CH_2)_3-$	A	**3b**	88	92 : 8
4	**1c**	**2b**	$-(CH_2)_3-$	A	**5b**	47 (46)	78 : 22
5	**1d**	**2b**	$-(CH_2)_3-$	A	**6b**	77 (50)	94 : 6
6	**1a**	**2b**	$-(CH_2)_3-$	B	**3b**	86 (42)	90 : 10
7	**1a**	**2c**	$-(CH_2)_3-$	B	**3c**	33	54 : 46
8	**1b**	**2b**	$-(CH_2)_3-$	B	**3b**	77 (40)	89 : 11
9	**1c**	**2b**	$-(CH_2)_3-$	B	**5b**	74 (71)	74 : 26
10	**1d**	**2b**	$-(CH_2)_3-$	B	**6b**	70 (61)	93 : 7
11	**1a**	**2b**	$-(CH_2)_3-$	C	**3b**	85	79 : 26

[a] All the reaction was carried out with polyfluoroalkylaldehyde ethyl hemiacetal (1 mmol) or hydrate (1 mmol) in the corresponding ketone (10 ml). [b] Method A ; piperidine (50 mol%) and CH_3CO_2H (50 mol%). Method B ; piperidine (50 mol%) and Wakogel C200 (silica gel). Method C; *n*-$PrNH_2$ (50 mol%) and CH_3CO_2H (50 mol%). [c] Measured by ^{19}F NMR using benzotrifluoride. Values in parentheses stand for the yields of isolated products. [d] Determined by ^{19}F NMR.

Cyclopentanone **2b** could also participate well in the direct aldol reaction of CF_3CHO ethyl hemiacetal **1a** to produce the corresponding β-hydroxy-β-trifluoromethyl ketone **5** in good yields with high *syn*-diastereoselectivities by the use of CH_3CO_2H (Method A) or silica gel (Method B) (entries 1 and 6). The reaction of CF_3CHO ethyl hemiacetal **1a** with cyclohexanone **2c** gave a 45% yield of the aldol product **3c** with low diastereoselectivity, because many unidentified by-products were produced (entries 2 and 7). *Syn*-selective direct aldol reaction of CF_3CHO ethyl hemiacetal **1a** with cyclopentanone **2b** by using *n*-propylamine in place of piperidine also occurred to produce the aldol product **3b** in 85% yield with slight reduction of diastereoselectivity (entry 11). CF_3CHO hydrate **1b** also reacted well with cyclopentanone **2b** to produce the aldol product **3b** in the similar yields with similar diastereoselectivities (entries 3 and 8). The direct aldol reaction of difluoroacetaldehyde ethyl hemiacetal **1c** as well as pentafluoropropionaldehyde hydrate **1d** with cyclopentanone **2b**, also successfully occurred to produce the corresponding β-difluoromethyl or pentafluoroethyl aldol adduct **5** or **6** in good yields with good to excellent *syn*-

diastereoselectivities in both cases of CH_3CO_2H and silica gel with piperidine (entries 4,5,9 and 10). Degree of *syn*-diastereoselectivities of the products **3b,5b,6b** derived from cyclopentanone, may depend on the bulkiness of the fluoroalkyl groups *(8)* in the following orders under the same conditions: the pentafluoroethyl (*syn* : *anti* = 94 : 6, 88% *de*) > trifluoromethyl (*syn* : *anti* = 93 : 7, 86% *de*) > difluoromethyl (*syn* : *anti* = 78 : 22, 56% *de*) group (entries 1,4 and 5), though the reason for this *syn*-selective outcome is not clear at this present.

L-Proline-catalyzed Asymmetric Direct Aldol Reaction of CF_3CHO Ethyl Hemiacetal with Ketones *(9,10,11)*

Optimization of the Reaction Conditions

The results of the *L*-proline-catalyzed reaction of CF_3CHO ethyl hemiacetal **1a** with acetone **2a** in various solvents are summarized in Table 3 (Figure 8).

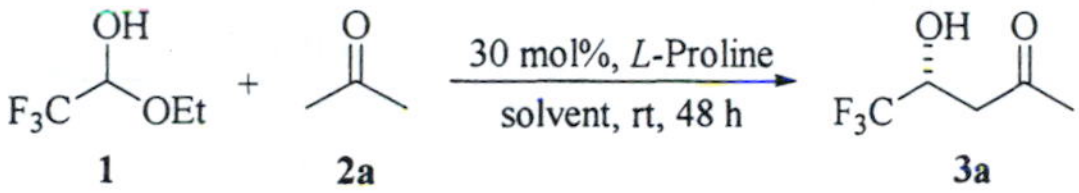

OH F$_3$C OEt **1** + O **2a** —30 mol%, *L*-Proline / solvent, rt, 48 h→ OH O F$_3$C **3a**

Figure 8. L-Proline-catalyzed Asymmetric Direct Aldol Reaction of CF_3CHO Ethyl Hemiacetal with Acetone.

The employment of DMSO or DMF resulted in a significant loss of enantioselectivities (entries 1 and 2). Other solvents, such as acetonitrile, dichloromethane and chloroform could be used to produce the good yields of the product **3a** with good enantioselectivities (entries 8-10). The reaction in THF or hexane was very sluggish and gave low yields of **3a** (entries 2 and 12). The reaction in water did not proceed at all (entry 3). The reaction in benzene resulted in best enatioselectivity of **3a** but with a decreased yield (entry 11). Importantly, acetone can be used as the solvent as well as a ketone donor with excellent yield in good enantioselectivity (entries 5 and 6). Lowering reaction temperature (0 °C) resulted in significant decrease of the product **3a** even in the presence of a stoichiometric amount of *L*-proline, together with slight increase of *ee* (entry 7).

Table 3. *L*-Proline-catalyzed Asymmetric Direct Aldol Reaction of CF_3CHO Ethyl Hemiacetal with Acetone

Entry[a]	*Solvent*	*Yield (%)*[b]	*R : S*[c]	*Ee*[c]
1	DMSO	96	50.3 : 49.7	2.8
2	DMF	94	52.0 : 48.0	4.0
3	H_2O	0	-	-
4	THF	19	68.7 : 31.3	37.4
5	Acetone	97	67.6 : 32.4	35.2
6	Acetone	96	69.2 : 30.8[d]	38.4[d]
7[e]	Acetone	18	71.3 : 28.7	42.6
8	MeCN	64	70.7 : 29.2	41.5
9	CH_2Cl_2	45	73.9 : 26.1	47.8
10	$CHCl_3$	31	72.8 : 27.2	45.6
11	Benzene	32	75.8 : 24.2	51.6
12	Hexane	19	73.0 : 27.0	46.0

[a] All the reaction was carried out with CF_3CHO ethyl hemiacetal **1a** (1 mmol) with *L*-proline (30 mol%) in the mixed solvent of dry acetone **2a** (2 ml) and solvent (8 ml) or dry acetone (10 ml). [b] Determined by ^{19}F NMR using benzotrifluoride as an internal standard. [c] Determined by HPLC analysis with DAICEL CHIRALCEL OD-H (hexane:*i*-PrOH=95/5) after *p*-chlorobenzoylation. [d] Determined by GC with InterCap CHIRAMIX (GL Science). [e] Carried out with *L*-proline (100 mol%) at 0 °C for 96 h.

Other CF_3CHO derivatives, such as hydrate (75 wt%) and 2,2,2-trifluoroethyl hemiacetal (88 wt%) as well as pentafluoropropionaldehyde hydrate **1d** were also examined, as shown in Figure 9. Hemiacetals are **1a,c** much effective than hydrate **1b** to produce almost quantitative yields of **3a**. However, there is only slight difference in *ee* of **3a**. Pentafluoropropionaldehyde hydrate **1d** also reacted with acetone to give 69% yield of aldol product **6a** with slight higher enantioselectivity.

The absolute configuration of the major aldol product **3a** generated by the reaction could be determined unambiguously as *R* by the comparision with the reported values of the optical rotation *(12)*.

L-Proline-catalyzed Asymmetric Direct Aldol Reaction with Ketones

The catalytic asymmetric direct aldol reaction of CF_3CHO ethyl hemiacetal with unmodified other ketones **2** are described in Figures 10 and 11.

Interestingly, cyclopentanone **2b** also nicely underwent the *L*-proline-catalyzed direct aldol reaction with CF_3CHO ethyl hemiacetal **1a** as well as pentafluoropropionaldehyde hydrate **1d** to give the corresponding aldol product **3b, 6b** in 77-96% yields with excellent *syn*-diastereoselectivities (*dr* = 94-98 : 2-6) as well as high enantioselectivities (*er* = 88.6-88.8 : 11.2-11.4). The reason for *syn*-selectivities in the reaction with cyclopentanone is unclear at this present.

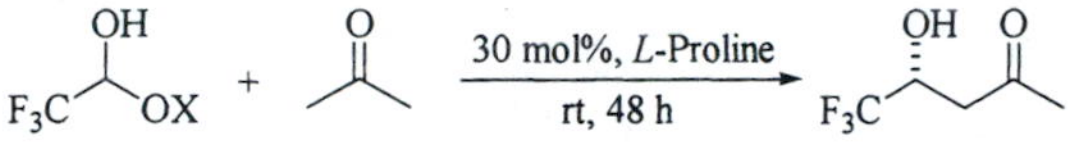

OH O
F3C OX + 30 mol%, L-Proline rt, 48 h F3C OH O
1 2a 3a

X = H . 64% yield, $R : S$ = 68.9 : 31.1, 37.8% *ee*
X = CH_2CF_3 . 97% yield, $R : S$ = 70.6 : 29.4, 41.2% *ee*

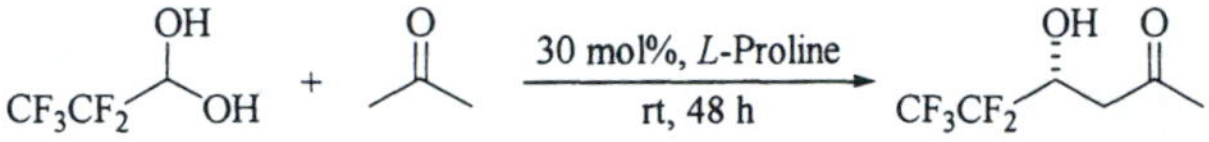

CF_3CF_2 OH OH + O 30 mol%, L-Proline rt, 48 h CF_3CF_2 OH O
1d 2a 6a

69% yield
$R : S$ = 72.0 · 28.0
44.0% *ee*

Figure 9. L-Proline-catalyzed Asymmetric Direct Aldol Reaction of Other CF_3CHO Derivatives and Pentafluoropropionaldehyde Hydrate with Acetone.

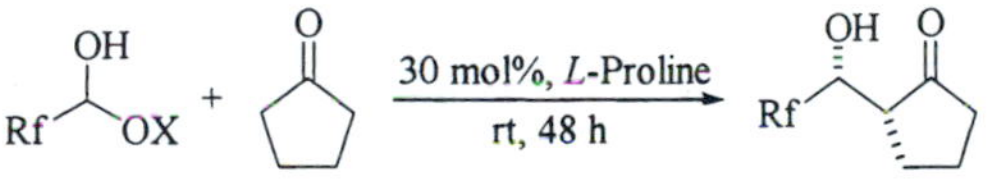

Rf OH OX + O 30 mol%, L-Proline rt, 48 h Rf OH O
1 2b 3b,6b

1a : Rf = CF_3, X = Et 96% yield[a], *Syn* : *Anti* = 98 . 2[b], $R . S$ = 88.8 : 11.2[c], 77.6% *ee*[c]
1d : Rf = CF_3CF_2, X = H 77 % yield[a], *Syn* : *Anti* = 94 : 6[b], $R : S$ = 88.6 : 11.4[c], 77.2% *ee*[c]

[a] Determined by ^{19}F NMR using benzotrifluoride as an internal standard.
[b] Determined by ^{19}F NMR.
[c] Determined by GC with InterCap CHIRAMIX (GL Science).

Figure 10. L-Proline-catalyzed Asymmetric Direct Aldol Reaction with Cyclopentanone.

Cyclohexanone **2c** reacted smoothly, *via* the proposed transition state by List and Houk *(13)*, with the CF_3CHO ethyl hemiacetal **1a** as well as pentafluoropropionaldehyde hydrate **1d** in the presence of 30 mol% of *L*-proline to produce **3c,6c** in 68-71% yields with high *anti*-selectivities (*dr* = 96-99 : 4-1) with excellent enantioselectivities (*er* = 95.5-96.7 : 3.3-4.5).

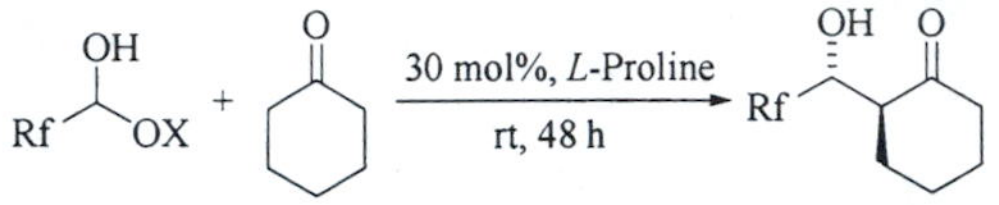

1a : Rf = CF_3, X = Et 68% yield[a], *Syn* : *Anti* = 4 : 96[b], *R* : *S* = 95.5 : 4.5[c], 91.0% *ee*[c]
1d : Rf = CF_3CF_2, X = H 71% yield[a], *Syn* : *Anti* = 1 : 99[b], *R* : *S* = 96.7 : 3.3[c], 93.4% *ee*[c]

[a] Determined by ^{19}F NMR using benzotrifluoride as an internal standard.
[b] Determined by ^{19}F NMR.
[c] Determined by GC with Chiralsil-Dex CB (Chrompack).

Figure 11. L-Proline-catalyzed Asymmetric Direct Aldol Reaction with Cyclohexanone.

Unfortunately, the reaction of diethyl ketone **2d** as a linear α,α'-disubstituted ketone with hemiacetal **1a** did not proceed at all.

Conclusions

In summary, we have achieved that the direct aldol reaction of CF_3CHO ethyl hemiacetal with unmidified ketones by the use of small amount of acids and amines without any strong acid and high reaction temperature, producing β-hydroxy-β-trifluoromethylated ketones in good yields. Furthermore, commercially available *L*-proline-catalyzed asymmetric aldol reaction of CF_3CHO ethyl hemiacetal with unmodified ketones also succeeded to produce β-hydroxy-β-trifluoromethylated ketones in good to excellent diastereo- and/or enatioselectivies.

Acknowledgements

This work was partially supported by a Grant-in-Aid from the Ministry of Education, Culture, Sports, Science and Technology of Japan, the OGAWA

Science and Technology Foundation and Gifu University. We are grateful to the Central Glass Co., Ltd., for the gift of CF_3CHO hemiacetal and hydrate as well as financial support. We also thank my coworkers, Messrs. H. Yamamoto, H. Nagaya, and Ms. M. Ishihara for their contribution.

References

1. For recent reviews, see. (a) Iseki, K. *Tetrahedron* **1998**, *54*, 13887-13914. (b) Soloshonok, V. In *Enantiocontrolled Synthesis of Fluoro-organic Compounds*, Soloshonok, V. A. Ed.; John Wiley & Sons, Chichester, **1999**, pp 230-262. (c) Mikami, K. In ref. 1b, pp 557-574. (c) Ishii, A.; Mikami, K. In *Asymmetric Fluoroorganic Chemistry: Synthesis, Application, and Future Directions*, Ramachandran, P. V. Ed.; ACS Symposium Series 746, Washington, DC, 1999, pp 60-73. (d) Mikami, K. In ref. 1c, pp 255-269. (e) Mikami, K.; Itoh, Y.; Yamanaka, M. *Chem. Rev.* **2004**, *141*, 1-16. (f) Mikami, K.; Itoh, Y.; Yamanaka, M. In *Fluorine-Containing Synthons*, Soloshonok, V. A. Ed.; ACS Symposium ries 911, Washington DC, **2005**, pp 356-367.
2. (a) Braid, M.; Iserson, H.; Lawlor, F. E. *J. Am. Chem. Soc.* **1954**, *76*, 4027. (b) Henne, A. L.; Pelley, R. L.; Alm, R. M. *J. Am. Chem. Soc.* **1950**, *72*, 3370-3371. (c) Shechter, H.; Conrad, F. *J. Am. Chem. Soc.* **1950**, *72*, 3371-3373.
3. (a) Funabiki, K.; Nojiri, M.; Matsui, M.; Shibata, K. *Chem. Commum.* **1998**, 2051-2052. (b) Funabiki, K.; Matsunaga, K.; Matsui, M.; Shibata, K. *Synlett* **1999**, 1477-1479. (c) Funabiki, K.; Matsunaga, K.; Nojiri, Hashimoto, W.; Yamamoto, H.; Shibata, K.; M.; Matsui, M. *J. Org. Chem.* **2003**, 68, 2853-2860. (d) Funabiki, K. In ref. 1f, pp 342-355.
4. (a) Funabiki, K.; Hashimoto, W.; Matsui, M. *Chem. Commun.* **2004**, 2056-2057. (b) Funabiki, K.; Hasegawa, K.; Murase, Y.; Nagaya, H.; Matsui, M. *J. Fluorine Chem.* (special issue) in press. And Ref. 3d.
5. Funabiki, K.; Honma, N.; Hashimoto, W.; Matsui, M. *Org. Lett.* **2003**, *5*, 2059-2061.
6. (a) Trost, B. M. *Science* **1991**, *254*, 1471-1477. (b) Trost, B. M. *Angew. Chem. Int. Ed. Engl.* **1995**, *34*, 259-281.
7. Funabiki, K.; Nagaya, H.; Ishihara, M.; Matsui, M. *Tetrahedron*, in press.
8. MacPhee, J. A.; Panaye, A.; Dubois, J. –E. *Tetrahedron* **1978**, *34*, 3553-3562.
9. (a) Funabiki, K.; Yamamoto, H.; Nagamori, M.; Matsui, M. US 2005119507 A1. (b) Funabiki, K.; Yamamoto, H.; Matsui, M. submitted for publication.

10. During our work, an example dealing with the catalytic generation of trifluoroacetaldehyde as well as only one proline-derived tetrazole catalyzed direct aldol reaction with cyclopentanone has been reported, see. Torii, H.; Nakadai, M.; Ishihara, K.; Saito, S.; Yamamoto, H. *Angew. Chem. Int. Ed.* **2004**, *43*, 1983-1986.
11. For *L*-proline-catalyzed asymmetric direct aldol reaction of trifluoropyruvate with aldehydes, see: Bøgevig, A.; Kumaragurubaran, N.; Jørgensen, K. A. *Chem. Commun.* **2002**, 620-621. For *L*-proline-catalyzed asymmetric Mannich reaction of trifluoromethylated imines with acetone, see: Funabiki, K.; Nagamori, M.; Goushi, S.; Matsui, M. *Chem. Commun.* **2004**, 1928-1929. For *L*-proline-catalyzed asymmetric Mannich reaction of trifluoromethylated imines with aldehydes, see: Fustero, S.; Jiménez, D.; Sanz-Cervera, J. F.; Sánchez-Roselló, M.; Estenban, E.; Simón-Fuentes, A. *Org. Lett.* **2005**, *7*, 3433-3436.
12. Bucciarelli, M.; Forni, A.; Moretti, I.; Prati F.; Torre, G. *Biocatalysis* **1994**, *9*, 313-320.
13. Bahmanyar, S.; Houk, K. N.; Martin, H. J.; List, B. *J. Am. Chem. Soc.* **2003**, *125*, 2475-2479.

Chapter 9

Cationic Cyclizations of Fluoro Alkenes: Fluorine as *a Controller* and *an Activator*

Junji Ichikawa

Department of Chemistry, Graduate School of Science, The University of Tokyo, Hongo, Bunkyo-ku, Tokyo 113–0033, Japan

A new function of fluorine as a synthetic tool has been discovered in cationic cyclizations. Ring-forming reactions via fluorine-containing carbocations have been achieved starting from fluoro alkenes by using the properties of fluorine, such as electronic effects and leaving-group ability. Fluoro alkenes with an alkenoyl or an aryl group undergo *fluorine-directed* and *-accelerated* Nazarov-type or Friedel–Crafts-type cyclizations, respectively. These cyclizations and their domino cyclizations allow the efficient construction of highly functionalized and fused ring systems. Throughout the reactions, fluorine functions as *a controller* over the reaction pathways and *an activator* of the substrates.

Introduction

Fluorine-containing compounds have attracted much interest in various fields, ranging from medicinal and agricultural chemistry to material science, because of the distinctive properties of the compounds. In contrast, the potential of fluorine as a tool in organic synthesis has not been fully realized, compared with other heteroatoms such as phosphorous, sulfur, and silicon. Synthetic methodologies that make extensive use of the unique properties of fluorine are yet to be developed. It is well known that fluorine has high electronegativity, which leads to the (i) β-cation-destabilizing effect, (ii) β-anion-stabilizing effect, and (iii) leaving-group ability as a fluoride ion (F^-). In addition, (iv) fluorine stabilizes α-cations and (v) destabilizes α-anions because of its lone pairs (*1*). These unique properties should provide a potent methodology in organic synthesis.

In order to generate fluorinated cations and anions, and to utilize the effects of fluorine on both of them, fluoro alkenes seem to be an ideal choice as a substrate. Based on these considerations, we have achieved two types of ring-forming reactions starting from fluorinated alkenes, one that proceeds via fluorine-containing carbocations (*2*) and another via fluorine-containing carbanions (*3*). In this chapter, we describe our investigations on the cationic cyclizations, Nazarov-type and Friedel–Crafts-type cyclizations, by the use of the properties shown in Scheme 1. These reactions are promoted and/or controlled by fluorine substituents to construct carbocycles starting from fluoro alkenes. Other cationic cyclization processes of fluoro alkenes provide steroids via polyene cyclization (*4*), lactones via halolactonization (*5*), and cycloalkanes via carbocyclization (*6*).

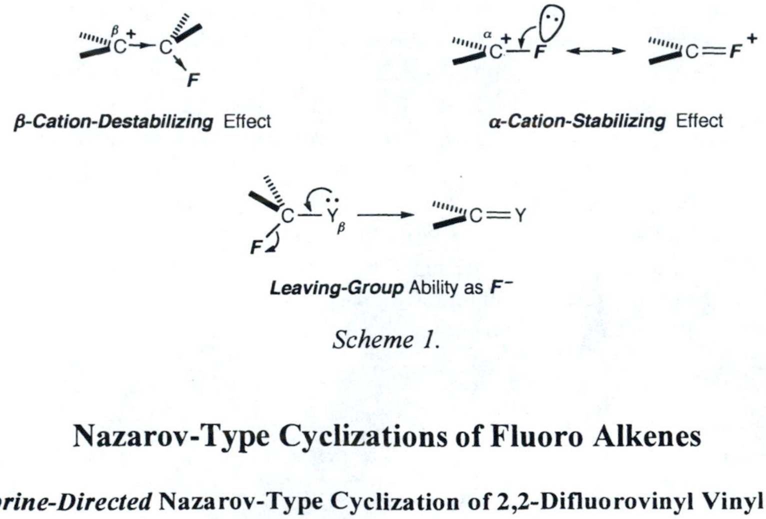

β-Cation-Destabilizing Effect

α-Cation-Stabilizing Effect

Leaving-Group Ability as ***F***⁻

Scheme 1.

Nazarov-Type Cyclizations of Fluoro Alkenes

Fluorine-Directed Nazarov-Type Cyclization of 2,2-Difluorovinyl Vinyl Ketones

The Nazarov cyclization is a versatile protocol for construction of the cyclopentenone framework, a structural motif found in a number of natural products (Scheme 2) (*7*). The classical reactions, however, had limited utility due to the lack of control over placement of the endocyclic double bond, which normally occupied the most substituted position because of its thermodynamic stability. Among a number of modifications of this reaction, it was the *silicon-directed* Nazarov cyclization (*7,8*) that overcame this drawback of regiochemistry by making use of two properties of silicon: (i) its *β-cation-stabilizing* effect and (ii) its leaving-group ability as a silyl *cation* (Si^+) (Scheme 3).

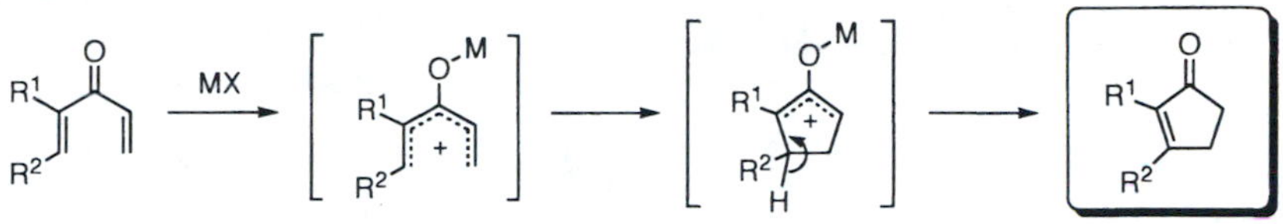

Scheme 2.

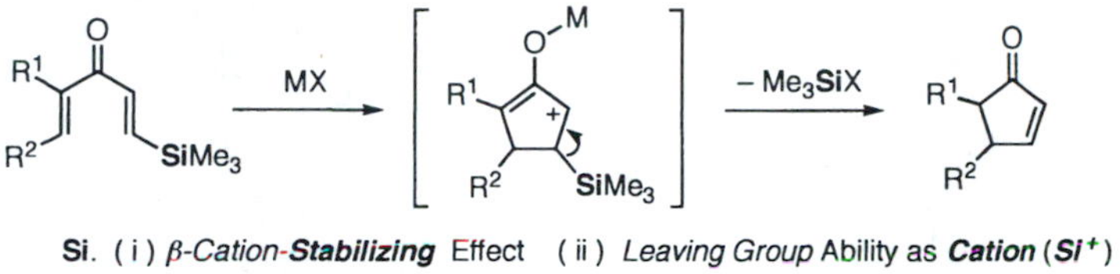

Si. (i) *β-Cation-**Stabilizing*** Effect (ii) *Leaving Group* Ability as ***Cation*** (***Si***$^+$)

Scheme 3.

In terms of these two properties, interestingly fluorine is a kind of negative image of silicon. Fluorine possesses a *β-cation-destabilizing* effect and also functions as a leaving group of an *anion* (a fluoride ion, F^-) (*1*). These facts suggest that not only silicon but also fluorine might be a controller of this electrocyclic reaction. Based on these considerations, 2,2-difluorovinyl vinyl ketones **1** were designed as the substrates for *fluorine-directed* Nazarov-type cyclizations.

Treatment of **1** with a silylating agent such as trimethylsilyl triflate (TMSOTf) and $TMSB(OTf)_4$ readily induced cyclization at room temperature to afford cross-conjugated cyclopent-2-en-1-ones **2** with defined placement of the two double bonds (Scheme 4) (*9*). The Nazarov-type ring closure occurs to generate the cyclopentenylic cation **3**, followed by its collapse in a fluorine-directed manner with the loss of a fluoride ion and a proton. These losses takes place under the strict control of fluorine regardless of the substrates, which causes the regioselective formation of the endocyclic and exocyclic double bonds in **2**. It should be noted that the addition of 1,1,1,3,3,3-hexafluoropropan-2-ol (HFIP) as a cosolvent dramatically promoted and accelerated the cationic cyclization, probably due to its high ionizing power and low nucleophilicity (*10*).

Thus, the fluorine-directed Nazarov-type cyclization provides a facile approach to highly functionalized cyclopentanones **2** (Scheme 5), which are versatile intermediates in the syntheses of cyclopentanoid natural products because of their readiness for further synthetic elaborations (*11*).

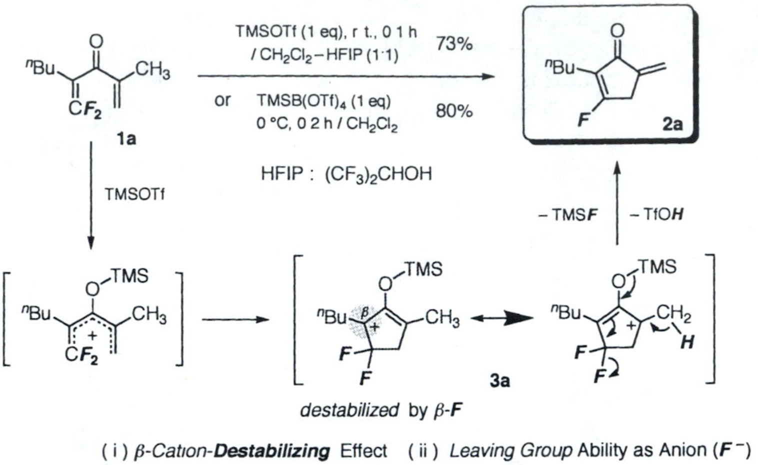

TMSOTf (1 eq), r.t., 0.1 h / CH_2Cl_2–HFIP (1:1) 73%

or $TMSB(OTf)_4$ (1 eq) 0 °C, 0.2 h / CH_2Cl_2 80%

HFIP : $(CF_3)_2CHOH$

destabilized by β-F

(i) β-Cation-**Destabilizing** Effect (ii) *Leaving Group* Ability as Anion (F^-)

Scheme 4.

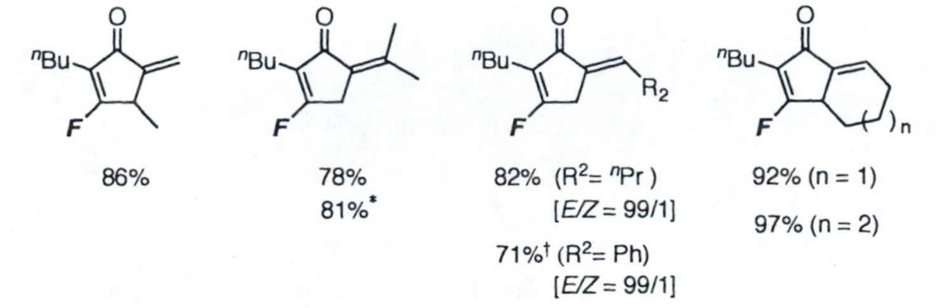

86% | 78%, 81%* | 82% (R^2= nPr) [E/Z = 99/1]; 71%† (R^2= Ph) [E/Z = 99/1] | 92% (n = 1), 97% (n = 2)

* $TMSB(OTf)_4$ (1 eq), 0 °C, 0.2 h / CH_2Cl_2. † TMSOTf (1 eq), 0 °C, 0.7 h / HFIP.

Scheme 5.

The effect of fluorine on the regiochemistry was confirmed by comparison of a monofluorinated and a fluorine-free substrate **4** and **5** (Scheme 6) with **1a**, which gave a single product as shown in Scheme 4. The monofluorinated substrate **4** gave a mixture of two compounds, which were derived from the same cation via deprotonation or attack by water. Concerning the endocyclic double bond in these compounds, the regiochemistry was perfectly controlled. In contrast, its fluorine-free counterpart **5** afforded a mixture of three regioisomers without control in the deprotonation step. Furthermore, in the case of the symmetrical (except for fluorine) substrate **1b**, the reaction proceeded through a defined pathway to give **2b**, not yielding the product **2b'** with an exocyclic double bond on the other side of the carbonyl group (Scheme 6). Therefore, fluorine plays a critical role in the regioselective cyclization (*9*).

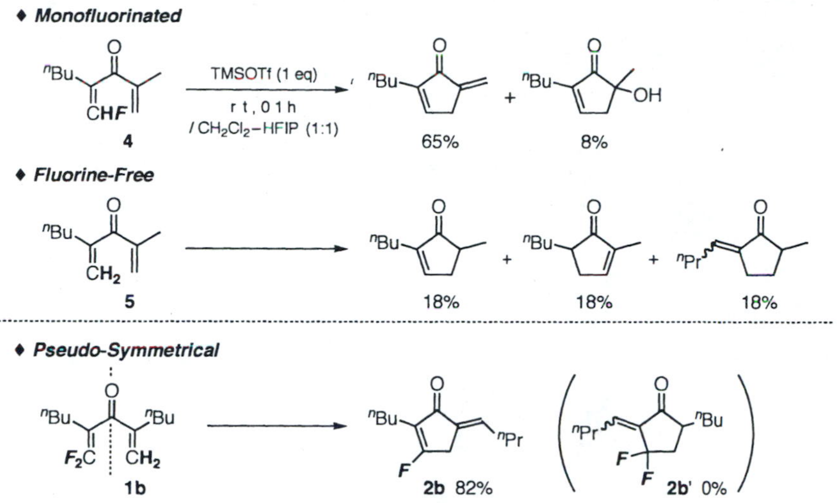

♦ *Monofluorinated*

TMSOTf (1 eq), r t, 0 1 h / CH_2Cl_2–HFIP (1:1): **4** → 65% + 8%

♦ *Fluorine-Free*

5 → 18% + 18% + 18%

♦ *Pseudo-Symmetrical*

1b → **2b** 82% (**2b'** 0%)

Scheme 6.

Fluorine-Directed Nazarov Cyclization of 1-(Trifluoromethyl)vinyl Vinyl Ketones

As the second generation of fluorine-directed Nazarov cyclizations, the introduction of fluorine at an alternative position in the substrates was examined. Divinyl ketones **6** bearing a vinylic trifluoromethyl group were designed as Nazarov substrates, so that the β-cation-destabilizing effect of fluorine could be exerted on the intermediary cyclopentenylic cation to control the reaction. When **6** were subjected to the same reaction conditions as above, 5-trifluoromethylcyclopent-2-en-1-ones **7** were selectively obtained in high to excellent yields (Scheme 7) (*12*). The directing effect of fluorine works well to place the double bond at the position distal to the CF_3 group, not at the proximal position. In contrast with 2,2-difluorovinyl vinyl ketones (*9*), the elimination of fluoride ion from **7** was not observed.

In order to confirm the effect of fluorine on the regiochemistry, the reaction of the pseudo-symmetrical substrate **6a** was tried. Cyclization proceeded through the defined pathway again to give the sole cyclized product **7a** with the double bond distal to the CF_3 group, which shows another strict control by fluorine (*12*).

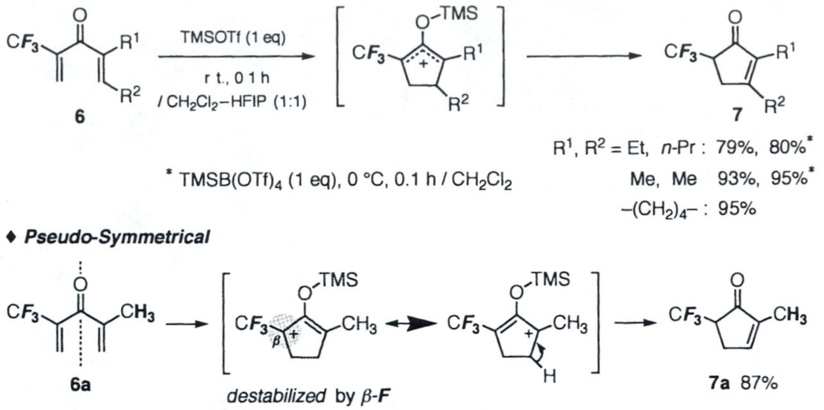

CF_3 … R^1 / R^2 **6** — TMSOTf (1 eq), r.t., 0.1 h / CH_2Cl_2–HFIP (1:1) → [O-TMS, CF_3, R^1, R^2, +] → CF_3 … R^1, R^2 **7**

R^1, R^2 = Et, *n*-Pr : 79%, 80%*
Me, Me 93%, 95%*
$-(CH_2)_4-$: 95%

* $TMSB(OTf)_4$ (1 eq), 0 °C, 0.1 h / CH_2Cl_2

♦ ***Pseudo-Symmetrical***

CF_3 … CH_3 **6a** → [O-TMS, CF_3, β, +, CH_3 ↔ O-TMS, CF_3, +, CH_3, H] → CF_3 … CH_3 **7a** 87%

destabilized by *β-F*

Scheme 7.

Fluorine-Directed **and** ***-Accelerated*** **Nazarov Cyclization of 1-Fluorovinyl Vinyl Ketones**

The first and the second generation of fluorine-directed Nazarov cyclizations mentioned so far utilized the β-cation-destabilizing effect. The rate-determining step in the Nazarov cyclization is supposed to be the electrocyclization process. Therefore, enhancement of the reactivity can be expected by stabilization of the cyclized cationic intermediate (*7a,7c*). In the third generation, we attempted to utilize the cation-stabilizing effect of fluorine, that is the *α-cation-stabilizing* effect. A fluorine should be introduced at the position α to the carbonyl group of divinyl ketone substrates **8** to stabilize the cyclopentenylic cation **10** (Scheme 8). Here, fluorine can exert two effects, an activating effect and an directing effect.

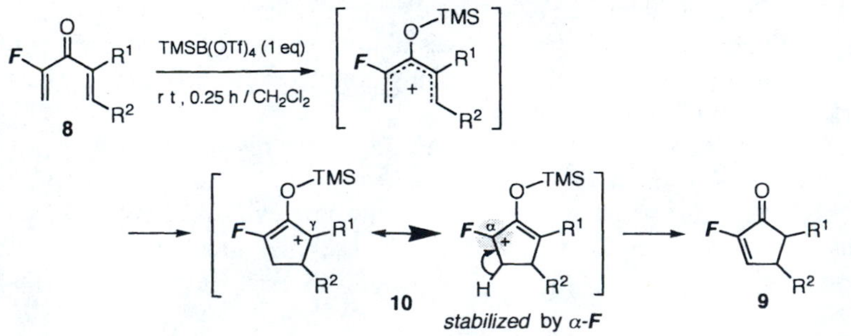

F … R^1, R^2 **8** — $TMSB(OTf)_4$ (1 eq), r.t., 0.25 h / CH_2Cl_2 → [O-TMS, F, R^1, R^2, +] → [O-TMS, F, +, γ, R^1, R^2 ↔ O-TMS, F, α, +, R^1, H, R^2] **10** → F … R^1, R^2 **9**

stabilized by *α-F*

Scheme 8.

Treatment of a 1-fluorovinyl vinyl ketone **8** with TMSOTf in CH_2Cl_2–HFIP afforded a 2-hydroxycyclopent-2-en-1-one, a hydrolyzed product of the expected compound, probably due to a trace amount of water in HFIP. However, a stronger silylating agent, $TMSB(OTf)_4$, successfully induced the cyclization in CH_2Cl_2 without HFIP to give the desired 2-fluorinated cyclopentenone **9**. All of the products **9** are exclusively obtained in high yield, and their position of the double bond does not depend on the substitution patterns of the substrates (Scheme 9) (*13*).

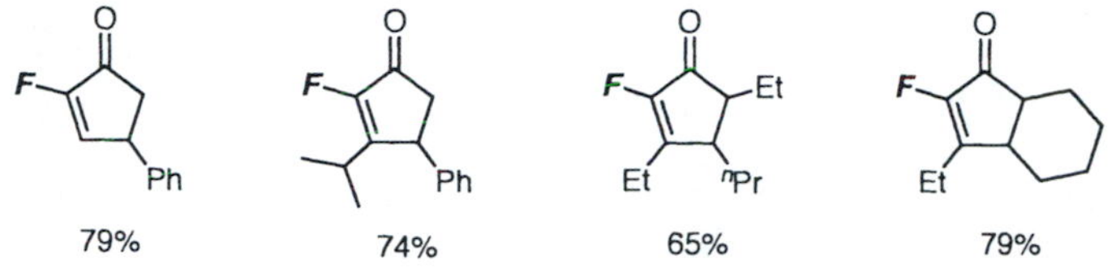

F Ph 79%
F Ph 74%
F Et Et ⁿPr 65%
F Et 79%

Scheme 9.

A competitive experiment was conducted to confirm the effect of fluorine on the reactivity. When a 1 to 1 mixture of the fluorinated and fluorine-free substrates **8a** and **11** was treated with 1 equiv of $TMSB(OTf)_4$, only the fluorine-containing cyclopentenone **9a** was obtained, and **11** was recovered (Scheme 10). This result obviously indicates that the fluorine activates the substrate by stabilizing the intermediary cyclopentenylic cation **10a**.

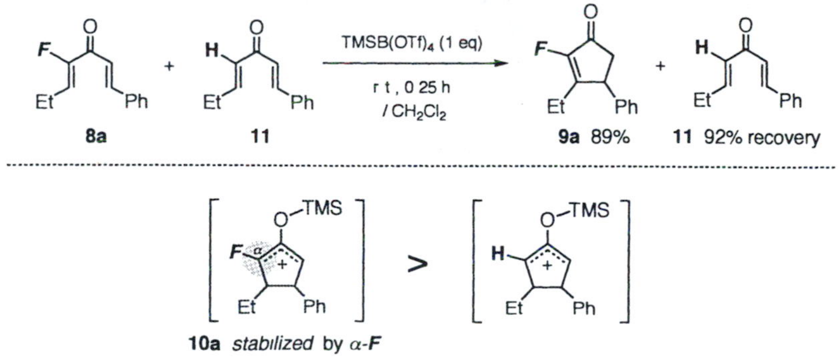

8a + 11 → $TMSB(OTf)_4$ (1 eq), r t, 0 25 h / CH_2Cl_2 → 9a 89% + 11 92% recovery

10a stabilized by α-F > [H analog]

Scheme 10.

While taking a much longer reaction time, the fluorine-free substrate **11** also underwent cyclization, which verifies the effect of fluorine on the regiochemistry. A mixture of two regioisomers, **12** and **13**, was obtained, the more stable isomer **13** with the longer conjugated system being the major isomer (Scheme 11). In contrast with **11**, the α-fluorinated substrate **8a** gave only one isomer, **9a**, which is the isomer with the shorter conjugated system. Thus, this reaction seems to be kinetically controlled by the electronic effect of fluorine. Furthermore, in the reaction of the pseudo-symmetrical substrate **8b**, a double bond was selectively introduced at the position proximal to the fluorine (Scheme 11). Thus, the fluorine functions as an activator as well as a controller in the third generation of the fluorine-directed Nazarov cyclizations (*13*).

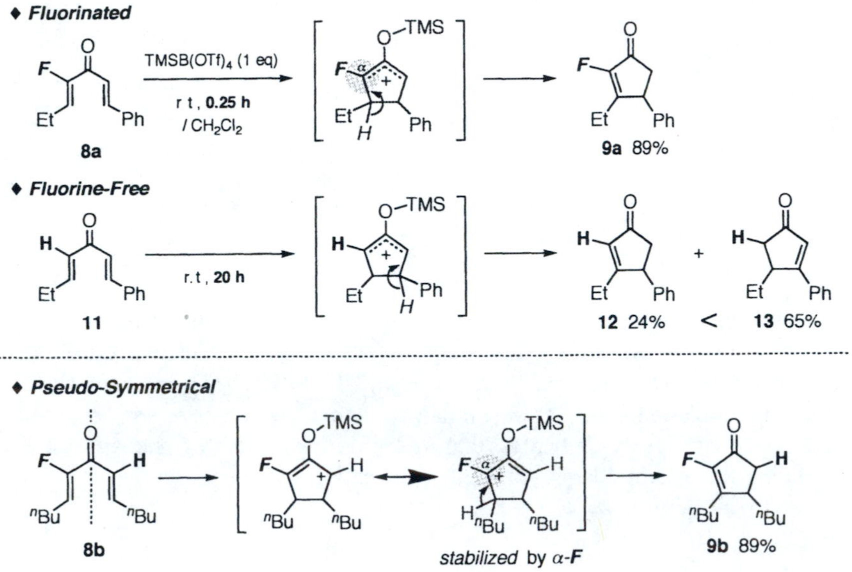

◆ *Fluorinated*

8a —TMSB(OTf)$_4$ (1 eq), r.t, 0.25 h / CH_2Cl_2→ 9a 89%

◆ *Fluorine-Free*

11 —r.t, 20 h→ 12 24% < 13 65%

◆ *Pseudo-Symmetrical*

8b → 9b 89%

stabilized by α-F

Scheme 11.

Now, three types of fluorine-containing cyclopentenones can be selectively prepared by the fluorine-directed Nazarov cyclizations.

Friedel–Crafts-Type Cyclizations of Fluoro Alkenes

Fluorine-Accelerated Friedel–Crafts-Type Cyclization of 2,2-Difluorovinyl Ketones

The results on the first generation of fluorine-directed Nazarov-type cyclizations imply that treatment of 2,2-difluorovinyl ketones with TMSOTf effectively generates the corresponding difluoroallyl cations (Scheme 12), which are reactive enough to induce electrocyclization. Another C–C bond forming reaction has been achieved by trapping these cations with an aryl group instead of a vinyl group, that is, Friedel–Crafts-type reactions (*14*).

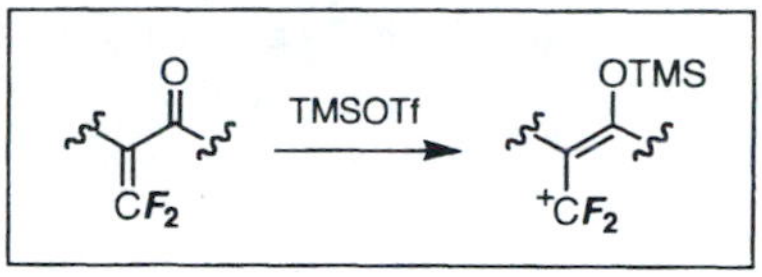

Scheme 12.

Difluorovinyl ketones **14** bearing an aryl group were designed as the Friedel–Crafts substrates to trap the generated cations with an intramolecular aryl moiety, leading to 4-fluorinated 3-acyl-1,2-dihydronaphthalenes **15**. When ketones **14** were treated with TMSOTf or $TMSB(OTf)_4$ under similar conditions, the expected cyclization proceeded smoothly to give fluorodihydronaphthalenes **15** in high yield (Scheme 13) (*15*). This cyclization was accompanied by the loss of a fluoride ion, which results in the substitution of an aryl group for the fluorine under *acidic conditions*. This is in contrast to the replacement of the vinylic fluorines by carbanions under *basic conditions* via an addition–elimination process (*11,16*).

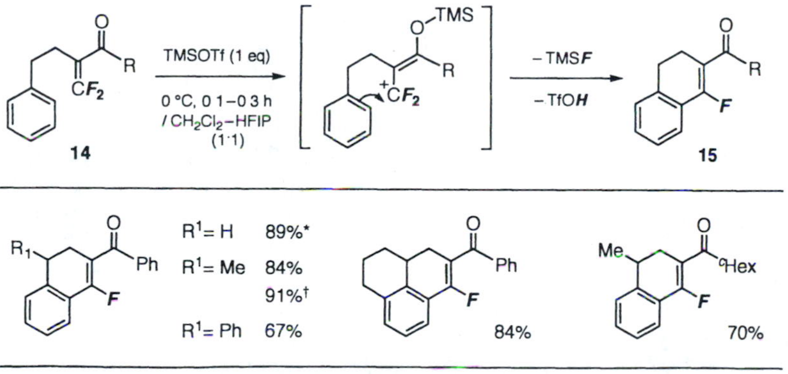

* TMSOTf (1 eq), 0 °C, 0 3 h / CH_2ClCH_2Cl–HFIP (1.1) † $TMSB(OTf)_4$ (1 eq), 0 °C, 0.2 h / CH_2Cl_2.

Scheme 13.

In order to confirm the effect of fluorine on the reactivity, the reactions of the corresponding monofluorinated and fluorine-free substrates **16** and **17** were studied (Scheme 14). While all of the substrates underwent the Friedel–Crafts-type cyclization, the conditions required for completion of the reaction were different: the monofluorinated substrate **16** needed a reaction time 10 times longer than **14a**. The fluorine-free substrate **17** gave no cyclized products until being heated under reflux. These results clearly show that the vinylic fluorine acts as an activating group. Its stabilizing effect on the α-carbocations is a key factor for the acceleration, because the rate of these Friedel–Crafts reactions seems to be controlled by the generation of carbocations from the alkenes (*17*).

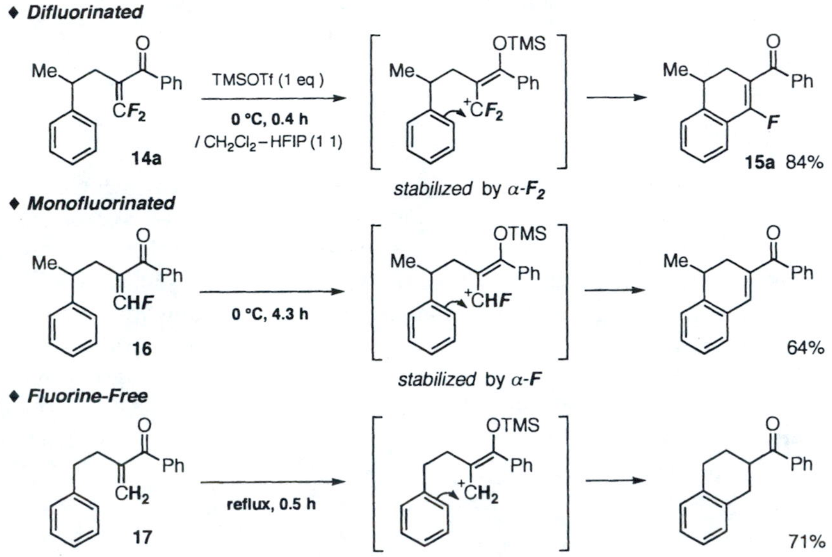

Scheme 14.

Domino Cyclizations for Fused Polycyclic Systems

Now that the Nazarov-type and the Friedel–Crafts-type cyclizations have been accomplished, the two vinylic fluorines can be used in a domino reaction of these cyclizations. 2,2-Difluorovinyl ketones **18** were designed to allow successive trapping of the intermediary carbocations with a vinyl and an aryl group. When **18** were exposed to TMSOTf in HFIP, the Friedel–Crafts-type and the Nazarov-type cyclizations proceeded sequentially via intermediary 2-fluorovinyl ketones. The reaction provided fused polycyclic systems **19** or **20a** bearing the steroid skeleton (*18*) in good yield in a one-pot operation (Scheme 15) (*15*).

Acylfluorodihydronaphthalenes **15** are a versatile class of compounds because of their 2-fluorovinylcarbonyl functionality, which readily reacts with nucleophiles to provide a variety of polysubstituted naphthalene derivatives (*11,16,19*). For example, **15a** was treated with hydrazines or amizines as a bifunctional nucleophile to construct heterocycles via consecutive addition–

elimination and cyclodehydration processes (Scheme 16) (*19b*). The reactions afforded pyrazole- or pyrimidine-fused ring systems, 4,5-dihydrobenzo[*g*]indazoles **21** (*20*) or 5,6-dihydrobenzo[*h*]quinazolines **22** (*21*), in excellent yield with excellent regioselectivity in the pyrazole formation (*15*).

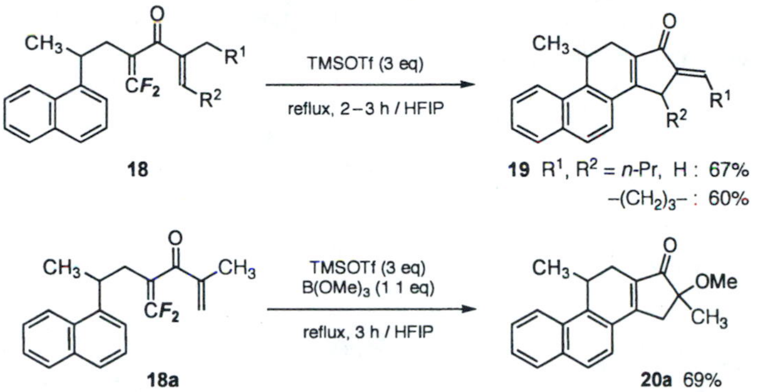

Scheme 15.

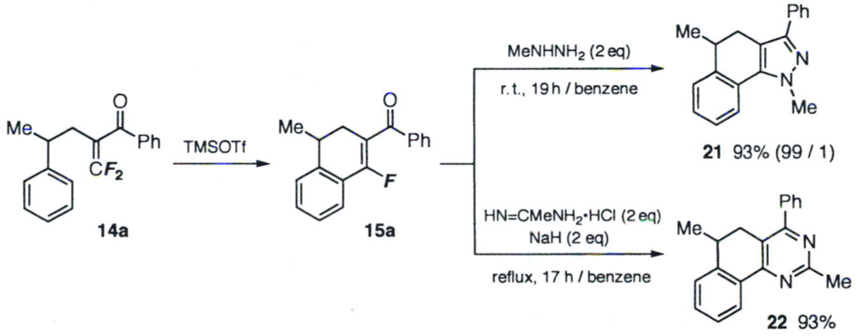

Scheme 16.

Fluorine-Directed Friedel–Crafts-Type Cyclization of 1,1-Difluoroalk-1-enes

After having accomplished the Friedel–Crafts-type cyclization of 2,2-difluorovinyl ketones, the generation of α-fluorocarbocations was investigated starting from simple 1,1-difluoroalk-1-enes **23** without a carbonyl group. Their protonation was expected to occur on the inner carbon β to fluorine, not the CF_2 carbon, because of the α-cation-stabilizing effect of fluorine. Thus, the reaction should generate a cation and allow bond formation at the terminal carbon (Scheme 17) .

On treatment with Magic acid ($FSO_3H{\bullet}SbF_5$) in HFIP, the cyclizaton of **23** readily proceeded to afford bicyclic ketones **24** in high yield (*10b*). 4,4-

Difluorohomoallylbenzene **23a** (n = 1, R = Ph) afforded the *six-membered* cyclic ketone **24a**, whereas the fluorine-free counterpart **27** was reported to give the *five-membered* carbocycle **28** (Scheme 17) (*22*). These results suggest that the Friedel–Crafts-type ring closure is regioselectively controlled by the vinylic fluorines to give the difluorocycloalkanes via the α-fluorocarbocations **25**. The second Magic acid abstracts fluoride ion to generate α-fluorocarbocations **26** again, whose hydrolysis gives bicyclic ketones **24**.

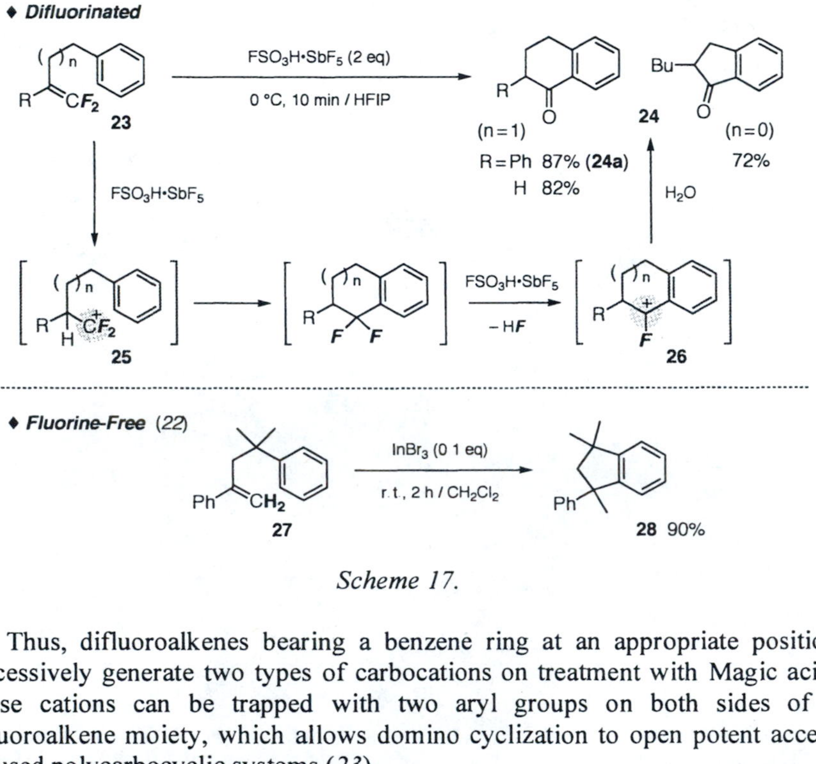

Scheme 17.

Thus, difluoroalkenes bearing a benzene ring at an appropriate position successively generate two types of carbocations on treatment with Magic acid. These cations can be trapped with two aryl groups on both sides of a difluoroalkene moiety, which allows domino cyclization to open potent access to fused polycarbocyclic systems (*23*).

Summary

The remarkable properties of fluorine, including (i) electronic effects on cations and (ii) leaving-group ability as a fluoride ion, can be utilized to activate substrates and control reaction pathways to provide new methodologies for synthesizing a variety of compounds inaccessible by other conventional methods. Thus, we have discovered a new function for fluorine as a versatile synthetic tool, a controller and an activator, opening a new area of fluorine chemistry in organic synthesis.

Acknowledgements

I thank my coworkers, Dr. M. Fujiwara, Ms. M. Yokota, Mr. S. Miyazaki, Mr. M. Kaneko, Mr. T. Yokoyama, Mr. H. Jyono, Mr. T. Kudo, Mr. D. Fujita, Mr. M. Ikemoto, and Mr. M. Itonaga for their contributions to this project.

References

1. (a) Smart, B. E. In *Organofluorine Chemistry, Principles and Commercial Applications*, Banks, R. E.; Smart, B. E.; Tatlow, J. C. Eds.; Plenum Press: New York, 1994; Chapter 3, pp 57–88. (b) Lee, V. J. In *Comprehensive Organic Synthesis*; Trost, B. M., Ed.; Pergamon: Oxford, 1991, Vol. 4, Chapter 1.2. pp 69–137.
2. A review on fluorinated carbocations: Allen, A. D.; Tidwell, T. T. In *Advances in Carbocation Chemistry*; Creary, X. Ed.; Jai Press: Greenwich, 1989; Vol. 1, pp 1–44.
3. Ichikawa, J. In *Fluorine-Containing Synthons*, Soloshonok, V. A. Ed.; Oxford University Press/American Chemical Society: Washington, D.C., 2005; Chapter 14, pp 262–275.
4. (a) Johnson, W. S.; Daub, G. W.; Lyle, T. A.; Niwa, M. *J. Am. Chem. Soc.* **1980**, *102*, 7800–7802. (b) Johnson, W. S.; Lyle, T. A.; Daub, G. W. *J. Org. Chem.* **1982**, *47*, 161–163. (c) Fish, P. V.; Johnson, W. S.; Jones, G. S.; Tham, F. S.; Kullnig, R. K. *J. Org. Chem.* **1994**, *59*, 6150–6152 and references cited therein.
5. (a) Morikawa, T.; Kumadaki, I.; Shiro, M. *Chem. Pharm. Bull.* **1985**, *33*, 5144–5146. See also: (b) Kendrick, D. A.; Kolb, M. *J. Fluorine Chem.* **1989**, *45*, 273–276.
6. Saito, A.; Okada, M.; Nakamura, Y.; Kitagawa, O.; Horikawa, H.; Taguchi, T. *J. Fluorine. Chem.* **2003**, *123*, 75–80.
7. For recent reviews of the Nazarov cyclization, see: (a) Harmata, M. *Chemtracts: Org. Chem.* **2004**, *17*, 416–435. (b) Pellissier, H. *Tetrahedron* **2005**, *61*, 6479–6517. (c) Frontier, A. J.; Collison, C. *Tetrahedron* **2005**, *61*, 7577–7606. (d) Tius, M. A. *Eur. J. Org. Chem.* **2005**, 2193–2206.
8. Silicon-directed Nazarov cyclizations: (a) Denmark, S. E.; Jones, T. K. *J. Am. Chem. Soc.* **1982**, *104*, 2642–2645. (b) Denmark, S. E.; Wallace, M. A.; Walker, Jr., C. B. *J. Org. Chem.* **1990**, *55*, 5543–5545. (c) Kang, K.-T.; Kim, S. S.; Lee, J. C.; U, J. S. *Tetrahedron Lett.* **1992**, *33*, 3495–3498. Tin-directed Nazarov cyclizations: (d) Peel, M. R.; Johnson, C. R. *Tetrahedron Lett.* **1986**, *27*, 5947–5950.
9. Ichikawa, J.; Miyazaki, S.; Fujiwara, M.; Minami, T. *J. Org. Chem.*, **60**, 2320–2321 (1995).
10. (a) Schadt, F. L.; Schleyer, P. V. R. *Tetrahedron Lett.* **1974**, *27*, 2335–2338. For recent reports on the cationic reactions conducted in HFIP, see: (b)

Ichikawa, J.; Jyono, H.; Kudo, T.; Fujiwara, M.; Yokota, M. *Synthesis* **2005**, 39–46 and references therein.

11. Ichikawa, J.; Fujiwara, M.; Miyazaki, S.; Ikemoto, M.; Okauchi, T.; Minami, T. *Org. Lett.*, **3**, 2345–2348 (2001).
12. Ichikawa, J.; Fujiwara, M.; Okauchi, T.; Minami, T. *Synlett*, **1998**, 927–929.
13. (a) Ichikawa, J. *Pure Appl. Chem.* **2000**, *72*, 1685–1689. (b) Ichikawa, J.; Fujiwara, M.; Yokota, M.; Yokoyama, T.; Miyazaki, S. Manuscript in preparation.
14. For recent reports on Friedel–Crafts-type reactions with conjugated enones and enals, see: (a) Imachi, S.; Onaka, M. *Chem. Lett.* **2005**, *34*, 708–709 and references therein. (b) Palomo, C.; Oiarbide, M.; Kardak, B. G.; García, J. M.; Linden, A. *J. Am. Chem. Soc.* **2005**, *127*, 4154–4155. (c) Evans, D. A.; Fandrick, K. R.; Song, H.-J. *J. Am. Chem. Soc.* **2005**, *127*, 8942–8943 and references therein.
15. Ichikawa, J.; Kaneko, M.; Yokota, M.; Itonaga, M.; Yokoyama, T. Manuscript in preparation.
16. Ichikawa, J.; Yokota, N.; Kobayashi, M.; Minami, T. *Synlett* **1993**, 186–188 and references therein.
17. For reviews on the alkylation mechanisms, see: Roberts, R. M.; Khalaf, A. A. *Friedel–Crafts alkylation chemistry: a century of discovery*, Marcel Dekker, New York, **1984**, Chapter 3, pp 122–227.
18. For recent reports on the synthesis of steroids, see: Ma, S.; Lu, P.; Lu, L.; Hou, H.; Wei, J.; He, Q.; Gu, Z.; Jiang, X.; Jin, X. *Angew. Chem. Int. Ed.* **2005**, *44*, 5275–5278 and references therein.
19. Sequential substitution of the two fluorines in 2,2-difluorovinyl ketones has been achieved under basic conditions with combinations of *C*, *O*, *S*, and *N*-nucleophiles. (a) Ichikawa, J.; Kobayashi, M.; Yokota, N.; Noda, Y.; Minami, T. *Tetrahedron* **1994**, *50*, 11637–11646. (b) Ichikawa, J.; Kobayashi, M.; Noda, Y.; Yokota, N.; Amano, K.; Minami, T. *J. Org. Chem.* **1996**, *61*, 2763–2769. (c) Xiao, L.; Kitazume, T. *J. Fluorine. Chem.* **1997**, *86*, 99–104. (d) Huang, X.; He, P.; Shi, G. *J. Org. Chem.* **2000**, *65*, 627–629.
20. For recent reports on 4,5-dihydrobenzo[*g*]indazole derivatives, see: (a) Peruncheralathan, S.; Khan, T. A.; Ila, H.; Junjappa, H. *J. Org. Chem.* **2005**, *70*, 10030–10035. (b) Murineddu, G.; Ruiu, S.; Mussinu, J.-M.; Loriga, G.; Grella, G. E.; Carai, M. A. M.; Lazzari, P.; Pani, L.; Pinna, G. A. *Bioorg. Med. Chem.* **2005**, *13*, 3309–3320.
21. 5,6-Dihydrobenzo[*h*]quinazoline derivatives are potent protein kinase inhibiors. Rapecki, S.; Allen, R. *J. Pharmacol. Exp. Ther.* **2002**, *303*, 1325–1333 and references therein.
22. Peppe, C.; Lang, E. S.; de Andrade, F. M.; de Castro, L. B. *Synlett* **2004**, 1723–1726.
23. Ichikawa, J.; Yokota, M.; Kudo, T.; Jyono, H. Manuscript in preparation.

New Technologies and Materials

Chapter 10

A Hydro–Fluoro Hybrid Compound, Hermaphrodite, Is Fluorophilic or Fluorophobic? A New Concept for the Fluorine-Based Crystal Engineering

T. Ono[1], H. Hayakawa[1], N. Yasuda[2], H. Uekusa[1], and Y. Ohashi[2]

[1]Research Institute of Instrumentation Frontier, National Institute of Advanced Industrial Science and Technology (AIST), 2266–98, Anagahora, Shimoshidami, Moriyama-ku, Nagoya, Japan
[2]Department of Chemistry, Tokyo Institute of Technology, 1–1, Ookayama, Meguro-ku, Tokyo, Japan

We proposed a new concept of the fluorine-based crystal engineering based on the hydro-fluoro hybrid compounds, R_H-X-R_F, of which structures were featured by the symbiotic name Hermaphrodite having a hydro-segment (R_H;Hermes) and perfluoro-segment (R_F;Aphrodite) connected by a junction (X). Hermaphrodites aligned in a way to face Hermes to Hermes and Aphrodite to Aphrodite, resulting in the segregated hydro- and fluoro-layers in the crystal packing. The origin of this nano-scale phase separation was discussed from both fluorophobic and fluorophilic interactions. A systematic study of the Hermaphrodites having functional groups with stronger intermolecular interactions revealed the universal nature of the Hermaphrodites for the high potential to the material design.

Introduction

The term crystal engineering was coined by Schmidt about three decades ago (*1*). His seminar work published in 1971 (*2*) gives an idea that too difficult task for the prediction of the crystal packing makes the solid state chemistry as empirical. The overwhelming difficulty has been emphasized in many ways and the simplest abstract composed of 'No' appeared in the well-known journal. (*3*). The famous example is the packing of trimesic acid which totally betray our expectation which is aesthetically driven by the molecular symmetry, but the reality is a complex supramolecular assemblage with an intertwined zigzag catenane-like structure with a very large Z of 48 (space group C2/c) (*4*). Even a simplest shape with perfect symmetry, sphere, packed into fcc or bcc in the pressurized core in the earth is just recently discussed (*5*).

As is stated by Schmidt, the crystal engineering has developed since then by mainly focusing on strong molecular interactions such as ionic or hydrogen bonds. To use such strong intermolecular interactions for the molecular assemblage is quite reasonable because the crystallinity of the concerned materials is essentially important and the substance having only weak intermolecular interactions is generally in poor of crystallinity. With the increasing attention of the supramolecular assemblage, the crystal engineering has been recognized as one of the branch of the supramolecular chemistry. The main player for the packing design has been hydrogen bonds due to a rather strong interaction and at the same time the directionality of the bond. The directionality of the hydrogen bond gives the bases for the life to do many sophisticated orthogonal reactions in a very complex soup of organic substances like a blood or a body fluid where the three dimensional architecture of the hydrogen bond doners and acceptors gives a molecule the ability of the highly sophisticated recognition in a high specific manner such as the bioactive molecule-to-receptor, enzyme-to-substarate, or antigen-to-antibody, the base-pair in DNA. This kind of beauty of recognition through hydrogen bond architecture has fascinated many researchers and driven the scientists toward the crystal engineering.

From the directionality point of view, one more directional bond, a coordinate bond, has been intensively investigated in connection with the supramolecular chemistry because this bond is strong enough to construct a molecular architecture. The latest addition to this category, directional and strong intermolecular interactions, is halogen bonds of which use for the crystal engineering was initiated by Resnati (*6*). This technology has been expanding rapidly into various fields of more general supramolecular chemistry (7). The nature of the halogen bonding is recently reviewed by Resnati as a world parallel to hydrogen bonding (*8*) and by others (*9*). This innovative supramolecular technology based on the halogen bonding is involved elsewhere in this book.

We described the new concept of the fluorine-based crystal engineering in the previous ACS series (*10*) in which how the new concept works is presented by an example of the design of the topochemically controlled solid state polymerization of the fluorine-containing hybrid diyne derivatives. With this successful example in hand, here we would like to shed a light on the phenomena, solid state nano-scale fluorous phase separation, which underlies in the new concept of the fluorine-based crystal engineering. Before entering into the crystal engineering, we will see the present technological circumstances in the next section, where we will confront with the wave of the revolution of X-ray instruments.

The technical revolution of X-ray analysis for the crystal engineering:

The X-ray crystallography is undoubtedly the essential technique for the crystal engineering. The unceasing efforts to develop this technology have shortened the time required for the analysis. The computational aspect is, of course, important but the most recent eye-opening development is found in the data collection method. The data collection time changed from months order in the film age to the minutes-to-hours order in the recent imaging plate (IP) or CCD detector age (*11*), and the latest feat comes in the second order by microstrip gas chamber (MSGC) (*12*). The first analysis by laboratory level equipment with an IP detector appeared in 1996, claiming only one hour data collection time for the standard cytidine crystal (*13*). The MSGC originally proposed in 1988 as a new type of a proportional gaseous detector as a neutron detector was applied for the rapid analysis of X-ray crystal structure. The champion data by this new MSGC detector is only 2 seconds which was obtained on the ammonium bitartrate crystal by using continuous rotation photograph method (*14*). The R value was a little bit large but in an acceptable level of 7.9 ($I>2\sigma(I)$).

As is mentioned above, the X-ray work was once a very time-consuming process, thus this process was apparently in the rate-determining step, hampering the crystal engineering as a practical approach. However, now both X-ray and synthetic processes are comparable in the time scale so that the X-ray results feed back to the design and synthesis of the molecules and the back-and-forth cycles are repeated in a reasonable time scale until the desired goal is attained. The appearance of the high sensitivity detector has one more important meaning in the crystal engineering because the size of the crystals amenable for the X-ray analysis becomes as small as 100 micron order which was not possible before. The advancement on these two factors, rapidness and a crystal size, are very important in a practical sense of crystal engineering. It has been increasingly recognized that the rational design for the functional materials by nano-level control with the supramolecular assemblage *(15)*. In this regards it should be again emphasized here that the time is ripe for the crystal engineering.

Supramolecular assemblage of the Hermaphrodite molecules

We have recently devised new fluorine-based crystal engineering which based on the hydro-fluoro hybrid compounds of which structures are represented by R_H-X-R_F. The hybrid compounds have both natures originated from hydro (R_H) and fluoro (R_F) segments at the same time so that the fluorophobic nature coming from the hydro segment and the fluorophilic nature from the fluoro segment. This kind of dichotomous combination of nature for which we call such hybrid compounds as Hermaphrodites meaning both sexes both anti in one body (Scheme 1).

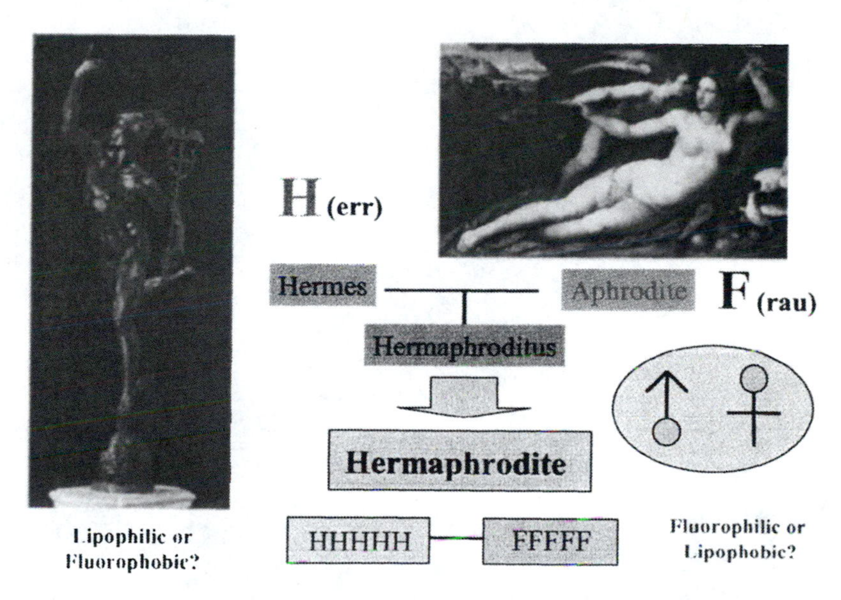

Scheme 1. The structure of the hydro-fluoro hybrid compounds, Hermaphrodites, R_H-X-R_F.

Due to these unique structures of the compounds, Hermaphrodites aligned in the crystal, resulting in the segregation of hydro and fluoro segments like a nano-scale fluorous phase. We have intensively investigated the topology-packing relationships by analyzing X-ray structures of a series of Hermaphrodites and showed the topochemical control of solid phase polymerization. We would like to propose that such a nano-scale fluorous phase formation could be used as a main driving force for the supramolecular alignment in the crystals.

The most essential schemes to explain our idea are recited here for convenience (Scheme 2,3).

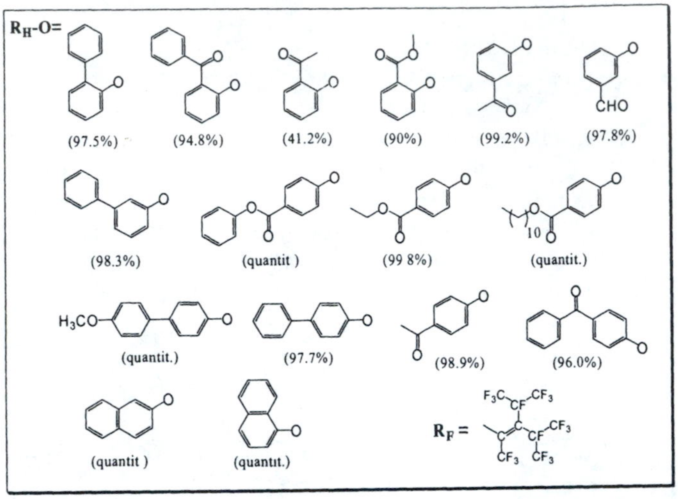

R_H-O=
(97.5%)
(94.8%)
(41.2%)
(90%)
(99.2%)
CHO
(97.8%)
(98.3%)
(quantit)
10
(99 8%)
(quantit.)
H3CO
(quantit.)
(97.7%)
(98.9%)
(96.0%)
(quantit)
(quantit.)
R_F =
F3C
CF3
CF
CF3
CF3
CF3

Scheme 2. Structures of the Hermaphrodies, R_F-O-R_H.

CF3
CF3
F
RH
CF3
F3C
CF3
F
Type 1
Type 2
Type 3
Type 4
Type 5
Perfluoro group
Hydrocarbon group

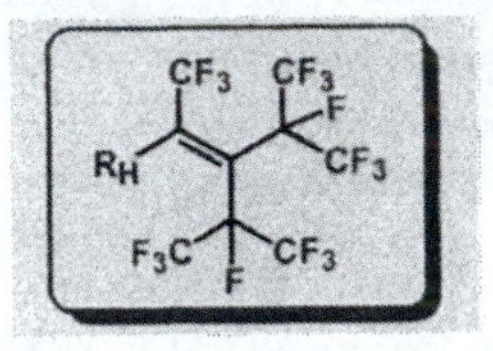

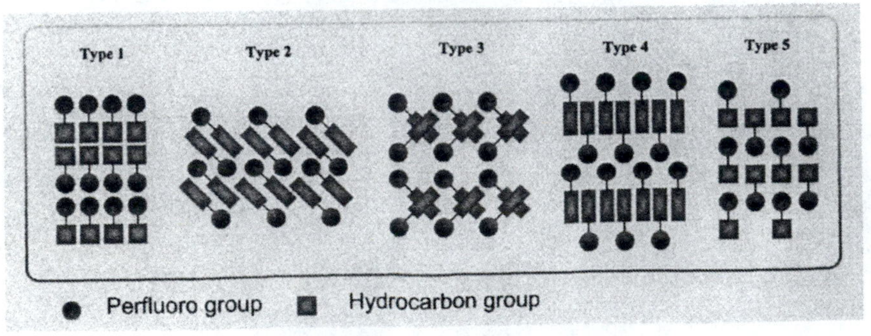

Scheme 3. Packing Motifs found in the Hermaphrodites, R_F-O-R_H, where the Aphrodite part is perfluoropropen trimers and the Hermes parts are described in Scheme 2.

One of the representatives Type 1 packing of the Hermaphrodites is shown below (Fig.1).

Predominant F-F intermolecular interactions in the Hermaphrodite molecules: Fluorophilic Nature is more important?

The intermolecular interaction is undoubtedly a main player for the supramolecular assemblage, and thus, no exception for the Hermaphrodites too. There are two parts in Hermaphrodites, H-segment (Hermes part) and F-segment (Aphrodite part) so that there are three combinations for the intermolecular interactions, two homolytic ones on H-segments and F-segments, and one heterolytic one between H- and F-segments. The terms, lipophilic and fluorophilic, are the terms for between the H-segments and for between the F-segments and both are favoruble for the nano-scale phase separation. These words focus on the philic nature of Hermes and Aphrodite, but there are reverse expressions from the phobia world, fluorophobic and lipophobic, in the heterolytic relations between Hermes and Aphrodite parts. Therefore, there could be a phobic relationship between the Hermes and Aphrodite. However, the electron distributions on the C-H and C-F bonds are polarized in a mirror image, thus, with a positive surface on the Hermes and a negative surface on the Aphrodite, suggesting the philic interctions. Moreover, the recent development of the understanding of the hydrogen bonding has diversified into a variety of unconventional hydrogen bondings, including X-H---F-C type (*16*). Therefore, the polar effect and new type hydrogen bonding suggest the circumstantial affairs between Hermes and Aphrodite, but still controversial and thus set aside at the moment and the conventional rather obscure phobic ideas of fluorophobicity and lipophobicity are taken up as they are. All in all, both philic and phobic ideas work in unison, thus Hermes and Aphrodite separate to create the hydro and fluoro layers, eventually occurring in the nano-scale fluorous phase separation.

Recently, intermolecular F-F interactions are attracting a wide attention because they play increasingly important roles in various ways for the molecular recognition (*17*) and the supramolecular assemblage of the molecules, especially in relation to the perfluoro system. The new crystal engineering developed by Resnati is based on the halogen bonding, but the technological base is in the perfluoro system so that the F-F interactions play an important role accordingly. The recent upsurging demand for understanding the F-F interactions in crystal engineering and also in the fluorous chemistry should be behind the scenes for the publication of the review titled “Fluorine in crystal engineering” (*18*). According to the review, there are 788 compounds with F-F distances within 0.30 nm in the CSD and only 13 of them show F-F contacts of type II which is claimed to be more attractive than type I. Furthermore, only two of them have short F-F contacts with distances smaller than the van der Waals radii (0.294 nm) but appear to be driven by the general packing and not by dispersion forces

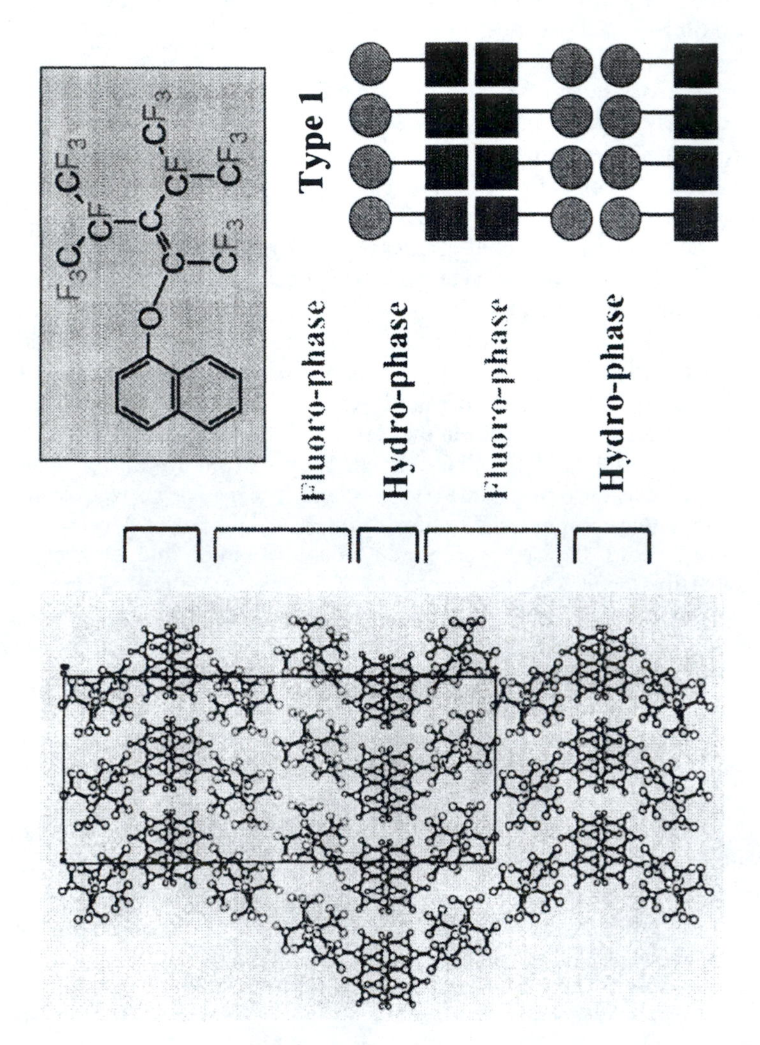

Figure 1. The Type 1 packing of the Hermaphrodites.

between F atoms. In summary, in contrast to the other heavier halogens (*19*), fluorine is more involved in X-H than in X-X interactions. The other recent view is also in this line (*20*).

In our new crystal engineering concept using Hermaphrodites, homolytic contact F-F and H-H are both philic, fluorophilic and lipophilic, and heterolytic contact H-F is phobic, fluorophobic viewed from H-side and lipophobic viewed from F-side, thus why not for nano-scale fluorous phase separation. But still remained is the question "which interactions of H-H and F-F is the main cause for the nano-scale phase separation phenomenon?" The answer may be very simple, both cooperatively, not large difference to be precise. However, in connection to this question we would like to draw your attention to an interesting reverse disorder structure found in the crystal packing of *N*-(4-methoxyphenyl)-*F*-2-(2,6-dimethylmorpholino)- propanamide (MPMPA) which gives a hint for understanding the underlying rules (Fig. 2).

The disorder structure is ubiquitous in the X-ray structures of fluoro compounds, especially notoriously in the trifluoromethyl group which is rarely ordered. However, in contrast with the usual fluoro compounds, no disorder is found in the fluoro segment of MPMPA, but surprisingly in the hydro segment of phenyl group. Thus, the scaffold for the packing of this molecule is apparently in the fluoro segment, but not in the hydro segment. To the most of our knowledge, this is the second example for the reverse order-disorder structure of fluoro compounds (*21*). These two reverse order-disorder structures are consistent with the idea of the predominancy of the F-F intermolecular interactions in the packing of the hybrid compounds, Hermaphrodites, at least in our cases.

But the story may be a little bit more complex. We believe the cumulative effect is the main cause of the segregation. It is very well known about the aggregation of the carbon nanotubes. The carbon nanotubes does not behave one by one, always act in the aggregated form due to a very strong intermolecular forces between the large surface of the carbon nanotubes. If the carbon nanotube molecule has a hydrophilic surface, then it can interact with hydrophilic solvent, thus, disperse into the hydrophilic solvent, but a carbon nanotube molecule has no ability to form such strong interactions with any other molecules so that they have no choice but aggregate each other. Fluoro- and hydro-segregation of the Hermaphrodite molecules is very much the case of this kind. Of course, there is no analogy between the molecular structures of the Hermaphrodites and carbon nanotubes, but the underlying rule for the aggregation of nanotube and segregation of Hermaphrodites are the same. If we select C_{60}, which is also notorious of aggregation, as an example instead of the carbon nanotubes, most people agree with our idea due to the very analogous shapes between the fluoro segment used in the Hermaphrodite and C_{60}, both soccer balls, one fluoro soccer

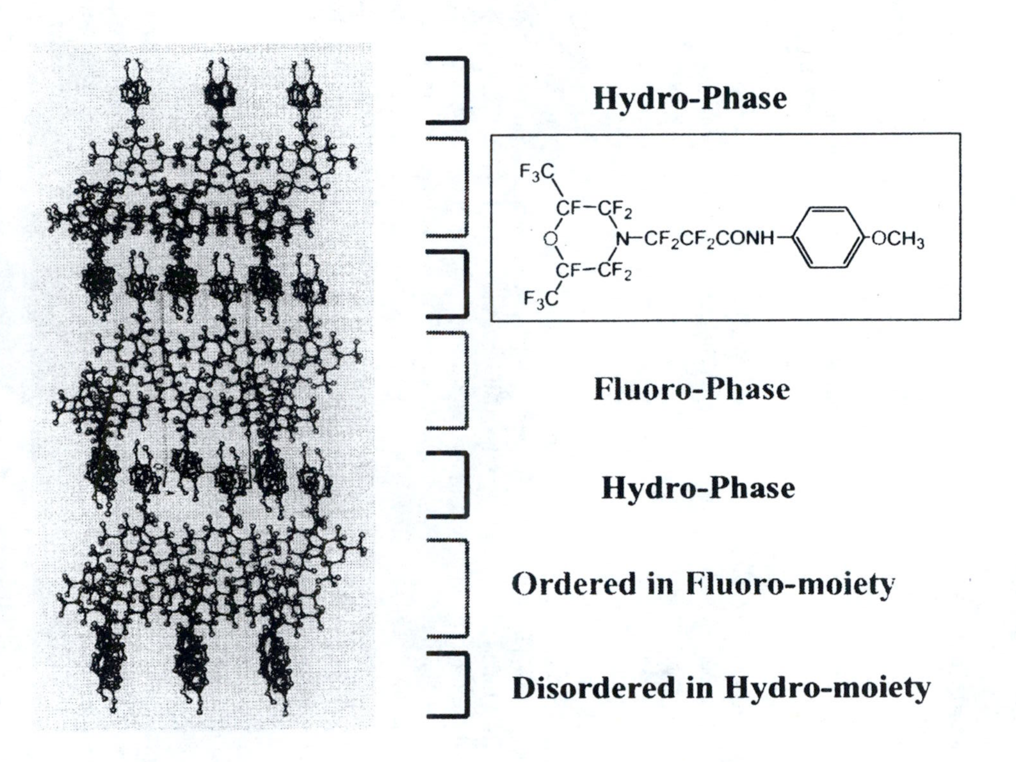

Figure 2. The unusual hydro-disordered fluoro-ordered structure.

ball, the other carbon soccer ball, as is seen in Fig. 3. We will see some evidence for this kind of mass effect of a large surface contact of F-F interactions in the later section of a cluster formation based on the F-F interactions.

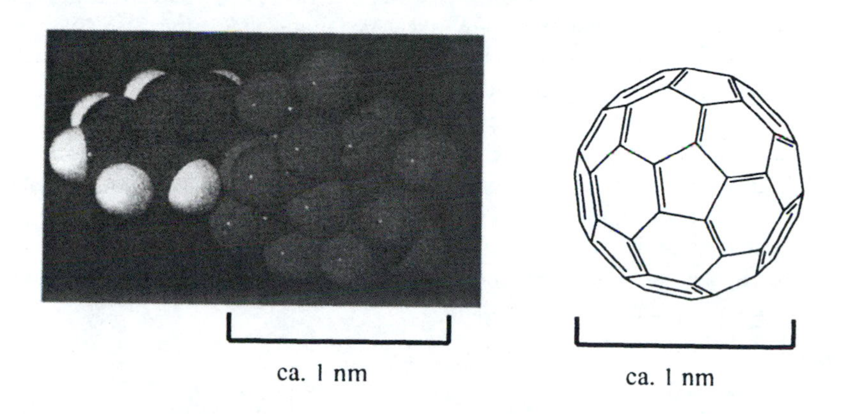

Figure 3. The similarity in shape and dimension between F-part of Hermaphrodite and C_{60}; the former is a fluoro soccer ball and the latter is a hydro soccer ball.

We have discussed on the predominacy of F-F interactions based on the phenomenological reverse order-disorder observations, but finally here presented is a very simple evidence for that the homolytic F-F intermolecular interactions are stronger than the H-H homolytic intermolecular interactions. The boiling point of tetrafluoromethane (b.p. -128.1°C) (*22*) is about 34°C higher than the one of methane (b.p. -162°C), supporting the F-F predominancy. Against such simplicity, here should be pointed out is that the phase separation should be understood as a whole by the combination of enthalpy and entropy terms.

Hermaphrodites with a hydrogen doner-acceptor functional group

We showed a unique packing nature of the Hermaphrodite molecules in the previous sections and claimed its uniqueness comes from the predominant F-F interactions with the emphasis on the analogy to the aggregational nature of the carbon nanotubes and C_{60}. We would like to see how the incorporation of the stronger-bondage-creating group(s) into the Hermaphrodite linkage takes effect on the packing patterns of the Hermaphrodites.

One example is already shown in the previous section as for the explanation of very rare reverse order-disorder structures about the F-segment and H-segment. The nano-scale F- and H-segregations are well preserved in this example in a consistent manner with the hydrogen bonding network. We would like to see if this coexistence, satisfying the demands for the segregation and hydrogen bond net work formation, could be found by the nature in general. Otherwise, we will come across something syrupy or intractable matters for X-ray analysis. We chose an amido group as the linkage between Hermes and Aphrodite because this functional group is very famous in the crystal engineering field to design the packing and at the same time from the curiosity due to the biological importance in the peptide or protein back bones.

The structures investigated are rather diverse in the F-parts and a little bit monotonous in the H-parts by the freak lack of consideration at the outset. Ten compounds, **1a-5a**, **1c**, **6a-6c**, and **7a**, out of the possible 21 compounds by the combination of seven F-parts and three H-parts, were prepared (Scheme 4). All the F-parts come from the perfluoro carboxylic acid fluorides which were prepared by the in-house electrochemical fluorination equipment (*23*). The other H-parts are commons.

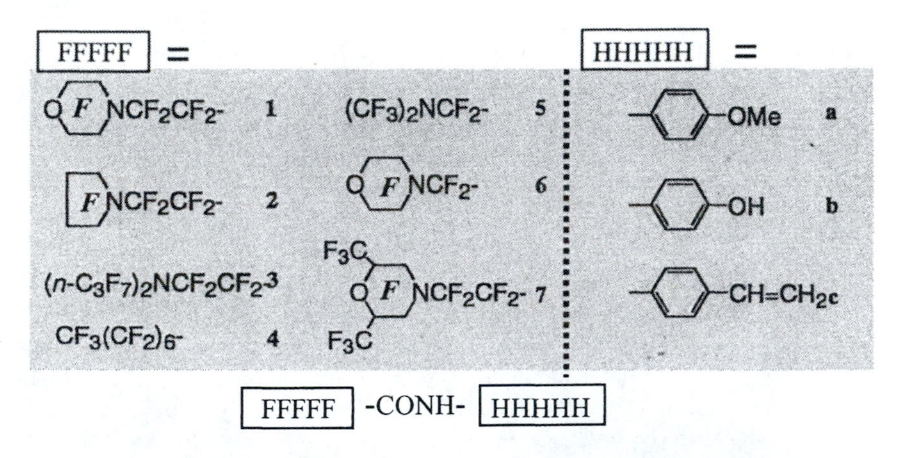

Scheme 4. Hermaphrodite with a polar functional group, –CONH-, having ability to make hydrogen bonding. F- inside the ring denotes the ring is perfluorinated.

Unfortunately all the crystals prepared have very thin needle-shapes. Therefore, the data collection was done by the high energy X-ray beam in SPring 8. Only **7a** gives somewhat fat needle crystals so that the IP detector was used for this crystal.

The crystal packing of **6b** was shown as the representative (Fig. 4). The amide group forms the hydrogen bond network perpendicular to the molecules. The direction of the hydrogen bond net work is along the b axis which is along the needle. One more hydrogen net work forms between the phenolic OH groups, parallel to the amide net work. The molecules are aligned head-to-head perpendicular to both hydrogen net works with the demand for the fluoro- and hydro-segregation structure satisfied. All the packing structures follows the same kind of packing patterns, thus both requirements for the nano-scale fluorous phase separation and the haydrogen bond net work formed by the amide linkage are satisfactorily found by nature in quite a general manner if considering of the diversity of the Hermaphrodite structures.

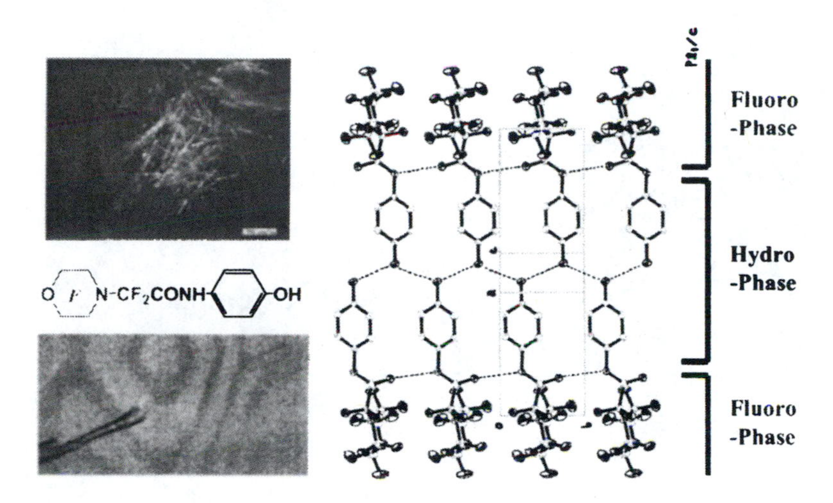

*Figure 4. The crystal packing of the representative Hermaphrodite, **6b**, R_F-CONH-R_H having an amide function as a linkage group.*

It should be noteworthy to point out the morphology of the crystals. It is easily expected that the crystal grows at much a higher rate in the direction along the the hydrogen net work formed by the amide groups than in the directions perpendicular to this hydrogen bond net work because the latter directions are only driven by weak dispersion forces. This is the reason for the needle shape seen for all the Hermaphrodites having the amide linkage. In one extreme case, **6b**, a few centimeters long needle-shape crystals are formed due to the two hydrogen bond net works paralleled each other, both driving the crystal growth in the same direction. This crystal morphology–Hermaphrodite structure relationship will be suggestive for designing the functional materials.

Finally, the crystal packing of a Hermaphrodite molecule which is known as perfluoroalkylating reagent, so-called FITS-8, is shown as an example of the Hermaphrodites having hypervalent cationic iodine function as the linkage part which is expected for stronger intermolecular interactions (Fig. 5). The type 2 packing, but in a reverse manner on the F- and H- parts due to the topological

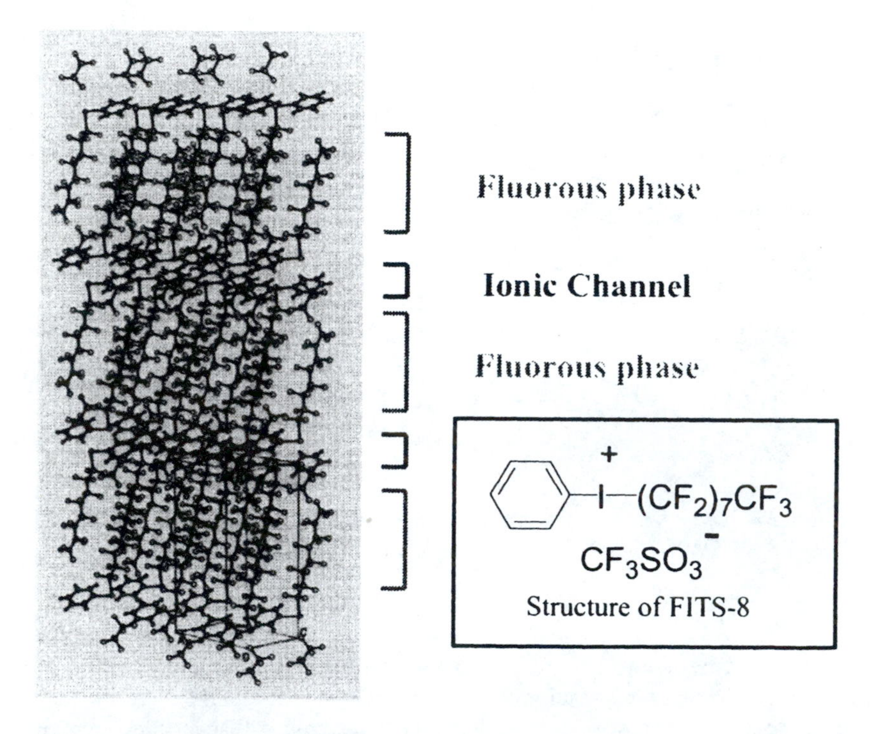

Figure 5. The crystal packing of FITS-8 Hermaphrodite with an ionic linkage.

reverse, the round shape in the H-part and rod shape in the F-part, appeared with the nano-scale fluorous phase separation. Interestingly, a channel like supramolecular structure was formed with the sulfonate anions aligned in the manner to face their fluoro-part toward the F-segregated layers.

Cluster formation based on the F-F intermolecular interactions

The supra-molecular assemblage of the Hermaphrodites was explained by emphasizing the predominance of the F-F intermolecular interactions over the H-H ones in the previous section. We would like to show an additional evidence for the F-F predominacy in the supramolecular cluster formation found in the Resnati's halogen bond based molecular assemblage.

We selected hexamethylenetetramine as the electron donor part because of its beautiful S_4 symmetry which gives some aesthetic intuitions to place the molecules in a space according to their symmetry (Scheme 5). Both are imaginations, but we prefer to the latticed packing (A) because there are voids in the center of the C_3 axis in the imaginative packing (B). The schemes are very much simplified to avoid the complexity but enough for explaining our aesthetic intuitions. If the centro-symmetric electron acceptor molecules with the acceptors in the both ends are placed at the center of the symmetry with the molecular center of symmetry co-centered, the space arrangement of the symmetry elements does not change. Perfluoro compounds, perfluoro-α,ω-diiodoalkylene molecules with C_{2v} symmetry, are exactly the things to satisfy the conditions. Therefore, we examined the system comprising of four fluorinated acceptor homologues, **A1-4**, and hexamethylenetetramine, **D** (Scheme 5). Rocky crystals appear, grow and fall down to the bottom from any places in the solution with beautiful glints immediately after the component molecules are mixed in chloroform for all combinations, **D-A1** to **D-A4**. All crystals are beautifully good enough for the X-ray analysis and analyzed.

The crystal packing of **D-A1** co-crystal (*24*) was exactly what we expected as is seen in Fig 6. We were delighted by the results and then the curiosity was growing to see if this packing maintains up to where. If the higher homologues follow this packing pattern, then, how the nature deals with the void problem because she does not like the void created by the longer rigid *F*-alkylene chains.

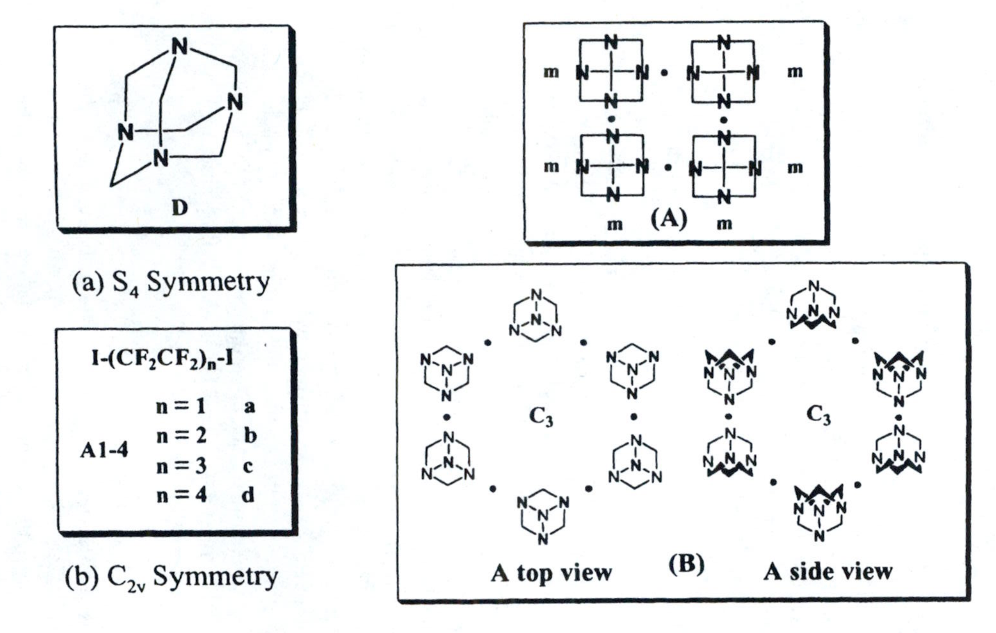

Scheme 5. A priori expectations (A) and (B) for the packing of co-crystals prepared by (a) hexamethylenetetramine and (b) α,ω-diiodo-perfluoroalkanes.

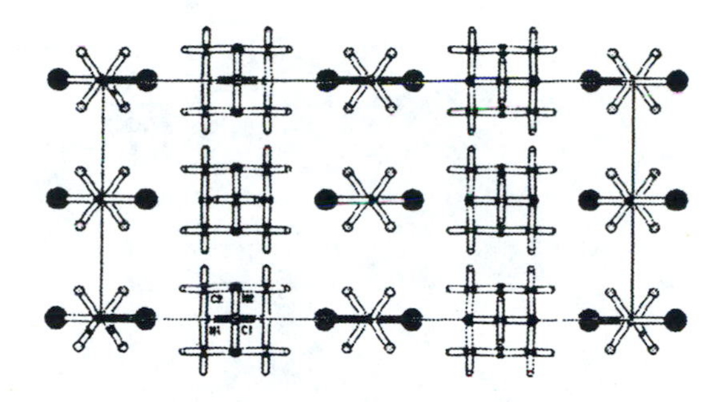

*Figure 6. The crystal packing of **D-A1** co-crystals prepared by hexamethylenetetramine and α, ω-diiodo-perfluoroethane, which was exactly the same as the expected packing (A).*

Soon technical difficulties appeared for the rest of the co-crystals, **D-A2** to **D-A4** (*25*), due partly to very large crysal lattices and due partly to the heavily disordered structures in the fluoro parts. The details will be reported elsewhere. The packing structure of **D-A2** crystal appears to be the supramolecular assemblage of the molecules like a screw with rotations of the acceptor molecules, perfluoro-α,ω-diiodoalkylenes, around the axis along the fluoro columns extending toward the c axis (Fig. 7).

Other **D-A3** and **D-A4** have the same kind of supramolecular clusters of perfluoro-α,ω-diiodoalkylene molecules with $Z > 11$ (Z = formula unit per asymmetric crystal unit). The Z values of three numbers 1/2, 1 and 2 make up 95.3% of all crystal structures (*26*) and there are only 7 reports up to now by the CSD search for the crystal lattice with $Z = 12$ or more than 12, and the highest number of Z is 16, thus these fluoro clusters are quite unusual (*27*). Here again, the analogy to the aggregated form of carbon nanotubes should be emphasized. The F-F interactions are undoubtedly the main player for these cluster formations.

The interesting supramolecular $(CF_3SO_3^-)_6$ cluster has been reported to be held together by twelve F-F interactions with two kinds of close contacts, one with the distance 0.2788 nm and another with 0.2822 nm (*28*).

We believe the fluoro clusters are the results of the F-F intermolecular interactions.

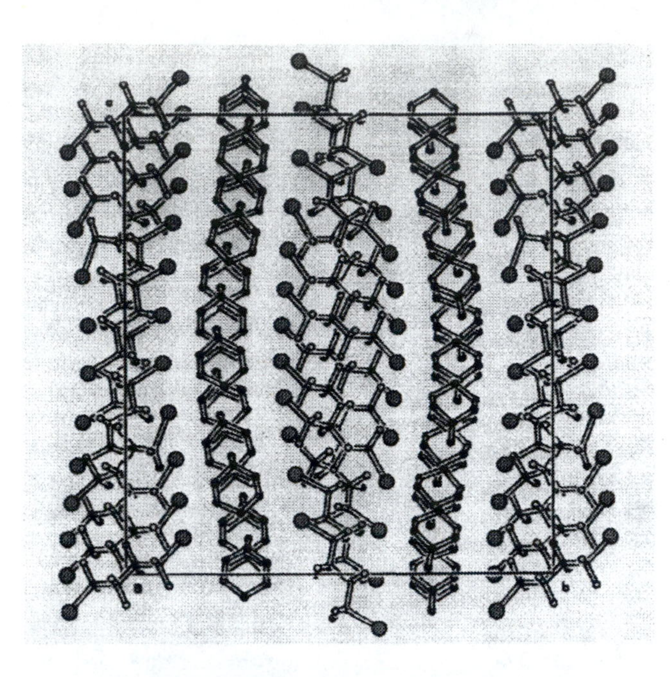

Figure 7. The crystal packing of the D-A2 crystal; a perfluoro layer is composed of the helical cluster formed by F-F intermolecular interactions.

Summary

We proposed new fluorine-based crystal engineering which based on the hydro-fluoro hybrid compounds of which structures are represented by R_H-X-R_F. The hybrid compounds have both natures originated from hydro (R_H) and fluoro (R_F) segments at the same time so that the fluorophobic nature coming from the hydro segment and the fluorophilic nature from the fluoro segment. This kind of dichotomous combination of nature for which we call such hybrid compounds as Hermaphrodites meaning both sexes both anti in one body. Due to this unique structures of the compounds, Hermaphrodites aligned in the crystal with the same kind of segments placed side by side, resulting in the segregation of hydro and fluoro segments like a nano-scale fluorous phase. This packing feature of the Hermaphrodites is universally found in various combinations of fluoro- and hydro-segments and explained by emphasizing on the predominant F-F intermolecular interactions for such fluoro-scaffold. The incorporation of a hydrogen doner-acceptor group as the functional group works in unison with various Hermaphrodite structures, suggesting this new Hermaphrodite technology will be very promising in the design of various functional materials.

Acknowledgement

This work was supported by an inner fund (Shin Houga Fund) from the Ministry of Economy, Trade and Industry. We greatly appreciate V. A. Soloshonok who named Hermaphrodites after Hermaphroditus in Greek mythology.

References

1. Desiraju, G. R.*Crystal Engineering, The Design of Organic Solids,* Material Science Monograph, 54, Elsevier, Amsterdam-Oxford-New York-Tokyo, **1989**; 3-10.
2. Schmidt, G. M. J. *Pure Appl. Chem.* **1971**, 27, 647.
3. Gavezzotti, A. *Acc. Chem. Res.* **1994**, *27*, 309-314.
4. Duchamp, D. J.; Marsh, R. E. *Acta Cryst.* **1969**, *B25*, 5-19.
5. Vocadlo, L.; Alfe, D.; Gillan, M. J.; Wood, I. G.; Brodholt, J. P.; Price, D. *Nature* **2003**, *424*, 536-539.
6. a) Casnati, A.; Liantonio, R.; Metrangolo, P.; Resnati, G.; Ungaro, R.; Ugozzoli, F. *Angew. Chem. Int. Ed.* **2006**, *45*, 1915-1918. b) Liantonio, R.; Metrangolo, P.; Meyer, F.; Pilati, T.; Navarrini, W.; Resnati, G *Chem. Commun.* **2006**, in print c) Boubekeur, K.; Syssa-Magale, J.-L.; Palvadeau, P.; Schollhorn, B. *Tetrahedron Lett.* **2006**, *47*, 1249-1252. d) Bertani, R.; Ghedini, E.; Gleria, M.; Liantonio, R.; Marras, G.; Metrangolo, P.; Meyer, F.; Pilati, T.; Resnati, G. *Cryst. Eng. Commun.* **2005**, *83*, 511-513. e) Neukirch, H.; Guido, E.; Liantonio, R.; Metrangolo, P.; Pilati, T.; Resnati, G. *Chem. Commun.* **2005**, 1534-1536. f) Mele, A.; Metrangolo, P.; Neukirch, H.; Pilati, T.; Resnati, G *J. Am. Chem. Soc.* **2005**, *127*, 14972-14973. g) Guido, E.; Metrangolo, P.; Panzeri, W.; Pilati, T.; Resnati, G.; Ursini, M.; Logothetis, T. *J. Fluorine Chem.* **2005**, *126*, 197-207. h) Goroff, N. S.; Curtis, S. M.; Webb, J. A.; Fowler, F. W.; Lauher, J. W. *Organic Letters* **2005**, *7*, 1891-1893. i) Zhu, S.; Jiang, H.; Zhao, J.; Li, Z. *Crystal Growth & Design* **2005**, *5*, 1675-1677. j) Lucassen, A. C. B.; Vartanian, M.; Leitus, G.; van der Boom, M. E. *Crystal Growth & Design* **2005**, *5*, 1671-1673. k) Takeuchi, T.; Minato, Y.; Takase, M.; Shinmori, H. *Tetrahedron Lett.* **2005**, *46*, 9025-9027. l) Xu, J.; Liu, X.; Lin, T.; Huang, J.; He, C. *Macromolecules* **2005**, 38, 3554-3557. m) Caronna, T.; Liantonio, R.; Logothetis, T. A.; Metrangolo, P.; Pilati, T.; Resnati, G. *J. Am. Chem. Soc.* **2004**, *126*, 4500-4501. n) Forni, A.; Metrangolo, P.; Pilati, T.; Resnati, G. *Crystal Growth & Design* **2004**, *4*, 291-295. o) Zhu, S.; Xing, C.; Xu. W.; Jin, G.; Li, Z. *Crystal Growth & Design* **2004**, *4*, 53-56. p) Nguyen, H. L.; Horton, P. N.; Hursthouse, M. B.; Legon, A. C.; Bruce, D. W. *J. Am. Chem. Soc.* **2004**, *126*, 16-17. q) Berski, S.; Ciunik, Z.; Drabent, K.; Latajka, Z.; Panek, J. *J. Phys. Chem.* **2004**, *108*, 12327-12332.; *For the theoretical:* r)

Zou, J.-W.; Jiang, Y.-J.; Guo, M.; Hu, G.-X.; Zhang, B.; Liu, H.-C.; Yu, Q.-S. *Chem. Eur. J.* **2005**, *11*, 740-751. s) Zordan, F.; Brammer, L.; Sherwood, P. *J. Am. Chem. Soc.* **2005**, *127*, 5979-5989. t) Wang, W.; Wong, N.-B.; Zheng, W.; Tian, A. *J. Phys. Chem. A* **2004**, *108*, 1799-1805. u) Glaser, R.; Chen, N.; Wu, H.; Knotts, N.; Kaupp, M. *J. Am. Chem. Soc.* **2004**, *126*, 4412-4419.

7. a) Espallargas, G. M.; Brammer, L.; Sherwood, P. *Angew. Chem. Int. Ed.* **2006**, *45*, 435-440. b) Jiang, Y.; Alcaraz, A. A.; Chen, J.-M.; Kobayashi, H.; Lu, Y. J.; Snyder, J. P. *J. Med. Chem.* **2006**, *49*, 1891-1899. c) Mugnaini, V.; Punta, C.; Liantonio, R.; Metrangelo, P.; Recupero, F.; Resnati, G.; Pedulli, G. F.; Lucarini, M. *Tetrahedron Lett.* **2006**, in print. d) Chandrasekhar, V.; Baskar, V.; Gopal, K.; Vittal, J. J. *Organometallics* **2005**, *24*, 4926-4932.
8. Metrangolo, P.; Neukirch, H.; Pilati, T.; Resnati, G. *Acc. Chem. Res.* **2005**, *38*, 386-395.
9. Fourmigue, M.; Batail, P. *Chem. Rev.* **2004**, *104*, 5379-5418.
10. Ono, T.; Hayakawa, Y. "*Fluorine-Based Crystal Engineering*" in *Fluorine-Containing Synthons*, ACS Symposium Series #911, Soloshonok, V. A. Ed., Oxford University Press/American Chemical Society, Washington, D. C., **2005**, Chapter 29, 498-513.
11. a) Ozawa, Y.; Terashima, M.; Mitsumi, M.; Toriumi, K.; Yasuda, N.; Uekusa, H. *Chem. Lett.* **2003**, 32, 62-63. b) Yasuda, N.; Uekusa, H.; Ohashi, Y. *Bull. Chem. Soc. Jpn.* **2004**, 77, 933-944.
12. Takeda, A.; Uekusa, H.; Kubo, H.; Miuchi, K.; Nagayoshi, T.; Ohashi, Y.; Okada, Y.; Orito, R.; Takada, A.; Tanimori, T. *J. Synchrotron Radiation* **2005**, *12*, 820-825.
13. Ohashi, Y.; Uekusa, H. *J. Molecular Strct.* **1996**, *374*, 37-42.
14. a) Ochi, A.; Uekusa, H.; Tanimori, T.; Ohashi, Y.; Toyokawa, H.; Nishi, Y.; Nishi, Y.;Nagayoshi, T.; Koishi, S. *Nucl. Instr. & Methods in Phys. Res., A: Accelerators, Spectrometers, Detectors, and Associated Equipment* **2001**, *467-468*, 1148-1151. b) Ochi, A.; Tanimori, T.; Nishi, Y.; Nishi, Y.; Nagayoshi, T.; Ohashi, Y.; Uekusa, H.; Toyokawa, H. *Nucl. Instr. & Methods in Phys. Res., A: Accelerators, Spectrometers, Detectors, and Associated Equipment* **2002**, *477*, 48-54.
15. a) *Frontiers in Crystal Engineering*, Tiekink, E. R. T.; Vittal, J. J. Eds,., John Wiley & Sons Inc., **2006**. b) *Crystal Engineering* in *Supramolecular Chemistry*, Steed, J. W.; Atwood, J. L. Eds., John Wiley & Sons, Ltd, Chichester/New York/Weinheim/Brisbane/Singapore/Toronto, **2000**., Chapter 6, 389-463. c) *Crystal Engineering: The Design and Application of Functional Solids*, Nato ASI Series, Series C: Mathematical and Physical Sciences, Vol.539, Seddon, K. R.; Zaworotko, M. Eds., Kluwer Academic Publishers, Dordrecht/Boston/London, **1999**. d) *Solid-State Supramolecular Chemistry: Crystal Engineering,* Mac.Nicol, D. D.; Toda, F.; Bishop, R., Eds. in *Comprehensive Supramolecular Chemistry*., Vol. 6, Atwood, J. L.;

Davies, J. E. D.; MacNicol, D. D.; Voegtle, F.; Lehn, J.-M. Eds., Pergamon, **1996** e) Dunitz, J. D. *Crystals as Supermolecules* in *Perspectives in Supramolecular Chemistry*, in *The Crystal as a Supramolecular Entity*, Vol. 2, Desiraaju, G. R. Ed., Wiley:Chichester, **1995**.

16. Belkova, N. V.; Shubina, E. S.; Epstein, L. M. *Acc. Chem. Res.* **2005**, *38*, 624-631.
17. a) Soloshonok, V. A. *Angewa. Chem. Int. Ed.* **2006**, *45*, 766-769. b) Yeston, J. S. *Science* **2006**, *311*, 16-17. For this paper, phobic idea seems to be prevailing.
18. Reichenbacher, K.; Suss, H. I.; Hulliger, *J. Chem. Soc. Rev.* **2005**, *34*, 22-30.
19. Sarma, J. A. R. P.; Desiraju, G. R. *Acc. Chem. Res.* **1986**, *19*, 222-228.
20. Zaman, M. B.; Udachin, K. A.; Ripmeester, J. A. *Crystal Growth & Design* **2004**, *4*, 585-589.
21. the previous one is also found in our hybrid compounds. See reference 10.
22. Smart, B. E.; Fernandez, R. E., *Fluorinated Aliphatic Compounds in Fluorine Chemistry: A Comprehensive Treatment*, Encyclopedia Reprint Series, Kroschwitz, J. I.; Howe-Grant, M.; Humphreys, L.; Punzo, C.; Altieri, L. Eds., John Wiley & Sons, New York/Chichester/Brisbane/ Toronto/Singapore, **1995**, 259-281.
23. Abe, T.; Hayashi, E.; Fukaya, H.; Hayakawa, Y.; Baba, H. *J.Fluorine Chem.* **1992**, *57*, 101-111.
24. **D-A1**: Cmcm, a = 9.4104(5), b = 7.0685(2), c = 20.8110(6), V = 1384.29(9), Z = 4
25. **D-A2**: P-1, a = 14.5178(7), b = 24.2689(11), c = 26.1782(10), α = 90.1893(12), β = 93.6602(15), γ = 90.083(2), V = 9204.5(7), Z = 22; **D-A3**: $P2_1/c$, a = 14.522(3), b = 28.825(6), c = 28.610(6), 103.71(3), V = 11635(4). Z = 24; **D-A4**: P-1, a = 14.6629(12), b = 30.131(3), c = 34.166(3), α = 92.9553(19), β = 99.098(6), γ = 95.470(4), V = 14803(2), Z = 23.
26. Steiner, T. *Acta Cryst.* **2000**, *B56*, 673-676.
27. Frequency of Z (formula units per asymmetric crystal unit): For Z = 16: a) Bruckner, S. *Acta Crystallogr.* **1982**, *B38*, 2405-2408. b) Hsu, L.-Y.; Nordman, C. E. *Science* **1983**, 220, 604-606. c) Bernal, J. D. *Philos. Trans. R. Soc. London*, **1940**, *A239*, 135. ; For Z = 15: d) Greiser, T.; Weiss, E. *Chem. Ber.* **1977**, *110*, 3388-3396.; For Z = 12: e) Bars-Combe, M. L.; Lajzerowicz, J. *Acta Crystallogr.* **1987**, *B43*, 386-393. f) Wingerter, S.; Pfeiffer, M.; Baier, F.; Stey, T.; Stalke, D. *Z. Anorg. Allg. Chem.* **2000**, *626*, 1121. g) Herbstein, F. H.; Marsh, R. E. *Acta Crystallogr.* **1977**, *B33*, 2358-2367.
28. Sun, C.-Y.; Kang, B.-S.; Wang, Q.-G.; Mak, T. C. W. *J. Chem. Soc., Dalton Trans.* **2000**, 1831-1833.

Chapter 11

Fluorous Nanoflow Microreactor: Nanoflow Microreactor with Fluorous Lanthanide Catalysts for Increase in Reactivity and Selectivity

Koichi Mikami and Takayuki Tonoi

Department of Applied Chemistry, Tokyo Institute of Technology, Meguro-ku, Tokyo 152–8552, Japan

The regioselectivity and reaction rate of the fluorous lanthanide complex-catalyzed Mukaiyama aldol and Baeyer-Villiger reactions were significantly increased using our nanoflow microreactor. A moderate level of kinetic resolution was also observed even in the lowest concentration of the fluorous lanthanide catalysts (catalyst-to-substrate (C/S) molar ratio is less than 1:1000 ($\ll$0.1 mol%)).

Introduction

Fluorous phase is the third phase orthogonal to aqueous and organic phases. Fluorous perfluoroalkane solvents are highly hydrophobic and their misibility to organic media is critically temperature dependent. The weaker van der Waals interaction results in the lower boiling points of perfluoroalkanes and their low miscibility with organic solvents. Fluorous biphase catalysis (FBC) (*1*) is based on this low miscibility with conventional organic solvents. Molecules can be rendered fluorous by attaching one to several perfluoroalkyl groups, which are referred to as fluorous ponytails (*2*). When fluorous ponytails are introduced in sufficient length and/or quantity, fluorous phase affinities (fluorophilicities) of fluorous catalysts can be increased. The electron-withdrawing properties of the perfluoroalkyl ponytails have a decreasing effect on the catalyst activity in comparison with the non-perfluoroalkyl catalysts. This is the reason why hydrocarbon spacers are introduced on ligands to minimize the electron-withdrawing effect of the fluorous ponytails on the late transition metal catalysts. We have, however, developed the lanthanide (III) bis(perfluoroalkanesulfonyl)amide and -methide complexes without any hydrocarbon spacer (*3,4,5*) rather than the late transition metal catalysts. The fluorous lanthanide catalysts are therefore super Lewis acidic (*6*) and soluble in both aromatic and aliphatic fluorocarbons. Fluoroaromatic solvents such as benzotrifluoride (BTF) are hybrid fluorous/organic solvents and miscible even with non-fluorous solvents (*7*).

Fluorous Nanoflow Microreactor

The miniaturisation of chemical analysis device using the micro total analysis system (μ-TAS) has afforded a new system in biochemical field (*8*). Recently, a miniaturized reaction system, namely a microreactor, has been attracting much attention of synthetic chemists (*9*). A microreactor is a microstructured chemical reactor, which is often fabricated from glass or silicon and attached with a series of microchannels. By virtue of the minimal structure, a microreactor has advantageous features; a high surface to volume ratio, small but detectable amounts of chemical products, strict fluidic control (sometime based on laminar flow), high heat and mass transfer rates, short residence times, and much improved safety even with explosive reagents, such as H_2O_2 in higher concentrations. For these reasons, various types of microreactors have been investigated.

These inherent benefits of microreactors can be applied in organic synthesis (*10*). Strict control of the flow rates of the reaction media is the key to exploit the fully integrated microreactors and thereby facilitate highly selective reactions that are difficult to achieve in conventional batch systems. However, usual flow rate in the recent micro flow pumping systems are still in the range of 10 – 500 μL/min (*9*). The development of nano-scale flow pumping system ("nanoflow system") has thus been required for further downsizing of the microreactors by three orders from microorder to nanoorder. Electro-osmotic flow (EOF) and hydrodynamic pumping technique are sometimes employed in microreactors (*11*). In EOF, however, the flow rate is proportional to the polarity of the solvent. Therefore, non-polar fluorous solvents cannot be used. By contrast, our nanoflow system can be employed even for the non-polar fluorous solvents by virtue of strict fluidic control in nano-ordered level. Nanofeeder *DiNaS* (*Direct Nanoflow System*) can now be delivered from KYA TECH Corp. as a high pressure (up to 20.0 MPa) syringe delivery system controlling the tunable flow rate of solution from 1 nL/min to 2 x 10^5 nL/min (*12*). The borosilicate microreactors are fabricated by a standard procedure (*13*) and supplied from Fuji Electric Co.

The total experimental system of nanoflow microreactor is illustrated in Scheme 1. The reaction path dimensions are 1, 2, and 3 cm lengths x 30 μm depth x 60 μm width. The experiments were also examined in a narrow (half-sized) cell of 30 μm width (1 and 3 cm lengths x 30 μm depth). Fick's second law of diffusion shows that the time of transportation T is reduced by shortening the width in the diffusion equation: $T \sim L^2/D$ (L: width, D: diffusion coefficient) (*14*). With decreasing cell dimensions, the surface area to volume ratio increases (*15*). Due to small dimensions and internal volumes, the reaction rate significantly increases.

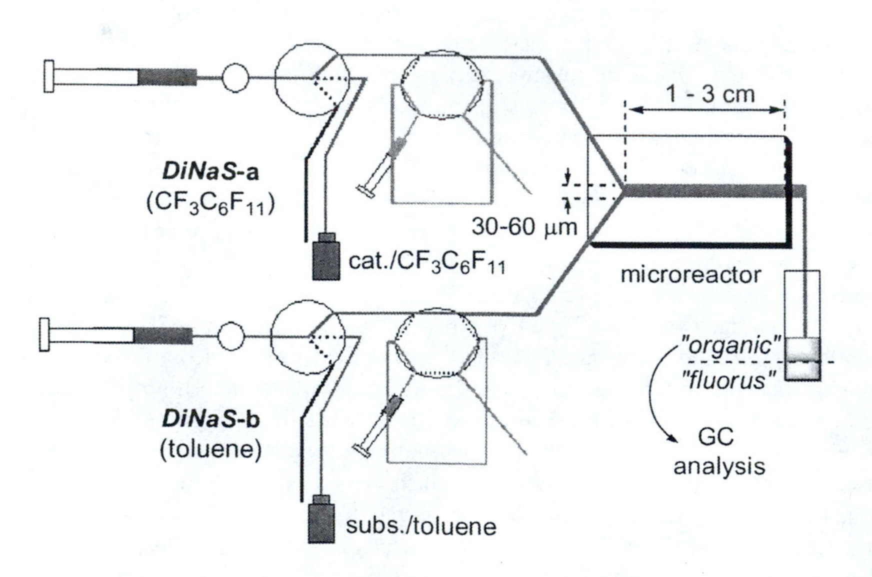

Scheme 1. The total experimental system of nanoflow microreactor.

The nature of fluid flow is of great influence on the mixing in the cell. In turbulent flow, a fluid portion does not stay within a certain layer but moves in a random manner across the flow path.. Under laminar flow conditions, molecular diffusion is the only mechanism for exchange of molecules.

Due to low flow rates (25-200 nL/min) and small characteristic dimensions, laminar flow conditions are sometimes encountered. Figure 1 exemplifies the introduction of two solutions into a microreactor at constant flow rates. Through two inlets of the nanofeeders, the substrates in hydrocarbon solvent and the fluorous catalyst in fluorocarbon solvent were introduced independently. Diffusion between the adjacent domains leads to a molecular mixing and larger interfacial areas lead to enhance mass and heat transfer, promotion of chemical reactions (*16*).

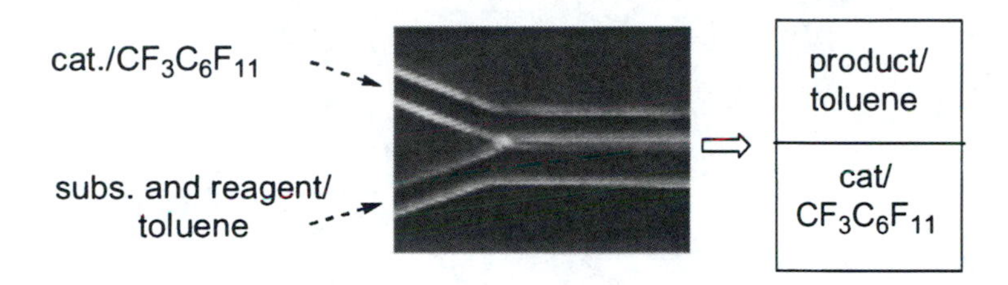

Figure 1. Biphasic laminar flow in a microreactor.

The Mukaiyama aldol reaction

The Mukaiyama aldol reaction is the most synthetically important carbon-carbon bond forming (CCF) reaction (*17*). Therefore, a wide variety of Lewis acid catalysts (*18*) have been developed for the CCF reaction involving storable silyl enol ether in batch system (round-bottomed flask) using 1-10 mol% of a catalyst for several hours under the Lewis acid catalysis usually in polar aprotic solvents. FBC has attracted a great deal of attention in view of environmentally friendly chemical process, however, only in batch systems, and been overlooked in microreactors. We have thus reported the great increase in reactivity of the Mukaiyama aldol reaction in the fluorous media in the microfabricated device controlled by *DiNaS* ("fluorous nanoflow" microreactor in Scheme 1) (*12*). Even by the lowest catalyst-to-substrate (C/S) molar ratio (C/S = 6.25:10000 (6.25×10^{-2} mol%)) in the non-polar fluorous solvents, the reaction completes within seconds as a biphasic residential time. Particularly, lanthanide complexes (*19*) with fluorous bis(perfluoroalkanesulfonyl)amide ponytails (*e.g.*

$Sc[N(SO_2C_8F_{17})_2]_3$) effectively form immobilized Lewis acid catalysts in the fluorous phase. Because these fluorous catalysts are virtually insoluble in usual hydrocarbon solvents at ambient temperature and soluble in fluorocarbon solvents. The significant increase in reactivity of the Mukaiyama aldol reaction is thus attained by the fluorous nanoflow microreactor (Scheme 2).

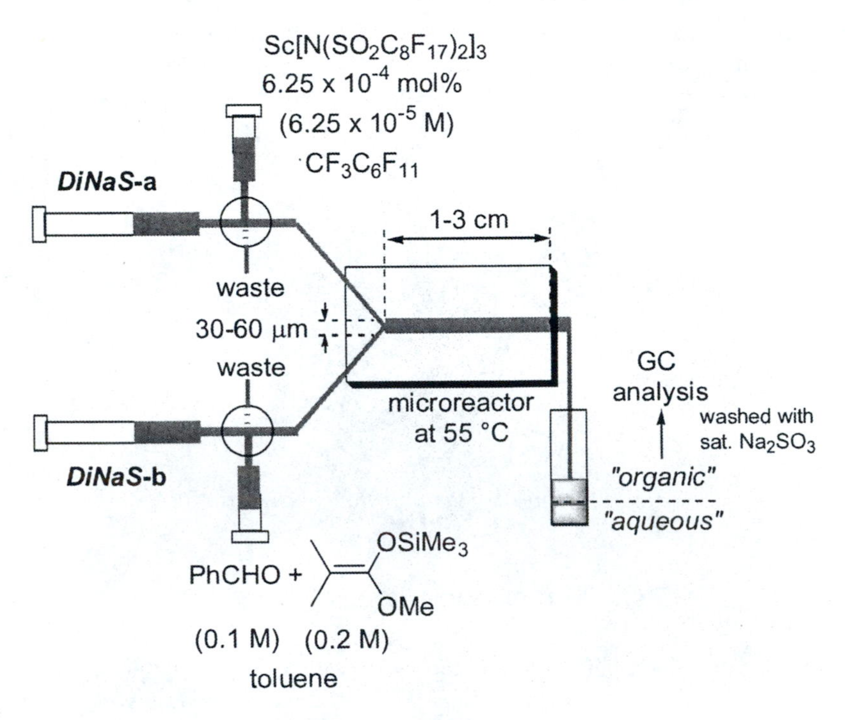

Scheme 2. Mukaiyama aldol reaction.

The concentration of $Sc[N(SO_2C_8F_{17})_2]_3$ in fluorous solvent could be controlled to be low enough (6.25×10^{-5} M) and therefore catalyst economical. The flow rates were continuously controlled by 25-200 nL/min; the length of the cell was 1, 2 and 3 cm along with 30 and 60 μm widths, and 30 μm width in different experiments. The Mukaiyama aldol reactions were thus examined for 5.4-43.2 seconds as residential times in the microreactor.

Significantly, high yields were obtained, in spite of the very short reaction time and the lowest C/S ratio, as calculated by GC analysis on the basis of the calibration curves for silylated and de-silylated aldol products using *n*-decane as an internal standard. Then, the heterogeneous two phases were separated into the upper toluene and the lower fluorous perfluomethylcyclohexane layers. The lanthanide catalyst could be recovered and reused in the fluorous phase without isolation.

Significantly, the aldol reaction of benzaldehyde (0.1 M) with methyl 2-methylpropanoate-derived trimethylsilyl enol ether (0.2 M) was completed within seconds even with only << 0.1 mol% of $Sc[N(SO_2C_8F_{17})_2]_3$ in perfluoromethylcyclohexane. The aldol products were obtained in up to quantitative yield as the silylated and de-silylated aldol forms (Table 1). We then examined macro-scale experiment using normal batch (round-bottomed flask) system for comparison. The product was obtained in only 11% yield after vigorous stirring at 55 °C for 2 hours using the same C/S ratio. In sharp contrast, virtually quantitative yield was obtained within seconds of residential time in the nanoflow microreactor system.

Table 1. The Mukaiyama aldol reaction in the nanoflow system.[a]

flow rate of both layers (nL/min)	residential time in micro cell (sec)	micro cell length (cm)	% combined yield[b]
200	16.2	3	88
200	10.8	2	67
200	5.4	1	50
100	10.8	1	76
50	43.2	2	94
50	21.6	1	82
25	43.2	1	97

[a] PhCHO (0.1 M), trimethysilyl ketene acetal (0.2 M) and cat. (6.25×10^{-5} M) was used [b] GC yield of the silylated and de-silylated aldol products

We also investigated our experiments in a half-sized microreactor of 30 μm width in order to examine the effect of widths of microreactors in separate experiments (Table 2). The reaction rates were significantly increased in the narrower cell as the surface area to volume ratio increases with decreasing cell dimension.

We then demonstrated the effect of the lengths of the cell (Figure 2) and the residential time (Figure 3) on the product % yields. In nanoflow microreactors with longer cell length and biphasic residential time, higher chemical yields of the aldol products were obtained in up to quantitative yield.

Table 2. The comparative yields in the nanoflow systems.[a]

flow rate of both layers (nL/min)	residential time in micro cell (sec)	micro cell width (μm)	% combined yield[b]
200	5.4	60	50
100	5.4	30	71
100	10.8	60	76
50	10.8	30	92

[a] PhCHO (0.1 M), trimethysilyl ketene acetal (0.2 M) and cat. (6.25x 10^{-5} M) was used [b] GC yield

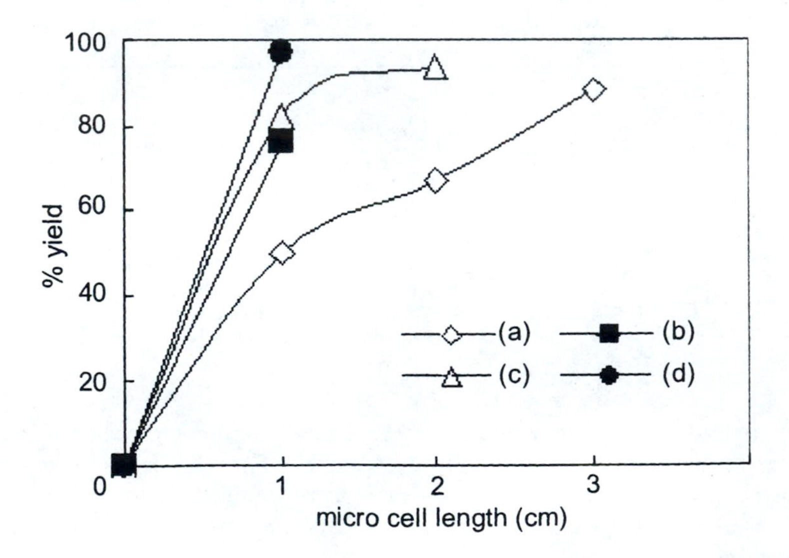

Figure 2. The effect of the lengths on reaction rate of the microreactor. Flow rates for both layers: (a) 200, (b) 100, (c) 50, and (d) 25 nL/min

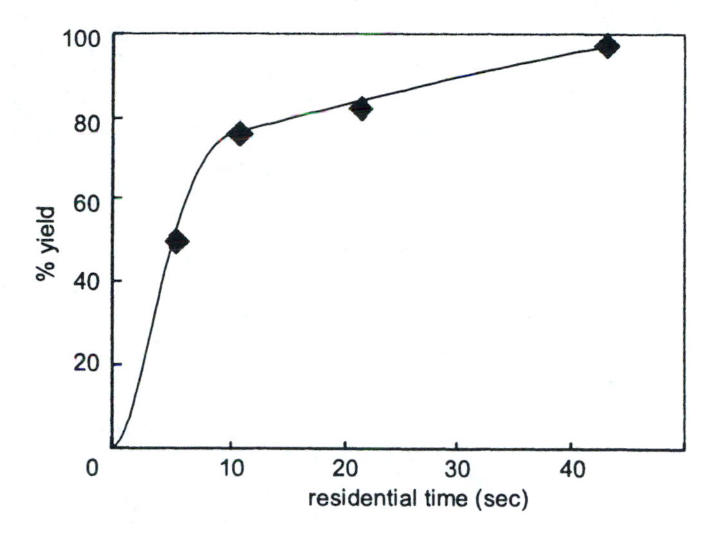

Figure 3. The effect of the residential time in the microreactor.

The reaction rate was significantly increased in narrower cell, although the residential time was the same as in the previous wider cell (Figure 4). These results indicate that in the narrower width and longer length of the microreactor, the rate of the Mukaiyama aldol reaction is significantly increased by the fluorous nanoflow microreactor.

The Baeyer-Villiger (B-V) reaction

The Baeyer-Villiger (B-V) reaction has been widely employed in organic synthesis (*20*). Among various oxidants, hydrogen peroxide is an environmentally friendly oxidant to waste only water. This environmental advantage can make hydrogen peroxide attractive for industrial use. Hydrogen peroxide has, in turn, disadvantageous features: less reactive, potentially exprosive in higher concentrations, formation of the regioisomeric mixtures particularly from cyclopentenone substrates, and hydrolysis of the ester or lactone products (*21*). To overcome these disadvantages, a wide variety of metal catalysts have been examined for the B-V oxidation (*20*). We have reported the great increase not only in the reaction rate but also in the regioselectivity of the B-V oxidation with hydrogen peroxide even in the lowest C/S ratio (*22*) (5:10000 (5 x 10^{-2} mol%)) in our nanoflow microreactor (*23*). The fluorous lanthanide catalyst can be fully employed in the highly regioselective B-V oxidation of cyclopentanones through the microfabricated flow device strictly controlled by nano feeder *DiNaS* (Scheme 3).

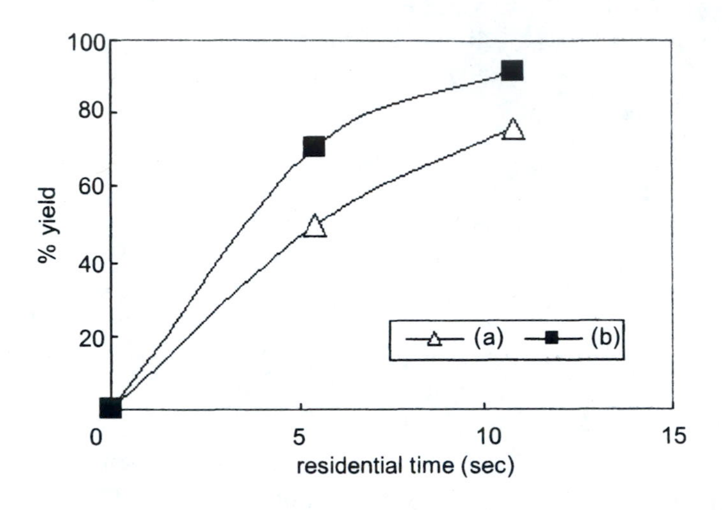

Figure 4. The effect of the width of the microreactor. For widths: (a) 60, and (b) 30 μm

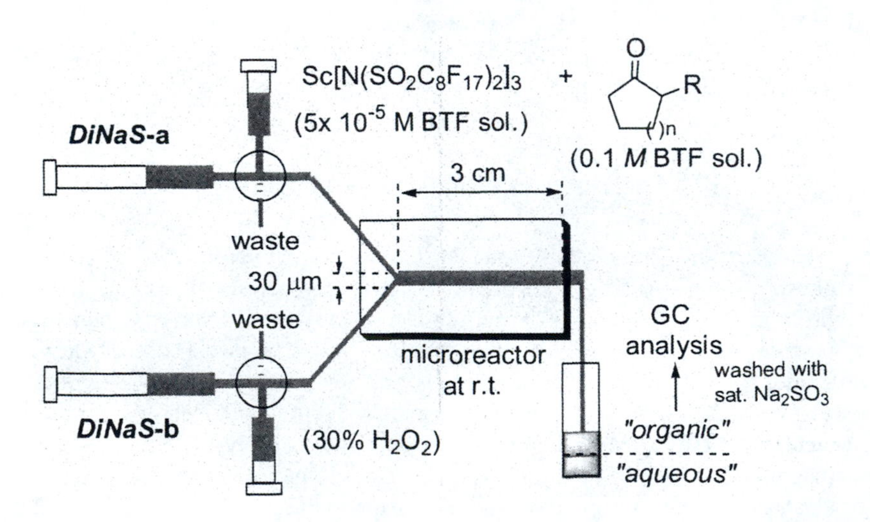

Scheme 3. Baeyer-Villiger reaction.

We conducted the B-V experiments in a narrow (half-sized) cell of 30 μm width (3 cm length x 30 μm depth). A batch system was preliminary examined for the B-V oxidation of 2-methylcyclopentanone in $CF_3C_6F_{11}$ biphasic system with 30% H_2O_2 but no product was observed even after 3 hours stirring. By replacing $CF_3C_6F_{11}$ with BTF as a fluorous-organic hybrid solvent, the lactone product was formed though in only low regioselectivity (67:33). When a substrate and $Sc[N(SO_2C_8F_{17})_2]_3$ in BTF and 30% H_2O_2 were separately introduced through two micro-inlets (Scheme 3). Significantly the B-V oxidation reactions of cyclic ketones were completed within seconds at room temperature even in the lowest C/S ratio (5:10000 (5 x 10^{-2} mol%)). The flow rate was continuously controlled by 100 nL/min for both phases and thus the reactions were examined for 8.1 seconds as biphasic residential time through 30 μm x 30 μm x 3 cm microreactor. High regioselectivity of the product as well as high chemical yield were obtained in contrast to lower yield and regioselectivity obtained in the batch system (Table 3). It should be noted here that the complete regioselectivity was achieved along with quantitative yield by the nanoflow microreactor even from cyclopentanone substrates.

Table 3. The B-V reaction in the nanoflow and batch systems.[a]

substrate	product	residential time (sec)	%yield	regioselectivity
O (2-methylcyclopentanone)	O, O : O, O	8.1	99	97:3
		(5 h	53	67:33)[b]
O, C_5H_{11}	O, O : O, O, C_5H_{11}, C_5H_{11}	8.1	92	99:1
		(5 h	55	70:30)[b]
O (2-methylcyclohexanone)	O, O	8.1	91	100:0
		(5 h	28	69:31)[b]
O, Ph	O, O, Ph	8.1	74	100:0
		(5 h	17	100:0)[b]

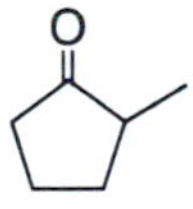

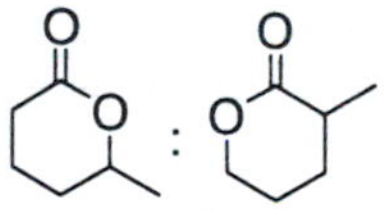

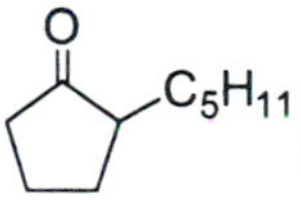

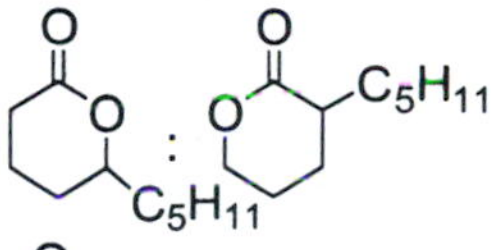

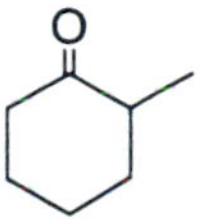

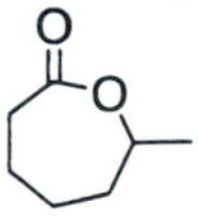

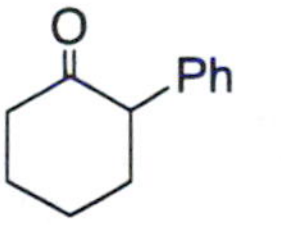

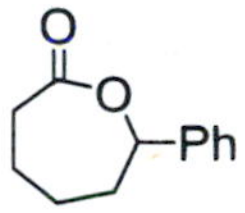

[a] Substrates (0.1 M), 30% H_2O_2 aqueous solution and a catalyst (5 x 10^{-2} mol%) in BTF were used in the nanoflow system and flow rates for both layer were maintained by 100 nL/min at r.t. [b] Catalyst (1 mol%) was used in batch system.

Such successful results in the nanoflow microreactor system indicate that the faster dispersion of H_2O_2 into fluorous-organic hybrid BTF including the substrate and the lanthanide catalyst can be achieved. The origin of the perfect regioselectivity and remarkably increased chemical yield in the nanoflow system can be rationalized by the formation of the metal peroxo species (*24,25*), which can increase the nucleophilicity of HO_2^-.

To shed right on the metal peroxo formation (*25*) in the nanoflow system, two experimental procedures were compared in the batch system (round bottomed flasks) for 2-methylcyclopentanone (Table 4).

In the standard batch experiment (Method a) under almost the same condition as the nanoflow system other than the higher lanthanide catalyst concentration (otherwise, oxidation did not proceed within reasonable reaction times). The lanthanide catalyst was dissolved in BTF (1×10^{-3} M) and the substrate was added (0.1 M) followed by 30% H_2O_2. The low regioselectivity was observed throughout the reaction in the batch Method a. In the premixing batch experiment (Method b), the Sc catalyst was pre-mixed with aq. H_2O_2 for about 30 min prior to substrate addition using the same concentrations of the substrate (0.1 M) and Sc catalyst (1×10^{-3} M) as in the Method a. In sharp contrast to the standard Method a, higher regioselectivity was achieved in the premixing Method b (Figure 5).

On the basis of these experimental observations in the batch system, the increased regioselectivity can be rationalized by efficient metal peroxo formation in the nanoflow microreactor (Scheme 4). It is crucial either the complex formation of the Sc catalyst with H_2O_2 affording more nucleophilic Sc peroxo species (path b) or the coordination of the Sc catalyst itself to the ketone followed by the addition of H_2O_2 herewith (path a). In path b, the intramolecular nucleophilic attack of the hydroperoxy moiety regioselectively from the Sc complex to the coordinated ketone can strictly control the direction of O-O bond *anti* to the migratory carbon bearing R group. Thus, high regioselectivity was obtained along with the formation of the Sc hydroxide complex. In the nanoflow system, the faster dispersion of H_2O_2 into BTF phase can efficiently lead to the re-formation of the Sc peroxo species from the Sc hydroxide complex. The highly regioselective path b is in contrast to the low regioselective path a, wherein the intermolecular H_2O_2 attack should take place from both sides.

Finally, we have examined the asymmetric B-V reaction by our nanoflow microreactor. The reaction rate is significantly increased with 60% H_2O_2 even by use of a chiral lanthanide complex with (*S,S*)-1,2-*N,N'*-bis-(trifluoromethanesulfonylamino)-1,2-diphenylethane (DPENTf) ligand (5×10^{-2} mol%) (Scheme 5). DPENTf itself showed no catalytic activity as a chiral Brønsted acid catalyst (*26*) even with 60% H_2O_2. Therefore, the fluorous Sc complex and DPENTf ligand were premixed to form the Sc-DPENTf complex in BTF before introducing into the nanoflow microreactor. The reaction proceeded within several sec (~50% conversion) in the microreactor to give moderate level of kinetic resolution (26% ee).

Table 4. The B-V oxidation with or without pre-formation of the Sc peroxo species.[a]

$$\text{2-methylcyclopentanone} \xrightarrow[\text{r. t.}]{30\%\ H_2O_2} \text{lactone} + \text{lactone}$$

time (h)	Method a		Method b	
	regioisomer ratio	%yield	regioisomer ratio	%yield
1	60:40	17	95:5	17
2	65:35	22	94:6	22
4	70:30	39	91:9	43

[a] Substrates (0.1 M) and catalyst (1 mol%) in BTF were used in batch system. Yield and regioselectivity were determined by GC analysis

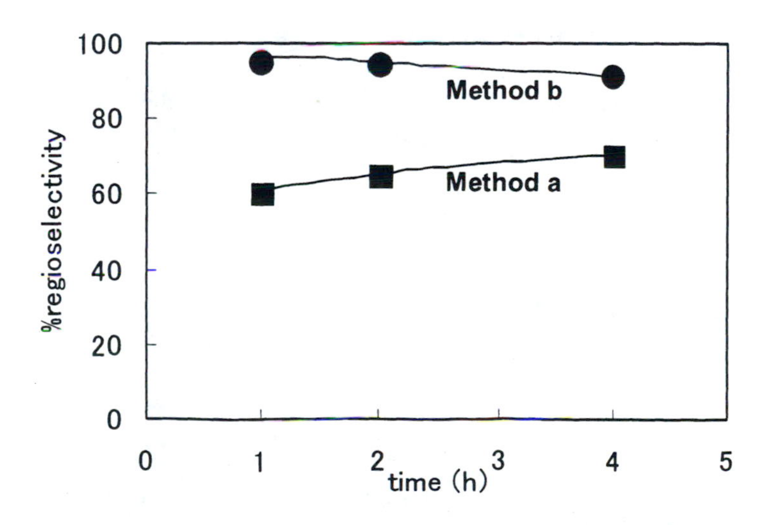

Figure 5. Regioselectivity of the two different methods compared in the batch system.

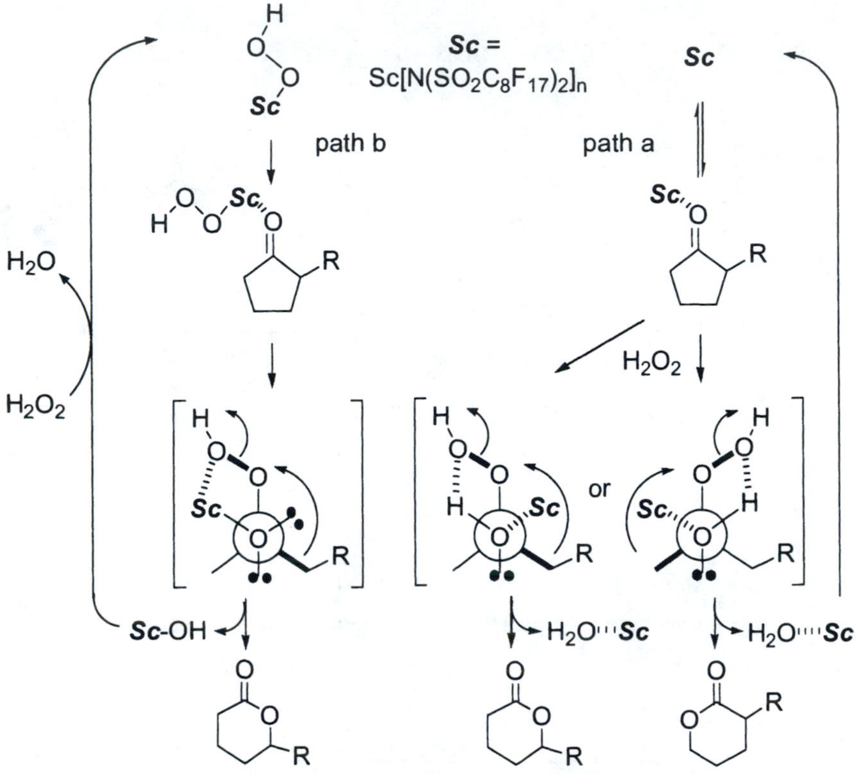

H O O Sc

Sc = $Sc[N(SO_2C_8F_{17})_2]_n$

Sc

path b

path a

H O O Sc O R

Sc O R

H_2O

H_2O_2

H_2O_2

H O O Sc O R

H O O H Sc O R

or

H O O H Sc O R

Sc-OH

H_2O···Sc

H_2O···Sc

O O R

O O R

O O R

Scheme 4. Metal peroxo formation followed by the highly regioselective oxygen atom insertion.

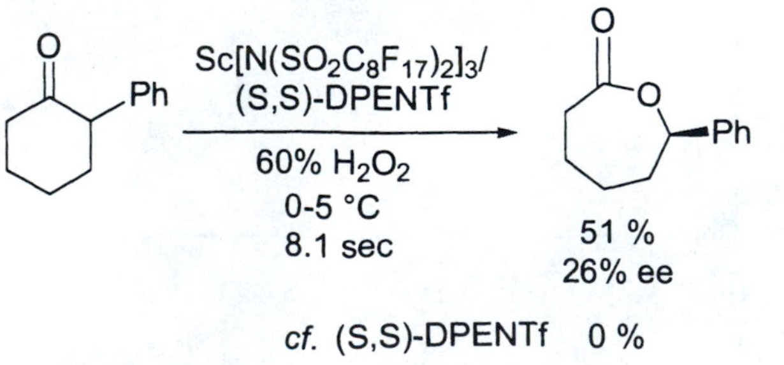

O Ph

$Sc[N(SO_2C_8F_{17})_2]_3$/
(S,S)-DPENTf

60% H_2O_2
0-5 °C
8.1 sec

O O Ph

51 %
26% ee

cf. (S,S)-DPENTf 0 %

Scheme 5. Asymmetric B-V reaction in nanoflow microreactor.

Conclusion

In summary, we have envisioned in our nanoflow microreactor that the reactivity of the fluorous lanthanide-catalyzed Mukaiyama aldol reactions is drastically increased and that not only the regioselectivity but also the reaction rate of the B-V reaction are significantly increased along with moderate level of kinetic resolution, even in the lowest C/S ratio (less than 1:1000 (<<0.1 mol%)).

References

1 a) Horvath, I. T.; Rabai, J. *Science* **1994**, *266*, 72. b) Gladysz, J. A. *Science* **1994**, *266*, 55. c) *Green Reaction Media in Organic Synthesis*; Mikami, K., Ed.; Blackwell, Oxford, 2005. d) Fish, R. H. *Chem. Eur. J.* **1999**, *5*, 1677. e) Horvath, I. T. *Acc. Chem. Res.* **1998**, *31*, 641. f) Cornils, B. *Angew. Chem. Int. Ed.* **1997**, *36*, 2057. g) Vogt, M. *The Application of Perfluorinated polyethers for Immobilization of Homogeneous Catalysts*, Ph.D. Thesis, Rheinisch-Wesfakischen Technischen Hochschule, Aachen, Germany, 1991. h) For historical view and definitions of fluorous chemistry, also see: Gladysz, J. A.; Curran, D. P. *Tetrahedron*, **2002**, *58*, 3823.

2 a) *Handbook of Fluorous Chemistry*; Gladysz, J.; Curran, D. P.; Horvath, I. T., Eds.; Wiley-VCH, 2004. b) Jiao, H.; Le Stang, S.; Soos, T.; Meier, R.; Kowski, K.; Rademacher, P.; Jafarpour, L.; Hamard, J. -B.; Nolan, S. P.; Gladysz, J. A. *J. Am. Chem. Soc.* **2002**, *124*, 1516. c) Curran, D. P. *Angew. Chem. Int. Ed.* **1998**, *37*, 1174.

3 a) Mikami, K.; Mikami, Y.; Matsuzawa, H.; Matsumoto, Y.; Nishikido, J.; Yamamoto, F.; Nakajima, H. *Tetrahedron* **2002**, 58, 4015. b) Mikami, K.; Mikami, Y.; Matsumoto, Y.; Nishikido, J.; Yamamoto, F.; Nakajima, H. *Tetrahedron Lett.* **2001**, *42*, 289. c) Nishikido, J.; Nakajima, H.; Saeki, T.; Ishii, A.; Mikami, K. *Synlett* **1998**, 1347. d) Ishii, A.; Kotera, O.; Saeki, T.; Mikami, K. *Synlett* **1997**, 1145.

4 Barrett, A. G. M.; Bouloc, N.; Braddock, D. C.; Catterick, D.; Chadwick, D.; White, A. J. P.; Williams, D. J. *Tetrahedron* **2002**, *58*, 3835, and references cited therein.

5 Unfortunately, bis(trifluorometanesulfonyl)methide is of too short hair to render the complexes in fluorous media. Earliest contributions on the lanthanide complex preparation: a) $La(NTf_2)_3$: Kobayashi, H.; Nie, J.; Sonoda, T. *Chem. Lett.* **1995**, 307. b) $Yb(NTf_2)_3$: Mikami, K.; Kotera, O.; Motoyama, Y.; Sakaguchi, H.; Maruta, M. *Synlett* **1996**, 171. c) $Sc(NTf_2)_3 \cdot 1.5\ CH_3CO_2H$: Ishihara, K.; Kubota, M.; Yamamoto, H. *Synlett* **1996**, 265.

6 Koppel and Taft et al. have reported the gas phase acidities of a variety of super acids: Bis(trifluoromethanesulfonyl)amine is stronger than

trifluofomethanesulfonic acid by ΔG = 7.7 kcal/mol and bis(trifluoromethanesulfonyl)methane is stronger than bis(trifluoromethanesulfonyl)amine by ΔG = 2.8 kcal/mol. They have also reported that bis(perfluorobutanesulfonyl)amine is stronger than bis(trifluoromethanesulfonyl)amine by ΔG = 7.7 kcal/mol: *J. Am. Chem. Soc.* **1994**, *116*, 3047.

7 a) Maul, J. J.; Ostrowske, P. J.; Ublackes, G. A.; Linclan, B.; Curran, D. P. *Top. Curr. Chem.* **1999**, *206*, 79. b) Ogawa, A.; Curran, D. P. *J. Org. Chem.* **1997**, *62*, 450.

8 a) Sato, K.; Hibara, A.; Tokeshi, M.; Hisamoto, H.; Kitamori, T. *Advanced Drug Delivery Reviews* **2003**, *55*, 379. b) *Microsystem Technology in Chemistry and Life Sciences*; Manz, A.; Becker, H., Eds.; Springer, Berlin, 1998. c) *Micro Total Analytical Systems*; Ache, H-J.; van den Berg, A.; Bergveld, P., Eds.; Kluwer, Dordrecht, 1995. d) Freemantle, M. *Chem. Eng. News.* **1999**, *77*, Feb. 22, 27. e) Manz, A.; Graber, N.; Widmer, H. M. *Sens. Actuators B 1* **1990**, 244.

9 a) Jensen, K. F. *Chem. Eng. Sci.* **2001**, *56*, 293. b) Jähnisch, K.; Hessel, V.; Löwe, H.; Baerns, M. *Angew. Chem.* **2004**, *116*, 410.; *Angew. Chem. Int. Ed.* **2004**, *43*, 406. c) Ehrfeld, W.; Hessel, V.; Lehr, H. *Top. Curr. Chem.* **1998**, *194*, 233. d) Ehrfeld, W.; Hessel, V.; Löwe, H. *Microreactors: New Technology for Modern Chemistry*; Wiley-VCH, Weinheim, Germany, 2000.

10 a) Ueno, M.; Hisamoto, H.; Kitamori, T.; Kobayashi, S. *Chem. Comm.* **2003**, 936. b) Lai, S. M.; Martin-Aranda, R.; Yeung, K. L. *Chem. Commun.* **2003**, 218. c) Zhang, W. X. *Tetrahedron* **2002**, *58*, 4735. d) Mello, A. D.; Wootten, R.; *Lab Chip 2* **2002**, 7N. e) Wiles, C.; Watts, P.; Haswell, S. J.; Pombo-Villar, E. *Chem. Commun.* **2002**, 1034. f) Fukuyama, T.; Shinmen, M.; Nishitani, S.; Sato, M.; Ryu, I. *Org. Lett.* **2002**, *4*, 1691. g) Haswell, S. J.; Middleton, R. J.; O'Sullivan, B.; Skelton, V.; Watts, P.; Styring, P. *Chem. Commun.* **2001**, 391. h) Taghavi-Moghadam, S.; Kleemann, A.; Golbig, K. G. *Org. Process Res. Dev.* **2001**, *5*, 652 i) DeWitt, S. H. *Curr. Opin. Chem. Biol.* **1999**, *3*, 350. j) Hisamoto, H.; Saito, T.; Tokeshi, M.; Hibara, A.; Kitamori, T. *Chem. Comm.* **2001**, 2662. k) Honicke, D. *Stud. Sur. Sci. Cat.* **1999**, *122*, 47. l) Salimi-Moosavi, H.; Tang, T.; Harrison, D. J. *J. Am. Chem. Soc.* **1997**, *119*, 8716.

11 a) Fletcher, P. D. I.; Haswell, S. J.; Paunov, V. N. *Analyst* **1999**, *124*, 1273. b) Manz, A.; Effenhauser, C. S.; Burggraf, N.; Harrison, D. J.; Seiler, K.; Fluri, K. *J. Micromech. Microeng.* **1994**, *4*, 257. c) Graveson, P.; Branebjerg, J.; Jensen, O. S. *J. Micromech. Microeng.* **1994**, *4*, 157.

12 a) Mikami, K.; Yamanaka, M.; Kudo, K.; Fuji Electric Systems Co. Ltd. JPC2004-195433 b) Our primary communication of the Mukaiyama aldol reaction in the nanoflow microreactor controlled by *DiNaS*: Mikami, K.;

Yamanaka, M.; Islam, M. N.; Kudo, K.; Seino, N.; Shinoda, M. *Tetrahedron Lett.* **2003**, *44*, 7545. c) Mikami, K.; Yamanaka, M.; Islam, M. N.; Kudo, K.; Seino, N.; Shinoda, M. *Tetrahedron* **2003**, *59*, 10593. d) The aldol reaction of silyl enol ethers using TBAF in the microreactor by EOF has been reported: Wiles, C.; Watts, P.; Haswell, S. J.; Pombo-Villar, E. *Lab on a Tip* **2001**, 100.

13 a) *Fundamentals of Microfabrication*; Madou, M., Ed.; CRC Press: Boca Raton, 2000. b) Jaeger, R. C. *Modular Series on Solid State Devices*; Neudeck, G. W.; Pierret, R. F., Eds.; Addison-Wesley Publishing Company, 1993. c) McCreedy, T. *Anal. Chim. Acta* **2001**, *427*, 39.

14 a) Crank, J. *The Mathematics of Diffusion*; Oxford University Press: Oxford, 1956. b) Carslaw, H. S.; Jaeger, J. C. *Heat Conduction in Solids*, Second Edition; Oxford University Press: Oxford, 1959.

15 a) Lowe, H.; Ehrfeld, W. *Electrochimica Acta* **1999**, *44*, 3679. b) Worz, O.; Jackel, K. P.; Richter, T.; Wolf, A. *Chemical Engineering Science* **2001**, *56*, 1029.

16 a) Kakuta, M.; Bessoth, F. G.; Manz, A. *The Chemical Record* **2001**, *1*, 395. b) DeWitt, S. H. *Current Opinion in Chemical Biology* **1999**, *3*, 350.

17 Reviews on Mukaiyama aldol Reactions: a) Carreira, E. M. *Comprehensive Asymmetric Catalysis;* Jacobsen, E. N.; Pfaltz, A.; Yamamoto, H., Eds.; Springer: Berlin, 1999; Vol. 3. pp 997. b) Nelson, G. *Tetrahedron: Asymmetry* **1998**, *9*, 357. c) Bach, T. *Angew. Chem. Int. Ed.* **1994**, *33*, 417. d) Mukaiyama, T. *Org. React.* **1982**, *28*, 203.

18 Reviews on Lewis acid catalysis: a) *Lewis Acids in Organic Synthesis*; Yamamoto, H., Ed.; Wiley- VCH: Weinheim, 2000; Vol. 1 & 2. b) Dias, L. C. *J. Braz. Chem. Soc.* **1997**, *8*, 289. c) Santelli, M.; Pons, J. M. *Lewis Acid and Selectivity in Organic Synthesis*; CRC Press: New York, 1996. d) Mikami, K.; Nakai, T. *Asymmetric Lewis Acid Catalysts;* Kagaku Zoukan, Kagaku Dojin: Tokyo, 1995; Vol. 124, p 177. e) Oh, T.; Reilly, M. *Org. Prep. Proced. Int.* **1994**, *26*, 129. f) Deloux, L.; Srebnik, M. *Chem. Rev.* **1993**, *93*, 763. g) Narasaka, K. *Synthesis* **1991**, 1. h) Shanbayati, S.; Schreiber, S. L. *Comprehensive Organic Synthesis*; Trost, B. M.; Fleming, I., Eds.; Pergamon: Oxford, 1991; Vol. 1, p 283. i) Reetz, M. T. *Organotitanium Reagents in Organic Synthesis*; Springer: Berlin, 1986.

19 Reviews on lanthanide complexes: a) Kagan, H. B.; Namy, J. L. *Tetrahedron* **1986**, *42*, 6573. b) Imamoto, T. *Comprehensive Organic Synthesis*; Trost, B. M.; Fleming, I., Eds.; Pergamon Press: 1991, Vol. 1, p 231. c) Molander, G. A. *Chem. Rev.* **1992**, *92*, 29. d) Imamoto, T. *Lanthanide in Organic Synthesis*; Academic Press: London, 1994. e) Molander, G. A.; Harris, C. R. *Chem. Rev.* **1996**, *96*, 307. f) *Lanthanides: Chemistry and Use in Organic Synthesis*; Kobayashi, S., Ed.; Springer: Berlin, 1999. g) Mikami, K.; Terada, M.; Matsuzawa, H. *Angew. Chem. Int. Ed.* **2002**, *41*, 3554.

20 a) Renz, M.; Meunier, B. *Eur. J. Org. Chem* **1999**, 737 b) Strukul, G. *Angew. Chem. Int. Ed.* **1998**, 37, 1198 c) Krow, G. R. *Org. React.* **1993**, 43, 251. d) Krow, G. R., *Comprehensive Organic Synthesis*; Trost, B. M., Ed.; Pergamon: Oxford, 1991, Vol. 7, p. 671.

21 Silbert, L. S.; Siegel, E.; Swern, D. *J. Org. Chem.* **1962**, *27*, 1336, and references cited therein.

22 The use of $Sc(OTf)_3$ in the Baeyer-Villiger oxidation of methylcyclohexanone by commercial grade m-CPBA provided high regioselectivity (96:4) for 12 minutes: Kotsuki, H.; Arimura, K.; Araki, T.; Shinohara, T. *Synlett* **1999**, 462.

23 a) Mikami, K.; Yamanaka, M.; Islam, M. N.; Tonoi, T.; Itoh, Y.; Shinoda, M.; Kudo, K. *J. Fluorine Chem.* in press. b) Mikami, K.; Islam, M. N.; Yamanaka, M.; Itoh, Y.; Shinoda, M.; Kudo, K. *Tetrahedron Lett.* **2004**, *45*, 3681.

24 a) Jacobson, S. E.; Tang, R.; Mares, F.; *Chem. Commun.* **1978**, 888. b) Jacobson, S. E.; Tang, R.; Mares, F. *Inorg. Chem.* **1978**, *17*, 3055. c) Herrmann, W. A.; Fischer, R. W. *J. Mol. Catal.* **1994**, *94*, 213. d) Huston, P.; Espenson, J. H.; Bakac, A. *Inorg. Chem.* **1993**, *32*, 4517. e) Yamazaki, S. *Chem. Lett.* **1995**, 127.

25 a) Todesco, M. D.; Pinna, F.; Strukul, G. *Organometallics* **1993**, *12*, 148. b) Strukul, G.; Varagnolo, A.; Pinna, F. *J. Mol. Catal. A* **1997**, *117*, 413. c) Roberta, G.; Maurizio, C.; Francesco, P.; Strukul, G. *Organometallics* **1998**, *17*, 661, and references cited therein.

26 Tonoi, T.; Mikami, K. *Tetrahedron. Lett.* **2005**, *46*, 6355.

Chapter 12

Fluorous Chemistry for Synthesis and Purification of Biomolecules: Peptides, Oligosaccharides, Glycopeptides, and Oligonucleotides

Wei Zhang

Fluorous Technologies, Inc., University of Pittsburgh Applied Research Center, 970 William Pitt Way, Pittsburgh, PA 15238

Fluorous chemistry has advanced well beyond the scope of biphasic catalysis and high-throughput synthesis of small molecules. This review describes the recent development of fluorous technologies in the synthesis and purification of biomolecules including peptides, oligosaccharides, glycopeptides, and oligonucleotides.

Highly fluorinated (fluorous) molecules are lipophobic and hydrophobic,[1] but they have strong interaction with fluorous separation media. This unique property has been exploited for phase separation of fluorous-tagged molecules from non-fluorous molecules by liquid-liquid extraction (LLE) with fluorous solvents or by solid-phase extraction (SPE) and chromatography with fiuorous silica gel.[2] In 1994 Horváth and Rábai introduced the concept of "fluorous biphasic catalysis" for catalyst recovery.[3,4] Since then, Curran group at the University of Pittsburgh, Fluorous Technologies, Inc., and others have developed a large collection of fluorous catalysts, reagents, scavengers, protecting groups for "fluorous synthesis" of small molecules.[5,8] Now fluorous technologies have been applied to synthesis arid separation of biomolecules.

Biomolecules such as peptides, oligosaccharides, glycopeptides, and oligonucleotides play pivotal roles in the life processes. They are the important subjects of molecular biology, cell biology, and medicinal chemistry research. Over the years, solid-phase synthesis (SPS) technologies for biomolecule preparation have been widely adopted for their efficiency and simplicity.[9,10] However, some issues associated with SPS still need to be addressed. For example, heterogeneous SPS has less favorable reaction kinetics than solution phase reactions, offers a limited range of chemical transformations, and prevents easy analysis of bound imtermediates. Most importantly, product purification, typically by HPLC, can be slow and costly, there are many examples of deletion sequences greatly complicate the separation.

Two general fluorous approaches can be used to enhance the SPS of biooligomers.[7] In the first approach (Scheme 1, A), organic capping agents are used to react with the deletion sequences, while the desired sequence is captured by a fluorous tag at the last coupling step. The mixture generated from resin cleavage is subjected to fluorous separation (F-SPE or F-HPLC) followed by the removal of the fluorous tag to give the oligomer product. The second approach (Scheme 1, B) employs fluorous reagents to cap the deletion sequences after each coupling reaction. At the end of synthesis, all the sequences are cleaved from the resin and the product is separated from the fluorous byproducts by fluorous separation; no detagging step is needed since the desired product is nonfluorous.

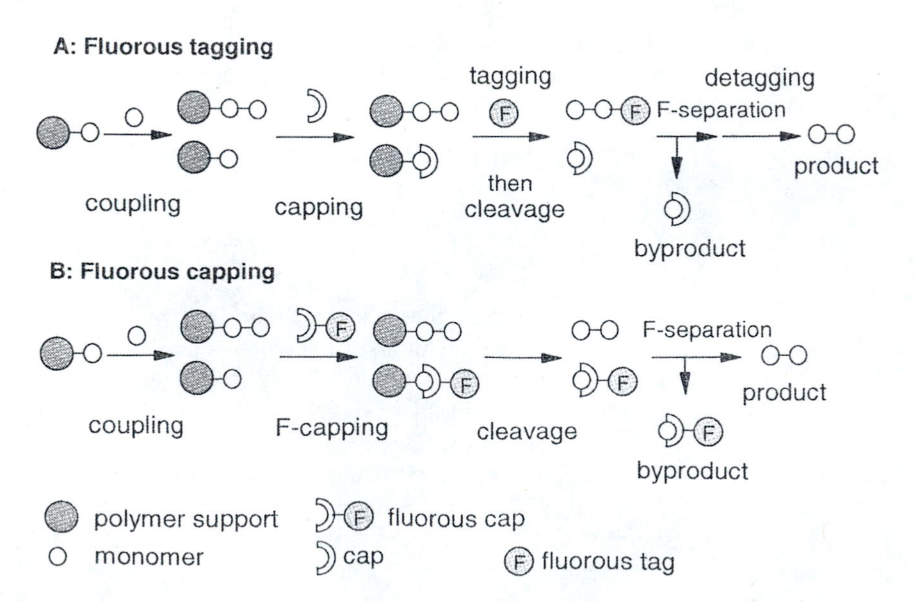

Scheme 1. Fluorous tagging and capping strategies for SPS of oligomers

In addition to fluorous-enhanced SPS, a fluorous support can be used to replace the solid support for solution-phase oligomer synthesis (Scheme 2). Compared to SPS, this approach takes advantages of solution-phase synthesis and it accelerates synthetic development by requing only one method for separation regardless of the characteristics of the product. However, it also requires high coupling efficiency since the deletion sequences with fluorous tags may not be easily separated from the desired sequences which are also attached to fluorous tags.

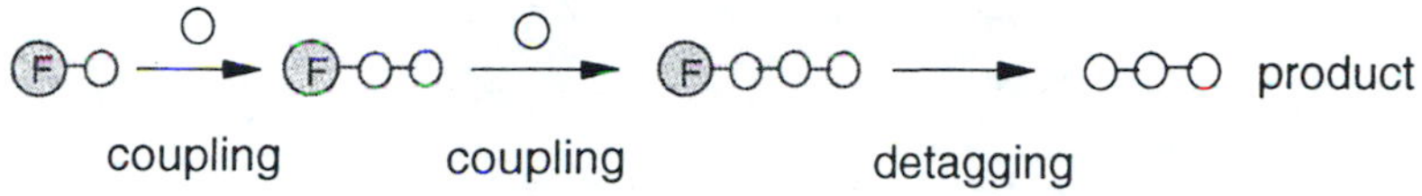

F-O —(O, coupling)→ F-O-O —(O, coupling)→ F-O-O-O —(detagging)→ O-O-O product

Scheme 2. Fluorous solution-phase synthesis of oligomers

Peptide Synthesis and Separation

SPS with Fluorous Tags

Fluorous adaptions of popular groups including Fmoc, Boc, Cbz and Msc have been developed for peptide synthesis. Many fluorous-protected amino acids are now available from Fluorous Technologies, Inc.[11] The van Boom group

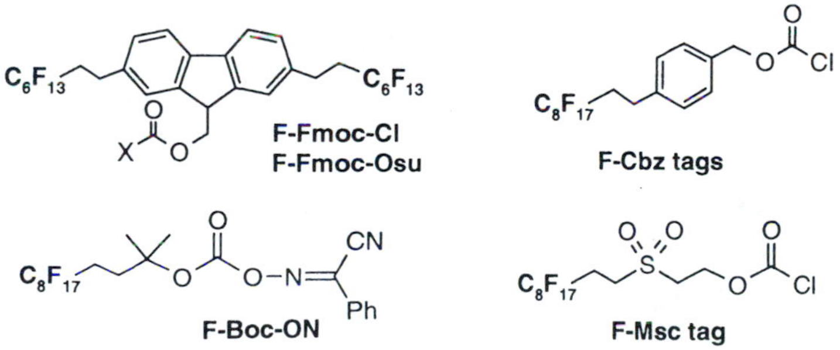

C_6F_{13} C_6F_{13} X O O — F-Fmoc-Cl, F-Fmoc-Osu

C_8F_{17} O O Cl — F-Cbz tags

C_8F_{17} O O O–N= CN Ph — F-Boc-ON

C_8F_{17} O O S O O Cl — F-Msc tag

employed three different fluorous Cbz tags in Fmoc-based solid-phase peptide synthesis.[12] At each coupling step any unreacted free amine was capped with an acetyl group. After the desired number of iterations, the deprotection of the final Fmoc group from the desired sequence was followed by tagging with a fluorous Cbz group. Resin cleavage gave a mixture containing desired F-tagged product and acetyl-capped byproducts. The mixture was separated by F-HPLC. Using this method, peptides up to 34 amino acids were readily synthesized. Similar to normal Cbz, the F-Cbz was found partially detagged during the acidic resin cleavage. To address this issue, Overkleeft and coworkers developed a base-labile fluorous methylsufonylethoxycarbonyl (F-Msc) as an alternative amine protecting group.[13] In this work, separations of resin cleaved mixtures were conducted by F-HPLC and also by F-SPE (Scheme 3). F-SPE gave good results for peptides **1**and **2**. F-HPLC succeeded on all three peptides **1-3**.

	F-HPLC		F-SPE	
	yield	purity	yield	purity
1, GCCSLPPCALNNPDYCamide	37%	98%	59%	91%
2, RQIKIWFQNRRMKWKKamide	10%	94%	7%	72%
3, LSELDDRADALQAGFSQFESS-AAKLKRKYWWKNLK	21%	99%		

Scheme 3. F-Msc tag for peptide synthesis

SPS with Fluorous Capping Agents

Montanari and Kumar developed a trivalent iodonium reagent **4** and used it as a capping reagent to remove deletion sequences up to 21 residues long (Scheme 4). Fluorous side-products cleaved from the resin were analyzed by reverse-phase HPLC and separated by fluorous flash chromatography or by precipitation with addition of water to the mixture. This fluorous capping reagent has been employed in both Boc- and Fmoc-based peptide syntheses.[14,15]

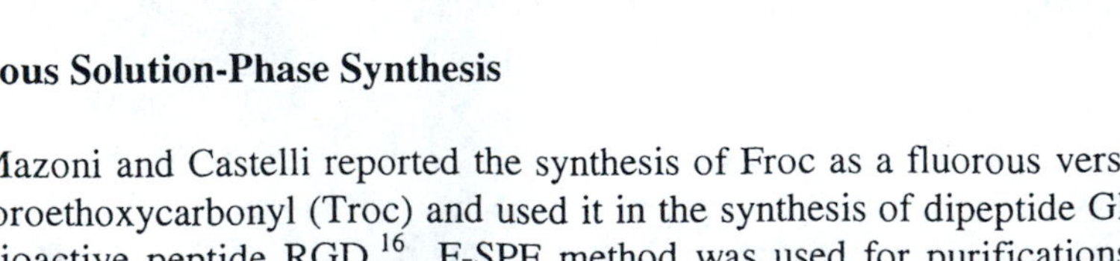

$$\text{(resin)}\text{-peptide-NH}_2 \xrightarrow[\text{collidine, } CH_2Cl_2]{\mathbf{4}\ \ Ph\text{-}\overset{+}{I}(CH_2C_6F_{13})\ {}^{-}N(SO_2CF_3)_2} \text{(resin)}\text{-peptide-NH-}CH_2C_6F_{13}$$

deletion sequence

*Scheme 4. Fluorous capping agent **4***

Fluorous Solution-Phase Synthesis

Mazoni and Castelli reported the synthesis of Froc as a fluorous version of trichloroethoxycarbonyl (Troc) and used it in the synthesis of dipeptide Gly-Gly and bioactive peptide RGD.[16] F-SPE method was used for purifications. The Fustero group used fluorous (trimethylsilyl)ethanol (F-TMSE) as a carboxylic acid protection group in the synthesis of tripeptide **5**. The fluorous tag was readily cleaved by treatment of TBAF (Scheme 5).[17]

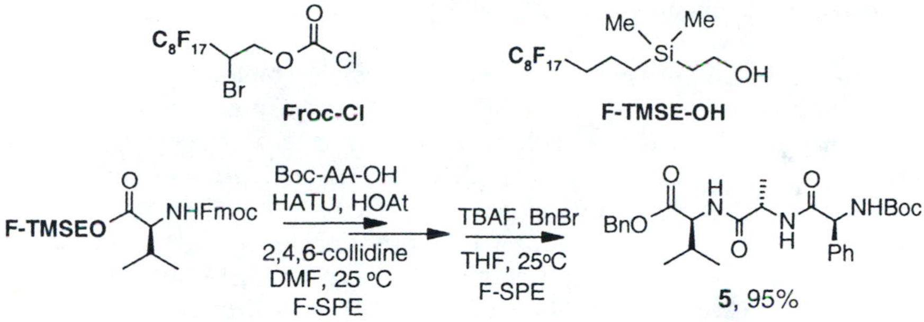

Scheme 5. *Fluorous tags and peptide synthesis with F-TMSE tag*

Inazu and coworkers developed a heavy fluorous support (Hfb) containing six C_8F_{17} chains and attached it to Rink-, Wang-, and *t*-butyl-type linkers. These heavy fluorous tags have been used in solution-phase peptide synthesis (Scheme 6).[18,19] In the synthesis of tripeptide pGlu-His-Pro-NH_2, an excess amount of amino acid derivative was used in each coupling step. The unreacted reagent and coupling agents were removed by extraction with MeCN from the perfluorohexane (FC-72) layer. At the last step, the fluorous tag was cleaved and recovered by F-LLE with FC-72, and the tripeptide product was produced in 62% yield. Several other heavy fluorous tags have also been developed and used for peptide and oligosaccharide synthesis (Scheme 9).

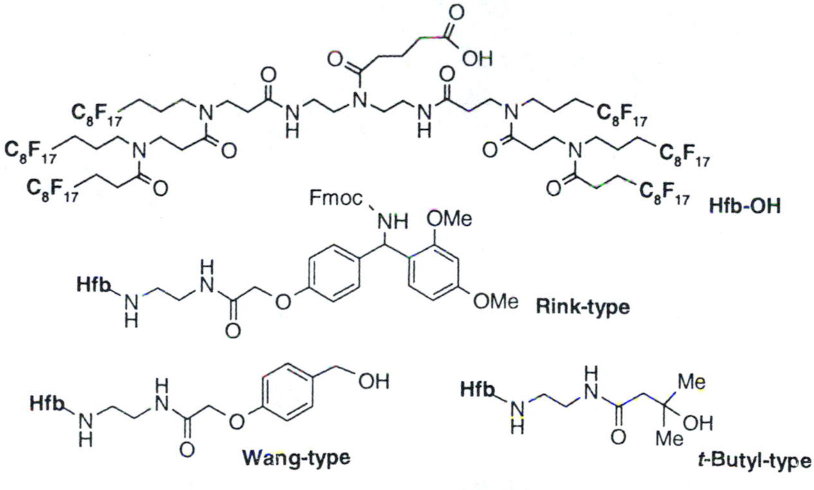

Scheme 6. *Heavy fluorous support Hfb-attached linkers*

Fluorous Enrichment of Proteomics Sample

The separation of peptide subsets from complex biological samples is an important and challenging task in molecular biology and proteomics. The Peters group has demonstrated the utility of fluorous tagging strategy for sample enrichment.[20] The overall process is shown in Figure 1. After peptide sample digestion, a specific class of peptides is labeled with a reagent bearing C_8F_{17} group through a selective chemical transformation. The mixture containing fluorous and non fluorous peptides is loaded onto an F-SPE cartridge. The initial wash with 50% MeCN-H_2O removes all the non-fluorous peptides, while the labeled-peptides are retained on the cartridge until they are washed with 100% methanol. The purification process is followed by MALDI-MS or ESI-MS analysis. This method avoids many problems associated with the classic biotin approach, such as fragmentation during MS/MS and high reagent cost.

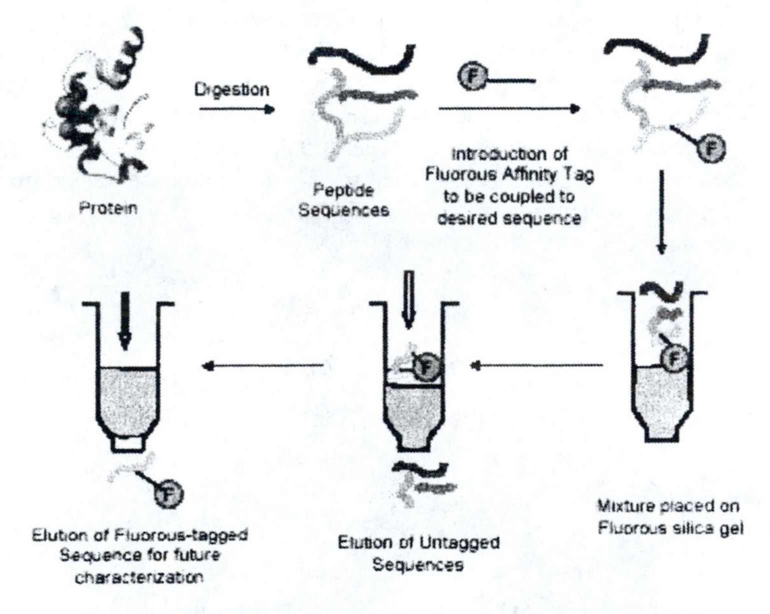

Figure 1. Fluorous tagging stratedgy for peptide sample enrichment

Fluorous Amide Coupling Agents

In addition to fluorous tagging and capping agents, fluorous amide coupling agents have also been developed for solution-phase peptide synthesis. The Dembinski group prepared the fluorous 2-chloro-4,6-dimethoxy-1,3,5-triazine (F-CDMT) and used it for the synthesis of di- and tripeptides (Scheme 7).[21] The coupling reactions were conducted in the presence of *N*-methylmorpholine.

After the reaction, the F-CDMT derivative was precipitated out from the reaction mixture with an appropriate organic solvent. No racemization was observed. Very recently, 1*H*,1*H*,2*H*,2*H*,3*H*,3*H*-perfluorodecanol displaced 2,4-dichloro-1,3,5-triazine (F-DCT) has been developed for amide coupling, which also has potential application for peptide synthesis.[22]

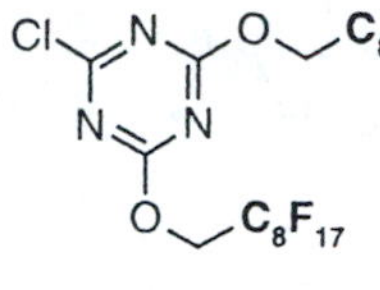

	yield
Cbz-Gly-Ala-OMe	92%
Cbz-Ala-Ala-OMe	98%
Cbz-Pro-Ser-OMe	96%
Cbz-Phe-Met-OMe	91%
Cbz-Phe-Aib-OMe	94%
Cbz-Aib-Aib-OMe	93%
Cbz-Ala-Ala-Ala-OMe	93%

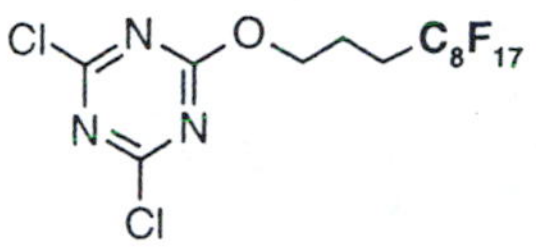

F-CDMT

F-DCT

Scheme 7. Peptide synthesis with F-CDMT

Oligosaccharide Synthesis

Solution-Phase Synthesis with Fluorous Tags

The Curran group first explored fluorous disaccharide synthesis using fluorous benzyl (F-Bn) as a hydroxyl protecting group (Scheme 21).[23] Tribenzyl-tagged D-glucal was coupled with excess diacetone galactose under standard reaction conditions using benzotrifluoride (BTF) as the solvent. Fluorous compound **6** was debenzylated by catalytic hydrogenation with H_2 and $Pd(OH)_2$. After F-LLE, product **7a** in the organic phase was isolated and then acylated to give disaccharide **7b**.

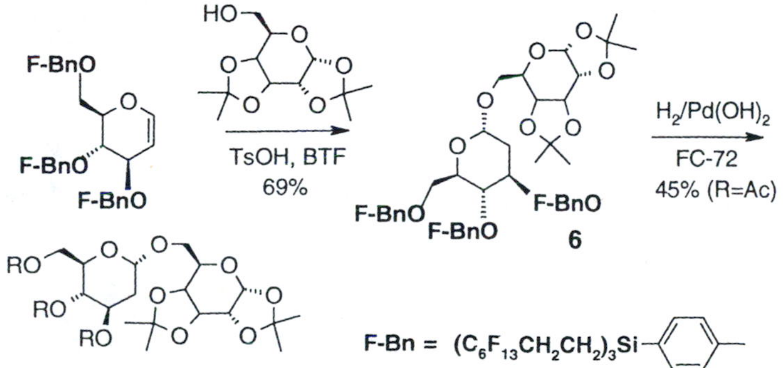

TsOH, BTF, 69%

H_2/$Pd(OH)_2$, FC-72, 45% (R=Ac)

7a, R = H → **7b**, R = Ac

F-Bn = $(C_6F_{13}CH_2CH_2)_3Si$–C_6H_4–

Scheme 8. Fluorous BnBr tag for disaccharide synthesis

Inazu, Mizuno and their coworkers have developed a series of heavy fluorous tags including Hfa, Hfb, Bfp, HfBn, and TfBz for F-LLE-based peptide and oligosaccharide synthesis.[24-30] Scheme 9 highlights the synthesis of disaccharide **10** using two Bfp-protected mannose derivatives.[25] The triphenylmethyl (Trt) group of **8** was selectively removed by treatment of 10-camphorsulfonic acid (CSA). The deprotected hydroxyl group was coupled with galactose derivative to give fluorous disaccharide **9**. Deprotection of both the acetyl and Bfp groups followed by FC-72/MeOH extraction gave disaccharide **10** in MeOH layer, while the protecting group was recovered from the FC-72 layer as a methyl ester. The Bfp tag has also been used in the synthesis of pentasaccharides.[25]

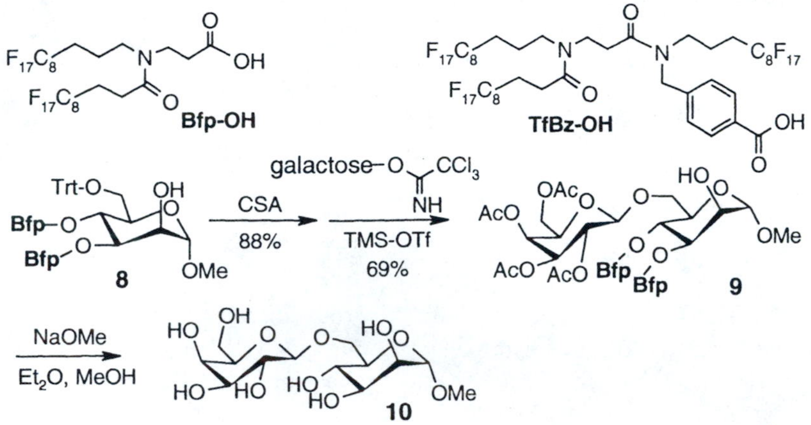

Scheme 9. Fluorous Bfp tag for disaccharide synthesis

The Huang group employed fluorous thiol **11** to displace the bromo group of a tetraacetyl-α-glucosyl bromide **12** (Scheme 10).[31] After removal of the

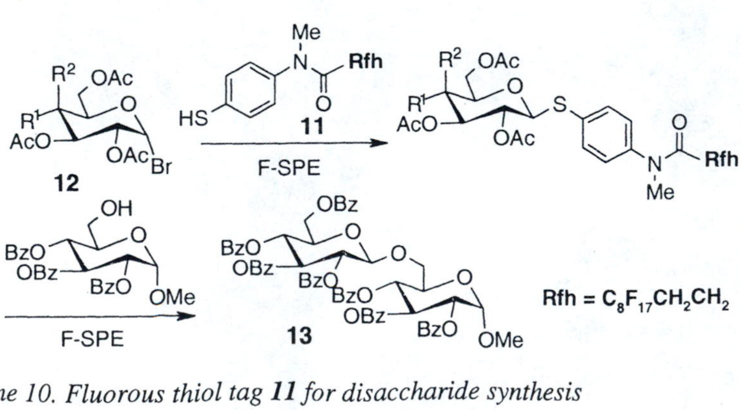

*Scheme 10. Fluorous thiol tag **11** for disaccharide synthesis*

acetate, the thiolglucoside was subjected to benzoylation and glycosylation to give disaccharide **13** in good yield and high purity after F-SPE.

Manzoni recently reported the use of a fluorous silyl tag to protect the anomeric position of sugar acceptors in the synthesis of trisaccharide Lewis a. F-SPE was employed for reaction mixture separation (Scheme 11).[32]

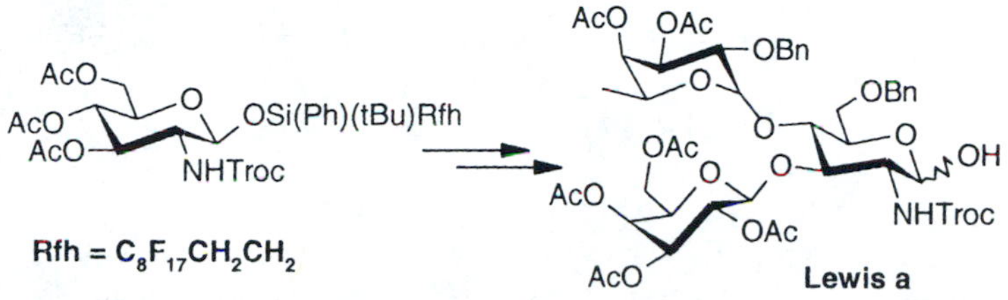

Scheme 11. Fluorous silyl tag for trisaccharide synthesis

Solid-Phase Synthesis with Fluorous Capping Agents

Seeberger and coworkers employed a fluorous silyl group (i-$Pr_2SiCH_2CH_2C_8F_{17}$) to cap the hydroxyl group of the undesired sequences in automated SPS of oligosaccharides.[33] After resin cleavage, the desired sequence was separated from fluorous byproducts by HPLC. A much cleaner trisaccharide **14** was prepared by the fluorous capping strategy as compared to a similar synthesis without fluorous capping, where a significant amount of deletion sequences (n-1 and n-2) was observed.

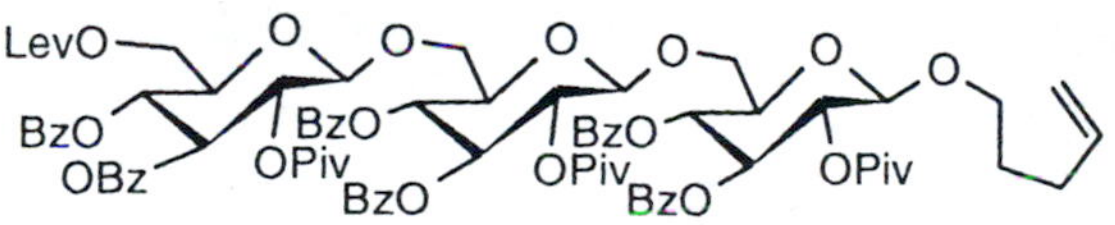

Enzymatic Synthesis with Fluorous Primers

The Hatanaka group synthesized a fluorous alcohol attached lactoside primer **15** and introduced it into muse B16 cells for an *in vivo* glycosylation to produce a GM3-type oligosaccharide (Scheme 12).[34,35] The product was isolated by F-LLE. The fluorous tag was found non-cytotoxic and increased hydrophobicity and membrane permeability of the primer. This unique cellular enzymatic approach is simple and also environmentally friendly since much less reaction solvent is involved in the production process.

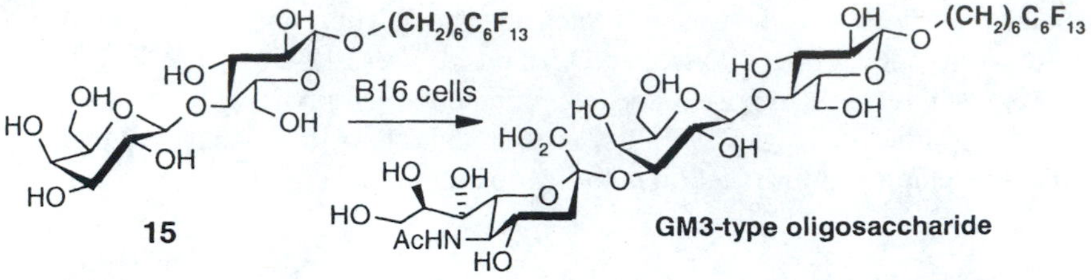

Scheme 12. Synthesis of GM3-type oligosaccharide with fluorous primer ***15***

Fluorous Microarrays

The Pohl group synthesized a series of fluorous alcohol ($C_8F_{17}(CH_2)_2O(CH_2)_4OH$)-attached saccharides and spotted them on a glass microscope slide whose surface had been pretreated with a fluorous silane. The fluorous affinity is strong enough to hold the substrates on the slide for screening with FITC-labeled lectins (Scheme 13).[36,37] This work demonstrates the concept of fluorous microarrays and also shows the potential for the development of affinity-based "spot-synthesis"[38] of fluorous small molecules or biomolecules on chips, and then use the microarrayed chips directly for biological screening.

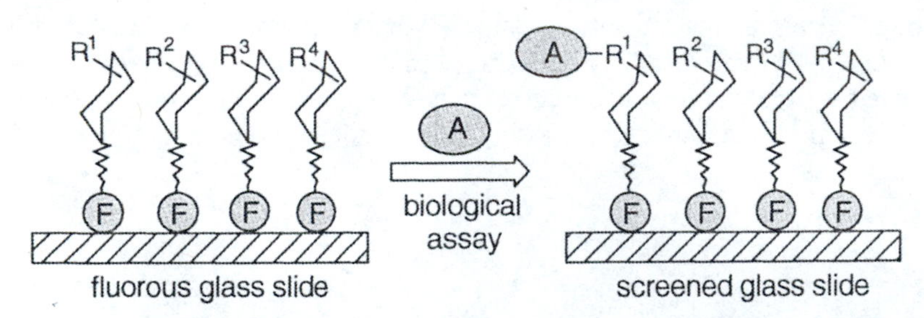

Scheme 13. Conceptual figure of fluorous microarray screening

Glycopeptide Synthesis

So far there is only one reported example of fluorous glycopeptide synthesis. The Mizuno group developed a hexakisfluorous support **16** and attached it to Fmoc-Leu-OH using 4-hydroxymethylphenoxyacetic acid (HMPA) as a linker. After steps of amide coupling and glycosylation the fluorous glycopeptide was cleaved from the support to give product **17** in an overall yield of 74% (Scheme 14).[39] Each reaction step was monitored by TLC and MALDI TOF mass spectroscopy. Intermediates and final product were purified by F-LLE using FC-72.

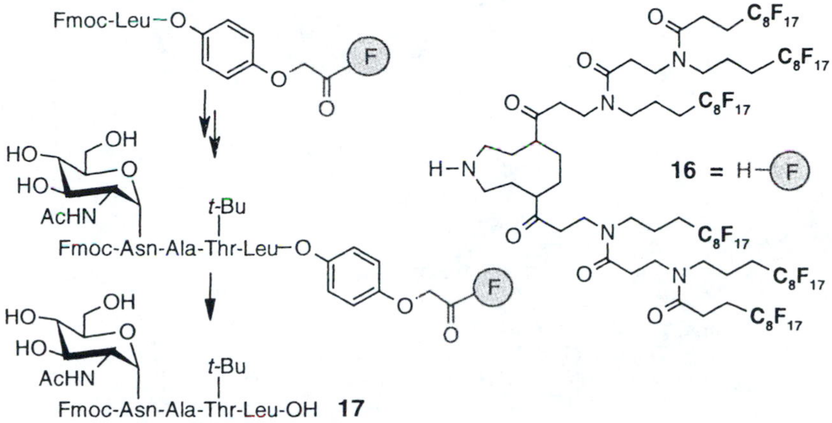

Scheme 14. First example of fluorous glycopeptide synthesis

Oligonucleotide Synthesis

Pearson and coworkers employed fluorous dimethoxytrityl (FDMT)-tagged nucleoside phosphoramidites **18** for solid-phase synthesis of oligonucleotides (Scheme 15).[40] The resin-cleaved mixture was subjected to F-SPE with a high pH-stable fluorinated sorbent. It efficiently retained the fluorous component and removed the failure sequences by washing with 10% MeCN in 0.1 M aqueous Et_3NHOAc followed by water. This step was followed by on-column detritylation with 3% aqueous trifluoroacetic acid. FDMT-off oligonucleotide product was eluted with 10% aqueous MeCN, and the cleaved FDMT tag was left on the cartridge. This fluorous affinity purification technique combines the tag cleavage and tag separation steps in a one-pass loading process without ammonia removal. It has high selectivity, high recovery (~70-100%), and capability to purify oligonucleotides up to 100-mers.

Beller and Bannwarth developed two fluorous trityl groups and used them to make 5'-*O*-attached thymidine-3'-phosphoramidite **19a** and **19b** for DNA synthesis.[41] Fluorous trityl-on purification (F-TOP) on a fluorous silica gel cartridge followed by tag cleavage provided 9- to 30-mer DNA sequences.

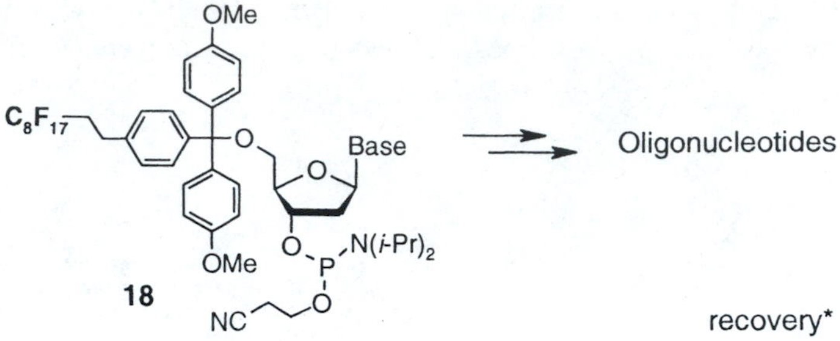

18 → Oligonucleotides

		recovery*
30-mer:	TTTCTCTAGACAATTGTGCAATACGTCTTT	75-88%
50-mer:	TTTCTGTTGACAATTTATCATCGGTCGTAT-AATGTGTGGAATTGGTCTTT	60-100+%
75-mer:	TTTCTGGTTAAGGTGTGTATATGCTCGGC-TACTAATTAGTGAGTATTATTCTCGCTACT-ATTAACAGTTGTCTTT	100+%
100-mer:	TTTCTGGTTAAGGTGTGTAATATGCTCAGC-TACTAATTAACAGTTGTCTAAGCTGGTTAA-CGTGAGTAATATGATCAGCTACTATTTAAC-AGTTGTCTTT	76-100%

* estimated by HPLC integration

Scheme 15. FDMT-nucleoside phosphoramidites for oligonucleotide synthesis

Tripathi and coworkers employed fluorous silyl-protected phosphoramidite **20** in the synthesis of oligonucleotides.[42] The enhanced hydrophobicity of fluorous oligonucleotide sequences made them precipitate when suspended in water. Fluorous tag deprotection was achieved by treatment with tetrabutylammonium fluoride (TBAF) in THF and the cleaved tag was isolated by F-LLE with FC-72 and H_2O. The products were further purified by HPLC.

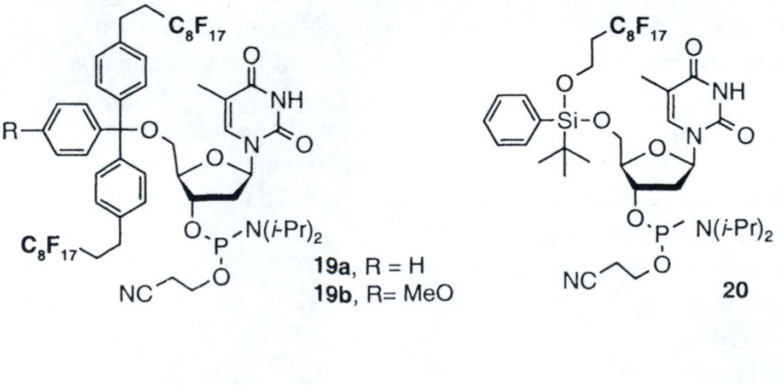

19a, R = H
19b, R= MeO

20

Summary and Outlook

Fluorous chemistry is broadly compatible with existing techniques in synthesis, separation, and immobilization of biomolecules. Fluorous tagging and capping strategies can be used to enhance long-established solid-phase technologies, while fluorous supports and reagents can be used for solution-phase synthesis, where reactions are easily monitored by conventional analytical methods. Small fluorous groups such as C_8F_{17} have proven effective in the separation of tagged oligomers up to 100-mers. Using fluorous molecules for microarrays and direct synthesis of fluorous molecules on fluorous chips could be active areas of research.

References

1. *Handbook of Fluorous Chemistry* Gladysz, J. A.; Curran, D. P.; Horváth , I. T. Eds.; Wiley-VCH: Weinheim, 2004.
2. Curran, D. P. In *Handbook of Fluorous Chemistry;* Gladysz, J. A.; Curran, D. P.; Horváth , I. T. Eds., Wiley-VCH: Weinheim, 2004, pp 101-127.
3. Horváth , I. T., Rábai; T. *Science* **1994**, 266, 72-76.
4. Horváth , I. T. *Acc. Chem. Res.* **1998,** *31*, 641-650.
5. Studer, A.; Hadida, S.; Ferritto, S. Y.; Kim, P. Y.; Jeger, P.; Wipf, P.; Curran, D. P. *Science* **1997**, *275*, 823-826.
6. W. Zhang, *Tetrahedron* **2003**, *59*, 4475-4489.
7. W. Zhang, *Chem. Rev.* **2004**, *104*, 2531-2556.
8. Curran, D. P. *Aldrichmica Acta* **2006**, *39*, 3-9.
9. Merrifield, R. B. *J. Am. Chem. Soc.* **1963**, 85, 2149-2154.
10. Caruthers, M. H. *Science* **1985**, *230*, 281-285.
11. Filippov, D. V.; van Zoelen, D. J.; Oldfield, S. P.; van der Marel, G. A.; Overkleeft, H. S.; Drijfhout, J. W.; van Boom, J. H. *Tetrahedron Lett.* **2002**, *43*, 7809-7812.
12. www.fluorous.com
13. de Visser, P. C.; van Helden, M.; Filippov, D. V.; van der Marel, G. A.; Drijfhout, J. W.; van Boom, J. H.; Noortc, D.; Overkleeft, H. S. *Tetrahedron Lett.* **2003,** *44*, 9013-9016
14. Montanari, V.; Kumar, K. *J. Am. Chem. Soc.* **2004**, 126, 9528-9529.
15. Montanari, V.; Kumar, K. *Eur. J. Org. Chem.* **2006**, 874-877.
16. Manzoni, L.; Castelli, R. *Org. Lett.* **2006**, *8*, 955-957.
17. Fustero, S.; Sancho, A. G.; Chiva, G.; Sanz-Cervera, J. F.; del Pozo, C.; Aceña, J. L. *J. Org. Chem.* **2006**, *71*, 3299-3302.
18. Mizuno, M.; Goto, K.; Miura, T.; Hosaka, D. Inazu, T. *Chem. Commun.*, **2003**, 972-973.

19. Mizuno, M.; Goto, K.; Miura, T.; Matsuura, T.; Inazu, T. *Tetrahedron Lett.* **2004,** 45, 3425–3428.
20. Brittain, S. M.; Ficarro, S. B.; Brock, A.; Peters, E. C. *Nat. Biotech.* **2005**, *23*, 463-469.
21. Markowicz, M. W.; Dembinski, R. *Synthesis* **2004**, 80-86.
22. Zhang, W.; Lu, Y. manuscript submitted to *QSAR Comb. Science*
23. Curran, D.P.; Ferritto, R.; Hua, Y. *Tetrahedron Lett.* **1998**, *39*, 4937-4940.
24. Miura, T.; Goto, K. Hosaka, D.; Inazu, T. *Angew. Chem. Int. Ed.* **2003**, *42*, 2047-2051.
25. Miura, T.; Goto, K.; Waragai, H.; Matsumoto, H.; Hirose, Y.; Ohmae, M.; Ishida, H.-k.; Satoh, A.; Inazu, T. *J. Org. Chem.* **2004**, *69*, 5348-5353.
26. Goto, K.; Miura, T.; Mizuno, M.; Takaki, H.; Imai, N.; Murakami, Y.; Inazu, T. *Synlett* **2004**, 2221-2223.
27. Goto, K.; Miura, T.; Hosaka, D.; Matsumoto, H.; Mizuno, M.; Ishida, H.-k.; Inazu, T. *Tetrahedron* **2004**, *60*, 8845-8854.
28. Miura, T.; Tsujino, S.; Satoh, A.; Goto, K.; Mizuno, M.; Noguchi, M.; Kajimoto, T.; Node, M.; Murakami, Y.; Imaib, N.; Inazua, T. *Tetrahedron* **2005,** *61*, 6518-6526.
29. Goto, K.; Miura, T.; Mizuno, M. *Tetrahedron Lett.* **2005**, *46*, 8293–8297.
30. Miura, T. *Trends Glycoscience Glycotech.* **2003**, *15*, 351-358.
31. Jing, Y.; Huang, X. *Tetrahedron Lett.* **2004**, *45,* 4615-4618.
32. Manzoni, L.; Castelli, R. *Org. Lett.* **2004**, *6*, 4195-4198.
33. Palmacci, E. R.; Hewitt, M. C.; Seeberger, P. H. *Angew. Chem. Int. Ed.* **2001**, *40*, 4433-4437.
34. Kasuya, M. C. Z.; Cusi, R.; Ishihara, O.; Miyagawa, A.; Hashimoto, K.; Sato, T.; Hatanaka, K. *Biochem. Biophy. Res. Commun.* **2004,** *316*, 599-604.
35. Kasuya, M. C. Z.; Ito, A.; Cusi, R.; Sato, T.; Hatanaka, K. *Chem. Lett.* **2005**, *34*, 856-857.
36. Ko, K.-S.; Jaipuri, F. A.; Pohl, N. L. *J. Am. Chem. Soc.* **2005**, *127*, 13162-13163.
37. Mamidyala, S. K.; Ko, K.-S.; Jaipuri, F. A.; Park, G.; Pohl, N. L. *J. F. Chem.* **2006**, article asap.
38. Frank,R. *Tetrahedron* **1992**, *48*, 9217-9232.
39. Mizuno, M.; Goto, K.; Miura, T. *Chem. Lett.* **2005**, *34*, 426-427.
40. Pearson, W. H.; Berry, D. A.; Stoy, P.; Jung, K.-Y.; Sercel, A. D. *J. Org. Chem.* **2005,** *70*, 7114-7122.
41. Beller, C.; Bannwarth W. *Helv. Chim. Acta* **2005**, 88, 171-179.
42. Tripathi, S.; Misra, K.; Sanghvi, Y. S. *Org. Prep. Proc.* **2005**, *37*, 257-263.

Chapter 13

Functional Compounds Based on Hypervalent Sulfur Fluorides

Peer Kirsch[1,*] and Gerd-Volker Röschenthaler[2,*]

[1]Merck Ltd. Japan, Technical Center at Atsugi, 4084 Nakatsu, Aikawa-machi, Aiko-gun, Kanagawa 243–0303, Japan
[2]Institute of Inorganic and Physical Chemistry, University of Bremen, Leobener Strasse, D–28334 Bremen, Germany

During the last decade hypervalent sulfur fluorides have been explored as building blocks for compounds with various functions. Primarily, this interest was driven by the unique property profile of the pentafluorosulfanyl and related groups. The development of more convenient synthetic procedures and the commercial availability of several synthetic precursors provided additional stimulus for the research on this fascinating class of compounds.

Introduction

Hypervalent sulfur fluorides, i.e., organic derivatives of SF_6, are showing a unique set of physicochemical properties. The combination of extreme kinetic stability with very strong electron acceptor capability (*-I* effect), high lipophilicity and high group electronegativity render in particular the SF_5 group a kind of "super-CF_3" function (*1*). This analogy to carbon-based fluoroorganic chemistry is further extended with the SF_4CF_3 group, which can be considered a linear analogue of the perfluoroethyl group (*2*). A fundamental difference between carbon-based and sulfur-based fluoroorganic systems is their geometry (tetrahedral *vs.* oktahedral), permitting, e.g., the design of linear SF_4 "bridge" structures which are of particular interest for highly fluorinated materials, such as liquid crystals for active matrix LCDs (*3*). The stability of radicals and radical ions generated from SF_6 and its derivatives makes these compounds attractive for application in electrical engineering and electronics.

The potential application as organic materials or bioactive compounds have provided a constant driving force for the delopment of industrial-scale access to SF_5 derivatives. Whereas hypervalent sulfur compounds have been practically inaccessible as "exotics" for several decades (*4*), progress in synthetic methodology (*5*) has now brought them on the way to become commodities like other fluorochemicals. The chemistry of SF_5 synthons until around 2003 has been subject of a recent review by G. L. Gard and coworkers (*6*). The following summary will cover more recent developments with emphasis on practical application, and it will provide a more in-depth review on some selected areas.

Useful Intermediates and Interesting Reactions

The synthesis of the long sought-for SF_5NO_2 was finally accomplished either by reacting $N(SF_5)_3$ with NO_2 or by the photolysis of a SF_5Br/NO_2 mixture. The solid compound (m.p. -78°C, extrapolated b.p. 9°C) was characterized by NMR, IR, Raman and UV spectroscopy, as well as by mass spectrometry. The molecular structure was elucidated in detail by gas phase electron diffraction (*7*).

Pentafluorosulfanylchloride (SF_5Cl) was found to add to 1,3-, 1,4-, and 1,5-alkadienes furnishing $F_5SCH{=}CH(CH_2)_nCH{=}CH_2$ (n = 0-2) and their epoxides, which are useful as monomers or intermediates for the preparation of polymers, surface coatings and SF_5 containing heterocyclic compounds (*8*). V. K. Brel and coworkers also treated 3-chloro-1-pentafluorosulfanylprop-1-ene with KCN to obtain the prototropic rearrangement product $ClCH{=}CHCH_2SF_5$, whereas reaction with NaN_3 and KSCN furnished the expected S_N2 products. High level *ab initio* calculations were used to explain the unexpected behavior of cyanide (*9*).

New pentafluorosulfanyl benzene derivatives were synthesized by dehydro(halo)genation of halogenated cycloaliphatic precursors which again were obtained by addition of SF_5Br to 1,4-cyclohexadiene, followed by bromination of the pentafluorosulfanyl cyclohexene intermediates (*10*). A similar addition reaction using SF_5Br provided 1-pentafluorosulfanyl-2-bromooctane in good yields (*11*). Other procedures based on SF_5Cl have been reported by the groups of W. R. Dolbier (*12*) and G. L. Gard (*13*).

A new fluorination method was applied in order to obtain (un)substituted phenyl sulfurpentafluorides. Reaction of $CF_2(OF)_2$ as an electrophilic fluorine source with bis(4-nitrophenyl) disulfide provided 4-nitropentafluorosulfanyl benzene in moderate yields (*14*). The practical relevance of this method lies in the large scale availability of $CF_2(OF)_2$ as an etching gas for microchip manufacturing.

New, versatile pentafluorosulfanyl arene building blocks carrying a substituent in *ortho* position of the SF_5 group were prepared by nucleophilic substitution of suitable halogeno precursors which were in turn obtained by fluorination of the corresponding disulfides with AgF_2 according to Sheppard's method (*4*). As a result, even some sterically demanding *N*- and *S*-nucleophiles could be introduced into the aromatic core system (Scheme 1), without affecting the neighboring SF_5 group (*15*).

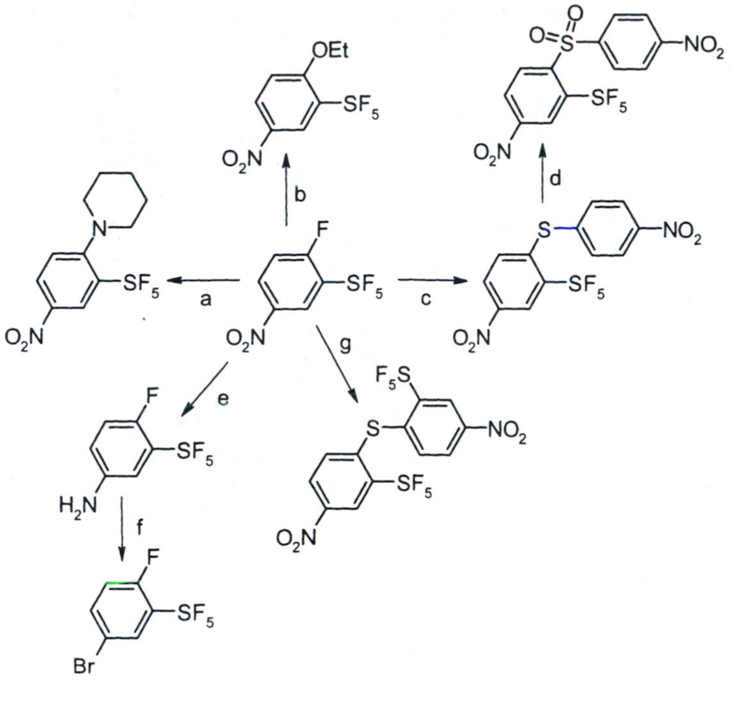

Scheme 1. Introduction of various ortho *substituents into pentafluorosulfanyl benzene derivatives by aromatic nucleophilic substitution: a) piperidine, EtOH; reflux (64%). b) KOH, EtOH; reflux (49%). c)* p-*NO_2PhSNa, EtOH; reflux (46%). d) $KMnO_4$, HOAc; 120°C (60%). e) Fe powder, EtOH, HCl; reflux (44%). f) 1. $NaNO_2$, 48% HBr; 0-5°C; 2. CuBr; 80°C (44%). g) EtOC(S)SK, EtOH; reflux (44%).*

Several preparatively useful pentafluoro-λ^6-sulfanyl perfluoroalkyl iodides were synthesized and reacted with ethylene and tetrafluoroethylene, furnishing, e.g., $SF_5(CF_2)_8I$ and $SF_5(CF_2)_{10}I$. The reaction conditions for the addition of ethylene were investigated in detail, and one of the products ($SF_5(CF_2)_4CH_2CH_2I$) was characterized by X-ray crystallography (*16*).

The reaction of trifluorostyrenes (*o*-, *m*-, *p*-$F_2C{=}CFC_6H_4X$; X = halo, CF_3) with SF_5Br provides the intermediates $F_5SCF_2CFBrC_6H_4X$, which were treated with $AgBF_4$ to furnish the corresponding SF_5 difluorostyrenes $F_5SCF_2CF_2C_6H_4X$. This method is the first high yield preparation of this substance class which shows some promise as a building block for SF_5 based materials (*17*).

Various sulfur fluorides (SF_6, SF_5Cl and CF_3SF_5) were reacted with C_6H_6, C_6H_5Br and C_6H_5Cl in a low temperature radio-frequency plasma. Due to stepwise dissociation of the sulfur fluorides, predominantly fluorination of the arenes was observed. As a minor product, $C_6H_5SF_5$ was found, and $BrC_6H_4SF_5$ or $ClC_6H_4SF_5$ along with numerous halogenated benzenes when C_6H_5Br or C_6H_5Cl were used as reactants, respectively (*18*).

The potentially very useful intermediate 1-pentafluorosulfanyl-2-trimethylsilyl acetylene was prepared by addition of SF_5Br to trimethylsilyl acetylene and subsequent dehydrohalogenation in 44% yield (*19*).

The compound SF_5CF_3 has recently drawn some attention due to its unexpected presence in the stratosphere (*20*), and due to its "super greenhouse gas" properties, with the largest global warming potential ever reported (*21*). The gas phase protonation of SF_5CF_3 was studied (*22*), and a theoretical investigation on the absorption and ionization spectra of the compound has been reported (*23*). Experimental evidence suggests that SF_5CF_3 might originate from the breakdown of SF_6 in high-voltage facilities via spark discharge reactions of SF_6 with some fluorocarbons, such as CHF_3 and CH_2F_2. However, it was also concluded that SF_5CF_3 is unlikely to be generated by attack of an SF_5 radical on a CF_3 group: no SF_5CF_3 was detected in reactions of SF_6 with CF_4 or CF_3Br. Reaction of SF_6 with CHF_3 and CH_2F_2 under spark discharge produced SF_5CF_3, and the yields of SF_5CF_3 were determined as a function of the initial fluorocarbon-to-SF_6 ratio, total gas pressure and discharge time (*24*).

Energetic Materials and Explosives

The pentafluorosulfanyl group has been employed in order to improve safety and effectiveness of energetic materials for use as explosives or rocket fuel (*6*). Most of the SF_5 based explosives are polynitroaliphatic compounds

(*25*), but one publication (*26*) also indicates the usefulness of pentafluorosulfanyl nitramides as oxidizing anions in energetic salts, such as $Q^+SF_5N(NO_2)^-$ (**1**) (Scheme 2).

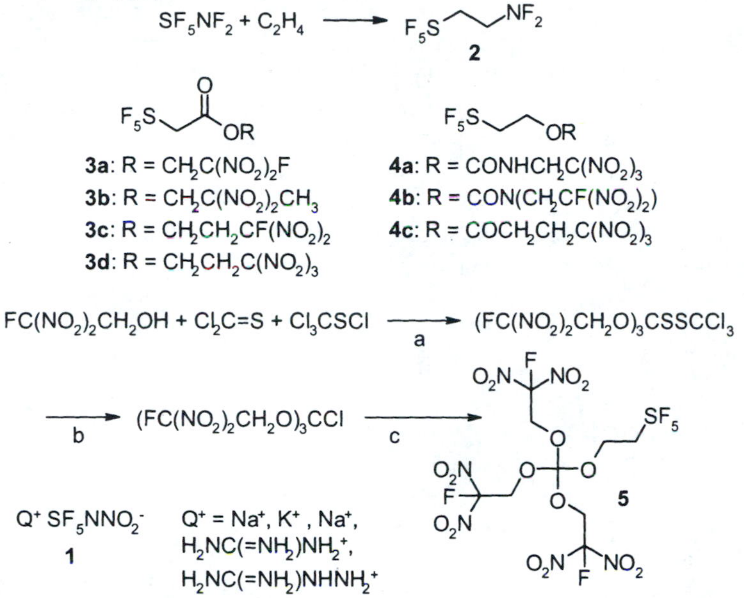

$SF_5NF_2 + C_2H_4 \longrightarrow F_5SCH_2CH_2NF_2$ **2**

3a: R = $CH_2C(NO_2)_2F$
3b: R = $CH_2C(NO_2)_2CH_3$
3c: R = $CH_2CH_2CF(NO_2)_2$
3d: R = $CH_2CH_2C(NO_2)_3$

4a: R = $CONHCH_2C(NO_2)_3$
4b: R = $CON(CH_2CF(NO_2)_2)$
4c: R = $COCH_2CH_2C(NO_2)_3$

$FC(NO_2)_2CH_2OH + Cl_2C{=}S + Cl_3CSCl \xrightarrow{a} (FC(NO_2)_2CH_2O)_3CSSCCl_3$

$\xrightarrow{b} (FC(NO_2)_2CH_2O)_3CCl \xrightarrow{c}$ **5**

$Q^+ SF_5NNO_2^-$ **1** $Q^+ = Na^+, K^+, Na^+, H_2NC(=NH_2)NH_2^+, H_2NC(=NH_2)NHNH_2^+$

*Scheme 2. Examples for SF_5 based energetic materials (**1-5**) (25). The anion $SF_5NNO_2^-$ serves either as an intramolecular oxidizer in the salts **1** which are containing a guanidinium cation, or as an oxidizer in explosive compositions: a) NaOH, H_2O, CH_2Cl_2 (65%). b) Cl_2, $(CH_2Cl)_2$; 60-65°C (95%). c) $SF_5CH_2CH_2OH$, $(CH_2Cl)_2$; reflux (74%).*

In particular, the very high density which can be achieved by inclusion of the SF_5 function results in a significant improvement of the detonation pressure (*27*). For example, the density of $(NO_2)_3CCH_2CH_2O(CO)NHSF_5$ is 1.96 $g{\cdot}cm^{-3}$, whereas the analogous $(NO_2)_3CCH_2CH_2O(CO)NH_2$ shows only 1.74 $g{\cdot}cm^{-3}$. The density of an explosive preparation is of decisive importance for energetic materials as they are used, e. g., for the first stage of nuclear weapons. Energetic materials based on SF_5 compounds also show high thermal stability and reduced shock-sensitivity (*25c*). The reasons for the - from a practical viewpoint highly desirable - decrease of shock-sensitivity have not been conclusively explained so far. As possible contributions to this effect, the formation of relatively stable SF_5 radicals inhibiting radical chain reactions has been discussed, as well as the differences in molecular volume of bound *vs.* free NO_2 and SF_5.

Initially, the SF_5 group also held some promise due to those explosives' potentially very high energy release on detonation, which is mostly determined by the large difference of the S-F bond energy (79 kcal·mol^{-1} on average) (*28*) and the heat of formation of hydrogen fluoride (136 kcal·mol^{-1}) (*29*). In metallized compositions the effect is supposed to be even further enhanced, e.g., by the formation of the very stable Al-F bond (158 kcal·mol^{-1}) (*29*). Thus, in order to achieve maximum energy release, the formation of highly stable fluorine-containing detonation products, such as CF_4, SF_6 or SF_4, is undesirable (*30*). Nevertheless, calorimetric studies showed that the energy release on detonation of $(F(NO_2)_2CH_2O)_3COCH_2CH_2SF_5$ (**5**) (1065-1105 cal·g^{-1}) does not meet these high expectations, although hydrofluoric acid was detected as the only fluorine-containing detonation product.

Ionic Liquids

Due to their lack of any vapour pressure, ionic liquids offer some very interesting features as solvents for various applications (*31*). The structural variation on the cationic moiety is still limited to relatively few basic structures, with even fewer published examples of fluorine-containing cations (*32*). On the other hand, the physico-chemical characteristics (e.g., miscibility with other solvents, viscosity, solvating power) of ionic liquids have been modified in a wide range by using a variety of complex fluorinated anions.

The most prominent feature of ionic liquids containing a pentafluorosulfanyl group as a part of the cation structure - compared to "conventional" analogues with, e. g., a CF_3 group - is their high density. The technical application of ionic liquids is currently just in the beginning, and so far no application has been identified where this property offers a unique advantage justifying the high cost of SF_5 derivatives. Another potential use for the SF_5 function is the modulation of reaction centers of "functional" ionic liquids playing an active role in chemical reactions. So far, there are very few examples of ionic liquids with SF_5 substituted cations in the literature (*33*) (Scheme 3).

The densities of the pentafluorosulfanylpolyfluoroalkyl imidazolium triflimide salts are around 2 g·cm^{-3}, making them the most dense non-chloroaluminate ionic liquids known so far. For comparison, the corresponding propyl and 3,3,3-trifluoropropyl salts exhibit densities of only 1.44-1.48 g·cm^{-3}. Even the highly fluorinated $CF_3(CF_2)_5(CH_2)_2$- derivative shows a significantly lower density of 1.77 g·cm^{-3}. A practical disadvantage of the SF_5 based ionic liquids is their limited thermal stability (>300°C) which is lower than for the other imidazolium triflimides (> 375°C).

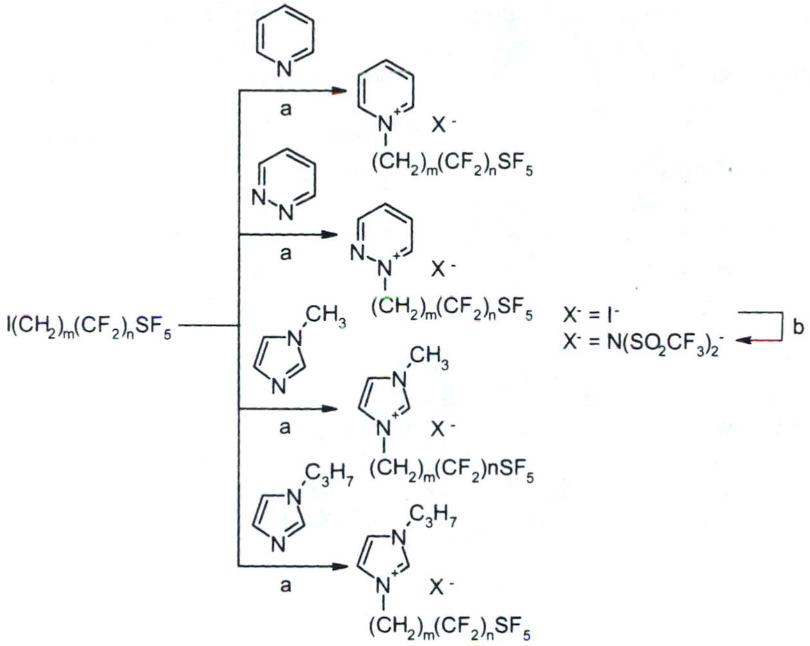

Scheme 3. Synthesis of some ionic liquids based on a SF_5 substituted cation (m = 2, 4; n = 2) (33): a) 1. neat; 65°C, 12 h; 2. washing with acetone/n-pentane 1:3 (typical yields 80-91%) . b) $LiN(SO_2CF_3)_2$ (1.2 equivs.), H_2O/acetone 2:1 (typical yields 85-96%).

Polymers

Several fluorinated monomers bearing an SF_5 group ($XYC=CZSF_5$; X, Y, Z = H or F) were co- and terpolymerized with 1,1-difluoroethylene, vinylidene fluoride and hexafluoropropene. Although the SF_5 monomers do not homopolymerize under radical initiation, they copolymerize in solution in the presence of radical initiators with various fluoroolefinic monomers, furnishing copolymers with SF_5 side groups. The degree of fluorination on the olefinic unit of the SF_5 monomers was found to have an influence on their reactivity, on their relative incorporation into the copolymers, and on the molecular weight of the resulting terpolymers: the higher the fluorine content in the SF_5 monomer, the higher the molecular mass of the resulting polymer and the higher the overall reaction yield. The thermal stability and glass transition temperature of the SF_5 fluoropolymers were also studied in detail (*34*).

A new fluorinated polymer surface comprising SF_5 groups was prepared by polymerization of $SF_5(CF_2)_6(CH_2)_2OC(O)CH=CH_2$ into films. The surface exhibits the typical characteristics of a highly apolar, non-wettable perfluorinated polymer. Depth-dependent XPS studies using angular-resolved methods as well as time-of-flight secondary ion mass spectrometry (TOF-SIMS) indicated the presence of the terminal SF_5 groups preferably in the outer 15 Å of the film surface (*35*).

Two new SF_5 terminated perfluoroalkylsilanes ($SF_5(CF_2CF_2)_nCH_2CH_2SiCl_3$; $n = 1$ and 2) were synthesized by hydrosilylation of the corresponding ω-SF_5 perfluoroalkenes. An SF_5 silane film prepared from the monomer with $n = 1$ shows a hydrophobic interface, with depth-dependent XPS analysis suggesting a poorly organized surface. The longer SF_5 silane analogue ($n = 2$) exhibits a high aqueous contact angle (106°) and angular-dependent XPS compositional variance consistent with an improved film assembly (fewer defects and more vertical chain orientation) (*36*).

As new monomers carrying an SF_5 group, β-SF_5 acrylates and methacrylates were synthesized. The SF_5 acrylate was prepared by addition of SF_5Br to the ethyl propynate, followed by reduction of the resulting $SF_5CH{=}CBrC(O)OEt$. The methacrylate $SF_5CH{=}CMeC(O)OMe$ was obtained by treatment of $SF_5CH_2CBrMeC(O)OMe$ with base (*37*).

Liquid Crystals

Since the beginning of their industrial production in the 1970s, liquid crystal displays (LCD) have developed into an indispensable utility of our daily lives. The high-resolution, full-color LCDs, which are used as PC monitors, for TV or in cellular phones, organizers (PDA) and video games, are based on active matrix (AM) technology (*38*). A key requirement for AM LCD technology is a high specific resistivity and a high voltage holding ratio of the liquid crystalline material. This can so far only be achieved by so-called super fluorinated materials (SFM) containing no other hetero atoms except oxygen and fluorine (Scheme 4) (*1, 39*).

The molecular basis for this requirement is the non-coordinating nature of such compounds precluding the mobilization of ionic trace impurities by coordinative solvation (*40*). Another requirement for many LCD devices is low energy consumption for a longer battery lifetime, which can be achieved by increasing the dielectric anisotropy ($\Delta\varepsilon$) of the liquid crystal which allows a lower operating voltage of the driving electronics. In general, it is very difficult to obtain really high dielectric anisotropies using only fluorine-containing polar groups as the source of molecular polarity, but the SF_5 group was found to offer a unique combination of a strong dipole moment and low ion-solvating capability.

Since the SF_5 group is a versatile and chemically highly stable structure element, a large variety of liquid crystals with a polar pentafluorosulfanyl phenyl terminal group has been prepared (Scheme 5) (*41*). The chemistry

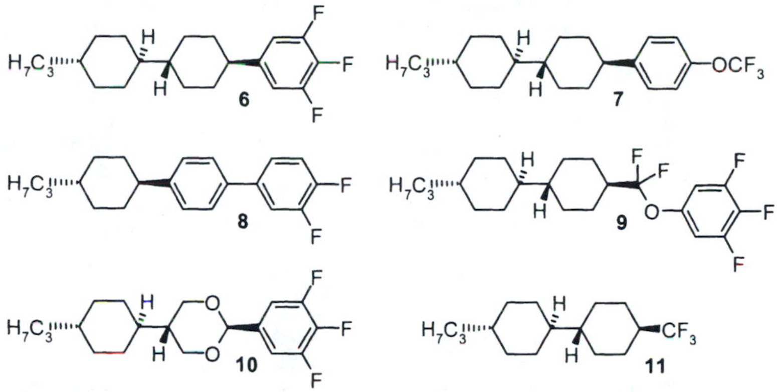

Scheme 4. Typical examples of super fluorinated materials (SFM) used in active matrix liquid crystal displays (1, 39).

employed is mainly based on the commercial availability of 4-nitropentafluorosulfanyl benzene (**12**) (*5*) and on the synthetic toolbox developed by W. A. Sheppard in the early 1960s (*4*) and more recently by J. S. Thrasher and coworkers (*42*).

In comparison to materials carrying a trifluoromethyl group as the closest "conventional" analog, the SF_5 compounds show significantly higher clearing points ($T_{NI,extr}$) and a higher dielectric anisotropy ($\Delta\varepsilon$) (Table 1). Disadvantageous with regard to technical application are a tendency towards a higher rotational viscosity (γ_1, basically proportional to the switching time of an LCD based on a liquid crystal) and relatively high melting points. However, materials based on a polar SF_5 terminal group are of practical interest as components of complex liquid crystal mixture preparations.

Recently, some aliphatic and olefinic SF_5 derivatives have also been prepared as liquid crystals with a combination of high dielectric anisotropy and low birefringence (Δn) (*41a*). In this case, no pre-formed building blocks (such as **12**) were available, and the SF_5 group had to be introduced via the radical addition of SF_5Br to carbon-carbon double bonds of suitable liquid crystalline precursors (Scheme 6) (*12, 13, 44*).

Also these cycloaliphatic compounds show significantly higher dielectric anisotropy than their CF_3 analogues (Table 2). Their low birefringences render them interesting key components for liquid crystal mixtures used in reflective or transflective LCDs for mobile applications, such as cellular phones, video games or mobile multi-media devices. The cholestane derivative **35** shows a helical twisting power of (HTP) of 7.7 μm^{-1} and is of interest as a chiral dopant for liquid crystal mixtures (*45, 46*).

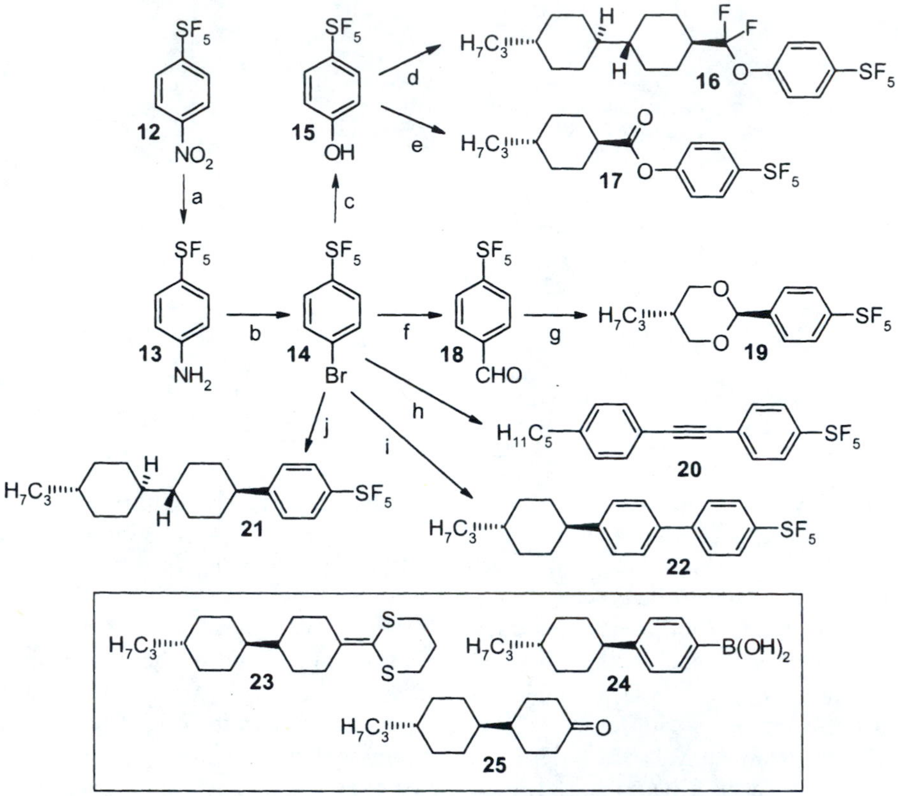

*Scheme 5. Synthesis of pentafluorosulfanyl based two-ring liquid crystals **16**, **17** and **19-22**: a) 1. H_2, 5% Pd-C, THF. b) 1. HBr, $NaNO_2$; -5°C; 2. CuBr; room temp. → 80°C (46%). c) 1.* tert-*BuLi, Et_2O; -70°C; 2. $B(OMe)_3$; -70°C → -20°C; 3. HOAc, H_2SO_4, H_2O_2 (30%); -20°C → 35°C, 1h (58%). d) 1. **23**, CF_3SO_3H, CH_2Cl_2; 0°C (5 min) → RT (30 min) → –70°C; 2. NEt_3, CH_2Cl_2; -70°C; 3. $NEt_3 \cdot 3HF$; -70°C; 4. DBH (1,3-dibromo-5,5-dimethylhydanthoin), CH_2Cl_2; -70°C → -10°C (25%). e) 4-Propylcyclohexylcarbonyl chloride, pyridine, 0.1 equiv. DMAP, CH_2Cl_2 (24%). f) 1.* tert-*BuLi, Et_2O; -78°C; 2.* N-*Formylpiperidine; -40°C → room temp. (76%). g) 2-(Trimethylsiloxymethyl)-1-trimethylsiloxypentane, cat. Me_3SiOTf, CH_2Cl_2; -78°C, 30 min. (67%). h) 1-Ethinyl-4-propylbenzene, cat. $Pd(PPh_3)_4$, pyrrolidine; room temp., 18 h (33%). i) **24**, cat. $Pd(PPh_3)_4$, toluene, 2 N NaOH; room temp., 2 d (42%). j) 1.* tert-*BuLi, Et_2O; -78°C; 2. **25**; -78°C → room temp.; 3. Cat. TsOH, toluene; azeotropic removal of water; 4. H_2, 5% Pd-C, THF (10% pure* trans-trans-*isomer).*

Table 1. Physical properties of some pentafluorosulfanyl derivatives and some selected conventional fluorinated liquid crystals (*43*).

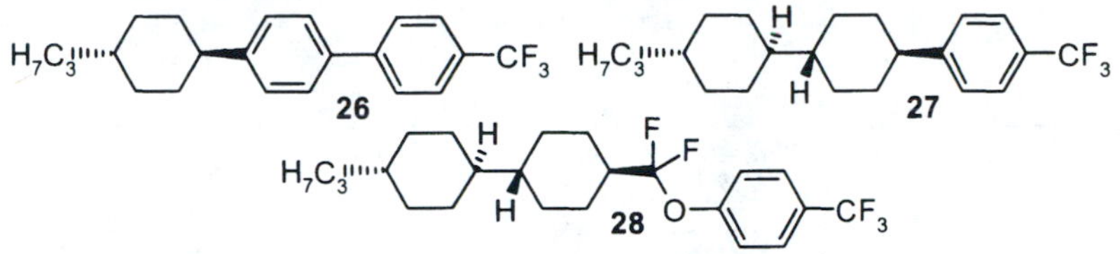

Compound	*Phase Sequence*	$T_{NI,extr}$	$\Delta\varepsilon_{virt}$	Δn_{virt}	$\gamma_{1,virt}$
26	C 134 I	108.6	13.0	0.1645	279
22	C 109 N (87.8) I	94.6	14.3	0.1537	634
27	C 133 I	112.2	9.5	0.0910	338
21	C 121 I	95.5	11.6	0.0933	612
28	C 78 S_G (66) N 125.0 I	114.2	9.2	0.0861	285
16	C 67 N 116.5 I	108.2	11.8	0.0800	488

NOTE: The temperatures are given in °C, the γ_1 values in mPa·s. C = crystalline, S_G = smectic G, N = nematic, I = isotropic. Numbers in parentheses denote monotropic phase transitions.

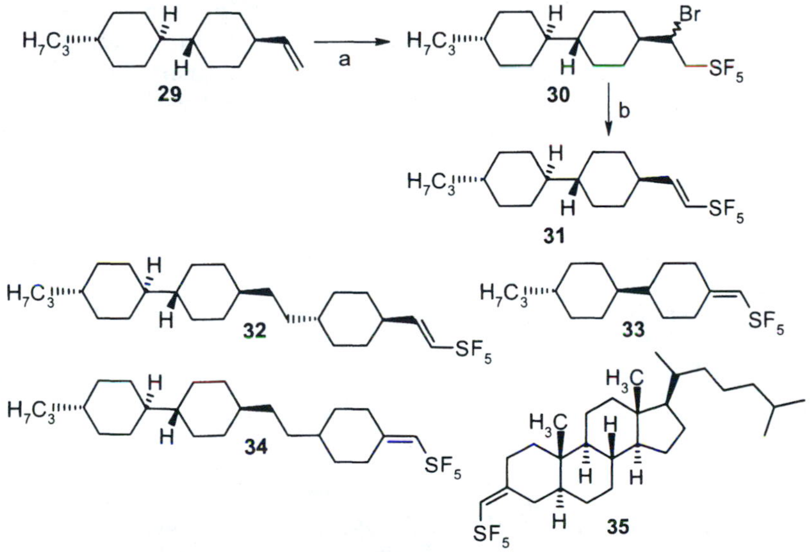

*Scheme 6. Syntheses of the liquid crystal **31**: a) SF_5Br, BEt_3 (0.15 equivs.),* n-*heptane; -40 to -20°C, 2 h (85%). b) KOH powder,* n-*heptane; 35°C, 18 h (74%). The liquid crystals **32-35** were synthesized by an analogous method.*

Table 2. The physical properties of olefinic SF_5 substituted liquid crystals in comparison to some of their CF_3 analogues (*43*).

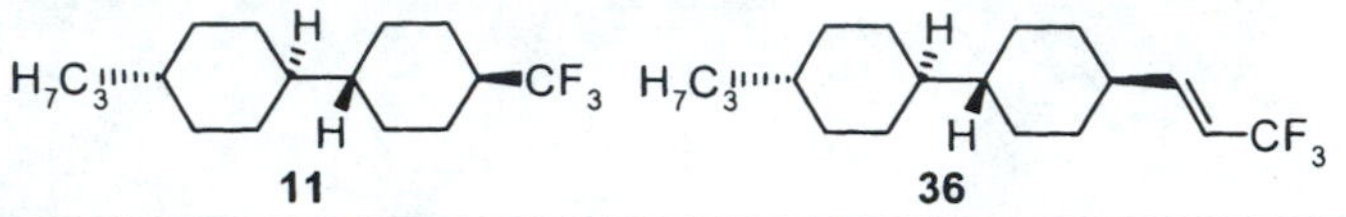

Compound	*Phase Sequence*	$T_{NI,virt}$	$\Delta\varepsilon_{virt}$	Δn_{virt}
11	C 19 S_H? (8) S_B? 41 I	-44.4	6.8	0.059
36	C 60 S_B 61 I	5.6	6.2[a]	0.051[a]
31	C 51 S_G? 65 I	8.3	10.5	0.070
33	C 41 I	-93.8	6.3	0.063
32	C 145 S_B 173 N 194.1 I	168.4	10.1	0.064
34	C 68 S_B 115 I	79.5	4.3	0.057
35	C 123 I	-	-	-

NOTE: The phase transition temperatures and the virtual clearing points ($T_{NI,virt}$) are cited in °C. C = crystalline, S_B = smectic B, S_G = smectic G, S_H = smectic H, N = nematic, I = isotropic. In the cases marked by a question mark the mesophase assignment is not clear. Values in parentheses denote monotropic phase transitions.

[a]Extrapolated from Merck mixture N.

Another type of SF_5 substituted liquid crystals (Scheme 7) has been presented recently by G. L. Gard and coworkers (*41c*). The compounds **37** and **38** carry a $CF_2CF_2SF_5$ function as their polar terminal group, and the structure and mesogenic properties of this class of liquid crystals was studied in detail. Due to the hydrolytic lability of their central Schiff base unit, these materials are not suitable for practical application in current active matrix LCD technology.

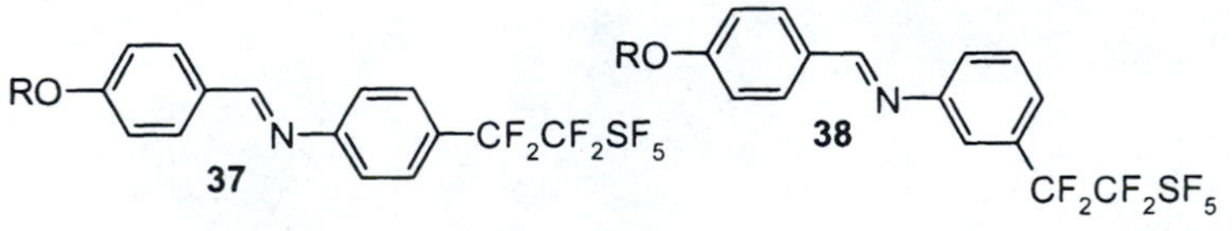

*Scheme 7. $SF_5CF_2CF_2$ substituted Schiff bases **37** and **38** ($R = C_7H_{15}$, $C_{10}H_{21}$).*

Two different approaches (Scheme 8) have been tried to further increase the dielectric anisotropy of SF_5 based liquid crystals – unfortunately with limited success due to the remarkable and unusual steric and dynamic properties of hypervalent sulfur fluorides.

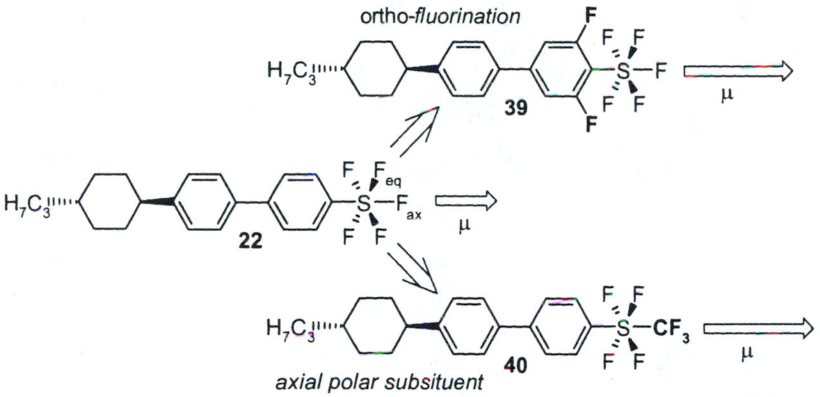

Scheme 8. Two strategies to increase the dielectric anisotropy ($\Delta\varepsilon$) of liquid crystals based on hypervalent sulfur fluorides as polar function: lateral ortho-*fluorination (**39**), and replacing the axial fluorine atom by a group wih stronger dipole moment (μ) (**40**).*

The first of these approaches (*41b*) is based on the introduction of additional aromatic *ortho*-fluorine substituents, a classic design principle for many liquid crystals in commercial use. The local dipole moments of the polar carbon-fluorine bonds are supposed to add to the SF_5 group dipole moment (μ), resulting in a dramatic increase of the dielectric anisotropy which is proportional to the square of the dipole moment which is oriented along the direction of the long molecular axis (*47*).

Another point in favour of *ortho*-fluorination is the flexibility of the F_{ax}-S-F_{eq} resp. C_{ar}-S-F_{eq} bond angle in the SF_5 group. DFT calculations (*48*) indicate that the C_{ar}-S-F_{eq} angle α (Figure 1) can be increased with a very small amount of strain energy: only about 1 kcal·mol^{-1} is required to extend this angle by 1°, resulting in an increase of the group dipole moment by ca. 0.3 D. Normally, the equatorial S-F_{eq} bonds are oriented perpendicularly to the long molecular axis, cancelling each other and thus not contributing significantly to the molecular dielectric anisotropy ($\Delta\varepsilon$). If *ortho*-fluorination pushes the ring of equatorial fluorine atoms "forward" (i.e., into the direction of the polar terminal group), this causes their local dipole moments to add to the overall molecular dipole moment pointed into the direction of the long molecular axis. This again is expected to result in a dramatic increase of the dielectric anisotropy.

Ortho-fluorinated SF_5 arenes and their chemistry have been explored first by J. S. Thrasher and coworkers (*15*), but the compounds reported by his group did not have the right substitution pattern to be useful as liquid crystal precursors. Therefore, the key intermediates for the synthesis of **39** and **47** were prepared by direct fluorination of the corresponding diaryl disulfides (Scheme 9).

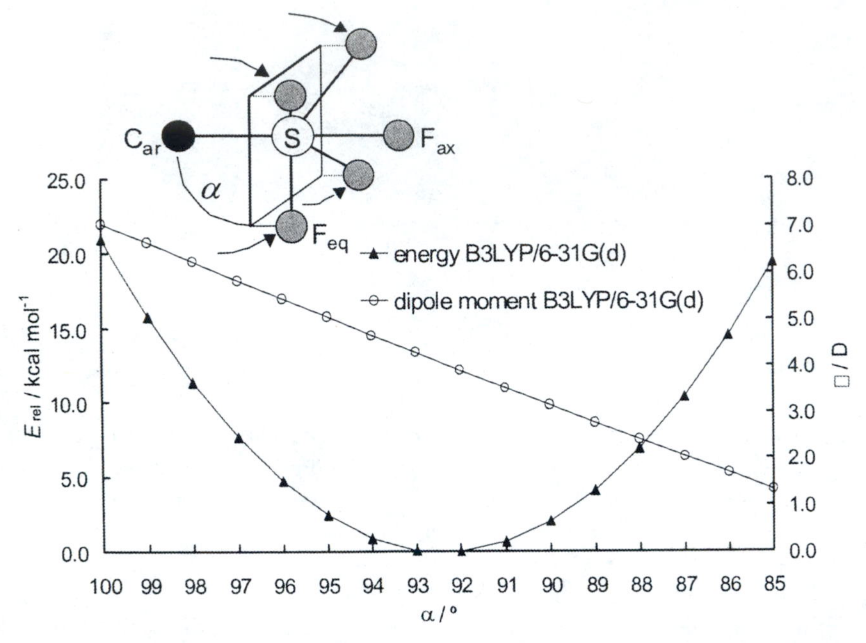

Figure 1. Correlation between dipole moment μ, the deformation energy and the twist angle α for pentafluorosulfanylbenzene, calculated on the B3LYP/6-31G(d) level of theory (48).

Surprisingly, the expectations in an extremely high dielectric anisotropy of *ortho*-difluorinated materials like **39** were not fulfilled (Table 3). The increase of $\Delta\varepsilon$ due to the two lateral *ortho*-fluorine substituents was much smaller than for the analogous trifluoromethyl system **26-48**. Thus, the SF_5 system **39** has about the same $\Delta\varepsilon$ as its CF_3 based analog **48**. On the other hand, in case of the *mono*-fluorinated system **47**, the value of $\Delta\varepsilon$ lies within the expected range.

A DFT study (*48*) showed that this less than satisfactory result could again be explained by the high conformational flexibility of the SF_5 group: the aromatic *ortho*-fluorine substituents are not pushing "forward" the equatorial fluorine atoms of the SF_5 function but "squeezing" between them, thereby reducing the C_{ar}-S-F_{eq} angle and thus resulting in a smaller group dipole moment by the slightly "backwards" oriented polar S-F_{eq} bonds.

The second strategy towards SFM liquid crystals with extremely high dielectric anisotropy is based on replacing the axial fluorine of the SF_5 group (which gives the strongest contribution to the group dipole moment) by the more polar CF_3 group (*2*). The synthesis of liquid crystals with a terminal *trans*-SF_4CF_3 group (**40**) is based on the direct fluorination of a suitable trifluoromethylthio precursor (Scheme 10). Since the *cis* isomer was formed as the major fluorination product, it was isomerized to the thermodynamically

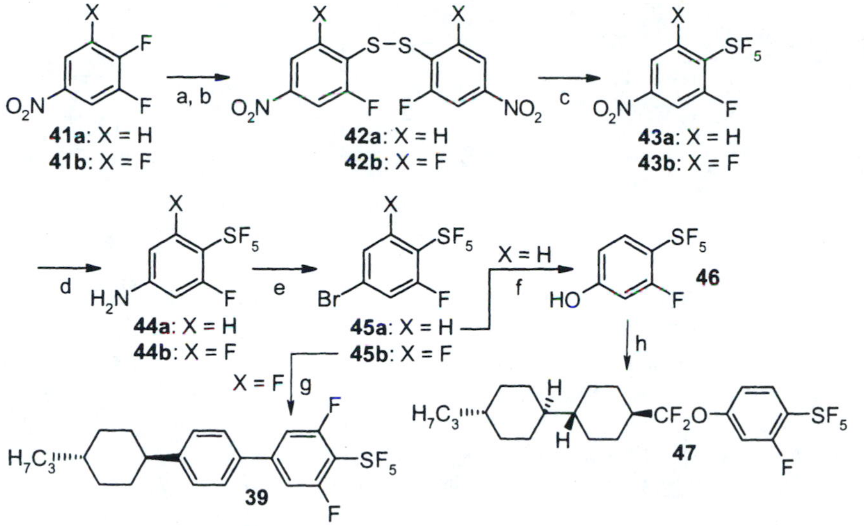

Scheme 9. Synthesis of the liquid crystals **39** *and* **47**: *a)* $Na_2S \cdot 9H_2O$, *DMSO; RT → 50°C. b)* $NaBO_4 \cdot 3H_2O$, *HOAc; RT (***42a***: 63%,* **42b***: 77%). c) 10%* F_2/N_2, CH_3CN*; -5°C → 0°C (***43a***: 9%,* **43b***: 4.6%). d) Cat. Raney-Ni,* H_2*, THF; RT, 1 bar (***44a***: 95%,* **44b***: 95%). e) 1.* $NaNO_2$*, 47% HBr; 0-2°C; 2. CuBr; 85°C (***45a***: 51%,* **45b***: 49%). f) 1.* tert-*BuLi,* Et_2O*; -70°C; 2.* $B(OMe)_3$*; 3. HOAc,* H_2SO_4*, 30%* H_2O_2*; -20°C → 35°C (crude product used for next step). g)* **24**, $NaBO_2 \cdot 8H_2O$, H_2O*, THF, cat.* $Pd(PPh_3)_4$*; reflux, 18 h (16%). h) 1.* **23**, CF_3SO_3H, CH_2Cl_2*; 0°C (5 min) → RT (30 min) → –70°C; 2.* **46**, NEt_3, CH_2Cl_2*; -70°C; 3.* $NEt_3 \cdot 3HF$*; -70°C; 4. DBH (1,3-dibromo-5,5-dimethylhydanthoin),* CH_2Cl_2*; -70°C → -10°C (7%).*

Table 3. The physical properties of *ortho*-fluorinated with a terminal SF_5 group (*43*).

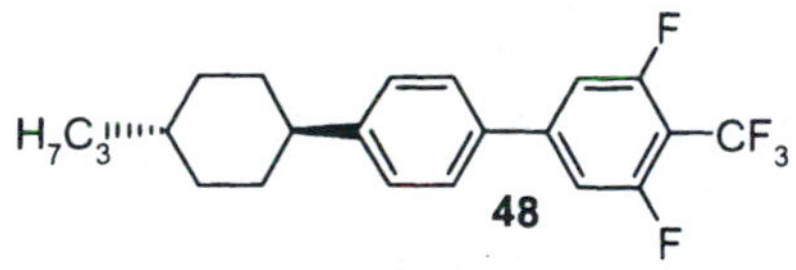

Compound	*Phase Sequence*	$T_{NI,virt}$	$\Delta\varepsilon_{virt}$	Δn_{virt}
48	C 86 I	47.1	21.5	0.149
39	C 103 I	49.6	21.4	0.132
47	C 50 N 101.9 I	87.6	14.3	0.080

NOTE: The phase transition temperatures are given in °C. C = crystalline, N = nematic, I = isotropic. Numbers in parentheses denote monotropic phase transitions.

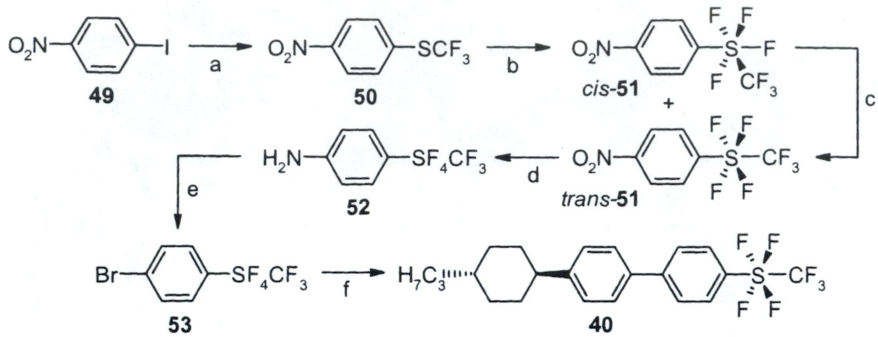

Scheme 10. *Synthesis of the liquid crystal* **40**: *a) $CuSCF_3$, DMF; 130°C (60%). b) 10% F_2/N_2, CH_3CN; -3 - 2°C (50% of a* cis/trans *85:15 mixture). c) $AlCl_3$ (0.8 equivs.), CH_2Cl_2; -10°C, 30 min (48% pure* trans*). d) Cat. Raney-Ni, H_2, THF; RT, 1 bar (87%). e) 1. $NaNO_2$, 47% HBr; 0-5°C; 2. CuBr; 85°C (52%). f)* **24**, *$NaBO_2 \cdot 8H_2O$, H_2O, THF, cat. $Pd(PPh_3)_4$; 60°C, 5 h (14%).*

preferred *trans* isomer by using the Lewis acid $AlCl_3$. The *trans*-SF_4CF_3 group has a hydrolytic stability comparable to the SF_5 function.

As already for the *ortho*-fluorination concept, also the introduction of the axial trifluoromethyl group in the liquid crystal **40** afforded a far lower value of $\Delta\varepsilon$ than expected (mesophase sequence of **40**: C 197 N 209.7 I; $T_{NI,virt}$ = 182.6°C; $\Delta\varepsilon_{virt}$ = 10.6; Δn_{virt} = 0.150), being even lower than for the analogous SF_5 derivative **22** with a $\Delta\varepsilon$ value of 14.3. Again, a DFT study (*48*) revealed that the sterically bulky axial CF_3 group pushed the ring of equatorial fluorine atoms "backwards" creating a local dipole moment opposed to the main dipole created by the trifluoromethyl group.

A first step on the way towards using organic derivatives of SF_6 not as a polar terminal group of liquid crystals but as a part of their mesogenic core structure is based on the *trans*-diaryl tetrafluorosulfurane moiety (*3*). The basic building block was synthesized by direct fluorination of bis(4-nitrophenyl)thioether (**54**) and subsequent catalytic isomerization of the isomer mixture (*cis*/*trans* 85:15) to pure *trans*-**51** (Scheme 11). The Lewis acid catalyzed isomerization proceeds presumably via a transient diaryltrifluorosulfuranonium cation (**56**$^+$) which – in contrast to hexavalent sulfur species - has a very low inversion barrier.

As a more direct approach towards liquid crystals with a *trans*-tetrafluorosulfurane bridge, the bis(triflate) **57** was synthesized by an analogous route (*49*). Further synthetic studies directed towards liquid crystals with a SF_4 bridge in their mesogenic core structure are ongoing.

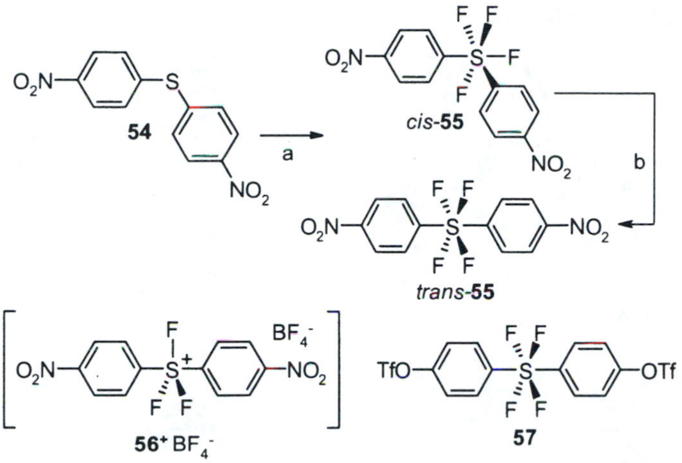

Scheme 11. Synthesis of the liquid crystal precursors **55** and **57**: a) 10% F_2/N_2, CH_3CN; -5 → 2°C (68%; mixture of trans- and cis-**55**, 15:85). b) 1. Cat. $BF_3 \cdot OEt_2$, CH_2Cl_2; room temp., 1 h; 2. Me_3SiOMe; room temp., 10 min (87% pure trans-**55**). The bis(triflate) **57** was synthesized by an analogous route.

Pharmaceuticals and Agrochemicals

The *trans*-SF_4CF_3 group has been characterized with regard to its physicochemical properties as a new substituent in biologically active compounds (*2*). The lipophilicity increment π_p was determined at +2.13, exceeding the already large values for the SCF_3 (+1.44) and SF_5 group (+1.23). The Hammett substituent parameter σ_p is +0.68, the same as for the SF_5 group but significantly higher than for CF_3 (+0.53) or CN (+0.66). Therefore, the *trans*-SF_4CF_3 group exhibits "polar hydrophobicity" (*50*) to an extreme extent.

In the last decade a number of SF_5 based bioactive compounds have been reported, mostly in the patent literature. A recent review by P. J. Crowley and colleagues (*51*) give a comprehensive overview on SF_5 pesticides, and provides some comparative data on their activity compared to CF_3 analogues. Unfortunately, to our knowledge there are still no published data on toxicity and metabolism of organic SF_5 derivatives.

Recently, the SF_5 compounds shown in Scheme 12 have been explored for their potential as anti-cancer drugs (**58**, **59**) (*52, 53*), and for the treatment of cardic arrhythmia (**60**) (*53*).

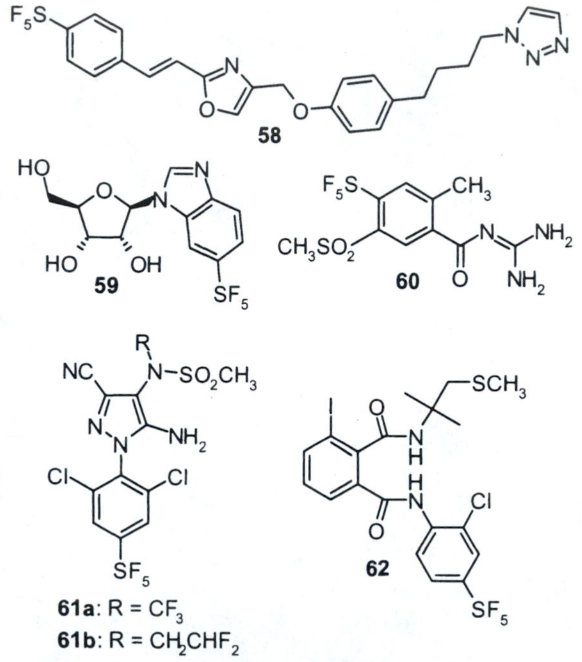

Scheme 12 structures: 58; 59; 60; 61a: R = CF_3; 61b: R = CH_2CHF_2; 62

Scheme 12. Recently published pharmaceuticals and pesticides containing a pentafluorosulfanyl group.

Pentafluorosulfanyl derivatives were found to offer some advantage over conventional fluoroorganic compounds as pesticides (Scheme 12). Thus, SF_5 substituted phenylpyrazoles **61a** (*51*), **61b** (*55*) and related compounds (*56-61*) show excellent activity against a variety of pests, including arthropods and helminths. An SF_5 analogue (**61a**) of the insecticide Fipronil has a significantly higher activity than the corresponding CF_3 compound (*51*). The pentafluorosulfanyl substituted phthalodiamide **62** (*62*) was identified as a highly effective agricultural or horticultural insecticide.

Summary and Outlook

Although functional groups based on hypervalent sulfur fluorides, such as SF_5 and *trans*-SF_4CF_3, offer a unique profile of highly unusual properties, they still have not found wide-spread use. In the past this was mostly due to the difficult access to this class of compounds. Nevertheless, the situation has changed in the last decade with the introduction of new synthetic methods opening up large scale access in particular to SF_5 arenes. Especially for biomedical and agrochemical applications it is still considered risky to have a closer look on hypervalent sulfur fluorides with so few data available on toxicity and metabolism of these compounds.

In the field of polymer and liquid crystal chemistry, using the SF_5 group provides access to physicochemical characteristics which cannot be easily achieved by other means. What is still missing for the real break-through, are key applications which require functional groups based on hypervalent sulfur fluorides as an essential necessity, and whose intrinsic value justifies the development risk and still high price for these compounds.

Acknowledgement

G.-V. R. thanks the Forchheimer Foundation for a fellowship as visiting professor at the Department of Medicinal Chemistry, The Hebrew University of Jerusalem (Israel) for 2005/2006.

References

1. Kirsch, P. *Modern Fluoroorganic Chemistry: Synthesis, Reactivity, Applications*, Wiley-VCH, Weinheim, Germany, **2004**.
2. Kirsch, P.; Hahn, A. *Eur. J. Org. Chem.* **2006**, 1125-1131.
3. Kirsch, P.; Bremer, M.; Kirsch, A.; Osterodt, J. *J. Am. Chem. Soc.* **1999**, *121*, 11277-11280.
4. a) Sheppard, W. A. *J. Am. Chem. Soc.* **1960**, *82*, 4751-4752. b) Sheppard, W. A. *J. Am. Chem. Soc.* **1962**, *84*, 3064-3071. c) Sheppard, W. A. *J. Am. Chem. Soc.* **1962**, *84*, 3072-3076.
5. a) Chambers, R. D.; Greenhall, M. P.; Hutchinson, J.; Moilliet, J. S.; Thomson, J. *Abstract of Papers*, in: *Proceedings of the 211th National Meeting of the American Chemical Society*, New Orleans, LA, March 24-26, **1996**, American Chemical Society, Washington, DC, **1996**, FLUO 11. b) Greenhall, M. P. *15th International Symposium on Fluorine Chemistry*, Vancouver, Canada, Aug. 2-7, **1997**, presentation FRx C-2. c) Bowden, R. D.; Greenhall, M. P.; Moillet, J. S.; Thomson, J. *WO 97/05106*, **1997**.
6. Winter, R. W.; Dodean, R. A.; Gard, G. L. in *ACS Symposium Series, 911: Fluorine-Containing Synthons* (Soloshonok, V. A.; ed.), ACS Press, Washington, DC, **2005**, pp. 87-118.
7. Lu, N.; Thrasher, J. S.; von Ahsen, St.; Willner, H.; Hnyk, D.; Oberhammer, H. *Inorg. Chem.* **2006**, *45*, 1783–1788
8. Brel, V. K. *Synthesis*, **2005**, 1245-1250.
9. Trushkov, I. V.; Brel, V. K. *Tetrahedron Lett.* **2005**, *46*, 4777-4779.
10. Lal, G. S.; Minnich, K. E. *US 20051076329*, **2005**.
11. Lal, G. S.; Minnich, K. E. *US 6870068*, **2005**.
12. a) Sergeeva, T. A.; Dolbier Jr., W. R. *Org. Lett.* **2004**, *6*, 2417-2419. b) Dolbier Jr., W. R.; Aït-Mohand, S. *WO 2004011422*, **2004**.

13. Winter, R. W.; Gard, G. L. *J. Fluorine Chem.* **2004**, *125*, 549-552.
14. Bailey III, W. H.; Casteel, W. J.; Pesaresi, R. J.; Prozonic, F. M. *EP 1484318*, **2004**.
15. Sipyagin, A. M.; Enshov, V. S.; Kashtanov, S. A.; Bateman, C. P.; Mullen, B. D.; Tan, Y.-T.; Thrasher, J. S. *J. Fluorine Chem.* **2004**, *125*, 1305-1316.
16. Nixon, P. G.; Mohtasham, J.; Winter, R.; Gard, G. L.; Twamley, B.; Shreeve, J. M. *J. Fluorine Chem.* **2004**, *125*, 553-560.
17. Winter, R. W.; Dodean, R.; Smith, J. A.; Anilkumar, R.; Burton, D. J.; Gard, G. L. *J. Fluorine Chem.* **2003**, *122*, 251-253.
18. Klampfer, P.; Skapin, T.; Kralj, B.; Zigon, D.; Jesih, A. *Acta Chim. Slov.* **2003**, *50*, 29-42.
19. Lal, G. S.; Minnich, K. E. *US 6479645*, **2002**.
20. Sturges, W. G.; Wallington, T. J.; Hurley, M. D.; Shine, K. P.; Shira, K.; Angel, A.; Oram, D. E.; Penkett, S. A.; Mulvaney, R.; Brenninkmeijer, C. A. M. *Science* **2000**, *289*, 611-613.
21. Baby, S. *Nature, Environment Pollution Tech.* **2005**, *4*, 171-178.
22. Pepi, F.; Ricci, A.; Di Stefano, M.; Rosi, M. *Phys. Chem. Chem. Phys.* **2005**, *7*, 1181-1186.
23. Turki, M.; Eisfeld, W. *Phys. Chem. Chem. Phys.* **2005**, *7*, 1700-1707.
24. Huang, L.; Zhu, L.; Pan, X.; Zhang, J.; Ouyang, B.; Hou, H. *Atmospheric Environment* **2005**, *39*, 1641-1653.
25. a) Witucki, E. F.; Frankel, M. B. *Energetic Aliphatic Sulfur Pentafluoride Derivatives*, UCRL-13809, **1978**. b) Bovin, J. L. *CA 1085875*, **1980**. c) Sitzmann, M. E.; Gilligan, W. H.; Ornellas, D. L.; Thrasher, J. S. *J. Energ. Mat.* **1990**, *8*, 352-374. d) Sitzmann, M. E. *US 5194103*, **1993**.
26. Sitzmann, M. E.; Gilarda, R.; Butcher, R. G.; Koppes, W. M.; Stern, A. G.; Thrasher, J. S.; Trivedi, N. J.; Yang, Z.-Y. *Inorg. Chem.* **2000**, *39*, 843.
27. Kamlet, M. J.; Jacobs, S. J. *J. Chem. Phys.* **1968**, *48*, 23.
28. Hildenbrand, D. L. *J. Phys. Chem.* **1973**, *77*, 897.
29. *Handbook of Chemistry and Physics* (Weast, R. C.; ed.), CRC Press, Inc., Boca Raton, **1985**.
30. Ornellas, D. L. *Propellants, Explosives, Pyrotechnics* **1989**, *14*, 122-123.
31. Recent review: Jain, N.; Kumar, A.; Chauhan, S.; Chauhan, S. M. S. *Tetrahedron* **2005**, *61*, 1015-1060.
32. Xue, H.; Shreeve, J. M. *Eur. J. Inorg. Chem.* **2005**, 2573-2580.
33. Singh, R. P.; Winter, R. W.; Gard, G. L.; Shreeve, J. M. *Inorg. Chem.* **2003**, *42*, 6142-6146.
34. Kostov, G.; Ameduri, B.; Sergeeva, T.; Dolbier Jr., W. R.; Winter, R.; Gard, G. L. *Macromolecules* **2005**, *38*, 8316-8326.
35. Winter, R.; Nixon, P. G.; Terjeson, R. J.; Mohtasham, J.; Holcomb, N. R.; Grainger, D. W.; Graham, D.; Castner, D. G.; Gard, G. L. *J. Fluorine Chem.* **2002**, *115*, 107-113.
36. Nixon, P. G.; Winter, R.; Castner, D. G.; Holcomb, N. R.; Grainger, D. W.; Gard, G. L. *Chem. Mat.* **2000**, 3108-3112.

37. Winter, R. W.; Dodean, R.; Holmes, L.; Gard, G. L. *J. Fluorine Chem.* **2004**, *125*, 37-41.
38. a) Pauluth, D.; Tarumi, K. *J. Mater. Chem.* **2004**, *14*, 1219-1227; b) Pauluth, D.; Tarumi, K. *J. SID* **2005**, *13*, 693-702.
39. Kirsch, P.; Bremer, M. *Angew. Chem.* **2000**, *112*, 4384-4405; *Angew. Chem. Int. Ed.* **2000**, *39*, 4216-4235.
40. Bremer, M; Naemura, S.; Tarumi, K. *Jpn. J. Appl. Phys.* **1998**, *37*, L88-L90.
41. a) Kirsch, P.; Binder, J. T.; Lork, E.; Röschenthaler, G.-V. *J. Fluorine Chem.* **2006**, in press [doi: 10.1016/j.fluchem.2005.11.015]. b) Kirsch, P.; Hahn, A. *Eur. J. Org. Chem.* **2005**, 3095-3100. c) Smith, J. A.; DiStasio Jr., R. A.; Hannah, N. A.; Winter, R. W.; Weakly, T. J. R.; Gard, G. L.; Rananavare, S. B. *J. Phys. Chem. B* **2004**, *108*, 19940-19948. d) Kirsch, P.; Bremer, M.; Taugerbeck, A.; Wallmichrath, T. *Angew. Chem.* **2001**, *113*, 1528-1532; *Angew. Chem. Int. Ed.* **2001**, *40*, 1480-1484. e) Kirsch, P.; Bremer, M.; Kirsch, A.; Manabe, A.; Poetsch, E.; Reiffenrath, V.; Tarumi, K. *Mol. Cryst. Liq. Cryst.* **2000**, *346*, 193-199. f) Kirsch, P.; Bremer, M.; Heckmeier, M.; Tarumi, K. *Mol. Cryst. Liq. Cryst.* **2000**, *346*, 29-33. g) Kirsch, P.; Bremer, M.; Heckmeier, M.; Tarumi, K. *Angew. Chem.* **1999**, *111*, 2174-2178; *Angew. Chem. Int. Ed. Engl.* **1999**, *38*, 1989-1992.
42. Sipyagin, A. M.; Bateman, C. P.; Tan, Y.-T., Thrasher, J. S. *J. Fluorine Chem.* **2001**, *112*, 287-295.
43. The application oriented evaluation of liquid crystals for use in LCDs is centered around „virtual“ clearing temperatures, electrooptical parameters and viscosities. These data are obtained by extrapolation from a standardized nematic host mixture: $T_{NI,virt}$, $\Delta\varepsilon_{virt}$, Δn_{virt} and $\gamma_{1,virt}$ were determined by linear extrapolation from a 10% w/w solution in the commercially available Merck mixture ZLI-4792 (T_{NI} = 92.8°C, $\Delta\varepsilon$ = 5.27, Δn = 0.0964). The values thus obtained are empirically corrected for changes in the order parameter. For the pure substances the mesophases were identified by optical polarization microscopy, and the phase transition temperatures by differential scanning calorimetry (DSC).
44. a) Hoover, F. W.; Coffman, D. D. *J. Org. Chem.* **1964**, *29*, 3567-3570. b) Case, J. R.; Ray, N. H.; Roberts, H. L. *J. Chem. Soc.* **1961**, 2066-2070. c) Wessel, J.; Hartl, H.; Seppelt, K. *Chem. Ber.* **1986**, *119*, 453. d) Henkel, T.; Klauck, A.; Seppelt, K. *J. Organomet. Chem.* **1995**, *501*, 1. e) Aït-Mohand, S.; Dolbier Jr., W. R. *Org. Lett.* **2002**, *4*, 3013-3015.
45. The helical twisting power (HTP) was measured at 20°C in a 1% w/w solution of the analyte in the Merck liquid crystal mixture MLC-6260 (T_{NI} = 103.5°C, $\Delta\varepsilon$ = 4.0, Δn = 0.088).
46. Pauluth, D.; Wächtler, A. E. F. *Synthesis an Application of Chiral Liquid Crystals* in *Chirality in Industry II* (Collins, A. N.; Sheldrake, G. N.; Crosby, J.; eds.), John Wiley & Sons Ltd., New York, **1997**, pp. 263-286.

47. The dielectric anisotropy is defined as $\Delta\varepsilon = \varepsilon_{\|} - \varepsilon_{\perp}$, the birefringence as $\Delta n = n_{\|} - n_{\perp}$, where $_{\|}$ stands for parallel and $_{\perp}$ perpendicular to the nematic phase director, which is approximated by the molecular orientation axis or the long molecular axis, respectively. The correlation between $\Delta\varepsilon$, the molecular dipole moment μ and the angle β between the dipole and the orientation axis is the following: $\Delta\varepsilon \propto \Delta\alpha - F\,(\mu^2/2k_BT)\,(1\text{-}3\cos^2\beta)\,S$; $\Delta\alpha$ is the anisotropy of the polarizability, F the reaction field factor, S the order parameter: a) Maier, W.; Meier, G. *Z. Naturforschg.* **1961**, *16a*, 262-267. b) Demus, D.; Pelzl, G. *Zeitschr. f. Chem.* **1982**, *21*, 1. c) Michl, J.; Thulstrup, E. W. *Spectroscopy with Polarized Light: Solute Alignment by Photoselection, in Liquid Crystals, Polymers, and Membranes*, VCH, Weinheim, Germany, **1995**, 171-221.
48. *Gaussian 98*, Revision A.6: Frisch, M. J.; Trucks, G. W.; Schlegel, H. B.; Scuseria, G. E.; Robb, M. A.; Cheeseman, J. R.; Zakrzewski, V. G.; Montgomery Jr., J. A.; Stratmann, R. E.; Burant, J. C.; Dapprich, S.; Millam, J. M.; Daniels, A. D.; Kudin, K. N.; Strain, M. C.; Farkas, O.; Tomasi, J.; Barone, V.; Cossi, M.; Cammi, R.; Mennucci, B.; Pomelli, C.; Adamo, C.; Clifford, S.; Ochterski, J.; Petersson, G. A.; Ayala, P. Y.; Cui, Q.; Morokuma, K.; Malick, D. K.; Rabuck, A. D.; Raghavachari, K.; Foresman, J. B.; Cioslowski, J.; Ortiz, J. V.; Stefanov, B. B.; Liu, G.; Liashenko, A.; Piskorz, P.; Komaromi, I.; Gomperts, R.; Martin, R. L.; Fox, D. J.; Keith, T.; Al-Laham, M. A.; Peng, C. Y.; Nanayakkara, A.; Gonzalez, C.; Challacombe, M.; Gill, P. M. W.; Johnson, B.; Chen, W.; Wong, M. W.; Andres, J. L.; Gonzalez, C.; Head-Gordon, M.; Replogle, E. S.; Pople, J. A. Gaussian, Inc., Pittsburgh, PA, USA, **1998**. The geometries were optimized on the B3LYP/6-31+G(d) level of theory.
49. Kirsch, P.; Bremer, M.; Kirsch, A.; Osterodt, J. *16th International Symposium on Fluorine Chemistry*, Durham, UK, July 16-21, **2000**, lecture A 36.
50. Biffinger, J. C.; Kim, H. W.; DiMagno, S. G. *ChemBioChem* **2004**, *5*, 622-627.
51. Crowley, P. J.; Mitchell, G.; Salmon, R.; Worthington, P. A. *Chimia* **2004**, *58*, 138-142.
52. Bossenmaier, B.; Friebe, W.-G.; Georges, G.; Rueth, M.; Voss, E. *US 2005197370*, **2005**.
53. Tatsuta, K.; Morisawa, Y. *JP 2004059452*, **2004**.
54. Kleemann, H.-W. *WO 2005047239*, **2005**.
55. Critcher, D. J.; Lauret, C.; Walshe, N. D. A. *WO 2005090313*, **2005**.
56. Billen, D.; Logan Chubb, N. A.; Gethin, D. M.; Hall, K. T.; Roberts, L. R.; Walshe, N. D. A. *WO 2005060749*, **2005**.
57. a) Schnatterer, S.; Maier, M.; Lochhaas, F.; Knauf, W.; Seeger, K. *WO 2006000313*, **2006**. b) Schnatterer, S.; Maier, M.; Lochhaas, F.; Knauf, W.; Seeger, K. *WO 2006000314*, **2006**.

58. Scribner, A.; Chou, D. T.-W.; Knauf, W.; Maier, M.; Lochhaas, F.; Schnatterer, S.; Seeger, K. *WO 2006000312*, **2006**.
59. Knauf, W.; Maier, M.; Lochhaas, F.; Seeger, K. *WO 2006000311*, **2006**.
60. Chou, D. T.-W.; Schnatterer, S.; Knauf, W.; Seeger, K. *WO 2005023776*, **2005**.
61. Doeller, U.; Maier, M.; Kuhlmann, A.; Jans, D.; Pinchuk, A. M.; Marchenko, A. P.; Koydan, G. N. *WO 2005082917*, **2005**.
62. Matsuzaki, Y.; Morimoto, M.; Fujioka, S.; Tohnishi, M. *WO 2003093228*, **2003**.

Chapter 14

Synthesis of Macrolactam Marine Natural Products Using Fluorous Protecting Groups

Yutaka Nakamura and Seiji Takeuchi*

Niigata University of Pharmacy and Applied Life Sciences, 265–1 Higashijima, Niigata 956–8603, Japan

Marine natural products, bistratamides, were synthesized by using heavy fluorous protecting group, ^{F}Teoc or light fluorous one, ^{F}Boc. The intermediates of both methods were easily isolated by fluorous liquid phase extraction and solid phase extraction, respectively. Precursors of bistratamide H and its unnatural diastereomer were simultaneously prepared by quasi-racemic mixture synthesis. Each component of the mixture was cleanly separated by fluorous flash column chromatography.

Bistratamides (*1,2*) and didmolamides (*3*) were isolated from ascidians *Lissoclinum bistratum* and *Didemnum molle,* respectively, while tenuecyclamides (*4*) were isolated from *cyanobacterium Nostoc spongiaeforme var. tenue.* They are macrolactams of amino acids that include thiazole, oxazole or oxazoline (Figure 1). The unique structures and their bioactivities such as moderate cytotoxic activity, anti-drug resistance and antimicrobial properties and

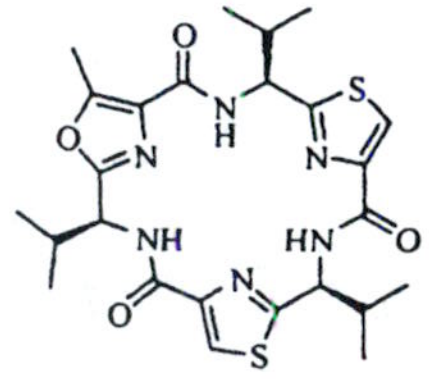

bistratamide H

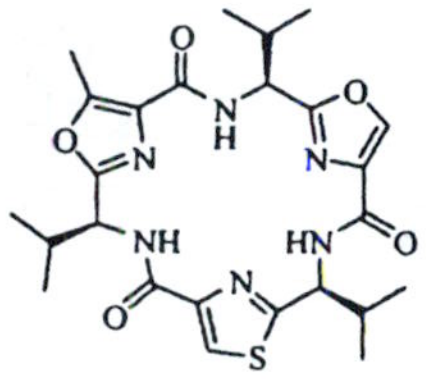

bistratamide G

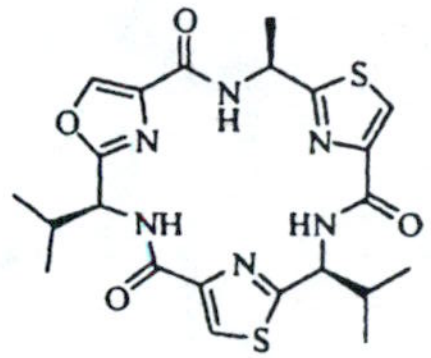

bistratamide C

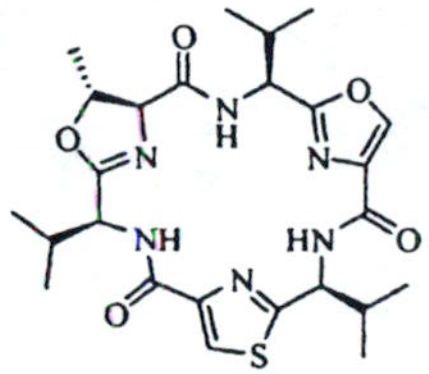

bistratamide E

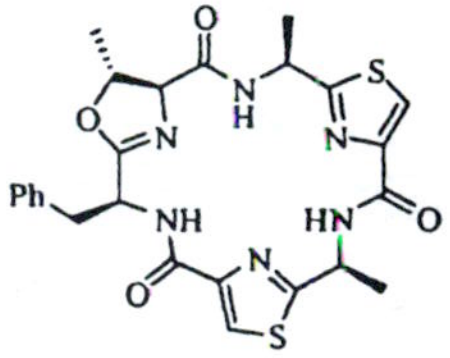

didmolamide A

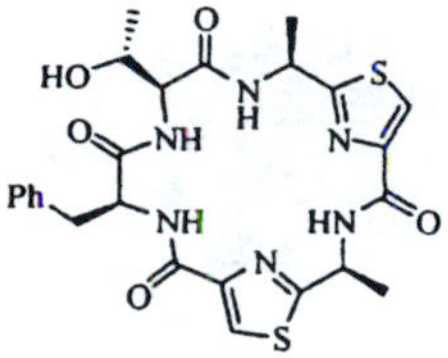

didmolamide B

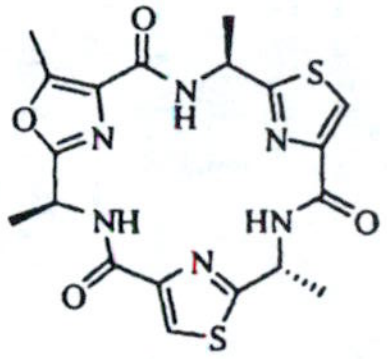

tenuecyclamide A

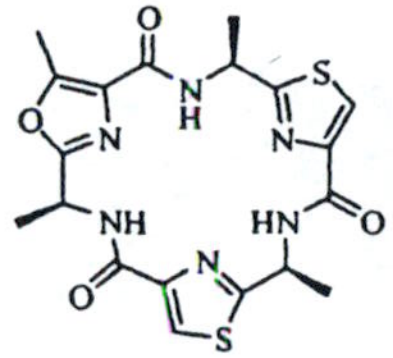

tenuecyclamide B

Figure 1. Bistratamide, didmolamide, and tenuecyclamide macrolactams.

inhibition of the division of sea urchin embryos have attracted synthetic organic chemists' interest. Among them, Kelly's and Shin's groups have recently reported syntheses of bistratamides according to their own methodologies of thiazole amino acid synthesis. Kelly stated in his reports that "Construction of the thiazoles in these macrolactams is central to their total synthesis" and discovered a facile and efficient biomimetic synthesis of thiazoles. The construction was accomplished by treating *N*-acylated cysteine with bis(triphenyl)oxodiphosphonium trifluoromethanesulfonate followed by oxidation of the thiazolines employing activated MnO_2. Utilizing this method, Kelly's group has synthesized most of bistratamides (*5*), didmolamides (*6*) and tenuecyclamides (*7*). Shin and his coworkers have synthesized bistratamide G in a high total yield by a modified Hantzsch's method using thioamide and bromoacyl derivatives of an oxazole amino acid as intermediates. The intermediates were prepared by a unique strategy via dehydropeptides (*8*). Coupling the thioamide and bromoacyl moieties of the intermediates led to the thiazole amino acid part of bistratamide G.

These interesting works prompted us to employ fluorous technologies for an expeditious synthesis of the macrolactams owing to the following reasons. Firstly, Kelly's group synthesized didmolamides and tenuecyclamides by solid phase assemblies of thiazole-containing amino acids. However they prepared the amino acids by the above biomimetic synthetic methods without using the solid support. Secondly, a modified Hantzsch's method used by Shin's group includes thiocarbonylation of α-amino acid amide with Lawesson's reagent. In the reaction, it is very hard to separate the products from unbearably bad smelling by-products by silica gel column chromatography. Partial racemization of the products during the reactions is another serious problem of this method.

Unlike Kelly's solid support synthesis, fluorous protecting groups can be used from the beginning of the modified Hantzsch's route for protecting amino groups of α-amino acids. If the fluorine atom content of the intermediates of the route is enough high (>60wt %), the fluorous intermediates can be clearly separated from organic by-products including the bad smelling ones by liquid phase extraction with fluorous and organic solvents (*9*). In the case of lower fluorine atom contents than 40 wt%, fluorous intermediates are separated from the organic by-products by solid phase extraction with a fluorous reverse phase silica gel column (*10*). Optimization of the reaction conditions is accomplished by checking the enantiomeric purity of the fluorous intermediates by HPLC with a chiral column to avoid the racemization of the intermediates. An excellent method using fluorous supports such as HfBz and HfBn has already been reported by Inazu and his coworkers for oligopeptide and ologosaccharide syntheses (*11,12*). However, we wanted to synthesize the macrolactams by using fluorous protecting groups *in order to apply our method also to a fluorous mixture synthesis* (*13,14*).

In an early stage of this work, we confirmed that Curran's light fluorous Boc-protecting group (*15*) and Bannwarth type heavy fluorous Cbz-protecting group (*16*) were really effective for a synthesis of thiazole amino acid derivative by a modified Hantzsch's method. However, the light fluorous Boc-protecting group is more practical for a small scale synthesis and *especially for a mixture synthesis.* We thought that a heavy fluorous protecting group was more suitable for establishing a long synthetic route as expeditiously as possible because the liquid phase extraction procedure was simpler and quicker than that of the solid phase one. Even if the fluorous solid phase extraction is needed in a later stage of the synthetic route, the heavy fluorous protecting group is considered to be more practical at least until the step. The heavy fluorous Cbz-protecting group, however, resisted being removed by hydrogenation over palladium on charcoal because of inactivation of the catalyst by sulfur atom of thiazole ring. Therefore we tried to prepare a new type of protecting group suitable for a modified Hantzsch's method and finally reached 2-tri(perfluorodecyl)silyl]ethoxycarbonyl (^{F}Teoc) group, a fluorous version of 2-(trimethylsilyl)ethoxycarbonyl group (*17*). The fluorous protecting group has a simple structure and thus high fluorine atom content and can be deprotected by tetrabutylammonium fluoride (TBAF).

In this paper, we would like to describe an expeditious synthesis of bistratamide H using the heavy fluorous protecting group ^{F}Teoc preparation of bistratamide H and C using the light fluorous Boc-protecting groups and an attempt of a mixture synthesis of precursors of bistratamide H and its unnatural diastereomer.

Synthesis of bistratamide H using heavy fluorous protecting group, ^{F}Teoc

Fluorous protecting agent, 2-[tris(perfluorodecyl)silyl]ethoxycarbonyl-O-succimide (^{F}Teoc-OSu) was prepared as shown in Scheme 1.

Fluorous hydrosilane (**1**) was reacted with bromine in FC-72 (perfluorohexane) and then the silylbromide (**2**) was reacted with vinyl magnesium chloride to give vinylsilane (**3**) in 87% yield via the two steps. The vinylsilane **3** was reacted with 9-borabicyclo[3,3,l]nonane (9-BBN) and then treated with hydrogen peroxide to give fluorous silylethanol (**4**) in 84% yield (*18*). The silylethanol **4** was treated with triphosgene and then reacted with *N*-hydroxysuccimide. After purification of the crude product by silica gel column chromatography, the fluorous protecting agent ^{F}Teoc-OSu (**5**) was obtained in 97% yield as white solid. The overall yield from the hydrosilane **1** was about 70%.

By using ^{F}Teoc-OSu **5**, thiazole amino acid derivative **10** was prepared from valine by a modified Hantzsch's method (Scheme 2).

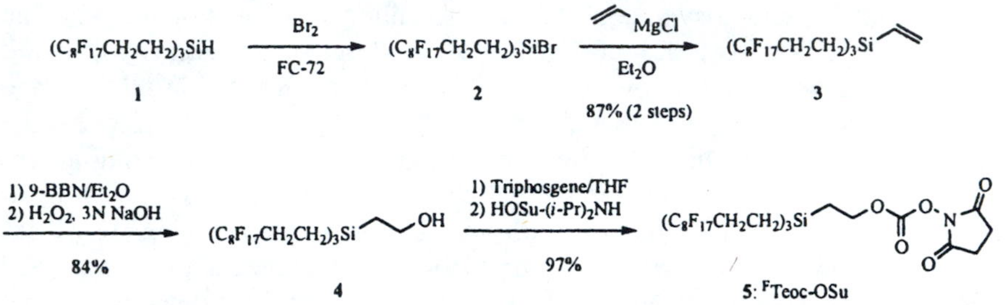

$(C_8F_{17}CH_2CH_2)_3SiH$ → (Br_2, FC-72) → $(C_8F_{17}CH_2CH_2)_3SiBr$ → (vinyl-MgCl, Et_2O) → $(C_8F_{17}CH_2CH_2)_3Si$-vinyl

1 — 2 — 3 (87% (2 steps))

1) 9-BBN/Et_2O
2) H_2O_2, 3N NaOH
84%

$(C_8F_{17}CH_2CH_2)_3Si$–CH_2CH_2–OH (4)

1) Triphosgene/THF
2) HOSu-$(i\text{-}Pr)_2NH$
97%

$(C_8F_{17}CH_2CH_2)_3Si$–CH_2CH_2–O–C(O)–O–N(succinimide)

5: ^{F}Teoc-OSu

Scheme 1. Preparation of fluorous protecting agent ^{F}Teoc-OSu

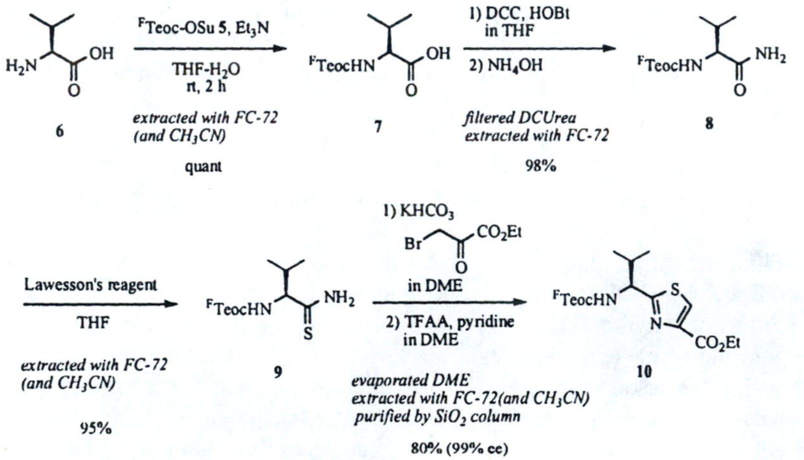

6 (valine, H_2N, OH, O) → ^{F}Teoc-OSu 5, Et_3N; THF-H_2O, rt, 2 h; *extracted with FC-72 (and CH_3CN)*; quant → 7 (^{F}TeocHN, OH, O)

7 → 1) DCC, HOBt in THF; 2) NH_4OH; *filtered DCUrea extracted with FC-72*; 98% → 8 (^{F}TeocHN, NH_2, O)

8 → Lawesson's reagent, THF; *extracted with FC-72 (and CH_3CN)*; 95% → 9 (^{F}TeocHN, NH_2, S)

9 → 1) $KHCO_3$, Br–CH_2–C(O)–CO_2Et in DME; 2) TFAA, pyridine in DME; *evaporated DME extracted with FC-72(and CH_3CN) purified by SiO_2 column*; 80% (99% ee) → 10 (^{F}TeocHN, S, N, CO_2Et)

Scheme 2. Preparation of N-^{F}Teoc-thiazole amino acid ester from valine.

Valine (**6**) was reacted with ^{F}Teoc-OSu **5** in the presence of triethylamine in aqueous THF. After the reaction, the reaction mixture was acidified and then extracted with FC-72 and acetonitrile. Concentration of the FC-72 layer gave *N*-^{F}Teoc-valine (**7**) in quantitative yield. The *N*-^{F}Teoc-valine **7** was reacted with DCC and 1-hydroxybenzotriazole (HOBt) in THF and then treated with an aqueous ammonia solution. The reaction mixture was filtered to remove DCU and then the filtrate was extracted with FC-72. Concentration of the FC-72 layer gave *N*-^{F}Teoc-Val-amide (**8**) in 98% yield. The amide **8** was reacted with Lawesson's reagent in THF and then the reaction mixture was extracted with FC-72 and acetonitrile. The bad smelling by-products and the reagent were completely partitioned into an organic layer. Therefore, *N*-^{F}Teoc-Val-thioamide (**9**) obtained from the FC-72 layer was almost odorless. The thioamide **9** was reacted with excess amount of ethyl bromopyruvate in DME and then treated with trifluoroacetic anhydride and pyridine. After evaporating DME, the reaction mixture was extracted with FC-72 and acetonitrile and the crude product obtained from FC-72 extract was purified by silica gel column chromatography to provide *N*-^{F}Teoc-thiazole amino acid ester (**10**) in 80% yield as waxy solid. The overall yield from valine was about 74%. Enantiomeric purity of the product **10** was determined to be higher than 99% ee by HPLC with a chiral column, DACEL CHIRALCEL OD-H.

Similarly, other *N*-^{F}Teoc-thiazole amino acid derivatives were prepared in good overall yields from alanine and phenylalanine as shown in Table 1.

Another component of bistratamide H, oxazole amino acid methyl ester (**14**) was prepared from the *N*-^{F}Teoc-valine **7** via 4 step reactions, coupling with threonine methyl ester, cyclization using PEG-Burgess reagent (*19*), dehydrogenation with DBU and $BrCCl_3$ (*20*) and deprotection with TBAF (Scheme 3).

Similarly, another *N*-^{F}Teoc-oxazole amino acid derivative was prepared in moderate overall yield by coupling with serine as shown in Table 2.

Finally bistratamide H was synthesized from the thiazole amino acid ester **10** and the oxazole amino acid methyl ester **14** (Scheme 4).

At first, the *N*-^{F}Teoc-thiazole amino acid **10** was treated with TBAF in THF to remove ^{F}Teoc protecting group. After evaporating THF, the residue was extracted with water, chloroform and FC-72. From chloroform layer, deprotected thiazole amino acid ester (**15**) was obtained in 82% yield. On the other hand, the thiazole amino acid ester **10** was treated with 1M LiOH aqueous THF solution and then the reaction mixture was acidified and extracted with FC-72. From the FC-72 extract, *N*-^{F}Teoc-thiazole amino acid (**16**) was obtained in 98% yield. The two thiazole amino acid derivatives **15** and **16** were coupled by using benzotriazol-1-yl-oxy-trispyrrolidinophosphonium hexafluorophophate (PyBOP) and *i*-Pr_2NEt in THF. After the work-up, thiazole amino acid dipeptide ester (**17**) was obtained in 96% yield. The thiazole dipeptide ester **17** was treated with 1M

Table 1. Preparation of *N*-[F]Teoc-thiazole amino acid esters

Entry	*Amino Acid*	*Yield(%)* **7**	**8**	**9**	**10**	*Total Yield(%)*	*% ee of* **10**[a]
1	Val	quant	98	95	80	74	99
2	Ala	quant	quant	93	69	64	92
3	Phe	quant	98	87	84	72	95

[a] Determined by HPLC analysis using DAICEL CHIRALCEL OD-H.

Table 2. Preparation of *N*-[F]Teoc-oxazole amino acid esters

Entry	*C-Terminus Amino Acid*	*Yield (%)* **11**	**12**	**13**	*Total Yield (%)*	*% ee of* **13**[a]
1	Thr	quant	70	79	55	95
2	Ser	96	83	83	66	N.D.[b]

[a] Determined by HPLC analysis using DAICEL CHIRALCEL OD-H.

[b] N.D. = Not determined.

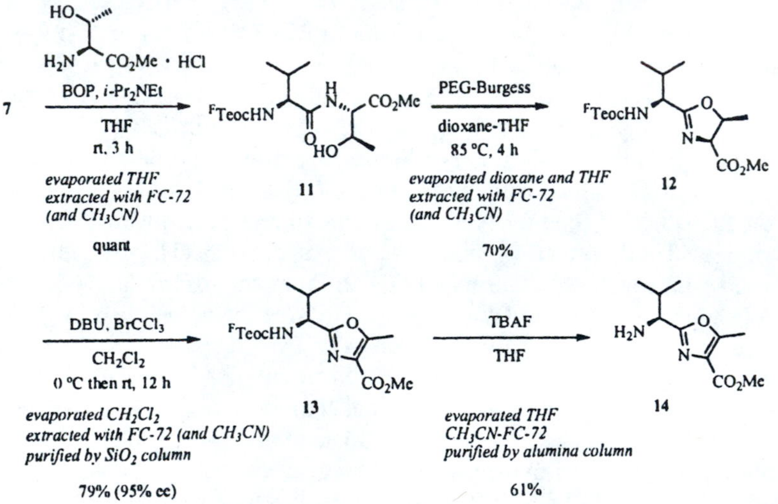

7 → 11: H_2N-Thr-CO_2Me · HCl, BOP, *i*-Pr_2NEt, THF, rt, 3 h; *evaporated THF, extracted with FC-72 (and CH_3CN)*; quant

11 → 12: PEG-Burgess, dioxane-THF, 85 °C, 4 h; *evaporated dioxane and THF, extracted with FC-72 (and CH_3CN)*; 70%

12 → 13: DBU, $BrCCl_3$, CH_2Cl_2, 0 °C then rt, 12 h; *evaporated CH_2Cl_2, extracted with FC-72 (and CH_3CN), purified by SiO_2 column*; 79% (95% ee)

13 → 14: TBAF, THF; *evaporated THF, CH_3CN-FC-72, purified by alumina column*; 61%

Scheme 3. Preparation of methyloxazole amino acid ester from N-[F]Teoc-valine.

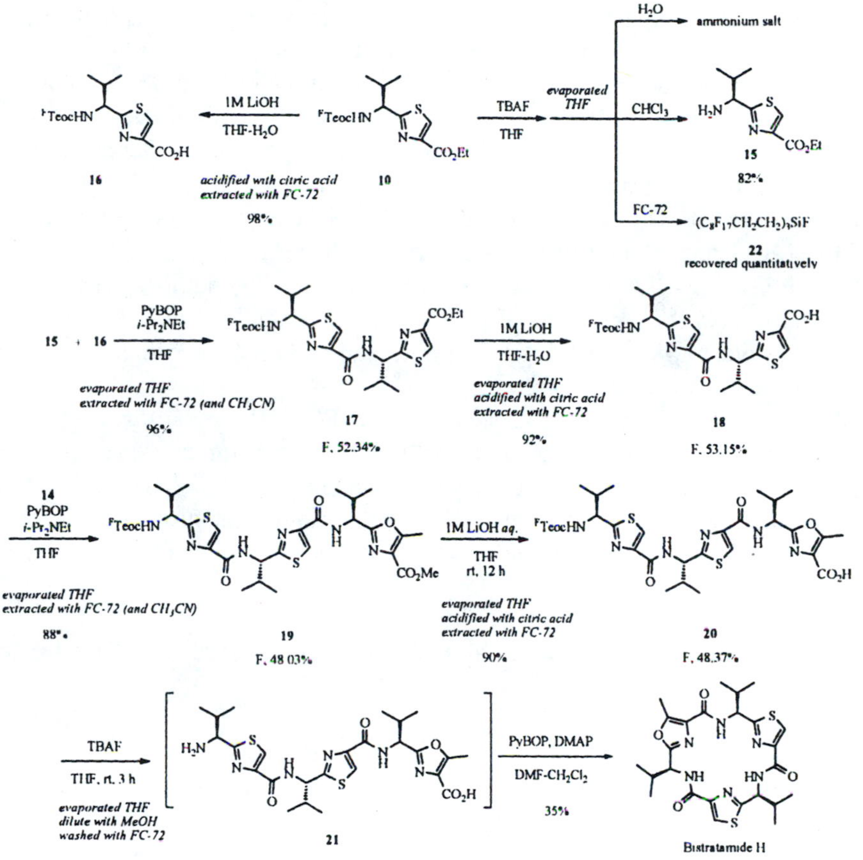

Scheme 4. Preparation of bistratamide H.

LiOH aqueous solution in THF. After evaporating THF, the aqueous solution was acidified and extracted with FC-72. From the FC-72 layer thiazole dipeptide amino acid (**18**) was obtained in 92% yield. The thiazole dipeptide **18** was coupled with the oxazole amino acid methyl ester **14** by using PyBOP and *i*-Pr_2NEt in THF. After the work-up, tripeptide methyl ester (**19**) was obtained in 88% yield as amorphous solid. The tripeptide methyl ester **19** was treated with 1M LiOH aqueous solution in THF. The reaction mixture was acidified, THF was evaporated and then the residue was extracted with FC-72. From the FC-72 layer tripeptide acid (**20**) was obtained in 90% yield. The tripeptide acid **20** was treated with TBAF to remove ^{F}Teoc protecting group in THF. After evaporating THF, the residue was extracted with FC-72 and methanol. *N,O*-Deprotected linear tripeptide **21** obtained from the methanol layer was dissolved in DMF-CH_2Cl_2 and the solution was added very slowly to a solution of PyBOP and DMAP in DMF-CH_2Cl_2 with maintaining high dilution conditions. The product was purified with a preparative TLC to give bistratamide H in 35% yield. The specific rotation, ^{1}H and ^{13}C NMR spectra and MS analysis demonstrated that the product was the desired compound.

The fluorous product (**22**) obtained from FC-72 layer on deprotection of ^{F}Teoc group with TBAF was reacted with vinyl magnesium chloride in ether to provide vinyl silane **3** in moderate yield (Scheme 5) (*21,22*). ^{1}H NMR of the vinyl silane product was the same as that of the sample obtained by the reaction of tris(perfluorodecyl)silylbromide **2** with vinyl magnesium chloride.

Preparation of bistratamide H and C using light fluorous Boc-protecting group

The new fluorous protecting group, ^{F}Teoc enabled us to prepare bistratamide H so expeditiously that the other macrolactams including didmolamides and tenuecyclamides were considered to be synthesized via a similar route to that of bistratamide H. Furthermore since the macrolactams have analogous structures, it was expected that they would be prepared simultaneously by using Curran's strategy of mixture synthesis. However, a mixture synthesis using the heavy fluorous protecting groups like ^{F}Teoc may cause a problem. That is, a large difference in partition coefficients to fluorous and organic solvents between the fluorous intermediates that bear fluorous Teoc groups of different size will limit the use of fluorous liquid phase extraction. If we use ^{F}Teoc protecting groups that have lower fluorine atom contents, we will have to use a fluorous solid phase extraction sooner or later. Therefore we thought that it was the best choice to use light fluorous Boc protecting groups for the mixture synthesis.

In order to confirm a generality of the synthetic route established above for bistratamide H, bistratamide H and C were prepared via similar synthetic routes except that light fluorous Boc-protected intermediates were separated from by-products by fluorous solid phase extraction with Fluoro*Flash* column (FSPE) (Scheme 6, 7, and 8).

Generally the FSPE was carried out by evaporating a solvent of a reaction, dissolving the residue in a fluorophobic solvent such as 80% aqueous methanol (H_2O:MeOH=20:80) and loading the solution onto the Fluoro*Flash* column. The column was eluted first with 80% aqueous methanol to remove organic by-products and remaining reagents and then washed with methanol to extract the fluorous product. In the case of thioamidation with Lawesson's reagent, however, the step of evaporation of the solvent was skipped to avoid scattering the unbearably bad smell over the laboratory and the building. In addition the amount of the solvent for the reaction was made as small as possible and 70% aqueous methanol was used as an eluting solvent to ensure the separation. The light fluorous thioamide ((*S*)-**25a**) was successfully obtained in extremely high yield without bad smell just as in the case of the fluorous liquid phase extraction of heavy fluorous ^{F}Teoc thioamide **9**.

Ethyl chloroformate was used for the amidation step instead of DCC and HOBt for the amidation of ^{F}Teoc valine **7**. However, there was no substantial difference between the two reagents for the reaction. In this step, the product was extracted simply with ethyl acetate, because the amide ((*S*)-**24a**) crystallized too easily to be loaded onto the Fluoro*Flash* column. However the product obtained by the extraction had enough high purity without further purification for the next reaction probably because of the quantitative yield. In the saponification step, the product ((*S*)-**27a**)) was also extracted with ethyl acetate after acidifying the reaction mixture. Since the reaction mixture was aqueous, the extraction was more adequate than FSPE in this case (Scheme 6).

Methyloxazole methyl ester (**30**) was purified by silica gel column chromatography in the middle of the synthetic route. The purification was necessary to provide the next intermediate (**31**) in pure state, because the product on the oxidation step was very messy due to low yield. The compound **31** was coupled with the thiazole dipeptide (**33**) to give the framework of bistratamide H, so it was inevitable to purify the intermediate **30** (Scheme 7).

The intermediate (**35a**) was also purified by silica gel column chromatography, because the compound was considered to be easier to purify with a silica gel column compared to the precursor of bistratamide H (**36a**).

In the final step, the yield was dramatically increased to 81% compared to that in the final step of the route using ^{F}Teoc protecting group. In the route using ^{F}Teoc protecting group, the solution of the precursor **21** in DMF and CH_2Cl_2 was slowly added to the solution of PyBOP and DMAP in DMF and CH_2Cl_2 with maintaining high dilution conditions. In contrast, in this case the solution of BOP

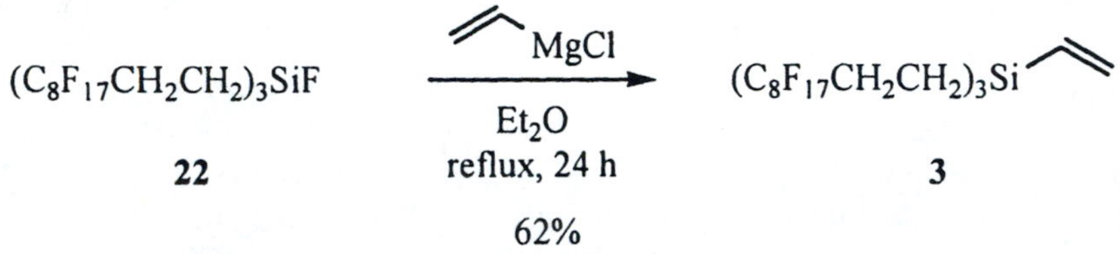

$(C_8F_{17}CH_2CH_2)_3SiF$ **22** → (vinyl-MgCl, Et_2O, reflux, 24 h, 62%) → $(C_8F_{17}CH_2CH_2)_3Si$-vinyl **3**

Scheme 5. Attempt to recycle the recovered fluorous silylfluoride.

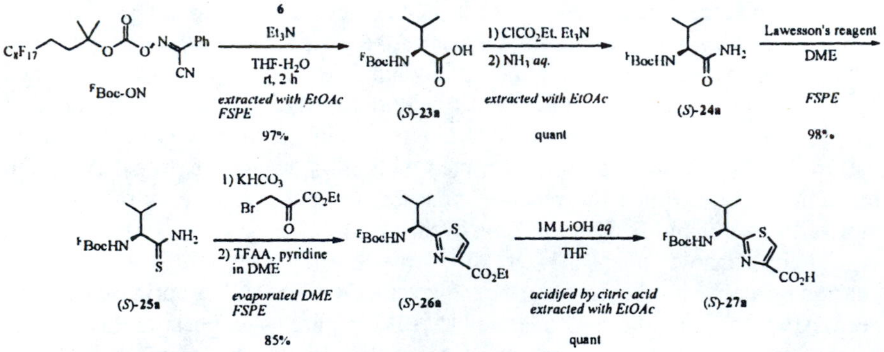

^{F}Boc-ON → (**6**, Et_3N, THF-H_2O, rt, 2 h; *extracted with EtOAc, FSPE*; 97%) → (*S*)-**23a** → (1) $ClCO_2Et$, Et_3N; 2) NH_3 *aq.*; *extracted with EtOAc*; quant) → (*S*)-**24a** → (Lawesson's reagent, DME; *FSPE*; 98%)

(*S*)-**25a** → (1) $KHCO_3$, Br-$CH_2C(O)CO_2Et$; 2) TFAA, pyridine in DME; *evaporated DME, FSPE*; 85%) → (*S*)-**26a** → (1M LiOH *aq*, THF; *acidifed by citric acid, extracted with EtOAc*; quant) → (*S*)-**27a**

Scheme 6. Preparation of ^{F}Boc-thiazole amino acid ester.

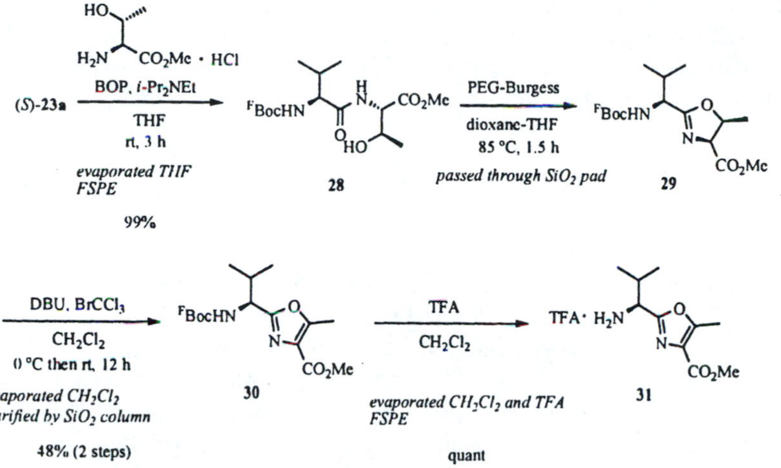

Scheme 7. Preparation of ^{F}Boc-oxazole amino acid ester.

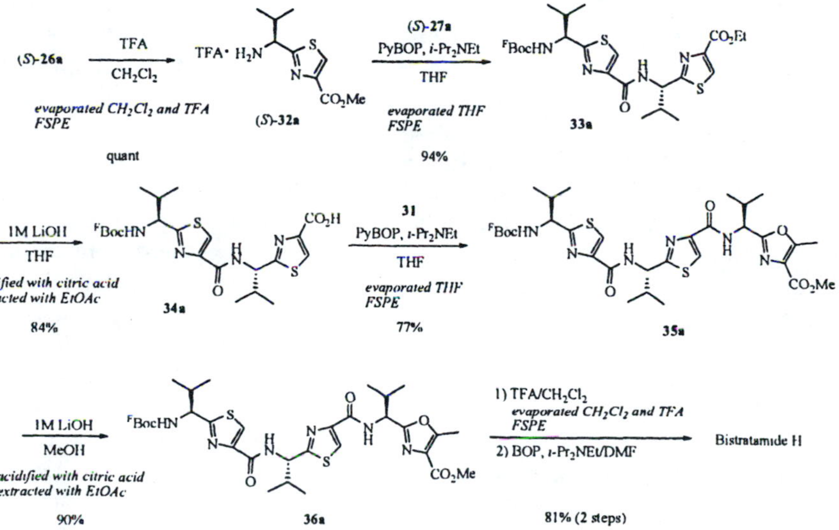

Scheme 8. Preparation of bistratamide H.

and i-Pr_2NEt in DMF was slowly added to the solution of the precursor **36a** in DMF. However, it is not clear whether or not the difference in a way of addition is actually the reason to lead the different yields (Scheme 8).

Bistratamide C was prepared by almost the same way from fluorous Boc protected alanine as the starting material.

Quasi-racemic mixture synthesis of the precursors of bistratamide H and its diastereomer

Very recently Curran and Zhang have reported a fluorous mixture synthesis of all 16 stereoisomers of Pinesaw fly sex pheromone (*14*). Starting from four stereoisomers that bore fluorous tags of different size, the target 16 stereoisomers were prepared via four lines of split parallel reactions. Each of the four stereoisomers in the four products was demixed by fluorous HPLC with Fluoro*Flash* column. In addition, Wilcox and Curran have found an amazing method to prepare all of 16 stereoisomer of Murisolin in one pot by combination of four different size each of fluorous and oligoethylene glycol (OEG) tags (*23*). The product that contained 16 stereoisomers was first separated into four fractions based on the properties of OEG tags and then each of the four fractions was further demixed by fluorous HPLC chromatography on a Fluoro*Flash* PF-C8 column. The simplest fluorous mixture synthesis of this kind is a quasi-racemic mixture synthesis in which a pair of enantiomers are tagged with different fluorous tags. In an example to prepare L- and D-phenylalanine tetrahydroisoquinoline amides, each of the quasi-enantiomers that contained $CH_2CH_2C_6F_{13}$ and $CH_2CH_2C_8F_{17}$ tags was separated by flash chromatography with standard fraction collection using Fluoro*Flash* column (*24*).

Since we had established the synthetic route of bistratamides H and C using the light fluorous protecting group ^{F}Boc we decided to prepare the precursors of bistratamide H and its unnatural diastereomer by the quasi-racemic mixture synthesis (exactly speaking "by quasi-diastereomer synthesis") to establish a method to prepare the macrolactams by the mixture synthesis (Scheme 9).

Starting from a mixture of C_8F_{17}-Boc-(*S*)-valine (*S*)-**23a** and $C_{10}F_{21}$-Boc-(*R*)-valine ((*R*)-**23b**) the precursor ^{F}Boc tripeptide methyl esters (M-**35ab**) were prepared via the route for bistratamide H. The mixture of the precursor compounds and the intermediates to the compounds all showed the same behaviors as those described in the literatures reported by Curran and Zhang. The mixtures M-**35ab** behaved as a single compound on a silica gel TLC but as a distinct mixture on a fluorous TLC as shown in Figure 2.

The two components of the product were clearly separated not only by fluorous HPLC with Fluoro*Flash* HPLC column (Figure 3; the difference in

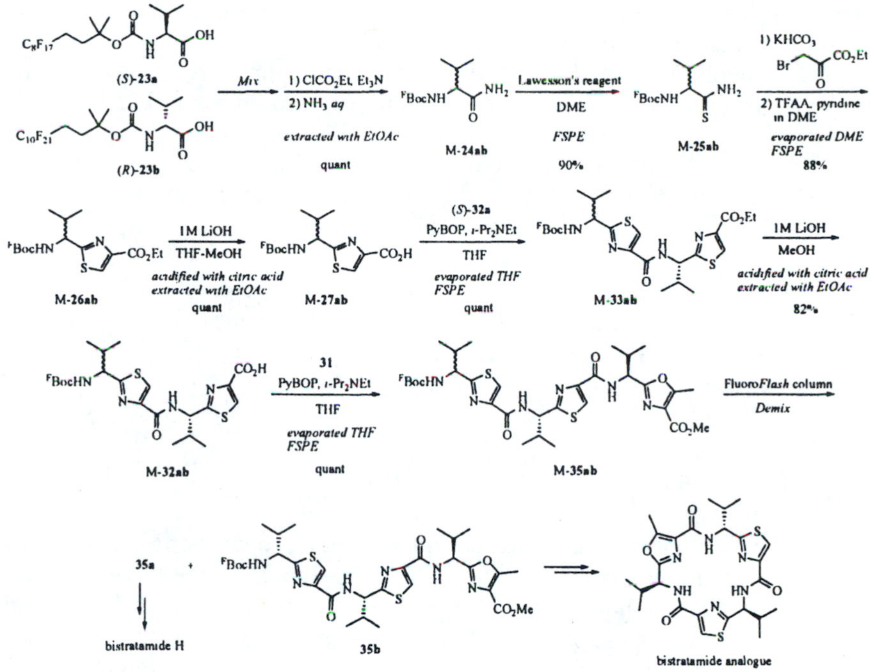

Scheme 9. Quasi-racemic mixture synthesis of bistratamide H and its analogue.

*Figure 2. TLC of fluorous mixture intermediates **35a** and **35b**.*
Left: silica gel, developed with 50% ethyl acetate in hexane;
*Right: Fluoro*Flash^{TM}*, developed with MeOH (top: **35a**, bottom: **35b**).*

retention time was larger than 9 min) but also by flash chromatography with Fluoro*Flash* cartridge column.

The samples separated by the flash chromatography were almost pure and gave ^{1}H NMR spectra in which only methine proton on the chiral carbon of *N*-terminal valine part exhibited a small difference in chemical shift (Figure 4).

The ^{1}H NMR of the product that bore C_8F_{17}-Boc group was exactly the same as that of the intermediate of bistratamide H prepared above. Therefore, bistratamide H and its unnatural diastereomer will be simultaneously synthesized by this method.

Natural tenuecyclamide A and B are real diastereomers in which the configurations of just one chiral center are different Therefore the quasi-racemic mixture synthesis is the best way for the simultaneous preparation of them and the work is now under way in our laboratory.

Conclusion

In conclusion, we synthesized bistratamide H expeditiously by using the new fluorous protecting group, ^{F}Teoc. The separation of the fluorous intermediates by fluorous liquid phase extraction was very easy and quick, although the fluorine atom content was just about 48% at the final stage. The product in thiocarbonylation was cleanly separated from the bad smelling by-products by the extraction. Optimization of the reactions was carried out as usual by monitoring the reactions with TLC. The enantiomeric purities of the fluorous intermediates were checked by HPLC with a chiral column to find optimal reaction conditions for a voiding racemization of the products. In addition, the fluorous fragment from the fluorous protecting group was demonstrated to be recycled. The synthetic route was applied to the preparation of bistratamide H and C using light fluorous protecting group, ^{F}Boc. The fluorous intermediates were separated from by-products by fluorous solid phase extraction with Fluoro*Flash* cartridge column. Fluorous quasi-racemic (quasi-diastereomer) mixture of the precursors of bistratamide H and its unnatural diastereomer was prepared by the same way as those of bistratamide H and C. The two fluorous diastereomers were easily separated by flash chromatography with a Fluoro*Flash* column.

Acknowledgement

The authors would like to thank Prof. Dennis P. Curran, University of Pittsburgh and Prof. Willi Bannwarth, Albert-Ludwigs-Universitat Freiburg for their helpful suggestions. They also would like to thank Dr. Nobuto Hoshi,

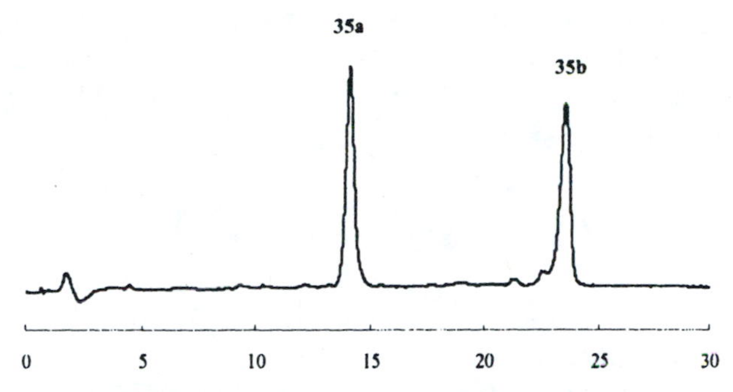

*Figure 3. HPLC analysis of fluorous mixture intermediates **35a** and **35b**. Conditions: Fluoro*Flash *PF-C8 column (4.6 × 50 mm) using a gradient from 70% CH_3CN in water to 100% CH_3CN in 30 min at 1.0 mL/min.*

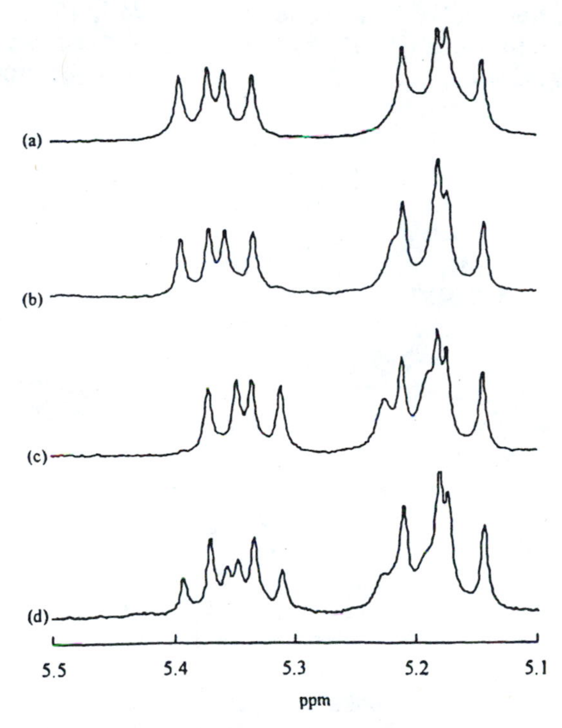

*Figure 4. ^1H NMR supectra of fluorous mixture intermediates **35a** and **35b** in $CDCl_3$. (a) product **35a** sample in Scheme 8. (b) **35a** sample separated from M-**35ab**. (c) **35b** sample separated from M-**35ab**. (d) mixture M-**35ab** sample.*

Noguchi Research Institute and Dr. Naoto Takada, Central Glass International, Inc. for their valuable advices about the fluorous silylfluoride and Noguchi Research Institute for the Noguchi Fluorous Project's Fund. This work has been done as a part of the project.

References and Notes

1. Foster, M. P.; Concepción, G. P.; Caraan, G. B.; Ireland, C. M. *J. Org. Chem.* **1992**, *57,* 6671-6675.
2. Perez, L. J.; Faulkner, D. J. *J. Nat. Prod.* **2003**, *66*, 247-250.
3. Rudi, A.; Chill, L.; Aknin, M.; Kashman, Y. *J. Nat. Prod.* **2003**, *66*, 575-577.
4. Banker, R.; Carmeli, S. *J. Nat. Prod.* **1998**, *61,* 1248-1251.
5. You, S.-L.; Kelly, J. W. *Tetrahedron* **2005**, *61,* 241-249.
6. You, S.-L.; Kelly, J. W. *Tetrahedron Lett.* **2005**, *46*, 2567-2570.
7. You, S.-L.; Deechongkit, S.; Kelly, J. W. *Org. Lett.* **2004**, *6*, 2627-2630.
8. Yonezawa, Y., Tani, N.; Shin, C. *Bull. Chem. Soc. Jpn* **2005**, *78,* 1492-1499.
9. Crich, D.; Neelamlavil, S. *Tetrahedron* **2002**, *58,* 3865-3870.
10. Curran, D. P. *Synlett* **2001**, 1488-1496.
11. Miura, T.; Satoh, A.; Goto, K.; Murakami, Y.; Imai, N.; Inazu, T. *Tetrahedron: Asymmetry* **2005**, *16,* 3-6.
12. Goto, K.; Miura, T.; Mizuno, M.; Takaki, H.; Imai, N.; Murakami, Y.; Inazu, T. *Synlett* **2004**, 2221-2223.
13. Fukui, Y.; Bruckner, A. M.; Shin, Y.; Balachandran, R.; Day, B. W.; Curran, D. P. *Org. Lett.* **2006**, *8,* 301-304.
14. Manku, S.; Curran, D. P. *J. Org. Chem.* **2005**, *70,* 4470-4473.
15. Luo, Z.; Williams, J.; Read, R. W.; Curran, D. P. *J. Org. Chem.* **2001**, *66,* 4261-4266.
16. Schwinn, D.; Bannwarth, W. *Helv. Chim. Acta* **2002**, *85*, 255-264.
17. Kocieński, P. J. P. J. *Protecting Groups, 3rd ed.;* Georg Thieme Verlag: Stuttgart, **2005**, pp 534-540.
18. Soderquist, J. A.; Hassner, A. *J. Organomet. Chem.* **1978**, *156,* C12-C16.
19. Wipf, P.; Hayes, G. B. *Tetrahedron* **1998**, *54,* 6987-6998.
20. Williams, D. R.; Lowder, P. D.; Gu, Y.-G.; Brooks, D. A. *Tetrahedron Lett.* **1997**, *38,* 331-334.
21. Ogi, K. Master's thesis, Tohoku University, 1973.
22. Boutevin, B.; Guida-Pietrasanta, F.; Ratsimihety, A.; Caporiccio, G. *J. Fluorine. Chem.* **1995**, *70,* 53-57.
23. Wilcox, C. S.; Gudipati, V.; Lu, H.; Turkyilmaz, S.; Curran, D. P. *Angew. Chem. Int. Ed.* **2005**, *44*, 6938-6940.
24. Curran, D. P.; Amatore, M.; Guthrie, D.; Campbell, M.; Go, E.; Luo, Z. *J. Org. Chem.* **2003**, *68,* 4643-4647.

Chapter 15

Carbohydrate Microarrays and Fluorous-Phase Synthesis: Interfacing Fluorous-Phase Tags with the Direct Formation of Glycoarrays

Nicola L. Pohl

Department of Chemistry and the Plant Sciences Institute, Iowa State University, Ames, IA 50011

Phase switching, such as the use of fluorocarbon tags, is a common technique to facilitate purification of intermediates in iterative biopolymer synthesis. After a review of the application of fluorous tags in carbohydrate synthesis, a new method to directly array sugars via noncovalent fluorous-fluorous interactions is discussed. Light-fluorous tags now can serve not only to simplify iterative oligosaccharide synthesis, but also to directly pattern sugars on surfaces for biological screening.

Introduction

Carbohydrates interact specifically with proteins to mediate biological processes that include inflammatory responses, pathogen invasion, cell differentiation, cell-cell communication, cell adhesion and development, and tumor cell metastasis (*1-3*). Information about these interactions would help illuminate the role of carbohydrates in the life cycles of organisms as well as foster the development of sugar-based therapeutics such as vaccines that intervene in these carbohydrate-protein interactions. Unfortunately, the molecular basis for many of these sugar-protein interactions is not understood, in part because homogeneous well-defined carbohydrates are extremely difficult to obtain (*3*). Whereas pure peptides and nucleic acids are readily commercially available by automated synthesis, an analogous commercial process for carbohydrate synthesis is still missing. The difficulty of carbohydrate synthesis also limits the range of structures that can be incorporated into microarrays for screening of carbohydrate-protein interactions.

Phase-Switching and Fluorous-Phase Carbohydrate Synthesis

Iterative biopolymer synthesis is often facilitated by the use of soluble or solid-phase supports to simplify the purification of intermediates (Figure 1). For example, solid-phase carbohydrate synthesis allows excess reagents required for reaction completion to be washed off between steps in a process that has been automated (*4*). Alternatively, to avoid poor reaction kinetics inherent to biphasic systems, soluble tags with unique physical properties can be attached to the growing chain to aid in purification of intermediates by tag precipitation, extraction into a liquid phase, or affinity chromatography/solid-phase extraction (*5*). Because tag precipitation is not quantitative, tags for extraction methods are attractive options.

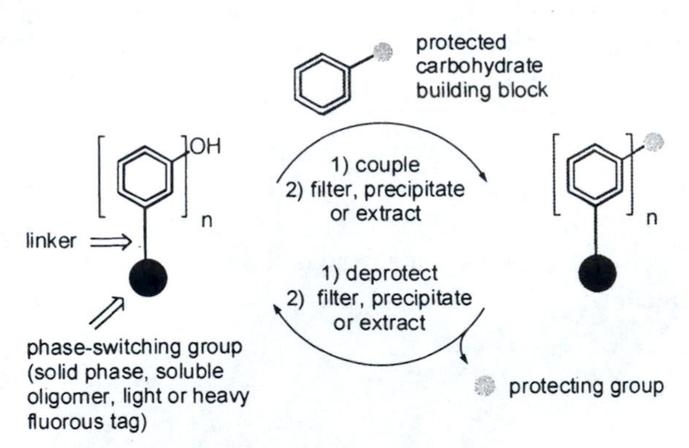

Figure 1. A basic scheme for phase-switching approaches to iterative carbohydrate synthesis.

Soluble fluorocarbon tags have been employed in the synthesis of a variety of carbohydrates. Fluorocarbons will phase separate from aqueous or conventional organic solvents. Several fluorocarbon chains can be incorporated into a protecting group to allow extraction of the compound containing the "heavy" fluorous tag into a liquid fluorocarbon layer or a single fluorocarbon chain, a "light" fluorous tag, can capture the tagged molecule by fluorous-derivatized silica gel in a solid-phase extraction process (*6-8*). Both "heavy" and "light" fluorous tags have been developed specifically for the challenges of oligosaccharide synthesis.

Fluorous Tags for Liquid-Liquid Extractions in Carbohydrate Synthesis

Nonenzymatic carbohydrate synthesis usually relies on protecting groups to permanently mask some hydroxyl and other functional groups and to temporarily block future reaction sites. These protecting groups are perfect locations for the introduction of fluorocarbon tags that allow liquid-liquid extraction of the reaction product away from excess reagents. The first protecting group introduced for this purpose to aid sugar synthesis was a variation of the commonly-used benzyl group. Three fluorocarbon chains attached to a silicon modify a benzyl group (Bnf, Figure 2) and allow extraction of molecules protected with the group into perfluorohexanes (*9*). After a monomer addition reaction to a triply Bnf-protected glucal, disaccharide **1** could be isolated by extraction into a liquid fluorocarbon phase. Two variations of the commonly-employed ester protecting group have also been designed for the synthesis of disaccharides **2** and **3** (Figure 2) with facile purification of intermediates by liquid-liquid extraction to create up to pentasaccharides (*10-12*). A benzoyl variation has also been reported recently (*13*).

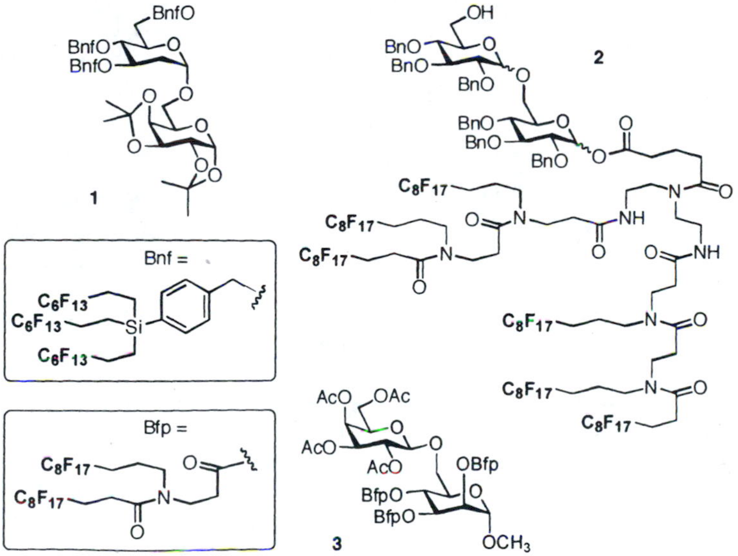

BnfO BnfO BnfO 1 Bnf = C_6F_{13} C_6F_{13} C_6F_{13} Si Bfp = C_8F_{17} C_8F_{17} N O O

OH BnO BnO BnO 2 BnO BnO BnO C_8F_{17} C_8F_{17} C_8F_{17} N NH NH C_8F_{17} C_8F_{17} C_8F_{17}

AcO OAc AcO AcO OBfp BfpO BfpO OCH_3 3

Figure 2. Fluorocarbon hydroxyl protecting groups for the synthesis of carbohydrates with liquid-liquid extraction of intermediates.

Unfortunately, the necessity to include multiple fluorocarbon tails for efficient liquid-liquid extraction also limits the solubility of the compounds in the nonfluorocarbon solvents required for a range of reaction types. Additionally, the large protecting groups can complicate spectral interpretation for characterization and a substantial amount of the molecular weight of the intermediates is accounted for by the fluorocarbon tags. Nonetheless, the tags allow iterative carbohydrate synthesis with minimal chromatography and with the benefits of solution phase reaction kinetics and reaction monitoring not possible by solid-phase approaches.

Fluorous Tags for Solid-Phase Extraction in Carbohydrate Synthesis

The addition of only one fluorocarbon chain to a protected carbohydrate renders the molecule separable from nontagged compounds not by liquid-liquid extraction but by solid-phase extraction (SPE) instead. The reaction mixture is loaded on fluorous silica gel, untagged compounds are eluted, and then a change of solvent allows elution of the pure tagged compound (*7*). Several carbohydrate protecting groups as well as an anomeric activating group for glycosylation reactions have been designed with single fluorous tags to simplify purification schemes.

Addition of a fluorous tag to a thiol anomeric activating group creates a glycosylation building block **4** (Figure 3) that could easily be purified by SPE (*14*). The thiol byproduct after glycosylation could be readily removed by SPE and recycled after reduction of any disulfide formed. In addition to the benefits of purification ease, the fluorous tag also rendered the thiol less repugnant.

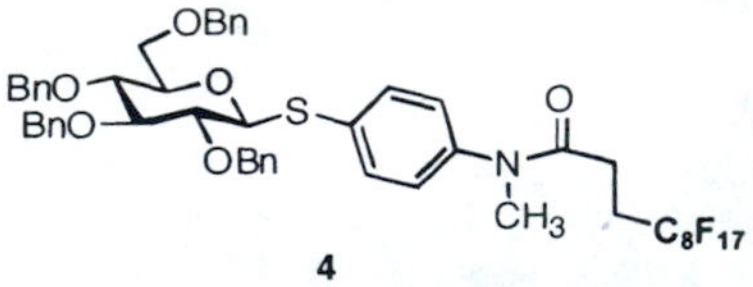

OBn
BnO
BnO
O
S
OBn
O
N
CH_3
C_8F_{17}
4

Figure 3. A single fluorous tag for a recyclable activating group for glycosylations.

Fluorous protecting groups have also been used to facilitate iterative carbohydrate synthesis by solid-phase extraction of growing chain intermediates. A fluorous version of a silicon protecting group (*15*) was used to protect the anomeric position of a glucosamine building block and build up the Lewis a trisaccharide **5** (Figure 4) with intermediates purified by fluorous SPE (*16-17*). However, unlike tags with multiple fluorocarbon chains, the fluorous group did not prevent the use of standard chromatography methods if necessary. A related fluorous silyl group has been used not in iterative synthesis but to cap oligosaccharides made on solid-phase for isolation of tagged sequences by SPE (*18*). More recently, a fluorous version of a carbamate nitrogen protecting group was developed and applied to the synthesis of a disaccharide **6** (Figure 4) (*19*). The group can be synthesized in three steps and removed with exchange to an acetyl group using zinc in acetic anhydride with triethylamine. Unfortunately, the new protecting group contains a stereogenic center, but at least in the cases reported this center does not overly complicate spectra. Finally, a fluorous version of the allyl protecting group has also been developed for facile purification of intermediates in the synthesis of polymannosides such as **7**, for example (Figure 4) (*20-21*). The fluorous allyl group allows fluorous SPE purification of intermediates and can be removed using standard palladium-mediated deallylation conditions.

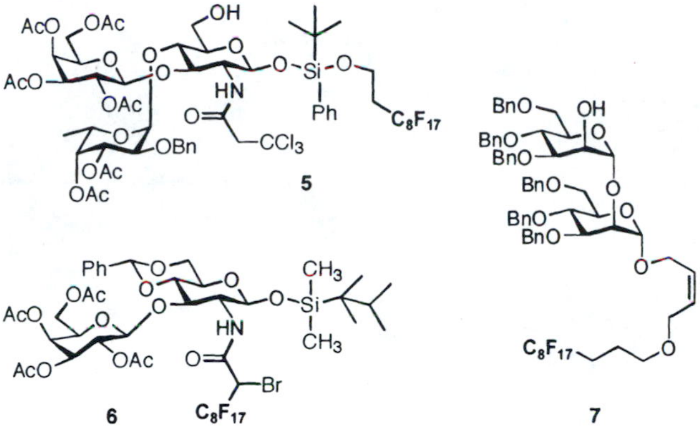

Figure 4. Protecting groups with single fluorous tags for the synthesis of carbohydrates with fluorous solid-phase extraction of intermediates.

Direct Formation of Fluorous-based Carbohydrate Microrrays

Like solid-phase approaches, fluorous-phase methods are showing promise in simplifying the iterative synthesis of carbohydrates with the added benefits of standard solution-phase reaction monitoring, reduced building block requirements at each coupling cycle, and the possibility of using other purification techniques as needed with intermediates. However, synthetic carbohydrates are still extremely precious and therefore, ideally, would be used sparingly to allow multiple bioassays. Microarrays such as DNA chips require minimal sample usage and therefore have spurred development of a range of technologies for protein-detection assays on glass slides (*22-23*). Some of these technologies have been applied to the synthesis of carbohydrate chips (*24-33*), but most of these microarray methods rely on covalent attachment of a compound to the slide and therefore require unique functional handles. Recently, a simpler method that relies on noncovalent fluorous-fluorous interactions has been developed that allows the direct formation of carbohydrate microarrays using a single fluorocarbon chain that can also facilitate oligosaccharide synthesis (Figure 5) (*34*).

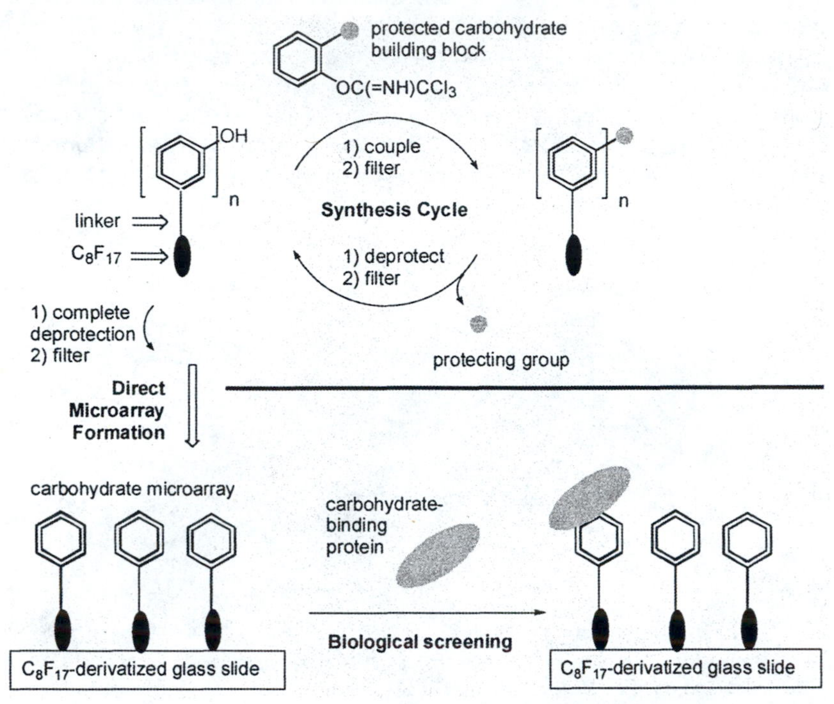

Figure 5. A combined strategy for use of a single fluorous tag to both facilitate iterative carbohydrate synthesis and allow direct formation of microarrays for biological screening. (Adapted with permission from reference 34. Copyright 2005 American Chemical Society.)

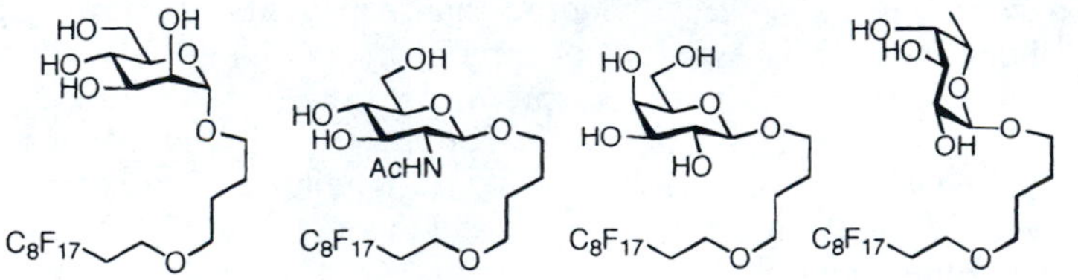

8 Man **9** GlcNAc **10** Gal **11** Fuc

Figure 6. A series of fluorous-tagged carbohydrates for microarray formation.

To test the feasibility of using noncovalent fluorous-fluorous interactions for surface patterning, a series of carbohydrates were synthesized with a fluorous-tagged allyl linker. The allyl linkers were then converted to more flexible and less reactive alkyl linkers by hydrogenation. After synthesis of the requisite fluorous-tagged sugars (Figure 6), a suitable fluorinated surface was needed. Initial microarray experiments used a commercially available glass microscope slide coated with a Teflon/epoxy mixture for compound spotting with a standard robot used for DNA arraying. The spotted slide was incubated for 20 minutes with a solution of the fluorescein isothiocyanate-labeled jack bean lectin concanavalin A (FITC-ConA) rinsed repeatedly with assay buffer and distilled water, and then scanned with a standard fluorescent slide scanner. The scan clearly showed binding of FITC-ConA only to the mannose-containing spots. The anomeric position also could be distinguished as the beta-linked GlcNAc **9** was not recognized by the lectin.

Although this initial lectin experiment demonstrated the ability of the C_8F_{17} tail to anchor the carbohydrates to the slide surface even after repeated washes, the intrinsic uneven fluorescence of the commercial slide at 488 nM, a wavelength that is commonly used to detect labeled analytes, necessitated another approach. An optically and fluorescently clear surface for the formation of compound microarrays then was obtained by reaction of a glass microscope slide with a fluoroalkylsilane.

The new microarray substrate also allowed spotting of fluorous-tagged sugars using an arraying robot, but after incubation with FITC-labeled lectins the protein-bound carbohydrates stood out clearly from the background in the fluorescent scan (Figure 7). A test of the ability of the array to withstand detergents often included in biological screens was carried out with the labeled plant lectin from the bush *Erythrina cristagalli* (FITC-ECA) and the hydrocarbon detergent Tween-20. Because hydrocarbons phase separate from fluorocarbons, the noncovalent fluorous-based array would be expected to be more stable to regular detergents than a noncovalent array approach based on hydrocarbon-hydrocarbon interactions. Indeed, the fluorous-based array withstood the 20 minute incubation time and repeated rinsing with this detergent-containing buffer.

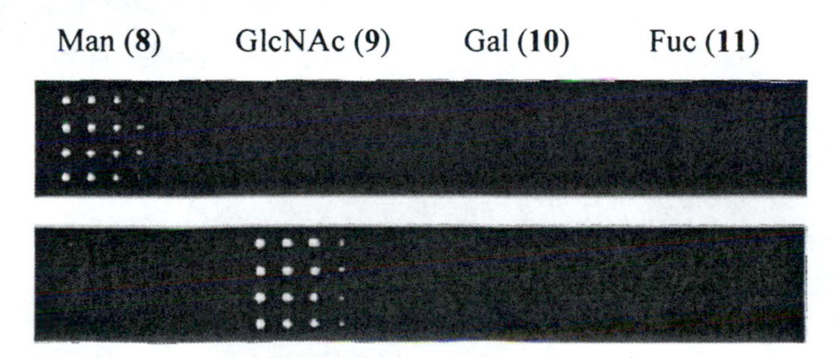

Figure 7. Fluorescence images of arrayed carbohydrates probed with FITC-labeled lectins. Columns of 4 spots each of 2, 1, 0.5 and 0.1 mM carbohydrates were incubated for 20 min with FITC-ConA (top) or FITC-ECA with 1% TWEEN-20 detergent (bottom) with BSA. (Reproduced with permission from reference 34. Copyright 2005 American Chemical Society.)

These initial results now have been expanded to the production of carbohydrate arrays that include disaccharides as well as a charged amino sugar with equal success (*20*). Applications of this fluorous-based strategy for microarray formation to larger saccharides as well as other molecules such as nucleic acids can easily be envisioned.

Future Directions

The ability to directly form compound microarrays with a fluorous-tail makes a fluorocarbon-based phase-switching approach to compound synthesis even more appealing. "Light" fluorous tags already have proven their utility in accelerating small molecule synthesis, including carbohydrate synthesis. Reactions can be monitored using traditional solution-phase techniques and, unlike solid-phase approaches, large excesses of reagents are not required for high yields. However, several challenges remain. Unlike solid-phase approaches, iterative fluorous-phase synthesis of any molecule class has never been automated. Automation is key to gaining the benefits of facile library synthesis seen with the automation of both peptide and nucleic acid synthesis. Chemistry amenable to automation for iterative oligosaccharide synthesis based on fluorous tags has been developed (*21*), but demonstration of its automation remains to be seen. The generality of using noncovalent fluorous-fluorous interactions for microarray formation also needs to be probed. Extensions to not only a variety of oligosaccharides but also other biopolymers and small molecules ultimately will test the robustness and utility of this new fluorous-based surface patterning approach.

References

1. Dube, D. H.; Bertozzi, C. R. *Nat. Rev. Drug Discov.* **2005**, *4*, 477-488.
2. Rudd, P. M.; Wormald, M. R.; Dwek, R. A. *Trends Biotechnol.* **2004**, *22*, 524-530.
3. Ratner, D. M.; Adams, E. W.; Disney, M. D.; Seeberger, P. H. *ChemBioChem*, **2004**, *5*, 1375-1383.
4. Plante, O. J.; Palmacci, E. E.; Seeberger, P. H. *Science*, **2001**, *291*, 1523-1527.
5. Ito, Y.; Manabe, S. *Chem. Eur. J.* **2002**, *8*, 3077-3084 and references therein.
6. Horváth, I. T. *Acc. Chem. Res.* **1998**, *31*, 641-650.
7. Curran, D. P. Separations with Fluorous Silica Gel and Related Materials. In *The Handbook of Fluorous Chemistry*; Gladysz, J.; Horváth, I.; Curran, D. P.; Wiley-VCH: Weinheim, 2004; pp 101-127.
8. Zhang, W. *Curr. Opin. Drug Discov. Develop.* **2004**, *7*, 784-797.

9. Curran, D. P.; Ferrito, R.; Hua, Y. *Tetrahedron Lett.* **1998**, *39*, 4937-4940.
10. Miura, T.; Hirose, Y.; Ohmae, M.; Inazu, T. *Org. Lett.* **2001**, *3*, 3947-3950.
11. Miura, T.; Goto, K.; Hosaka, D.; Inazu, T. *Angew. Chem. Int. Ed.* **2003**, *42*, 2047-2051.
12. Miura, T.; Goto, K.; Waragai, H.; Matsumoto, H.; Hirose, Y.; Ohmae, M.; Ishida, H.-k.; Satoh, A.; Inazu, T. *J. Org. Chem.* **2004**, *69*, 5348-5353.
13. Miura, T.; Satho, A.; Goto, K.; Murakami, Y.; Imai, N.; Inazu, T. *Tetrahedron: Asymmetry* **2005**, *16*, 3-6.
14. Jing, Y.; Huang, X.; *Tetrahedron Lett.* **2004**, *45*, 4615-4618.
15. Röver, S.; Wipf, P. *Tetrahedron Lett.* **1999**, *40*, 5667-5670.
16. Manzoni, L. *Chem. Commun.* **2003**, 2930-2931.
17. Manzoni, L.; Castelli, R. *Org. Lett.* **2004**, *6*, 4195-4198.
18. Palmacci, E. E.; Hewitt, M. C.; Seeberger, P. H. *Angew. Chem. Int. Ed.* **2001**, *40*, 4433-4437.
19. Manzoni, L.; Castelli, R. *Org. Lett.* **2006**, *8*, 955-957.
20. Mamidyala, S. K.; Ko, K.-S.; Jaipuri, F. A.; Park, G.; Pohl, N. L. *J. Fluor. Chem.* **2006**, *127*, ASAP.
21. Jaipuri, F. A.; Pohl, N. L. *Abstracts of Papers*. 230th National Meeting of the American Chemical Society, Washington, DC, Aug 28-Sept 1, 2005; American Chemical Society: Washington, DC, 2005, ORGN 185.
22. Tomizaki, K.; Usui, K.; Mihara, H. *ChemBioChem* **2005**, *6*, 782-799.
23. Uttamchandani, M.; Wang, J.; Yao, S. Q. *Mol. BioSyst.* **2006**, *2*, 58-68.
24. Dyukova, V. I.; Shilova, N. V.; Galanina, O. E.; Rubina, A. Y.; Bovin, N. V. *Biochim. Biophys. Acta* **2006**, ASAP.
25. Huang, G. L.; Zhang, H. C.; Wang, P. G. *Bioorg. Med. Chem. Lett.* **2006**, *16*, 2031-2033.
26. Shin, I.; Lee, M. *Angew. Chem. Int. Ed.* **2005**, *44*, 2881-2884.
27. Biskup, M.; Muller, J. U.; Weingart, R.; Schmidt, R. R. *ChemBioChem* **2005**, *6*, 1007-1015.
28. Adams, E. W.; Ratner, D. M.; Bokesch, H. R.; McMahon, J. B.; O'Keefe, B. R.; Seeberger, P. H. *Chem. Biol.* **2004**, *11*, 875-881.
29. Bryan, M.; Fazio, F.; Lee, H.; Huang, C.; Chang, A.; Best, M.; Paulson, J. C.; Burton, D.; Wilson. L.; Wong, C.–H. *J. Am. Chem. Soc.* **2004**, *126*, 8640-8641.
30. Ratner, D. M.; Adams, E. W.; Su, J.; O'Keefe, B. R.; Mrksich, M.; Seeberger, P. H. *ChemBioChem* **2004**, *5*, 379-383.
31. Blixt, O.; Head, S.; Mondala, T.; Scanlan, C.; Huflejt, M. E.; Alvarez, R.; Bryan, M. C.; Fazio, F.; Calarese, D.; Stevens, J.; Razi, N.; Stevens, D. J.; Skehel, J. J.; van Die, I.; Burton, D. R.; Wilson, I. A.; Cummings, R.; Bovin, N.; Wong, C.-H.; Paulson, P. *Proc. Natl. Acad. Sci., USA* **2004**, *101*, 17033-17038.

32. Köhn, M.; Wacker, R.; Peters, C.; Schröder, H.; Soulère, L.; Breinbauer, R.; Niemeyer, C. M.; Waldmann, H. *Angew. Chem. Int. Ed.* **2003**, *42*, 5830-5834.
33. Schwarz, M.; Spector, L.; Gargir, A.; Shevti, A.; Gortler, M.; Altstock, R. T.; Dukler, A. A.; Dotan, N. *Glycobiology* **2003**, *13*, 749-754.
34. Ko, K.-S.; Jaipuri, F. A.; Pohl, N. L. *J. Am. Chem. Soc.* **2005**, *127*, 13162-13163.

Chapter 16

Fluorous Carboxylates as Useful Metal Ligands for Catalysts–Products Recovery Procedures and Sensing

Mounir El Bakkari and Jean-Marc Vincent*

Laboratoire de Chimie Organique et Organométallique, (UMR-CNRS 5802), Université Bordeaux 1, 351 Cours de la Libération, 33405 Talence Cedex, France

A range of fluorous carboxylates has been used as ligating counter-anions to increase the fluorophilicity of transition metal complexes. In this chapter, applications using fluorous carboxylates in catalyst recovery procedures, organic synthesis and sensing will be highlighted.

Introduction

Fluorous chemistry, based on molecules bearing perfluoralkyl chains, has recently found numerous applications in catalysis, organic synthesis, polymer sciences, supramolecular chemistry and medicine (*1*). In particular, highly efficient purification strategies through liquid/liquid or solid/liquid extraction procedures have been developed to improve the recycling efficiency of catalysts (*2*), and facilitate the purification of organic compounds including complex natural molecules (*3*), polypeptides (*4*), oligosaccharides (*5*), and oligonucleotides (*6*). These methodologies rely on the chemical modification of a molecule with perfluoroalkyl chains, called fluorous "ponytails" or fluorous tags. The extremely apolar nature of these tags strongly affects the overall polarity of the tagged compound which allows its purification by straightforward hydrocarbon (HC)/perfluorocarbon (PFC) liquid/liquid phase separation, elution on fluorous silica, adsorption on fluorous supports (Silica, Teflon) or precipitation from organic solvents. While only one perfluoroalkyl chain is enough to impart a very efficient separation of the fluorous molecules from the non-fluorous molecules on fluorous silica, at least three of these tags are necessary to obtain high partition coefficient for a PFC, thus ensuring a high recovery of the fluorous compound when using liquid/liquid separation protocol. The requirement of highly fluorophilic derivatives is a critical issue, particularly when considering the recovery of catalysts for which a partition coefficient higher than 99% is desirable. However, transition metal complexes are rather polar molecules for which solubilization in PFCs can be challenging. In this context, the use of fluorous carboxylates as anionic ligands has emerged as an attractive strategy to obtain highly fluorophilic transition metal complexes. In this review we will focus on the use of metal-fluorous carboxylates complexes and their application in catalyst recovery procedures, but also in organic synthesis and sensing.

Fluorous carboxylic acids: Structures, properties

Carboxylates are widely encountered transition metal anionic ligands in both natural and synthetic systems. In proteins and enzymes, the carboxylate ligand is provided by the side chain of glutamate or aspartate amino acids. A remarkable property of these anionic ligands is that they can adopt a variety of coordination modes, thus allowing fine tuning of the electronic properties of the metal center and generation of free coordination sites, for example by shifting from a bidentate to a monodentate coordination mode (*7*). In synthetic catalysts, carboxylates also represent ligands of choice, the transition metal-acetate complexes being among the most useful pre-catalysts or catalysts for many reactions (*8*).

Commercially available acids (**1a-i**) (Aldrich, ABCR...) with only one perfluoroalkyl chain, are less fluorophilic than acids (**2**) and (**3**) having two fluorous tags, and thus a higher fluorinated Van der Waals surface. When employing (**1a-j**) under their carboxylate form, additional ligands modified by fluorous tags are generally necessary to obtain highly soluble metal complexes in PFCs. It should be noted that very few data are known concerning the solubility of carboxylic acids in PFCs (*9*). To the best of our knowledge, a partition coefficient (between the perfluoro(methylcyclohexane) and toluene) was reported only for the highly fluorophilic acid (**2**), 97.1% of the acid being recovered in the fluorinated phase (*10*). To obtain highly fluorophilic metal-carboxylate complexes it is thus preferable to use the basic form of (**2**) or (**3**). Acids (**2**) and (**3**) can be prepared in good overall yields in three steps from commercially available compounds. The carboxylate of (**5**) was never employed as ligand, but should prove an interesting candidate for future applications. Indeed, this commercial acid (ABCR) has a long perfluoropolyether chain and several trifluoromethyl groups ensuring a significant fluorophilicity as elegantly

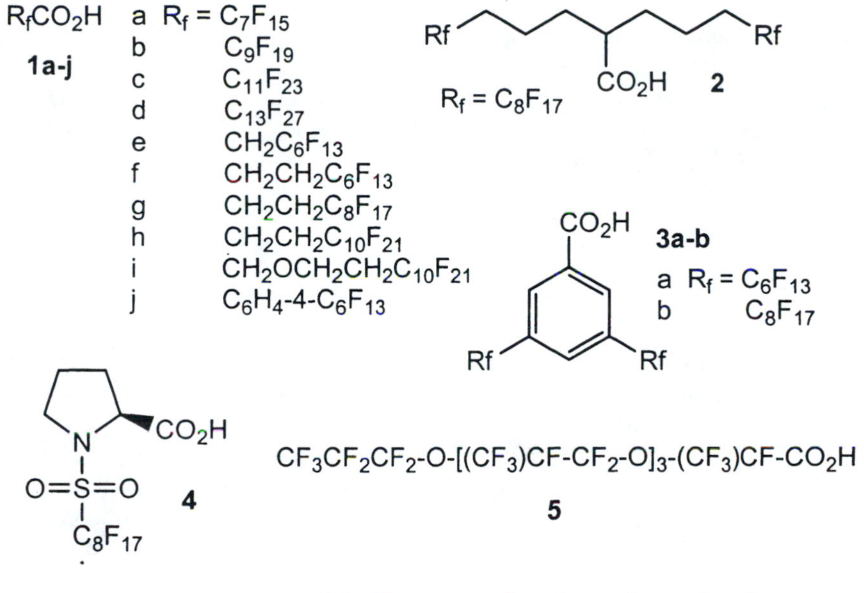

R_fCO_2H **1a-j**: a $R_f = C_7F_{15}$; b C_9F_{19}; c $C_{11}F_{23}$; d $C_{13}F_{27}$; e $CH_2C_6F_{13}$; f $CH_2CH_2C_6F_{13}$; g $CH_2CH_2C_8F_{17}$; h $CH_2CH_2C_{10}F_{21}$; i $CH_2OCH_2CH_2C_{10}F_{21}$; j C_6H_4-4-C_6F_{13}

2: Rf, Rf, CO_2H; $R_f = C_8F_{17}$

3a-b: CO_2H, Rf, Rf; a $R_f = C_6F_{13}$; b C_8F_{17}

4: N, CO_2H, O=S=O, C_8F_{17}

5: $CF_3CF_2CF_2$-O-[$(CF_3)CF$-CF_2-O]$_3$-$(CF_3)CF$-CO_2H

Figure 1. Structures of the fluorous carboxylic acids employed in fluorous chemistry.

shown by Crooks et al. (*11*). Thus, a poly(amidoamine) (PAMAM) dendrimer in which nanoparticules of Pd were encapsulated was made highly soluble in PFCs by complexing the final dendrimer amine groups with the carboxylic end groups of (**5**). Interestingly, long perfluoroalkanoic acids $CF_3(CF_2)_nCO_2H$ (n = 11, 17) failed to solubilize the dendrimers in the PFCs (*11*).

Finally, an important issue that one has to keep in mind when manipulating these fluorous carboxylic acids deals with their potential toxicity. Indeed, the Environmental Protection Agency (EPA) has identified potential human health concerns from exposure to perfluorooctanoic acid (PFOA) (**1a**), and its salts, which are used as essential processing aids in the production of fluoropolymers and fluoroelastomers such as polytetrafluoroethylene (PTFE). Studies indicated that PFOA can cause developmental anomalies and other adverse effects in laboratory animals. PFOA also appears to remain in the human body for a long time. Considerable scientific uncertainty regarding potential risks remains (for more details see, www.epa.gov, where a draft risk assessment of the potential human health effects associated with exposure to PFOA and its salts is available). For these reasons, these fluorous acids should be handled with great care until full and accurate toxicity data are available.

Alkanes/alkenes oxidations

The first application of fluorous carboxylates ligand, with the objective of facilitating catalyst recovery, was reported in 1997 by Fish and coworkers for oxidation reactions (*2d,12*). As depicted in Figure 2, Mn^{2+} and Co^{2+} complexes bearing both fluorous-1,4,7-triazacyclononane ($R_f$tacn) and carboxylate ligands derived from (**1g**) were employed as precatalysts for oxidation of alkanes and alkenes in the presence of the necessary oxidants, TBHP and O_2. Catalyst recycling by a straightforward liquid/liquid separation protocol was made possible because the fluorous carboxylate (**1g**) was used as counteranion leading to highly fluorophilic metal complexes. The limited scope of the substrates studied also showed that allylic oxidation, for example cyclohexene to cyclohexenol and cyclohexenone, was more favourable than alkane functionalization, based on thermodynamic grounds. Interestingly, spectroscopic evidence shows that these FBC oxidation reactions occurred via a classical autoxidation mechanism through the intermediacy of fluorous $Mn^{3+}Mn^{4+}$ mixed valent species. Indeed, a characteristic and intense transient 16-line EPR spectrum was observed in the fluorous phase during the course of the reaction (*12*).

Subsequently, the usefulness of fluorous carboxylates was also demonstrated by Quici and coworkers for similar oxidation reactions, but using [R_fLCo(**1a**)$_2$] associated with a tetraazacyclotetradecane macrocycle (R_fL) modified with four perfluoroether chains (*13*).

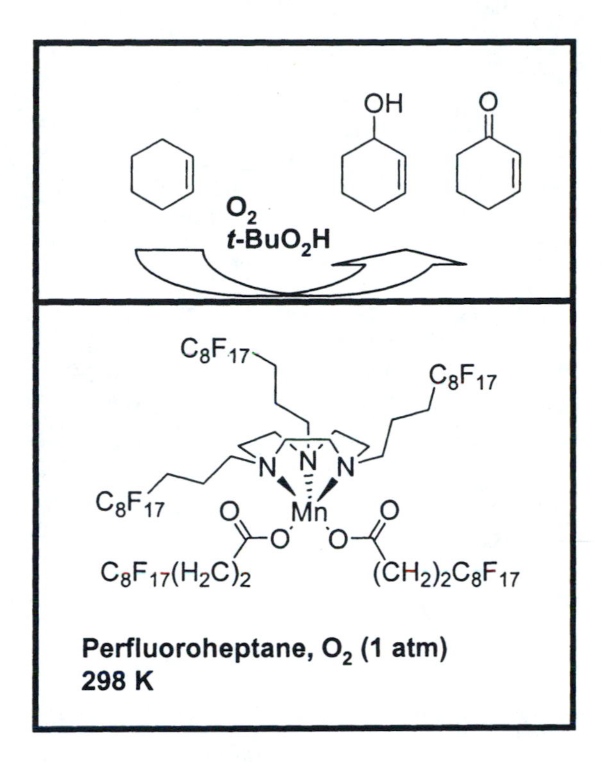

Figure 2. Fluorous biphasic oxidation catalysis with the tentative structure of the Mn^{2+} precatalyst formed in situ from the $R_f$tacn ligand and [Mn(1g)$_2$].

Another interesting application for fluorous carboxylates was reported by Pozzi and coworkers with the development of recyclable fluorous chiral (salen)manganese complexes for enantioselective epoxidations (*14*). The highly fluorous complexes (**6**) and (**7**) featuring either chloride or carboxylate ligands were prepared (Figure 3). They found that the chloride complexes, whilst being soluble in *n*-perfluorooctane tend to aggregate on standing, possibly through mutual interactions among the extended organophilic frameworks of the molecules. A solution to this problem was provided by the exchange of the chloride counterion for the fluorophilic anion (**1a**), leading to a partition coefficient $P > 100$ for (**6**) between *n*-perfluorooctane and toluene or acetonitrile.

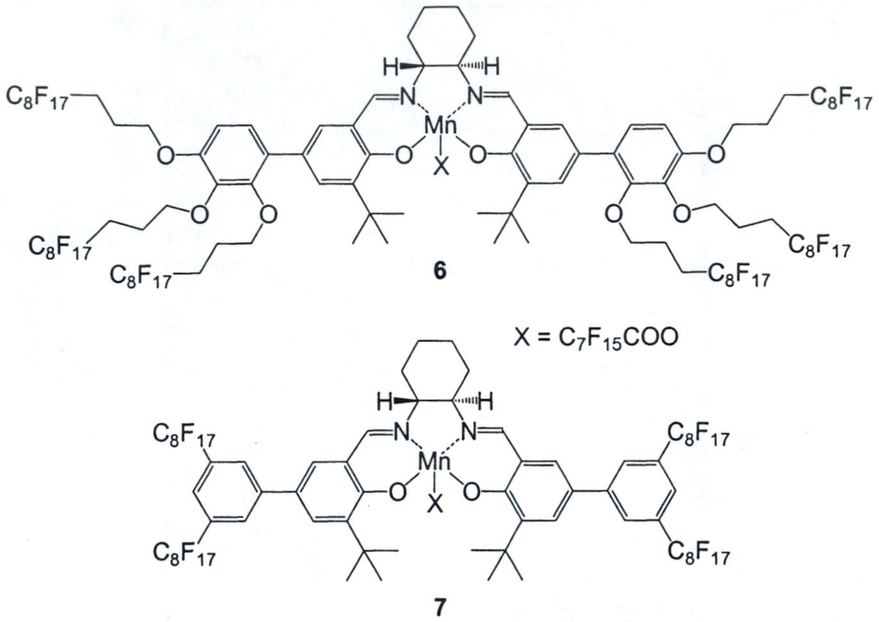

Figure 3. Fluorous Mn-(salen)manganese complexes prepared by Pozzi et al.

Using these crowded second-generation (salen)manganese complexes, highly effective and enantioselective epoxidations for a range of alkenes were achieved, with the best results obtained with a biphasic *n*-perfluorooctane/CH_3CN solvent system at 100 °C (Figure 4) in the presence of PhIO/pyridine *N*-oxide (PNO) (*14b*). Under such conditions, good to excellent epoxide yields (68-98%,) and *ee*s (50-92%) were obtained with values being close to those reported by Janda and Reger using gel-type resins. The catalysts were recycled two times without significant drop in enantioselectivities and epoxide yields. Only after the fourth run a decrease was observed.

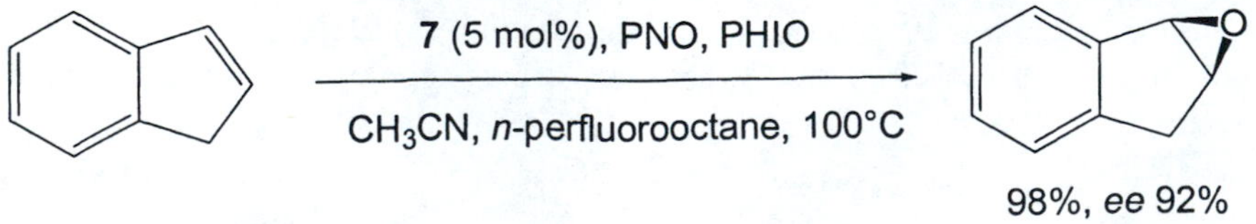

Figure 4. Asymmetric epoxidation of indene under fluorous biphasic conditions.

Alcohol oxidations

Fish, Contel, Laguna and coworkers reported the oxidation of 4-nitrobenzyl alcohol to 4-nitrobenzaldehyde using the fluorous mononuclear copper(II)-carboxylate precatalysts [Cu(**1g**)$_2$R$_f$bpy] and [Cu(**1g**)$_2$R$_f$tacn] (*15*). Due to its high fluorophilicity, [Cu(**1g**)$_2$R$_f$tacn] was employed in a perfluoroheptane/ chlorobenzene solvent mixture at 90 °C in the presence of TEMPO/O_2. At elevated temperature a single homogeneous phase was obtained leading to the benzaldehyde in excellent yields (> 90%). The catalyst was separated by liquid/liquid decantation and was recycled 4 times without significant loss of activity.

More recently, in collaboration with Fish, Contel and coworkers (*16*), we developed a perfluorcarbon-free catalytic system based on the highly fluorous dimeric copper(II) complex (**8**), a complex that we initially employed for the reversible fluorous phase-switching of pyridyl tagged compounds (*17*). Because of the high fluorophilicity of carboxylate (**2**) bearing branched perfluoroalkyl chains, the mononuclear complex [Cu(**2**)$_2$Me$_3$tacn], (**9**), which does not have fluorous tags on the polyamino ligands was found to be fully soluble in PFCs (Figure 5).

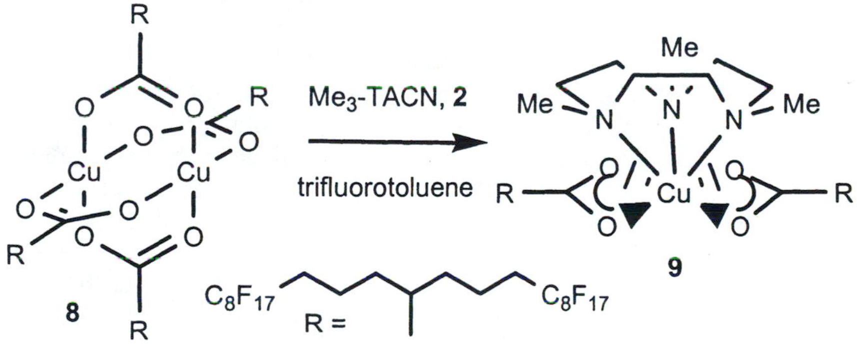

Figure 5. Structures of highly fluorous Cu(II)-carboxylate complexes.

Interestingly, reactions can be conducted under thermomorphic perfluorocarbon-free conditions (Figure 6), as demonstrated by Gladysz and coworkers (*2j*). Reactions were performed in a mixture of toluene/chlorobenzene using solid (**9**) as precatalyst, the precatalyst becoming soluble after 1 h of reaction at

90 °C. After the reaction had taken place, ~ 90% of the catalyst could be recovered via filtration techniques and reused.

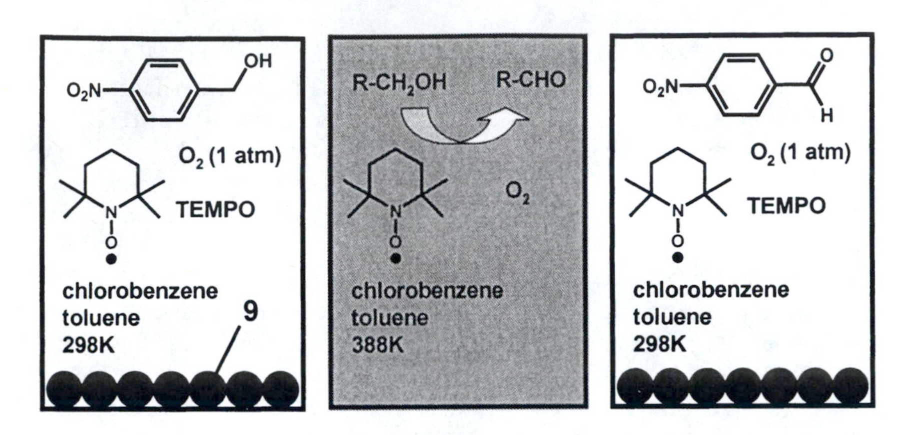

Figure 6. Oxidation of p-nitrobenzyl alcohol using (9) in a solid/liquid thermomorphic mode.

Fluorous rhodium-carboxylate catalysis

In 1999, Maas and coworkers reported on rhodium catalyzed carbenoid reactions with diazoacetates (*18*). Dimeric [Rh_2(**1a**)$_4$] and [Rh_2(**1j**)$_4$] (2 mol%) were used to achieve cyclopropanation of alkenes with methyl diazoacetate. Reactions were conducted in homogeneous CH_2Cl_2/alkene mixtures, leading to cyclopropanes in good yields, with better yields observed using the less electron-deficient benzoate derivative (71-83% against 54-79%). The catalysts were recovered by extracting with perfluoro(methylcyclohexane) the liquid residue obtained after solvent/excess alkene evaporation. Additionally, the recovered complexes have been used for four consecutive runs without significant loss of activity. They also prepared the dirhodium complexes [Rh_2(**1e-i**)$_4$] and found that all complexes with more than one methylene spacer were not soluble in PFMC (*19*). Alternatively, using the benzoates derived from (**3a-b**), the complexes [Rh_2(**3a-b**)$_4$] were prepared and found to be highly soluble in PFMC but insoluble in standard organic solvents (*20*). Cyclopropanations of styrene and α-methylstyrene were conducted with [Rh_2(**3a-b**)$_4$] (2 mol%) in a 1,1,2-trichloro-1,2,2-trifluoroethane (FC-113) fluorous/organic hybrid solvent. Good yields were obtained (70-83%) and the catalysts were recovered in a fluorous phase by replacing the FC-113 with a PFMC/CH_2Cl_2 mixture at the end of the reaction (Figure 7). By this protocol the catalysts have been reused for 5 consecutive runs affording essentially the same product yields. Additionally, compound $[Rh_2(\mathbf{3b})_4]$ showed promising results as catalyst for inter- and intramolecular carbenoid insertion into non-activated C-H bonds.

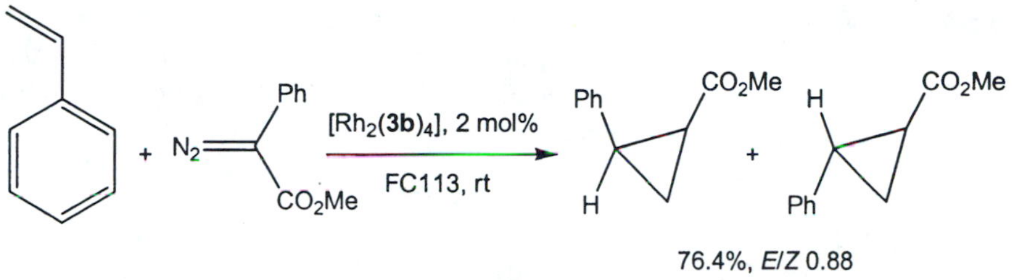

+ N_2 Ph CO_2Me $[Rh_2(\mathbf{3b})_4]$, 2 mol% FC113, rt Ph H CO_2Me + H Ph CO_2Me

76.4%, *E*/*Z* 0.88

Figure 7. Rh-catalyzed cyclopropanation of styrene.

Recently, a fluorous asymmetric version of the well-known highly selective dirhodium-prolinate complexes was described by Biffis and coworkers (*21*). The complex $[Rh_2(\mathbf{4})_4]$ was found to exhibit a good catalytic activity and chemoselectivity when the reaction was run with the catalyst dissolved in PFMC, the second phase being formed by the excess alkene (Figure 8). The catalyst can be separated by decantation and was shown to retain its activity for three consecutive runs.

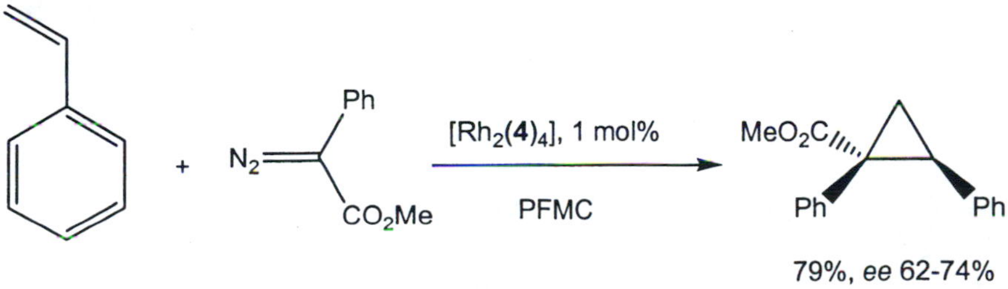

+ N_2 Ph CO_2Me $[Rh_2(\mathbf{4})_4]$, 1 mol% PFMC MeO_2C Ph Ph

79%, *ee* 62-74%

Figure 8. Asymmetric Rh-catalyzed cyclopropanation of styrene.

The dirhodium complexes $[Rh_2(\mathbf{1a\text{-}d})_4]$ were also applied by Biffis and coworkers as catalysts for silane alcoholysis under solventless conditions (*22*).

Of particular interest was their discovery of the "bonded fluorous phase catalysis" (BFPC) strategy in which the fluorous catalyst is non-covalently grafted on a solid support covalently modified with perfluoroalkyl chains, typically fluorous silica gel. Under the optimal reaction conditions triethylsilane alcoholysis was conducted using heterogeneous catalyst BFP-[Rh_2(**1a**)$_4$] (0.01 mol%) at 80 °C with an alcohol/silane solvent mixture (Figure 9). In these conditions, after 16 hrs of reaction and filtration of the catalyst, the silyl ether product was obtained in almost quantitative yield and > 95% purity. The catalyst was efficiently recycled (81%) once without decrease in catalytic activity.

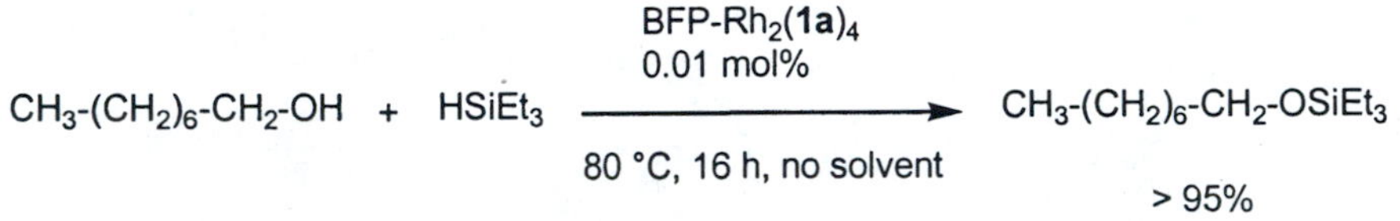

$$CH_3\text{-}(CH_2)_6\text{-}CH_2\text{-}OH + HSiEt_3 \xrightarrow[\text{80 °C, 16 h, no solvent}]{\text{BFP-Rh}_2(\mathbf{1a})_4\text{, 0.01 mol\%}} CH_3\text{-}(CH_2)_6\text{-}CH_2\text{-}OSiEt_3 \quad (> 95\%)$$

Figure 9. Rh-catalyzed silane alcoholysis under solventless conditions.

Reversible HC/PFC phase-switching process

As described above, most of the applications using fluorous carboxylates deals with catalysis. In our group in Bordeaux, we have been engaged in a program aiming at the development of new strategies to reversibly switch non-fluorinated molecules between hydrocarbon (HC) and perfluorocarbon phases. In a preliminary communication we described an unprecedented example of reversible HC/PFC phase-switching of a pyridyl-tagged prophyrin and C_{60} fullerene (*17*). This "catch-and-release" approach took advantage of both pyridyl and fluorous tags, the pyridyl tag activation being realized by pyridine coordination to the highly fluorophilic dicopper(II)-carboxylate complex (**8**). As shown in Figure 10, highly efficient liquid/liquid extraction of the 5,10,15,20-tetrapyridylporphyrine (~ 99.9%) in the fluorous phase can be achieved by simply stirring the biphasic system. Perhaps more importantly, due to the high lability of the monopyridyl ligands, instantaneous and quantitative release of TPyP was carried out by adding THF in excess, a competitive ligand for the copper ion. Additionally, the fluorous phase can be easily recycled and was shown to retain its extraction efficiency.

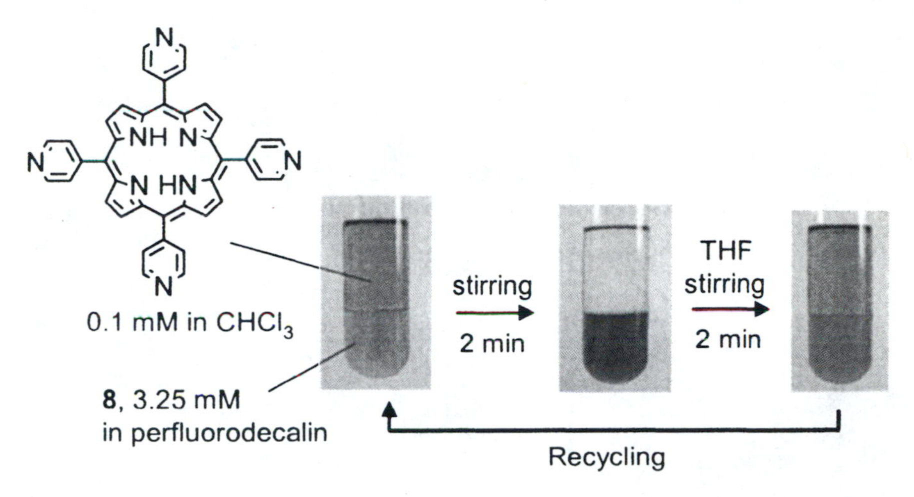

Figure 10. Photographs of reversible hydrocarbon/perfluorocarbon phase-switching of TPyP.

The 5,10,15-tripyridyl-20-phenylporphyrin (TPyMPP), 5,15-dipyridyl-15,20-diphenylporphyrin (*trans*-DPyDPP) were also extracted while the 5-monopyridyl-10,15,20-triphenylporphyrin (MPyTPP) was not (*23*). The peculiarity of this reversible phase-switching system is its high sensitivity. Accordingly, coordinating solvents such as MeOH and EtOH released the TPyP with high efficiency while the less coordinating acetone, and diethyl ether solvents were found to be much less efficient. More important, it was found that TPyP was only slightly released by water, while similar affinity for copper(II) ions was expected for water compared to alcohols (*23*). This result was attributed to the very low maximum solubility of water in C_6F_{14}, which was estimated by molecular simulations at around 2.4 mM at 20 °C (*24*). In this system, the perfluorocarbon can thus play the role of a liquid membrane impervious to water; a property could be exploited in the sensing/titration of analytes through an Indicator-Displacement-Assay (IDA) (*25*).

As a first attempt to demonstrate this concept, histamine sensing in water solution (*23*) was studied under the triphasic C_6F_{14}/DCM/H_2O system represented below (Figure 11). The histamine in solution in water binds to the copper(II) releasing the TPyP in the DCM, the amount of porphyrin released being titrate by visible light absorption measurement at 415 nm (Soret band). Using this triphasic IDA, histamine was detected in water at a concentration as low as 10 μM. We are now studying this system for the detection/titration of other analytes.

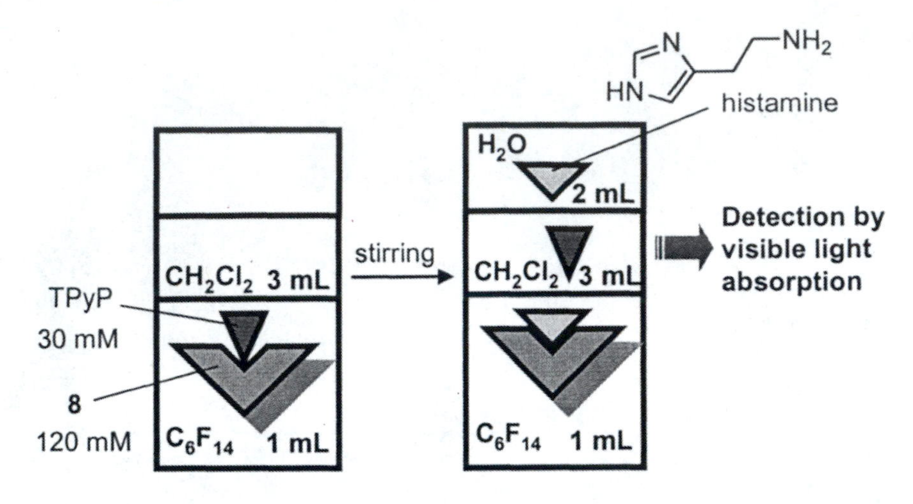

Figure 11. Fluorous triphasic IDA protocol for the sensing/titration of histamine.

Finally, the reversible phase-switching of pyridyl tagged procedure (Figure 12) was applied to the purification of substrates/products in the multi-step synthesis of a hydantoin (*26*).

For comparison, hydantoin (**10**) was prepared via the aforementioned tagging process using analogous reactions reported in a classical synthetic

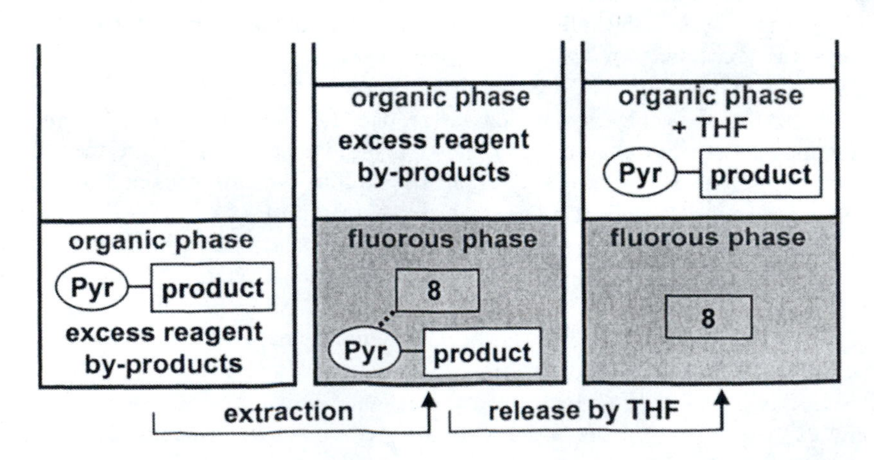

Figure 12. Fluorous phase-switching protocol applied to organic synthesis.

procedure described by Ley, Pritchard and coworkers (*27*). Indeed, in 2001 they reported the synthesis of benzodiazepins and hydantoins assisted by a 2,2'-bipyridine tag. A resin functionalized with iminodiacetic groups and loaded with copper(II) ions was employed as a solid support to trap substrates/products linked to the 4,4'-bis(hydroxymethyl)-2,2'-bipyridine tag. A limitation to this system is associated with the release of the strongly coordinating bipyridine-linker-product moiety from the support. This was achieved by shaking the suspension of the beads for 8 h in the presence of TMEDA ligand (N,N,N',N'-tetramethylethylenediamine). Moreover, the problem of the recovery of the copper-carboxylate resin was not addressed.

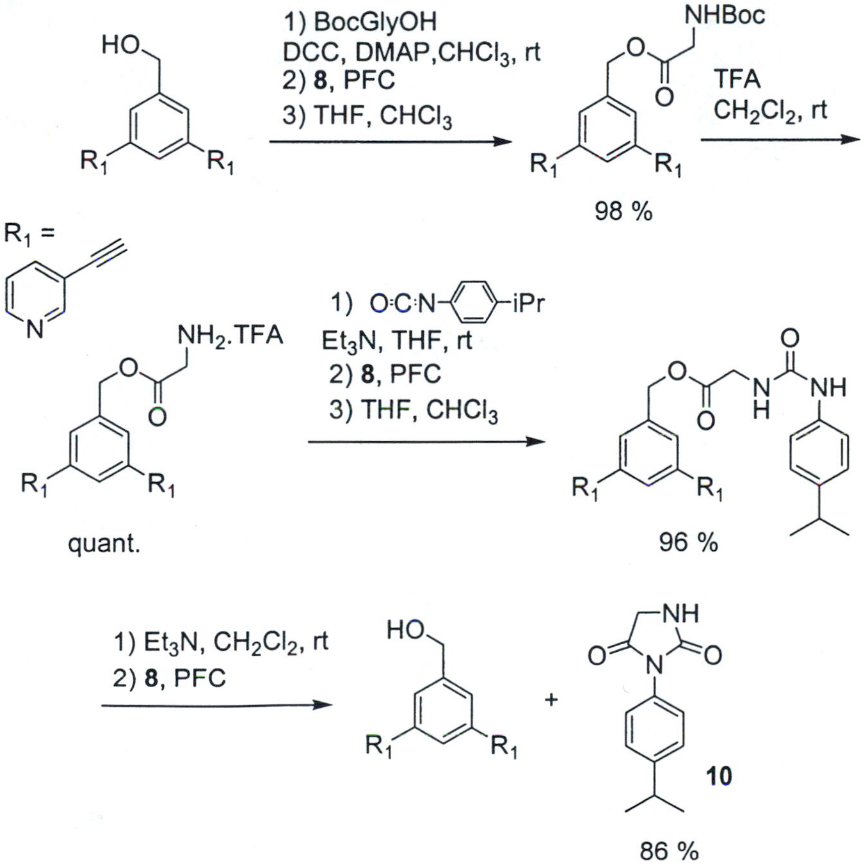

Figure 13. Synthesis of a hydantoin assisted by a reversible fluorous phase-switching process.

Using the fluorous liquid/liquid protocol in conjunction with a benzyl alcohol bis-monopyridyl tag, the hydantoin (**10**) was prepared in 81% overall yield by a four-step reaction sequence (Figure 13). The "extraction-release" process was achieved with high efficiency for the first and third steps of the synthesis, while in the last step, the fluorous phase containing **8** was used to selectively remove the bis-monopyridyl tag.

Conclusion

During last ten years fluorous carboxylates have been shown to be extremely useful ligands to increase the fluorophilicity of cationic catalysts, thus facilitating their recovery. As many highly fluorous carboxylic acids, such as (**5**), are now commercially available, there is little doubt that new applications for fluorous carboxylate-based catalysts will appear. We are engaged in a program focusing on the development of new and practical applications based on the reversible fluorous phase-switching processes. As shown in this review, the use of metal-carboxylates is emerging as a strategy of choice and, consequently, we are pursuing our investigations with such compounds and will report new applications in due course.

References

1 a) Special Tetrahedron issue on Fluorous Chemistry: Gladysz, J. A.; Curran, D. P., Eds.; *Tetrahedron* **2002**, *58*, 3823-4132. b) *Handbook of Fluorous Chemistry*; Gladysz, J. A.; Curran, D. P.; Horváth I. T., Eds., Wiley-VCH, Weinheim, 2004.

2 Selected examples: a) Horváth, I. T.; Rábai, J. *Science* **1994**, *266*, 72. b) Juliette, J. J. J.; Rutherford, D.; Horváth, I. T.; Gladysz, J. A. *J. Am. Chem. Soc.* **1999**, *121*, 2696. c) Klement, I.; Lütjens, H.; Knochel, P. *Angew. Chem. Int. Ed. Engl.* **1997**, *36*, 1454. d) Vincent, J.-M.; Rabion, A.; Yachandra, V. K.; Fish, R. H. *Angew. Chem. Int. Ed. Engl.* **1997**, 36, 2346. e) Cavazzini, M.; Manfredi, F.; Montanari, S.; Quici, S.; Pozzi, G. *Eur. J. Org. Chem.* **2001**, 4639. f) De Campo, F.; Lastécouères, D.; Vincent, J.-M.; Verlhac, J.-B. *J. Org. Chem.* **1999**, *64*, 4969. g) van den Broeke, J.; de Wolf, E.; Deelman, B.-J.; van Koten, G. *Adv. Synth. Catal.* **2003**, *345*, 625. h) Maayan, G.; Fish, R. H.; Neumann, R. *Org. Lett.* **2003**, *5*, 3547. i) Dinh, L. V.; Gladysz J. A. *Angew. Chem. Int. Ed.* **2005**, *44*, 4095. j) Wende, M.; Gladysz, J. A. *J. Am. Chem. Soc.* **2003**, *125*, 5861.

3 a) Zhang, W. *Chem. Rev.* **2004**, *104*, 2531. b) Wipf, P.; Reeves, J. T.; Balachandran, R.; Giuliano, K. A.; Hamel, E.; Day, B. W. *J. Am. Chem. Soc.* **2000**, *122*, 9391. c) Zhang, W.; Luo, Z.; Hiu-Tung Chen, C.; Curran, D. P. *J. Am. Chem. Soc.* **2002**, *124*, 10443. d) Wilcox, C. S.; Gudipati, V.; Lu, H.; Turkyilmaz, S.; Curran, D. P. *Angew. Chem. Int. Ed.* **2005**, *44*, 6938.

4 a) Montanari, V.; Kumar, K. *J. Am. Chem. Soc.* **2004**, *126*, 9528. b) Filippov, D. V.; van Zoelen, D. J.; Oldfield, S. P.; van der Marel, G. A.; Overkleeft, H. S.; Drijfhout, J. W.; van Boom, J. H. *Tetrahedron. Lett.* **2002**, *43*, 7809.

5 a) Miura, T.; Goto, K.; Hosaka, D.; Inazu, T. *Angew. Chem. Int. Ed.* **2003**, *42*, 2047. b) Miura, T.; Inazu, T. *Tetrahedron. Lett.* **2003**, *44*, 1819.

6 a) Pearson, W. H.; Berry, D. A.; Stoy, P.; Jung, K.-Y.; Sercel, A. D. *J. Org. Chem.* **2005**, *70*, 7114. b) Andrushko, V.; Schwinn, D.; Tzschucke, C. C.; Michalek, F.; Horn, J.; Mössner, C.; Bannwarth, W. *Helv. Chim. Acta* **2005**, *88*, 936. c) Beller, C.; Bannwarth, W. *Helv. Chim. Acta* **2005**, *88*, 171.

7 An impressive illustration of the coordination flexibility of the carboxylates is found in the catalytic cycle of the methane monooxygenase enzyme, see: Merkx, M.; Kopp, D. A.; Sazinsky, M. H.; Blazyk, J. L.; Müller, J.; Lippard, S. J. *Angew. Chem. Int. Ed.* **2001**, *40*, 2782.

8 Reviews: a) Merlic, C. A.; Zechman, A. L. *Synthesis* **2003**, 1137. b) Ley, S. V.; Thomas, A. W. *Angew. Chem. Int. Ed.* **2003**, *42*, 5400.

9 For extensive fluorous/organic liquid/liquid partition coefficient data see: Gladysz, J. A.; Emnet, C.; Rabai, J. In *Handbook of Fluorous Chemistry*; Gladysz, J. A.; Curran, D. P.; Horváth, I. T.; Eds., Wiley-VCH, Weinheim, 2004, pp 56-100.

10 Loiseau, J.; Fouquet, E.; Fish, R. H.; Vincent, J.-M.; Verlhac, J.-B. *J. Fuor. Chem.* **2001**, *108*, 195.

11 Chechik, V.; Crooks, R. M. *J. Am. Chem. Soc.* **2000**, *122*, 1243.

12 Vincent, J.-M.; Rabion, A.; Yachandra, V. K.; Fish, R. H. *Can. J. Chem.* **2001**, *79*, 888.

13 Pozzi, G.; Cavazzini, M.; Quici, S.; Fontana, S. *Tetrahedron Lett.* **1997**, *38*, 7605.

14 a) Cavazzini, M.; Manfredi, F.; Montanari, S.; Quici, S.; Pozzi, G. *Chem. Commun.* **2000**, 2171. b) Cavazzini, M.; Manfredi, F.; Montanari, S.; Quici, S.; Pozzi, G. *Eur. J. Org. Chem.* **2001**, 4639.

15 Contel, M.; Izuel, C.; Laguna, M.; Villuendas, P. R.; Alonso, P. J.; Fish, R. H. *Chem. Eur. J.* **2003**, *9*, 4168.

16 Contel, M.; Villuendas, P.R.; Fernández-Gallardo, J.; Alonso, P. J.; Vincent, J.-M.; Fish, R. H. *Inorg. Chem.* **2005**, *44*, 9771.

17 El Bakkari, M.; McClenaghan, N.; Vincent, J.-M. *J. Am. Chem. Soc.* **2002**, *124*, 12942.

18 Endres, A.; Maas, G. *Tetrahedron Lett.* **1999**, *40*, 6365.

19. Endres, A.; Maas, G. *Organomet. Chem.* **2002**, *174-180*, 643.

20 Endres, A.; Maas, G. *Tetrahedron* **2002**, *58*, 3999.
21 Biffis, A.; Braga, M.; Cadamuro, S.; Tubaro, C.; Basato, M. *Org. Lett.* **2005**, *7*, 1841.
22 a) Biffis, A.; Castello, E.; Zecca, M.; Basato M., *Tetrahedron* **2001**, *57*, 10391. b) Biffis, A.; Zecca, M.; Basato, M. *Green Chem.* **2003**, *5*, 170. c) Biffis, A.; Braga, M.; Basato, M. *Adv. Synth. Catal.* **2004**, *346*, 451.
23 El Bakkari, M.; Fronton, B.; Luguya, R.; Vincent, J.-M. *J. Fluorine Chem.* **2006**, in press.
24 a) Deschamps, J.; Costa Gomes, M. F.; Pádua, A. A. H. *J. Fluor. Chem.* **2004**, *125*, 409. b) Costa Gomes, M. F. personal communication.
25 Selected examples of IDAs: a) Fabbrizzi, L.; Marcotte, N.; Stomeo, F.; Taglietti, A. *Angew. Chem. Int. Ed.* **2002**, *41*, 3811. b) Zhong, Z.; Anslyn, E. V. *J. Am. Chem. Soc.* **2002**, *124*, 9014. c) Wright, A. T.; Zhong, Z.; Anslyn, E. V. *Angew. Chem. Int. Ed.* **2005**, *44*, 5679. d) Maue, M.; Schrader, T. *Angew. Chem. Int. Ed.* **2005,** 44, 2265. e) Buryak, A.; Severin, K. *J. Am. Chem. Soc.* **2005**, *127*, 3700. f) Fabbrizzi, L.; Foti, F.; Taglietti, A. *Org. lett.* **2005**, *7*, 2603. g) Buryak, A.; Severin, K. *Angew. Chem. Int. Ed.* **2004**, *43*, 4771. h) Hilderbrand, S. A.; Lim, M. H.; Lippard, S. J. *J. Am. Chem. Soc.* **2004**, *126*, 4972.
26 El Bakkari, M.; Vincent, J.-M. *Org. Lett.* **2004**, *6*, 2765.
27 Ley, S. V.; Massi, A.; Rodriguez, F.; Horwell, D. C.; Lewthwaite, R. A.; Pritchard, M. C.; Reid, A. M. *Angew. Chem. Int. Ed.* **2001**, *40*, 1053.

New Biological Applications

Chapter 17

Organofluorine Chemistry at the Biomedical Interface: A Case Study on Fluoro-Taxoid Anticancer Agents

Iwao Ojima, Larissa V. Kuznetsova, and Liang Sun

Institute of Chemical Biology and Drug Discovery and Department of Chemistry, State University of New York at Stony Brook, Stony Brook, NY 11794–3400

New and effective biochemical tools as well as medicinal and therapeutic agents have been successfully developed through strategic incorporation of fluorine(s) based on the rational design by taking into account the special properties of fluorine. Fluorinated analogues of biologically active molecules can also serve as excellent probes for the investigation of biochemical mechanisms. ^{19}F-NMR can provide unique and powerful tools for the structural and mechanistic investigations in chemical biology. This article describes our recent studies on the use of fluorine in the medicinal chemistry and chemical biology of taxane anticancer agents, exploiting the unique nature of this element, as a showcase of the "current fluoroorganic chemistry" at the biomedical interface.

Introduction

Fluorine is one of the smallest atoms with the highest electronegativity. Thus, the introduction of fluorine(s) to a bioactive molecule causes minimal steric alteration, while significantly affects the physico-chemical properties of the molecule.[1] The replacement of an oxidizable C-H group by a C-F group increases the metabolic stability of the molecule. The presence of fluorine(s) enhances lipophilicity of drugs and thus increases hydrophobic interactions and membrane permeability.[1] New and effective biochemical tools as well as medicinal and therapeutic agents have been successfully developed through rational design by these special properties of fluorine.[1,2] Fluorinated analogues of biologically active molecules can also serve as excellent probes for the investigation of biochemical mechanisms. ^{19}F-NMR can provide unique and powerful tools for the mechanistic investigations in chemical biology.[2-5]

This article describes an account of our research on the use of fluorine, exploiting the unique nature of this element mentioned above, in the medicinal chemistry and chemical biology of taxane anticancer agents as a showcase in the "current fluoroorganic chemistry".

Paclitaxel, docetaxel, taxoids, and second-generation taxoids

Paclitaxel (Taxol) and docetaxel (Taxotère) are two of the most important "taxane" anticancer drugs approved for clinical use in chemotherapy against various human tumors including metastatic breast cancer, advanced ovarian cancer, non-small cell lung cancer, and Kaposi's sarcoma.[6,7] These anticancer drugs have been expanding their clinical applications for the treatment of other cancers such as head and neck, prostate, and cervical cancers. Effective chemotherapy in combination with other anticancer agents like cisplatin, carboplatin or doxorubicin has also been developed.[6] Docetaxel is the first "taxoid" ("Taxol-like" compound), which became a FDA-approved anticancer drug, produced through semisynthesis, i.e., chemical synthesis building upon a readily available naturally occurring key intermediate.[8,9]

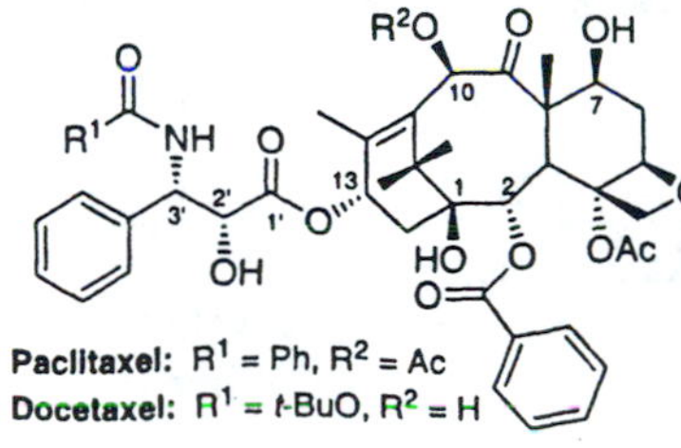

Paclitaxel: R^1 = Ph, R^2 = Ac
Docetaxel: R^1 = *t*-BuO, R^2 = H

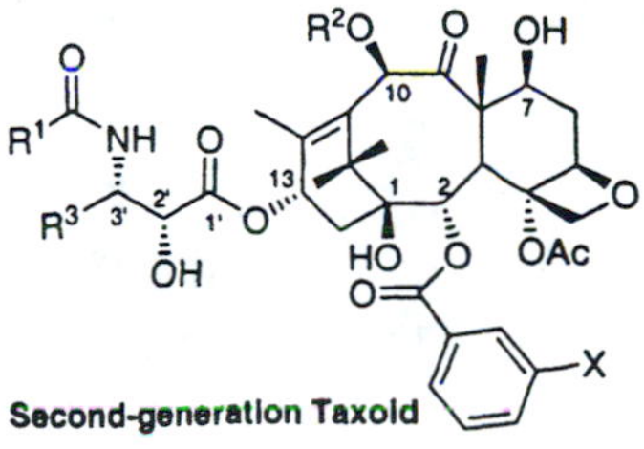

Second-generation Taxoid

These "taxane" anticancer drugs bind to the β-tubulin subunit, accelerate the polymerization of tubulin, and then stabilize the resultant microtubules. The stabilized microtubules lead to the inhibition of their depolymerization, which results in the arrest of the cell division cycle mainly at the G2/M stage, inducing the programmed cell death (apoptosis) through the cell-signaling cascade.[10,11]

Although both paclitaxel and docetaxel possess potent antitumor activity, chemotherapy with these drugs encounters undesirable side effects as well as drug resistance.[6,8,9,12] Therefore, it is important to develop new taxane anticancer agents or efficacious drug delivery systems with fewer side effects, superior pharmacological properties, and improved activity against various classes of tumors, especially against drug-resistant cancers.

Second-generation taxoids bearing a non-aromatic substituent at the C3' position were designed and developed in our laboratory. These taxoids possess one order of magnitude better activity against drug-sensitive cell lines and more than two orders of magnitude better activity against drug-resistant cell lines (Table 1)[13] It was found that *meta*-substitutions at the benzene ring of the benzoate moiety at the C-2 position dramatically increase cytotoxicity of taxoids against drug-resistant cancer cell lines.[14] Thus, this class of the second-generation taxoids virtually overcomes MDR at least for the cell level, which means that the cytotoxicity of these taxoids is three orders of magnitude stronger than paclitaxel and docetaxel against drug-resistant cance cell lines, expressing MDR.[14]

Table 1. Cytotoxicity (IC_{50})[a] of a few seconds-generation taxoids

Taxoid	R^1	R^2	R^3	X	MCF7-S[b] (breast)	MCF7-R[c] (breast)	LCC6-WT[d] (breast)	LCC6-MDR[e] (breast)
Paclitaxel					1.7	300	3.1	346
Docetaxel					1.0	215	---	---
SB-T-1213	*t*-BuO	EtCO	$Me_2C{=}CH$	H	0.18	2.2	---	---
SB-T-1103	*t*-BuO	EtCO	Me_2CHCH_2	H	0.35	2.8	---	---
SB-T-121303	*t*-BuO	EtCO	$Me_2C{=}CH$	MeO	0.36	0.33	1.0	0.9
SB-T-11033	*t*-BuO	EtCO	Me_2CHCH_2	MeO	0.36	0.43	0.9	0.8

[a]The concentration of compound which inhibits 50% (IC_{50}, nM) of the growth of a human tumor cell line after 72 h drug exposure. [b]human breast carcinoma. [c]multidrug resistant human breast carcinoma. [d]human breast carcinoma. [e]multidrug resistant human breast carcinoma.

In the course of our extensive studies on the design, synthesis and the structure-activity relationship (SAR) of taxane anticancer agents, we synthesized fluorine-containing paclitaxel and docetaxel analogs to investigate the effects of fluorine–incorporation on the cytotoxicity and the blockage of known metabolic pathways. Our earlier studies have been reported in several publications.[15,18] Thus, we describe here our recent studies on the synthesis and SAR of fluorine-containing second-generation taxoids.

Second-generation taxoids with strategic incorporation of fluorine

Synthesis and biological evaluation of 3'-difluoromethyl- and 3'-trifluoromethyl-taxoids

A series of the second-generation 3'-CF_2H- and 3'-CF_3-taxoids was synthesized through *β-Lactam Synthon Method.*[19,20] Enantiopure β-lactams bearing CF_2H and CF_3 groups at the C-4 position were synthesized through [2+2] ketene-imine cycloaddition followed by enzymatic optical resolution (PS Amano lipase, 50 °C) and functional group transformation (for the CF_2H group).[21] Enantiopure (3*R*,4*S*)-3-TIPSO-4-CF_2H-β-lactam **9** was obtained using the method reported from our laboratory[21] via difluoromethylation of 4-formyl-β-lactam **8** with DAST[22] (Scheme 1). Synthesis of enantiopure (3*R*,4*S*)-3-TIPSO-4-CF_3-β-lactam **13** was carried out via enzymatic optical resolution (PS-Amano lipase, 3 °C) of racemic 3-AcO-4-CF_3- β-lactam **11,** which was prepared from benzyloxyacetyl chloride and *N*-PMP-CF_3-aldimine through [2+2] cycloaddition followed by debenzylation and acetylation (Scheme 2).[21]

The ring-opening coupling of 4-CF_2H- and 4-CF_3- β-lactams **9** and **13** thus obtained with 2,10-modified baccatins **1** was carried out at -40 °C in THF using LiHMDS as base followed by HF/pyridine deprotection of silicon protecting groups to give the corresponding second-generation 3'-CF_2H- and 3'-CF_3-taxoids (Scheme 3)[17,21,23] 2,10-Modified baccatins **1** were prepared by the methods reported previously by our laboratory.[14]

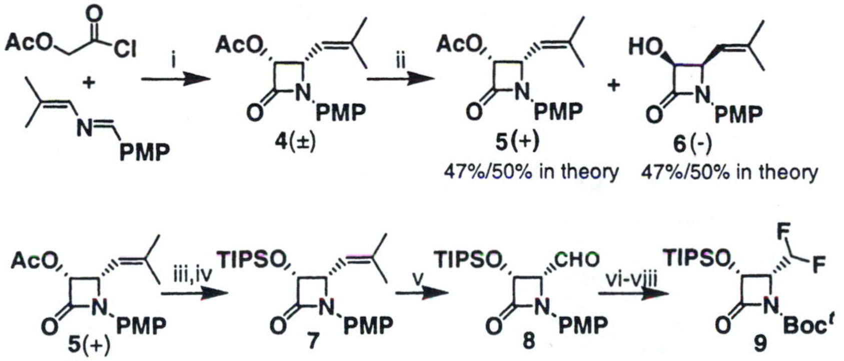

4(±) 5(+) 6(-)

47%/50% in theory 47%/50% in theory

5(+) 7 8 9

i) Et_3N,CH_2Cl_2, -78 °C ~ r.t., 70%; ii) PS-Amano, buffer pH 7.0, 10% CH_3CN, 50 °C; iii) KOH, THF, 0 °C, 100%; iv) TIPSCl, Et_3N, DMAP, CH_2Cl_2, 85%; v) O_3, MeOH/CH_2Cl_2, Me_2S, -78 °C, 73%; vi) DAST,CH_2Cl_2, 86%; vii) CAN, H_2O/CH_3CN, -15 °C, 68%; viii) *t*-Boc_2O, Et_3N, DMAP, CH_2Cl_2, 80%.

Scheme 1

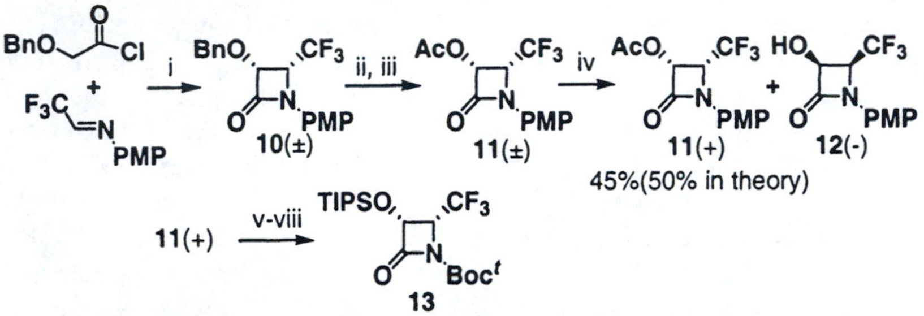

i) Et_3N, CH_2Cl_2, 40 °C, 83%; ii) H_2, Pd, MeOH, 45 °C, 98%; iii) Ac_2O, DMAP, Py, CH_2Cl_2, 74%; iv) PS-Amano, buffer pH 7, 10 % CH_3CN, 0-5 °C; v) KOH, THF, -5 °C, 100%; vi) TIPSCl, Et_3N, CH_2Cl_2, 95%; vii) CAN, CH_3CN/H_2O, -10 °C, 84%; viii) Boc_2O, Et_3N, DMAP, CH_2Cl_2, 87%.

Scheme 2

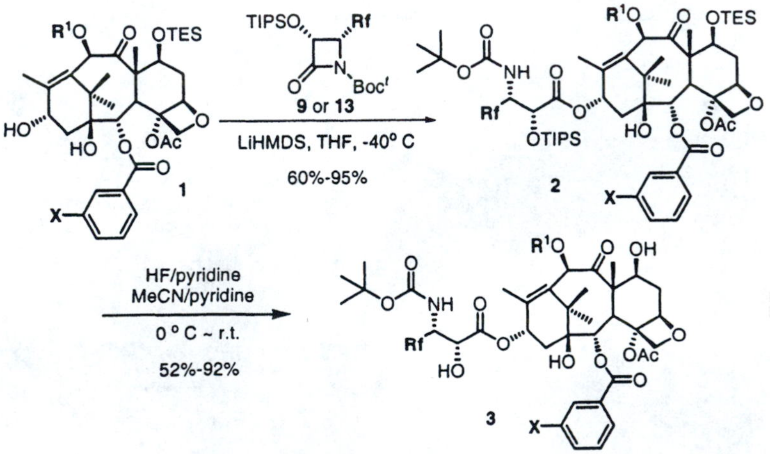

R^1 = MeCO, EtCO, MeOCO, Me_2NCO, TES; X= MeO, F, Cl, N_3; Rf = CF_2H, CF_3

Scheme 3

Cytotoxicities of these second-generation fluoro-taxoids **3** were evaluated *in vitro* against human breast cancer cell lines MCF7-S and LCC6-WT cancer cell lines and their corresponding drug-resistant cell lines MCF7-R and LCC6-MDR as well as human ovarian (H 460) and colon (HT-29) cancer cell lines. Results are summarized in Table 2 (3'-CF_2H-taxoids) and Table 3 (3'-CF_3-taxoids). The IC_{50} values were determined through 72 h exposure of the fluorotaxoids to the cancer cells according to the protocol developed by Skehan *et al.*[24]

Table 2. Cytotoxicity (IC_{50} nM)[a] of 3'-CF_2H-taxoids

Taxoid	R¹	x	MCF7-S[b] (breast)	MCF7-R[c] (breast)	R/S	LCC6-WT[d] (breast)	LCC6-MDR[e] (breast)	R/S	H460[f] (ovarian)	HT-29[g] (colon)
Paclitaxel			**1.7**	**300**	**176**	**3.1**	**346**	**112**	**4.9**	**3.6**
Docetaxel			**1.0**	**215**	**215**	---	---	---	---	**1.0**
SB-T-1284-1	**Ac**	**MeO**	0.34	4.16	12	0.26	5.57	21	0.38	0.52
SB-T-1284-2	**Ac**	**F**	0.44	5.53	13	0.52	10.03	19	0.2	0.35
SB-T-1284-3	**Ac**	**N_3**	0.32	1.68	**5.3**	0.22	1.57	**7.1**	0.48	0.57
SB-T-12841-3	**Me_2N-CO**	**N_3**	0.37	1.44	**3.9**	0.29	1.69	**5.8**	0.52	0.40
SB-T-12841-4	**Me_2N-CO**	**Cl**	0.31	2.96	**9.5**	0.21	3.87	**18**	0.36	0.58
SB-T-12842-3	**Et-CO**	**N_3**	0.32	0.96	**3**	0.39	1.15	**2.9**	0.27	0.37
SB-T-12842-3	**MeO-CO**	**N_3**	1.69	2.56	**1.5**	0.26	2.06	**7.9**	0.23	0.36

[a]See footnote a in Table 1. [b-e]See footnotes b-e in Table 1. [f]human ovarian carcinoma. [g]human colon carcinoma.

Table 3. Cytotoxicity (IC_{50} nM)[a] of 3'-CF_3-taxoids

Taxoid	R¹	x	MCF7S[b] (breast)	MCF7-R[c] (breast)	R/S	LCC6WT[d] (breast)	LCC6-MDR[e] (breast)	R/S	H460[f] (ovarian)	HT-29[g] (colon)
Paclitaxel			1.7	300	**176**	3.1	346	**112**	4.9	3.6
Docetaxel			1.0	215	**215**	---	---	---	---	1.0
SB-T-1281-3	**H**	**N_3**	0.57	7.45	13	1.18	12.3	10	0.2	0.7
SB-T-1282-3	**Ac**	**N_3**	0.47	3.85	**8.2**	1.18	4.0	**3.4**	0.2	0.5
SB-T-12821-1	**Me_2N-CO**	**MeO**	0.57	1.84	**3.2**	0.28	4.48	16	0.35	0.68
SB-T-12821-2	**Me_2N-CO**	**F**	0.32	2.64	**8.3**	0.32	5.57	17	0.5	0.76
SB-T-12821-3	**Me_2N-CO**	**N_3**	0.47	2.61	**5.6**	1.27	3.52	**2.8**	0.3	0.5
SB-T-12822-1	**Et-CO**	**MeO**	0.19	2.16	11	0.45	4.24	**9**	0.41	0.54
SB-T-12822-3	**Et-CO**	**N_3**	0.38	1.61	**4.2**	1.09	2.56	**2.3**	0.2	0.4
SB-T-12823-1	**MeO-CO**	**MeO**	0.17	2.88	17	0.27	399	15	0.38	0.53
SB-T-12823-2	**MeO-CO**	**F**	0.31	4.88	16	0.39	5.81	15	0.61	0.85
SB-T-12823-3	**MeO-CO**	**N_3**	0.47	2.92	**6.2**	1.09	4.0	**3.7**	0.2	0.4

[a-g]See footnotes in Tables 1 and 2.

As Table 2 shows, all second-generation 3'-CF_2H-taxoids exhibit sub-nanomolar IC_{50} values against drug-sensitive MCF7-S, LCC6-WT, H460, and HT-29 cancer cell lines (except for one case), which are several to ten times more potent than paclitaxel. Their cytotoxicity against drug-resistant MCF7-R and LCC6-MDR cell lines is all in the single digit nanomolar IC_{50} value (except for one case), which is two orders of magnitude on average more potent than paclitaxel. Among these 3'-CF_2H-taxoids examined, SB-T-12842-3 (R^1 = *n*-propanoyl; X = N_3) appears to be the most potent compound with the R/S ratio of only 2.9-3.0 against two sets of human breast cancer cell lines.

The cytotoxicity of second-generation 3'-CF_3-taxoids is very close to that of the 3'-CF_2H-taxoids, as shown in Table 3. Their potency against MCF7-S appears slightly higher and uniform with different substitution patterns.

Synthesis and biological evaluation of second-generation 3'-difluorovinyl-taxoids

Prior to the introduction of the CF_2H and CF_3 groups to the 3'-position of taxoids, replacing the phenyl group of paclitaxel and docetaxel, highly potent second-generation taxoids have been developed in our laboratory, which beared isobutyl or isobutenyl group at the C3' position. Recently, metabolism studies on these 3'-isobutyl- and 3'-isobutenyl-taxoids were performed in collaboration with Dr. Gut's laboratory at the National Institute of Public Health, Czech Republic. These studies disclosed that the metabolism of these second-generation taxoids (SB-T-1214, SB-T-1216, and SB-T-1103) is markedly different from that of docetaxel and paclitaxel.[25] For example, as Figure 1 illustrates, the primary metabolic sites of these taxoids by CYP 3A4 of the cytochrome P-450 family are the two allylic methyl groups of the 3'-isobutenyl moiety and the methyne group of the 3'-isobutyl group.[25] The results form a sharp contrast to that of decetaxel, in which the *tert*-butyl group of the 3'*N*-*t*Boc moiety is the single predominant site.[26]

These results prompted us to synthesize novel 3'-difluorovinyl-taxoids, which should be able to block the allylic oxidation by CYP 3A4 mentioned above, hence should have enhanced metabolic stability and *in vivo* activity.

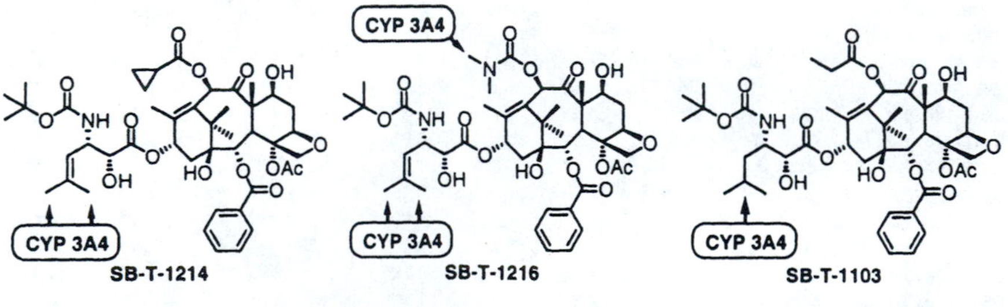

Figure 1. Primary sites of hydroxylation on second-generation taxoids by cytochrome P450

For the synthesis of a series of 3'-difluorovinyl-taxoids, enantiopure (3*R*,4*S*)-1-*t*-Boc-3-TIPSO-4-difluorovinyl-β-lactam **17**(+) is necessary. We successfully prepared this key intermediate in three steps from 4-formyl-β-lactam **14**(+), using the Wittig reaction of the formyl moiety with difluoromethylphosphorus ylide generated *in situ* from $(Me_2N)_3P/CF_2Br_2/Zn$ (Scheme 4).

The Ojima-Holton ring-opening coupling[3,27,28] of β-lactam **17**(+) with 2-modified, 10-modified or 2,10-modified baccatins[14] and the subsequent removal of the silyl protecting groups gave the corresponding 3'-difluorovinyl-taxoids in good to excellent yields (Scheme 5).

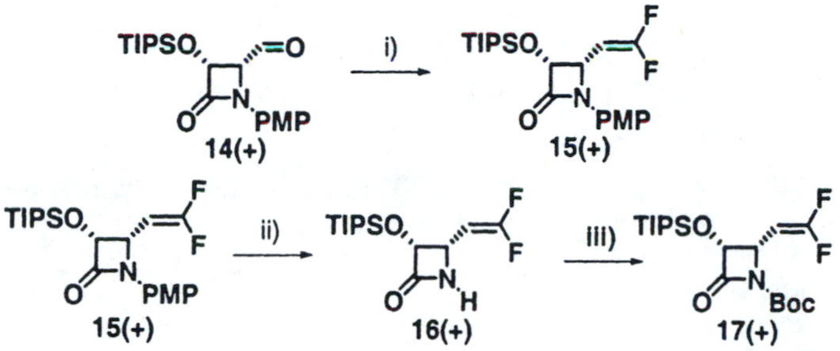

i) CBr_2F_2, HMPT, Zn, THF, 84%; ii) CAN, H_2O/CH_3CN, -15 °C, 92%; iii) *t*-Boc_2O, Et_3N, DMAP, CH_2Cl_2, 100%.

Scheme 4

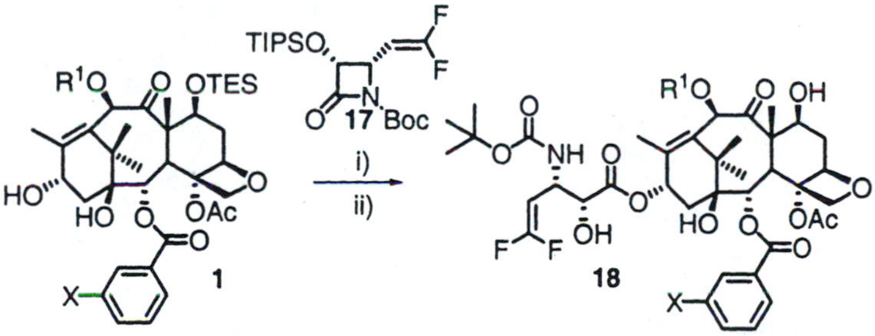

R^1= MeCO, EtCO, c-PrCO, $(Me)_2NCO$, MeOCO,
X= a) H, b) MeO, c) F, d) Cl, e) N_3,
i) LiHMDS, THF, -40° C, ii) HF/Py, Py/CH_3CN, overnight, 0° C ~ r.t., 57-91%.

Scheme 5

Cytotoxicities of the 3'-difluorovinyl-taxoids were evaluated *in vitro* against human breast cancer cell lines, MCF7-S and MCF7-R.[24] Results are summarized in Table 4. As Table 4 shows, these novel taxoids exhibit exceptional potency against MCF7-S, i.e., most of these taxoids possess less than 100 pM IC_{50} values, exceeding the potency of the highly potent second-generation taxoids previously developed in our laboratory. Cytotoxicities against MCF7-R are all subnanomolar level IC_{50}'s except for one, which are three orders of magnitude more potent than paclitaxel. The two most potent taxoids in this series against MCF7-S are **SB-T-12855-1** (IC_{50} 78 pM) and **SBT-12855-3** (IC_{50} 76 pM), in which a MeOCO group is at C10 for both taxoids, and *meta*-MeO and *meta*-N_3 groups, respectively, at the C2-benzoate moiety.

Table 4. Cytotoxicity (IC_{50} nM)[a] of 3'-difluorovinyl-taxoids

Taxoid	R[1]	X	MCF7-S[b] (breast)	MCF7-R[c] (breast)	R/S
Paclitaxel			1.2	300	250
SB-T-1213	Et-CO	H	0.13	2.2	17
SB-T-121303	Et-CO	MeO	0.25	0.33	1.3
SB-T-121304	Et-CO	N_3	0.64	1.1	1.7
SB-T-12851	Ac	H	0.099	0.95	9.6
SB-T-12853	Et-CO	H	0.17	1.2	7.06
SB-T-12852-1	c-Pr-CO	MeO	0.092	0.48	5.2
SB-T-12853-1	Et-CO	MeO	0.34	0.57	1.7
SB-T-12855-1	MeO-CO	MeO	0.078	0.50	6.4
SB-T-12851-3	Ac	N_3	0.092	0.34	3.7
SB-T-12852-3	c-Pr-CO	N_3	0.092	0.45	4.9
SB-T-12855-3	MeO-CO	N_3	0.076	0.40	5.3

[a-c]See the footnotes of Table 1.

Possible Bioactive Conformations of Fluoro-Taxoids

The wide dispersion of the ^{19}F NMR chemical shift makes it particularly amenable to the observation of dynamic conformational equilibria through the freezing of conformers at low temperature. In fact, we have successfully used fluorine-containing taxanes as probes for NMR analysis, in conjunction with molecular modeling, of the conformational dynamics of paclitaxel.[4] Furthermore, in order to estimate the F-F distance in the microtubule-bound docetaxel, we have extended the fluorine-probe protocol to solid-state magic-angle spinning (SSMAS) ^{19}F NMR analysis with radio-frequency driven dipolar recoupling (RFDR) (Figure 2).[3] Schaefer and co-workers used the rotational echo double resonance (REDOR) to investigate the structure of the microtubule-bound paclitaxel by determining the ^{19}F-^{13}C distances of a fluorine-probe of paclitaxel (Figure 2).[5] These SS NMR studies have provide critical information on the bioactive conformation of paclitaxel. Recently, we proposed a new bioactive conformation of paclitaxel, "REDOR-Taxol",[29] based on the ^{19}F-^{13}C distances determined by the REDOR experiment,[5] photoaffinity labeling of microtubules,[30], the crystal structure of α,β-tubulin dimmer model determined by cryo-electron microscopy,[31] and molecular modeling. In this computational biology analysis, we docked a paclitaxel-photoaffinity label molecule to the position identified by our photoaffinity labeling study first and then optimized the position with a free paclitaxel molecule in the binding space.[29]

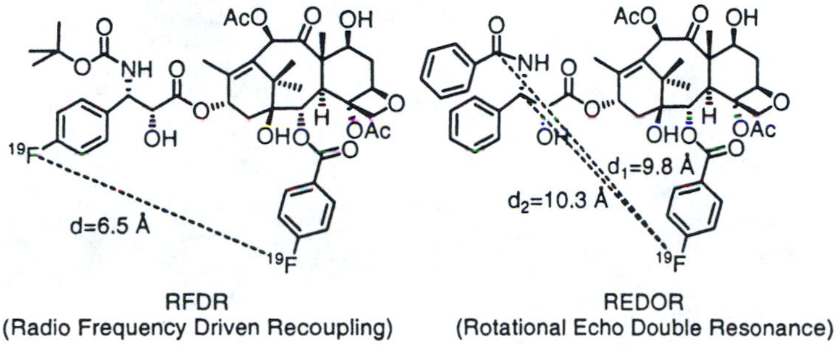

RFDR
(Radio Frequency Driven Recoupling)

REDOR
(Rotational Echo Double Resonance)

Figure 2. Solid state NMR studies on microtubule-bound flouorine-probes

We used the same protocol to investigate the microtubule-bound structures of the 3'-CF_2H-, 3'-CF_3-, and 3'-$CF_2C{=}CH$-taxoids, which should be the bioactive conformations of these fluoro-taxoids. The computational analyses have revealed that there are substantial differences between the "REDOR-Taxol" and fluoro-taxoids due to the lack of 3'-phenyl group as well as *3'N*benzoyl group in the fluoro-taxoids. For example, there is a very strong hydrogen bond between the 2'-OH of "REDOR-Taxol" and His227, while the strong hydrogen-bonding is observed between the 2'-OH and Arg359 for all fluoro-taxoids.

We examined the chemical spaces that these 3'-CF_2H-, 3'-CF_3-, and 3'-$CF_2C{=}CH$-taxoids occupy in the β-tubulin binding site by using 2-unmodified fluoro-taxoids, **SB-T-1284** (Rf = CF_2H, R_1 = Ac, X = H), **SB-T-1282** (Rf = CF_3, R_1 = Ac, X = H), and **SB-T-12853** (Rf = $CF_2C{=}CH$, R_1 = EtCO, X = H). The computer generated binding structures of these fluoro-taxoids are shown in Figure 3 (A, B, and C). The baccatin part occupies virtually the same space in all cases, as expected. Also, it is not surpising that the CF_2H and CF_3 moieties fill essentially the same space. However, the $CF_2C{=}CH$ moiety occupies substantially different space from the CF_2H and CF_3 moieties. This extended hydrophobic space is likely contributing to the exceptional cytotoxicity of difluorovinyl-taxoids. The overlay of **SB-T-12853** with a representative second-generation taxoid, SB-T-1213, shows very good fit, but with appreciable difference (Figure 3, D). Difluorovinyl group is in between vinyl and isobutenyl groups in size, but two fluorine atoms may mimic two hydroxyl groups rather than two methyl groups. Accordingly, the difluorovinyl group can be regarded as "magic vinyl", like "magic methyl" for the trifluoromethyl group, in drug design, including its anticipated metabolic stability against P-450 family enzymes.

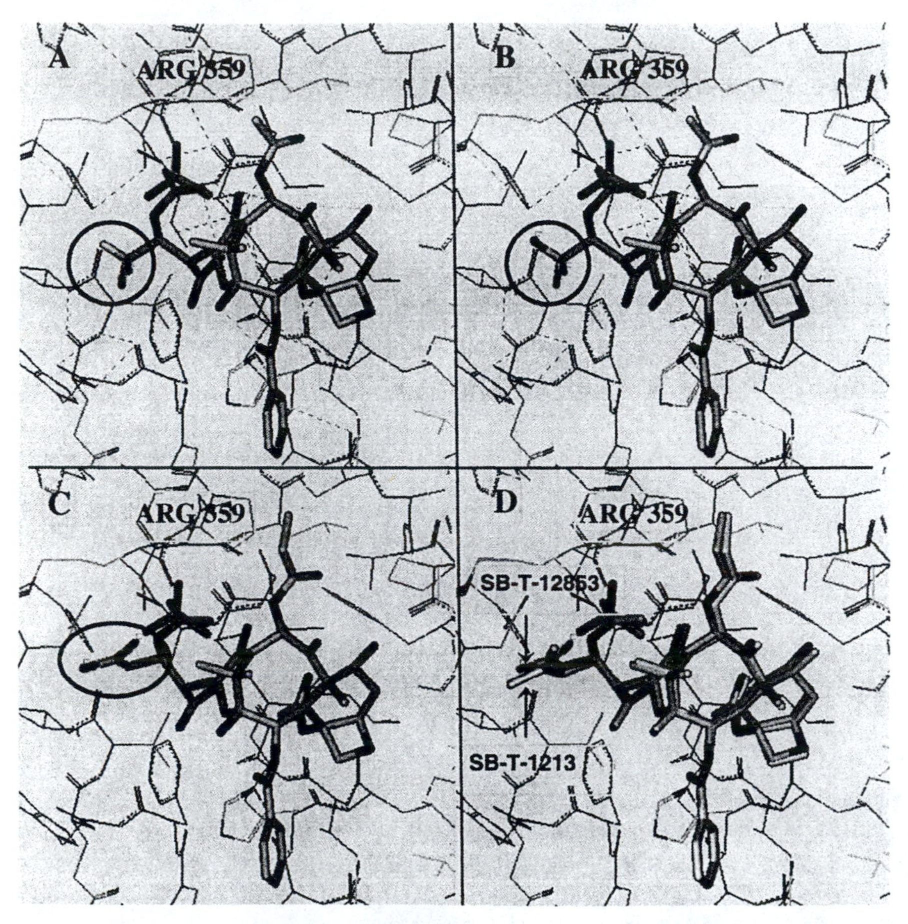

Figure 3. Computer-generated binding structures of fluoro-taxoids to β-tubulin: (A) SB-T-1284 (3'-CF_2H); (B) SB-T-1282 (3'-CF_3); (C) SB-T-12853 (3'-CF_2=CH); (D) Overlay of SB-T-12853 and SB-T-1213 (3'-isobutenyl).

Use of Fluorine in Tumor-Targeting Anticancer Agents

Current cancer chemotherapy is based on the premise that rapidly proliferating tumor cells are more likely to be killed by cytotoxic drugs. However, the difference in activity of current drugs against tumor tissues in comparison to primary tissues is relatively small. Consequently, the amount of a drug required to achieve clinically effective level of activity against the targeted tumor cells often causes severe damage towards actively propagating non-malignant cells such as cells of the gastrointestinal tract and bone marrow, resulting in a variety of undesirable side effects. Therefore, it is very important to develop new chemotherapeutic agents with improved tumor specificity.

The most promising approach to addressing tumor specificity is the development of tumor-targeting anticancer drug conjugates, consisting of a tumor-targeting molecule, a functional linker and a highly potent cytotoxic agent. As the tumor-targeting molecules, monoclonal antibody (mAb), hyaluronic acid, folic acid, biotin, somatostatin peptide mimic, etc., can be used.[32,33] For example, monoclonal antibodies (mAb's) specifically bind to antigens that are overexpressed on the surface of tumor tissues or cells, distinguishable from normal tissues. Therefore, in principle, mAb-cytotoxic drug conjugates can be specifically delivered to the tumor, internalized *via* receptor-mediated endocytosis and release the parent drug.[32,34,35] Mylotarg® (gemtuzumab-ozogamicin)[36] was approved by the Food and Drug Administration (FDA) for the treatment of acute myelogenous leukemia (AML), providing the first mAb-drug immunoconjugate for the treatment of cancer in clinic. Several other mAb-drug conjugates have advanced to human clinical trials.[35,37-40]

For the development of efficacious tumor-targeting drug conjugates, efficient mechanism-based linkers are essential since the conjugates should be stable during circulation in the blood, but readily cleavable in the tumor. Recently, we reported novel mAb-taxoid conjugates as tumor-targeting anticancer agents, which exhibited extremely promising results in human cancer xenografts in SCID mice, clearly demonstrating tumor-specific delivery of a taxoid anticancer agent without any noticeable toxicity to the animals, curing all animals tested.[41] As the linker for these mAb-taxoid conjugates, we used a disulfide linker, which was stable in blood circulation but efficiently cleaved by glutathione or other thiols in the tumor (glutathione level is known to be 1,000 times higher in tumor tissues than blood plasma).[32] However, in this first-generation mAb-taxoid conjugates, the original taxoid molecule was not released because of the compromised modification of the taxoid molecule to attach the disulfide linker. Accordingly, the cytotoxicity of the taxoid released in these conjugates was 8-10 times weaker than the parent taxoid.[41]

In order to solve this problem, we have been developing the second-generation mechanism-based disulfide linkers. One of our approaches is the glutathione-triggered cascade drug release, forming a thiolactone as a side product (Scheme 6). This mechanism-based drug release concept has nicely been proven in a model system by monitoring the reaction with ^{19}F NMR using fluorine-labeled compounds (Figure 4)[42] The strategic incorporation of a fluorine substituent at the *para* position to the disulfide linkage would direct the cleavage of this linkage by a thiol to generate the desirable thiophenolate or sulfhydrylphenyl species for thiolactonization.

Also, as described above, the incorporation of a fluorine substitution may increase the metabolic stability of the conjugate. This type of linkers is readily applicable to any tumor-targeting drug conjugates, including fluoro-taxoids. Moreover, a combination of a fluorine-containing linker and a fluoro-taxoid can be used as fluorine-probes for monitoring the internalization and drug release of these conjugates in the tumor cells and tissues.

Further applications of the strategic incorporation of fluorine(s) for the exploration of fluoroorganic chemistry at the biomedical interface are actively underway in these laboratories.

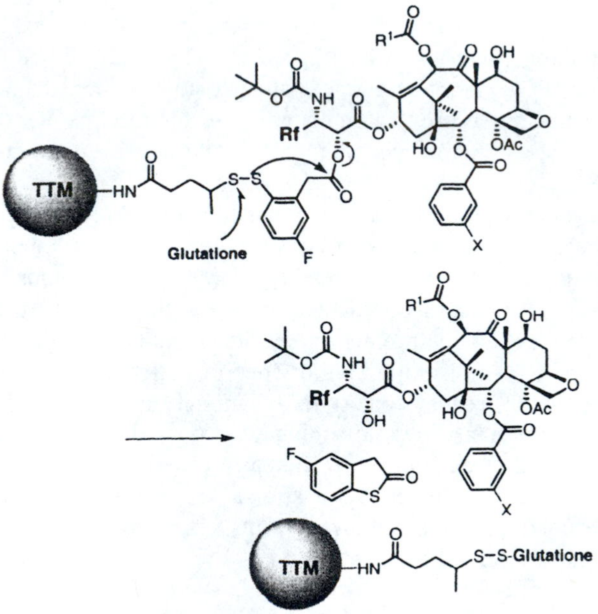

Scheme 6. Glutathione-triggered cascade drug release for fluorine-containing tumor-targeting anticancer agents

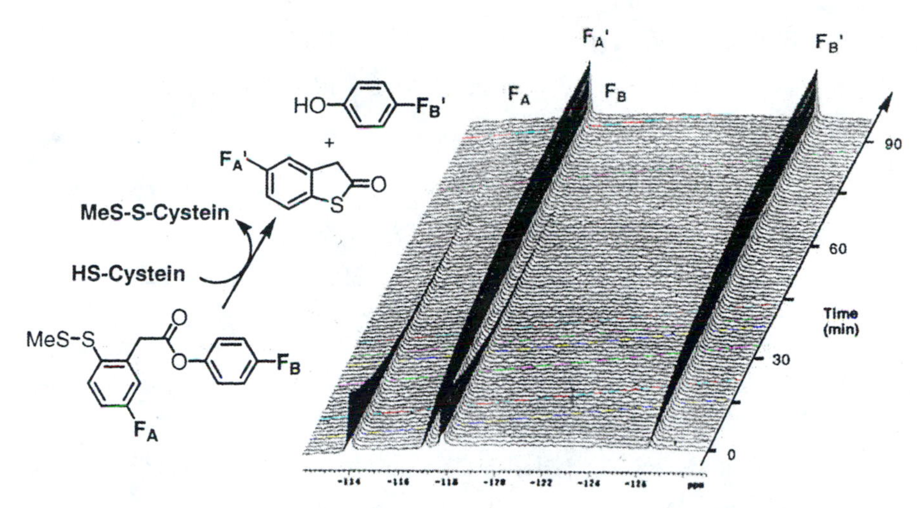

Figure 4. A proof of concept model for the mechanism-based drug release

References

1. Bohm, H.-J.; Banner, D.; Bendels, S.; Kansy, M.; Kuhn, B.; Muller, K.; Obst-Sander, U.; Stahl, M. *Chembiochem* **2004**, *5,* 637-43.

2. Ojima, I.; McCarthy, J. R.; Welch, J. T. (Eds.) *Biomedical Frontiers of Fluorine Chemistry; ACS Symp. Ser. 639;* American Chemical Society: Washington, D. C., 1996.

3. Ojima, I.; Kuduk, S. D.; Chakravarty, S. In *Adv. Med. Chem.;* Maryanoff, B. E., Reitz, A. B., Eds.; JAI Press: Greenwich, CT, 1998; Vol. 4, p 69-124.

4. Ojima, I.; Kuduk, S. D.; Chakravarty, S.; Ourevitch, M.; Bégué, J.-P. *J. Am. Chem. Soc.* **1997**, *119*, 5519-5527.

5. Li, Y.; Poliks, B.; Cegelski, L.; Poliks, M.; Gryczynski, Z.; Piszcek, G.; Jagtap, P. G.; Studelska, D. R.; Kingston, D. G. I.; Schaefer, J.; Bane, S. *Biochemistry* **2000**, *39,* 281-291.

6. Rowinsky, E. K. *Ann. Rev. Med.* **1997**, *48*, 353-374.

7. FDA 2004, http://www.fda.gov/cder/approval/t.htm.

8. Suffness, M. *Taxol, Science and Applications;* CRC Press: New York, 1995.

9. Georg, G. I.; Chen, T. T.; Ojima, I.; Vyas, D. M. (Eds.) *Taxane Anticancer Agents: Basic Science and Current Status; ACS Symp. Ser. 583;* American Chemical Society: Washington, D. C., 1995.

10. Schiff, P. B.; Fant, J.; Horwitz, S. B. *Nature* **1979**, *277*, 665-667.

11. Jordan, M. A.; Toso, R. J.; Wilson, L. *Proc. Natl. Acad. Sci. USA* **1993**, *90,* 9552-9556.

12. Bristol-Myers Squibb 2003, http://www.bms.com/cgi-bin/anybin.pl?sql=select%20PPI%20from%20TB_PRODUCT_PPI%20where%20PPI_SEQ=19&key=PPI.

13. Ojima, I.; Slater, J. C.; Michaud, E.; Kuduk, S. D.; Bounaud, P.-Y.; Vrignaud, P.; Bissery, M.-C.; Veith, J.; Pera, P.; Bernacki, R. J. *J. Med. Chem.* **1996**, *39,* 3889-3896.

14. Ojima, I.; Wang, T.; Miller, M. L.; Lin, S.; Borella, C. P.; Geng, X.; Pera, P.; Bernacki, R. J. *Bioorg. Med. Chem. Lett.* **1999**, *9,* 3423-3428.

15. Ojima, I.; Kuduk, S. D.; Slater, J. C.; Gimi, R. H.; Sun, C. M. *Tetrahedron* **1996**, *52,* 209-24.

16. Ojima, I.; Kuduk, S. D.; Slater, J. C.; Gimi, R. H.; Sun, C. M.; Chakravarty, S.; Ourevitch, M.; Abouabdellah, A.; Bonnet-Delpon, D.; Begue, J.-P.; Veith, J. M.; Pera, P.; Bernacki Ralph, J. In *Biomedical Frontiers of Fluorine Chemistry, ACS Symp. Ser. 639;* Ojima, I.; McCarthy, J. R.; Welch, J. T. (Eds.); American Chemical Society: Washington, D. C., 1996, p pp 228-243.

17. Ojima, I.; Slater, J. C.; Pera, P.; Veith, J. M.; Abouabdellah, A.; Begue, J.P.; Bernacki, R. J. *Bioorg. Med. Chem. Lett.* **1997**, *7,* 133-138.

18. Ojima, I.; Inoue, T.; Slater, J. C.; Lin, S.; Kuduk, S. C.; Chakravarty, S.; Walsh, J. J.; Gilchrist, L.; McDermott, A. E.; Cresteil, T.; Monsarrat, B.; Pera, P.; Bernacki, R. J. In *"Asymmetric Fluoroorganic Chemistry: Synthesis, Application, and Future Directions"*; *ACS Symp. Ser. 746;* Ramachandran, P. V., Ed.; American Chemical Society: Washington, D. C., 1999, p pp 158-181.

19. Ojima, I. *Acc. Chem. Res.* **1995**, *28,* 383-389 and references cited therein.

20. Ojima, I. In *Advances in Asymmetric Synthesis;* Hassner, A., Ed.; JAI Press: Greenwich, 1995; Vol. 1, p 95-146.

21. Kuznetsova, L.; Ungureanu, I. M.; Pepe, A.; Zanardi, I.; Wu, X.; Ojima, I. *J. Fluor. Chem.* **2004**, *125,* 487-500.

22. Lin, S.; Geng, X.; Qu, C.; Tynebor, R.; Gallagher, D. J.; Pollina, E.; Rutter, J.; Ojima, I. *Chirality* **2000**, *12,* 431-441.

23. Ojima, I.; Lin, S.; Slater, J. C.; Wang, T.; Pera, P.; Bernacki, R. J.; Ferlini, C.; Scambia, G. *Bioorg. Med. Chem.* **2000**, *8,* 1619-1628.

24. Skehan, P.; Storeng, R.; Scudiero, D.; Monks, A.; McMahon, J.; Vistica, D.; Warren, J. T.; Bokesch, H.; Kenney, S.; Boyd, M. R. *J. Nat'l. Cancer Inst.* **1990**, *82,* 1107-1112.

25. Gut, I.; Ojima, I.; Vaclavikova, R.; Simek, P.; Horsky, S.; Linhart, I.; Soucek, P.; Kondrova, E.; Kuznetsova, L., V.; Chen, J. 2006, unpublished results.

26. Vuilhorgne, M.; Gaillard, C.; Sanderlink, G. J.; Royer, I.; Monsarrat, B.; Dubois, J.; Wright, M. In *Taxane Anticancer Agents: Basic Science and Current Status, ACS Symp. Ser. 583;* Georg, G. I., Chen, T. T., Ojima, I., Vyas, D. M., Eds.; American Chemical Society: Washington, D. C., 1995, p98-110.

27. Ojima, I.; Habus, I.; Zhao, M.; Zucco, M.; Park, Y. H.; Sun, C. M.; Brigaud, T. *Tetrahedron* **1992**, *48,* 6985-7012.

28. Ojima, I.; Sun, C. M.; Zucco, M.; Park, Y. H.; Duclos, 0.; Kuduk, S. D. *Tetrahedron Lett.* **1993**, *34,* 4149-4152.

29. Geney, R.; Sun, L.; Pera, P.; Bernacki Ralph, J.; Xia, S.; Horwitz Susan, B.; Simmerling Carlos, L.; Ojima, I. *Chemistry & biology* **2005**, *12,* 339-48.

30. Rao, S.; He, L.; Chakravarty, S.; Ojima, I.; Orr, G. A.; Horwitz, S. B. *J. Biol. Chem.* **1999**, *274,* 37990-37994.

31. Nogales, E.; Wolf, S. G.; Downing, K. H. *Nature* **1998**, *391,* 199-203.

32. Chen, J.; Jaracz, S.; Zhao, X.; Chen, S.; Ojima, I. *Exp. Opin. Drug Deliv.* **2005**, *2,* 873-890.

33. Jaracz, S.; Chen, J.; Kuznetsova, L. V.; Ojima, I. *Bioorg. Med. Chem.* **2005**, *13,* 5043-5054.

34. Chari, R. V. J. *Adv. Drug Deliv. Rev.* **1998**, *31,* 89-104.

35. Liu, C.; Tadayoni, B. M.; Bourret, L. A.; Mattocks, K. M.; Derr, S. M.; Widdison, W. C.; Kedersha, N. L.; Ariniello, P. D.; Goldmacher, V. S.; Lambert, J. M.; Blättler, W. A.; Chari, R. V. J. *Proc. Nat. Acad. Sci. USA* **1996**, *93,* 8618-8623.

36. Hamann, P. R.; Hinman, L. M.; Hollander, I.; Beyer, C. F.; Lindh, D.; Holcomb, R.; Hallett, W.; Tsou, H.-R.; Upeslacis, J.; Shochat, D.; Mountain, A.; Flowers, D. A.; Bernstein, I. *Bioconjugate Chem.* **2002**, *13,* 47-58.

37. Lam, L.; Lam, C.; Li, W.; Cao, Y. *Drugs of the Future* **2003**, *28,* 905-910.

38. Saleh, M. N.; LoBuglio, A. F.; Trail, P. A. *Basic Clin. Oncol.* **1998**, *15,* 397-416.

39. Chan, S. Y.; Gordon, A. N.; Coleman, R. E.; Hall, J. B.; Berger, M. S.; Sherman, M. L.; Eten, C. B.; Finkler, N. J. *Cancer Immun. Immunother.* **2003**, *52,* 243-248.

40. Gillespie, A. M.; Broadhead, T. J.; Chan, S. Y.; Owen, J.; Farnsworth, A. P.; Sopwith, M.; Coleman, R. E. *Ann. oncol.* **2000**, *11,* 735-41.

41. Ojima, I.; Geng, X.; Wu, X.; Qu, C.; Borella, C. P.; Xie, H.; Wilhelm, S. D.; Leece, B. A.; Bartle, L. M.; Goldmacher, V. S.; Chari, R. V. J. *J. Med. Chem.* **2002**, *45,* 5620-5623.

42. Ojima, I. *ChemBioChem* **2004**, *5,* 628-635.

Chapter 18

Synthesis of *gem*-Difluoromethylated Sugar Nucleosides

Feng-Ling Qing[1,2] and Xiao-Long Qiu[1]

[1]Key Laboratory of Organofluorine Chemistry, Shanghai Institute of Organic Chemistry, Chinese Academy of Sciences, 354 Fenglin Lu, Shanghai 200032, China
[2]Institute of Biological Sciences and Biotechnology, Donghua University, 2999 North Renmin Lu, Shanghai 201620, China (flq@mail.sioc.ac.cn)

Of all the fluorinated nucleosides, *gem*-difluoromethylated sugar nucleosides currently attract more and more attention because of their highly antiviral and antitumor activities. Herein described are the recent achievements in synthesis of *gem*-difluoromethylated sugar nucleosides, which are grouped into four main types: *gem*-difluoromethylated furanyl (pyranyl) nucleosides, *gem*-difluoromethylated thionucleosides, *gem*-difluoromethylated azanucleosides and *gem*-difluoromethylated carbocyclic nucleosides.

Generally speaking, analogues of natural nucleosides are classed into carbocyclic nucleosides (*1*), thionucleosides (*2*), phosphanucleosides (*3*), and azanucleosides (*4*) when the oxygen atom of pyranose or furanose ring is replaced by carbon, sulfur, phosphorus, and nitrogen, respectively. Nucleoside analogues have achieved considerable success in the fight against virus and tumor. Many highly biological nucleosides and their analogues have been synthesized (*5*). Among all the highly bioactive nucleosides, what should be greatly mentioned are fluorinated nucleosides, which are usually defined as nucleosides containing fluorine atom(s) or fluorine-containing group(s) in the sugar moiety or the base moiety of nucleosides (*6*). The best known fluorinated nucleoside is 1-(2-deoxy-2,2-difluoro-*β*-D-arabino-furanosyl)cytosine (Gemcitabine, Gemzar or dFdC), which has been approved by FDA for treatment of inoperable pancreatic cancer and of 5-fluorouracial resistant pancreatic cancer (*7*). The highly antitumor activities of Gemcitabine reveal the special influences of *gem*-difluoromethylated group on biological activities of nucleosides. Thus, a number of nucleosides containing CF_2 group at the sugar moiety have been synthesized and biologically evaluated. Herein described are the recent achievements of synthesis of *gem*-difluoromethylated nucleosides, which are grouped into four main types: *gem*-difluoromethylated furanyl (pyranyl) nucleosides, *gem*-difluoromethylated thionucleosides, *gem*-difluoromethylated azanucleosides and *gem*-difluoromethylated carbocyclic nucleosides.

Synthesis of *gem*-difluoromethylated furanyl (pyranyl) nucleosides

Among all the reported *gem*-difluoromethylated nucleosides, 2'-deoxy-2',2'-difluoro-furanosyl pyrimidine nucleosides and their analogues grasped more and more eyes from organic, biological and pharmaceutical chemists. Originally, 2'-deoxy-2',2'-difluorocytidine, known as Gemcitabine (or Gemzar, dFdC) and currently marketed for the treatment of non-small cell lung and pancreatic cancer, was synthesized by Hertel et. al (*8*). Their synthetic strategy started from *D*-mannitol, which was first converted to (*R*)-2,3-*O*-isopropylideneglyceraldehyde **1** (Scheme 1). Reformatskii reaction between compound **1** and $BrCF_2CO_2Et$ provided the mixture of diastereomers **2a** and **2b**. Hydrolytic removal of the isopropylidine group in major isomer **2a** with concomitant closure gave γ-lactone **3**. Protection of hydroxyl groups as *tert*-butyldimethylsilyl ethers followed by reduction with DIBAL-H afforded the disilyl lactol **5**. Mesylation of the lactol **5** gave the mesylate derivative **6**, which was subjected to condensation with silylated cytosine followed by removal of protecting groups to deliver *β*-isomer Gemcitabine **7a** as minor isomer (β/α = 1:4). Coupling of the mesylate **6** with silylated thymine gave the blocked thymidine derivative **8**, which was treated with 1,2,4-triazol/$ClPhOPOCl_2$/pyridine to afford the protected nucleoside **9.** Exposure of the compound **9** to aqueous ammonia and subsequent removal of the protecting groups provided the nucleoside **10**. In addition, they also reported that hydrolytic deamination of Gemcitabine with $AcOH/H_2O$ gave the corresponding 2',2'-difluorouridine. In view of the fact that above synthetic route needed separation of isomers **2a**/**2b** and **7a**/**7b** by HPLC, Chou et. al. (*9*) improved the synthetic method by utilizing the same synthetic route but selecting benzoyl over *tert*-butyldimethylsilyl as the protecting group for hydroxyl group in compound **3**. Once this modification, three distinctive improvements were available. First and foremost, crystallization of the desired ribonolactone **12a** from a diastereomeric mixture consisting of **12a** and **12b** was easily realized. Besides, ratio of desirable β-isomer **7a** against α-isomer **7b** was increased to 1:1 from 1:4 which was afforded when *tert*-butyldimethylsilyl was used as the protecting group. Finally, crystallization of Gemicabine **7a** from a 1:1 anomeric mixture was also accomplished. In addition, starting from lactol **5** and **12a**, a series of 1-(2-deoxy-2,2-difluororibofuranosyl) purine nucleosides (*10*) and 4-amino-8-(2,2-difluoro-2-deoxy-β-D-ribofuranosyl amino)pyrimido[5,4-d]pyrimidine (*11*) were also synthesized.

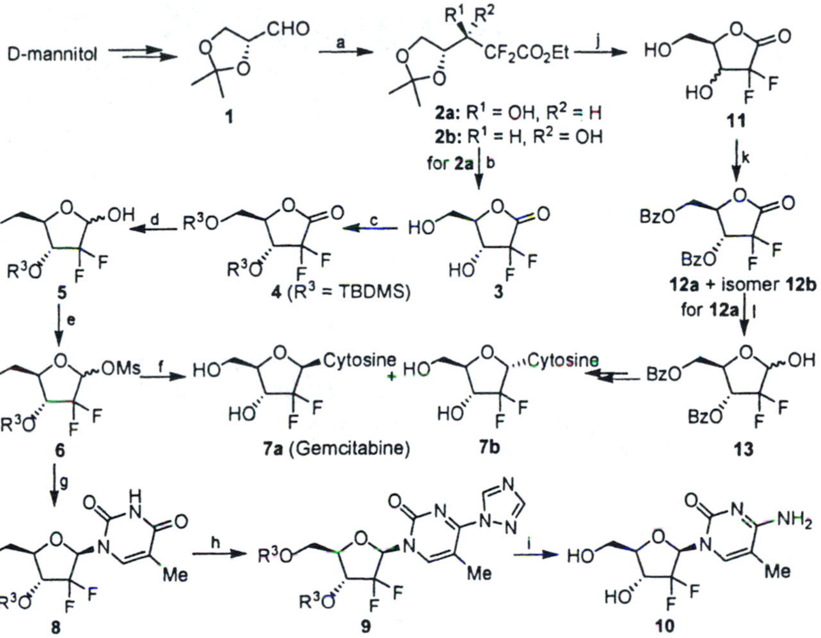

Scheme 1. *Reagent and Conditions:* (a) Zn, $BrCF_2CO_2Et$, THF, Et_2O; (b) Dowex 50, MeOH, H_2O; (c) TBDMSCl, lutidine, CH_2Cl_2; (d) DIBAL-H; (e) MsCl, Et_3N, CH_2Cl_2; (f) i. silylated cytosine, TMSOTf, $ClCH_2CH_2Cl$, reflux; ii. NH_3, MeOH; iii. MeOH, AG 50W-X8; (g) silylated thymidine, TMSOTf, $ClCH_2CH_2Cl$, reflux; (h) 1,2,4-triazol, $ClPhOPOCl_2$, pyridine; (i) i. NH_4OH, dioxane; ii. Dowex 50; (j) TFA, H_2O, MeCN; (k) BzCl, lutidine, DMAP, CH_2Cl_2; (1) $LiAl(OBu^t)_3H$, Et_2O, THF.

Recently, 3'-Deoxy-3',3'-difluoro-D-arabinofuranosyl nucleosides were also synthesized (*12*). The synthesis featured a novel and efficient strategy to prepare the versatile intermediate, 3-deoxy-3,3-difluoro-D-arabinofuranose **16** (Scheme 2). Indium-mediated reaction of 1-(*R*)-glyceraldehyde acetonide **1** with 3-bromo-3,3-difluoropropene gave the difluorohomoallyl alcohol **14**. Protection of hydroxyl group in the compound **14** followed by Os-catalyzed dihydroxylation gave the diol **15**, which was subjected to a series of transformation of protecting groups to furnish the furanose **16 in** 35:1 *dr.* Acetylation of the lactol **16** provided the single isomer **17**, which was condensed with silyated N^4-benzoylcytosine followed by removal of protecting group gave N^1-(3-Deoxy-3,3-difluoro-D-arabinofuranosyl)cytosine **18a** and **18b.**

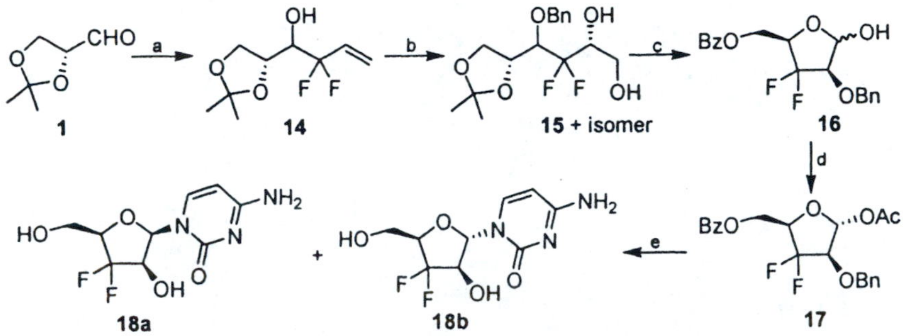

Scheme 2. *Reagent and Conditions:* (a) In powder, CH_2=$CHCF_2Br$; (b) i. NaH, BnBr, TBAI, THF; ii. OsO_4, NMMO; (c) i. BzCl, pyridine; ii. 75% AcOH, 50°C; iii. $NaIO_4$, acetone/H_2O; (d) Ac_2O, DMAP; (e) i. N^4-benzoylcytosine, *N,O*-bis(trimethylsilyl)acetamide, TMSOTf, CH_3CN, 80°C; ii. Sat. NH_3/MeOH; iii. H_2 (1 atm), 20% $Pd(OH)_2$ on charcoal, MeOH/cyclohexane.

Chu and co-workers synthesized the *L*-counterpart of Gemcitabine 2'-deoxy-2',2'-difluoro-L-*erythro*-pentofuranosyl nucleosides, starting from L-gulono γ-lactone (*13*) (Scheme 3). Chu's synthetic strategy was almost identical to that of synthesizing Gemcitabine. After preparation of key intermediates **21a** and **21b**, treatment of them with MsCl/Et_3N gave the crude mesylates **22a** and **22b**, respectively. Coupling of **21a**, **21b**, **22a**, **22b** with various persilylated pyrimidines, 6-chloropurine and silylated-protected 2-amino-6-chloropurine followed by some standard procedures (desilylation with TBAF, treatment with ammonia or treatment with 2-mercaptoethanol, etc.) provided 2'-deoxy-2',2'-difluoro-L-*erythro*-pentofuranosyl nucleosides **23a** and **23b**.

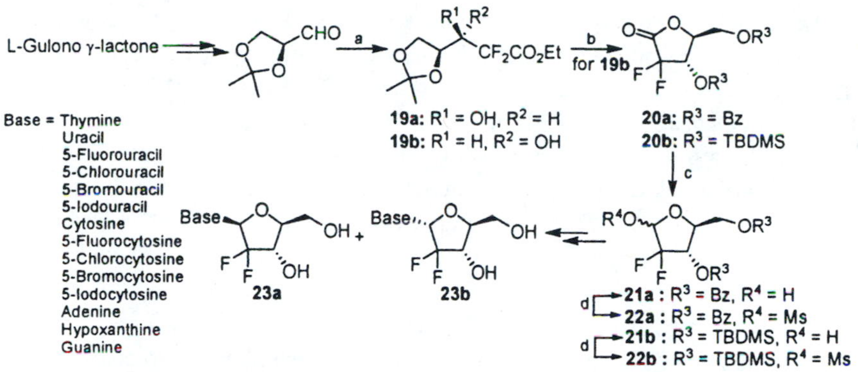

Scheme 3. *Reagent and Conditions:* (a) Zn, $BrCF_2CO_2Et$, THF, Et_2O; (b) i. 5% HCl, EtOH; or Dowex 50, MeOH, H_2O; ii. BzCl, 2,6-lutidine, CH_2Cl_2 or TBDMSCl, imidazole, DMF; (c) Li(OBu')$_3$AIH, THF, -78°C or DIBAL-H, toluene, -85°C; (d) MsCl, Et_3N, CH_2Cl_2.

Starting from alcohols **19a/19b,** 2',3'-dideoxy-2',2'-difluoro-L-*erythro*-pentofuranosyl nucleosides **28a** and **28b** were also prepared by Chu group (*14*). The key step was the removal of hydroxyl groups in alcohols **19a/19b** to give the compound **24** via radical reaction (Scheme 4). Treatment of compound **24** with acid followed by concomitant ring closure provided the lactone **25**. Benzoylation of the compound **25** gave the lactone **26**, which was subjected to reduction and protection of resultant hydroxyl group to furnish mesylate **27**.

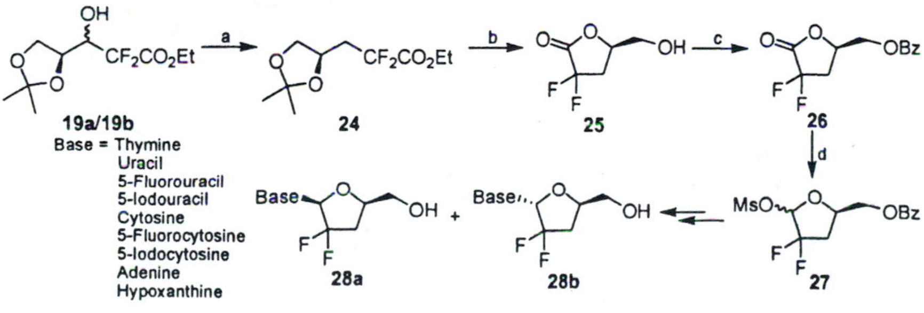

Scheme 4. *Reagent and Conditions:* (a) i. 1,1'-thiocarbonyldiimidazole, Me_2NCHO, 50°C; or CS_2, NaH, imidazole, Me_2NCHO; ii. Bu_3SnH, AIBN, toluene; (b) i. 5% HCl, EtOH; ii. Benzene, 80°C, Dean-Stark apparatus; (c) BzCl, 2,6-lutidine; (d) i. Li(OBu')$_3$AIH, THF, -78°C; ii. MsCl, Et_3N, CH_2Cl_2.

Condensation of the mesylate **27** with the various persilylated pyrimidines or 6-chloropurine followed by removal of protecting groups and (or) treatment with 2-mercaptoethanol delivered 2',3'-dideoxy-2',2'-difluoro-L-*erythro*-pentofuranosyl nucleosides **28a** and **28b**.

In 2003, a process for synthesis of 2'-deoxy-2',2'-difluoro-*β*-nucleocytidine **35** was described by Chen.(*15*). His synthetic strategy started from 1 ,6-anhydro-*β*-L-glucopyranose **29**, which was first treated with Me_3SiCl/Et_3N/DMAP to afford the alcohol **30** (Scheme 5). Oxidation of the compound **30** with Dess-Martin oxidant followed by protection the resulting hydroxyl groups with methyl groups gave the ketone **31.** Difluorination of compound **31** with DAST provided the *gem*-difluoromethylated derivative **32**, which was subjected to acid hydrolysis and oxidation with $NaIO_4$ to deliver the lactol **33**. Acetylation of the compound **33** provided the acetate **34**. Condensation of the compound **34** with (*Z*)-$H_2NC(O)NHCH{=}CHCN$ followed by removal of protecting group with NaOEt provided 2'-deoxy-2',2'-difluoro-*β*-nucleocytidine **35**. Simple process and high output rate were the advantages of the synthetic route.

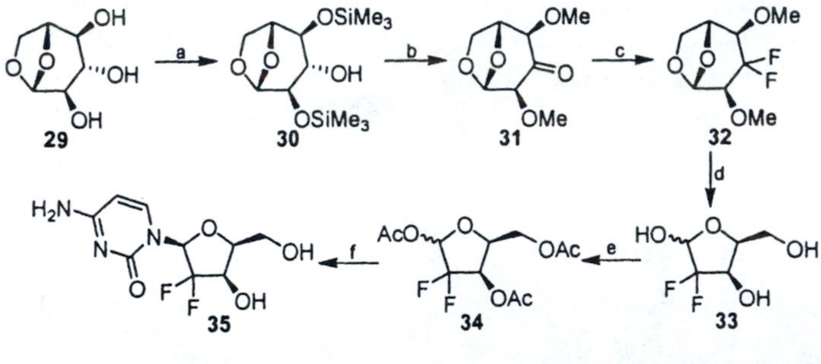

Scheme 5. *Reagent and Conditions:* (a) Me_3SiCl, DMAP, Et_3N, CH_2Cl_2; (b) i. Dess-Martin oxidant, CH_2Cl_2; ii. $Na_2S_2O_3$, $NaHCO_3$; iii. MeI, Et_3N, NaH, MeCN; (c) DAST, CH_2Cl_2, HF, DMPU; (d) i. HCl, reflux; ii. KOH, H_2O, dioxane; iii. $NaIO_4$; (e) AcCl, Et_3N, DMAP, CH_2Cl_2; (f) i. (*Z*)$H_2NC(O)NHCH{=}CHCN$, LiCl, Me_3SiCl, CH_2Cl_2; ii. $CdCl_2$, CH_2Cl_2; iii. $SnCl_4$, CH_2Cl_2, iv. NaOEt, ^{i}PrOH.

Although the synthesis of D-2',3'-deoxy-3',3'-difluoronucleoside **38** was first reported by Jarvi group in 1989 (*16*), detailed experiment procedure, the structure determination and the data on antiviral activity were described by Bergstrom group (*17*) (Scheme 6). That is, tritylation of thymidine and the oxidation of 5'-*O*-tritylthymidine gave 3'-ketothymidine **36,** which was subjected to difluorination with DAST to afford 5'-*O*-trityl-3',3'-difluoro-3'-deoxythymidine **37** in low yield. Removal of trityl protecting group by heating compound **37** in 80% AcOH gave D-3'-deoxy-3',3'-difluorothymidine **38** in 66% yield.

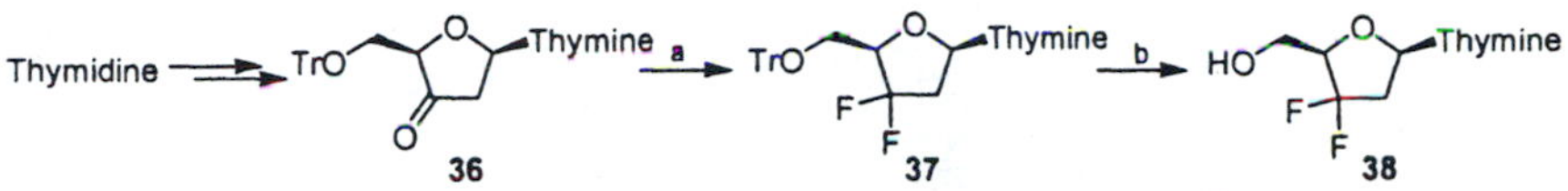

Thymidine → TrO–(O)–Thymine **36** –a→ TrO–(O)–Thymine (F, F) **37** –b→ HO–(O)–Thymine (F, F) **38**

Scheme 6. *Reagent and Conditions:* (a) DAST, CH_2Cl_2; (b) 80% AcOH, 100°C.

More recently, Chu et. al. synthesized a series of *L*-2',3'-dideoxy-3',3'-difluoronucleosides (*18*) and D-2',3'-dideoxy-3',3'-difluoronucleosides (*19*) *L*-xylose **39** was first converted to the protected *L*-ribose **40** in 73% overall yield (Scheme 7), which was further subjected to a series of transformation of protecting groups to furnish the ketone **41.** Treatment of the compound **41** with DAST provided the difluorinated intermediate **42** in 66% yield. Transglycosylation of the compound **42** by treatment with $Ac_2O/AcOH/H_2SO_4$ gave the acetate **43** in high yield. Condensation of the compound **43** with the various pyrimidine heterocylic base, 6-chloropurine and 2-fluoro-6-chloropurine followed by removal of protecting groups and (or) amination (NH_4OH) delivered the different base-containing *L*-2',3'-dideoxy-3',3'-difluoronucleosides **44a** and **44b**. In addition, starting from 2,3-*O*-isopropylidene-D-glyceraldehyde **1**, *D*-2',3'-dideoxy-3',3'-difluoronucleosides were also been synthesized in a straightforward fashion (*19*). Difluorination of ketone **45**, prepared from aldehyde **1** in two steps, gave the *gem*-difluoromethylated compound **46**, which was subjected to hydrolysis followed by selective protection of the resultant diol to provide alcohol **47**. Compound **47** was converted to epimeric acetates **48** by treatment with HCl/MeOH followed by H_2SO_4(conc.)/$AcOH/Ac_2O$. Condensation of the key intermediate **48** with the various persilylated pyridimine and purines gave the corresponding *D*-2',3'-dideoxy-3',3'-difluoronucleosides **49a** and **49b.**

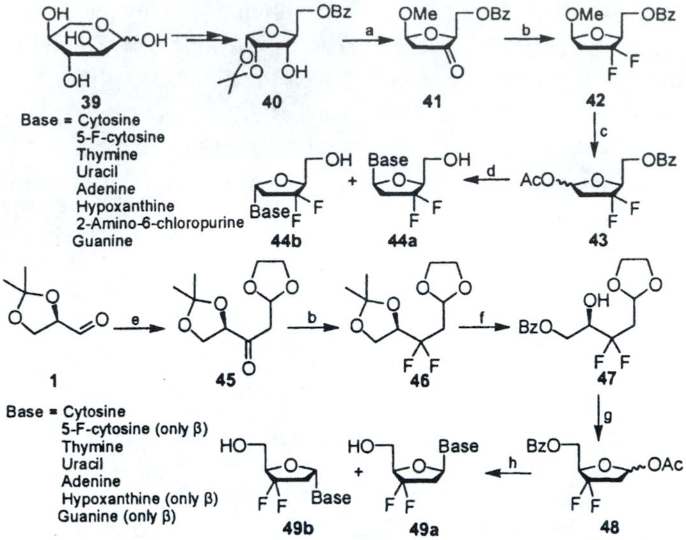

Base = Cytosine
5-F-cytosine
Thymine
Uracil
Adenine
Hypoxanthine
2-Amino-6-chloropurine
Guanine

Base = Cytosine
5-F-cytosine (only β)
Thymine
Uracil
Adenine
Hypoxanthine (only β)
Guanine (only β)

Scheme 7. *Reagent and Conditions:* (a) i. NaH, THF, then BnBr, TBAI; ii. 1:2 (4 M HCl/dioxane)/MeOH; iii. PhOC(S)Cl, DMAP; iv. Bu_3SnH, AIBN, reflux; v. H_2, 10% Pd/C; vi. CrO_3, Ac_2O, pyridine, CH_2Cl_2; (b) DAST, CH_2Cl_2, reflux; (c) conc. H_2SO_4, Ac_2O, AcOH; (d) i. persilylated pyrimidine, TMSOTf, CH_3CN; ii. NH_3, MeOH or i. HMDS, $(NH_4)_2SO_4$, 6-chloropurine (or 2-fluoro-6-chloropurine), TMSOTf, CH_3CN; ii. NH_4OH, 60°C ($HSCH_2CH_2OH$, NaOMe, 60°C or NH_3 bubbling, DME), (e) i (1,3-dioxolan-2-ylmethyl)magnesium bromide, THF, reflux; ii DMSO, $(ClCO)_2$, TEA; (f) i. 1:1 (5% HCl/dioxane); ii. BzCl, pyridine; (g) conc. H_2SO_4, Ac_2O, AcOH; (h) i. persilylated pyrimidine or purine, TMSOTf, CH_3CN, ii NH_3, MeOH ($HSCH_2CH_2OH$, NaOMe).

In view of high bioactivities against cancer cells of Gemcitabine, Castillón and co-worker (*21*) synthesized 2',3'-dideoxy-3',3'-difluoro and 2',3'-dideoxy-2',2'-difluoro pyranosyl nucleosides (Scheme 8), regarded as pyransyl analogues of Gemcitabine. Synthesis of 2',3'-dideoxy-3',3'-difluoro and 2',3'-dideoxy-2',2'-difluoro pyranosyl nucleosides started from D-mannose and D-glucose, respectively. The difluorination of ketone carbonyl with DAST was used as key steps. After preparation of *gem*-difluoromethylated pyranoses **50** and **55**, thioglycoside **51**, 1-*O*-acetylpyranose **52** and 1-mesyl pyranose 56 were obtained via a series of transformation of protecting groups. Finally, glycosylations of all the three glycosyl donors **51**, **52** and **56** gave the 2',3'-dideoxy-3',3'-difluoro pyranosyl nucleosides **53**, **54** and 2',3'-dideoxy-2',2'-difluoro pyranosyl nucleosides **57a**, **57b**. In addition, Castillón et. al. found that glycosyl donors had the important effect on introduction of the different bases. For example, synthesis of the cytosine derivative from thioglycoside 51 was unsuccessful, however, it proceeded well in the case of 1-O-acetylpyranose 52.

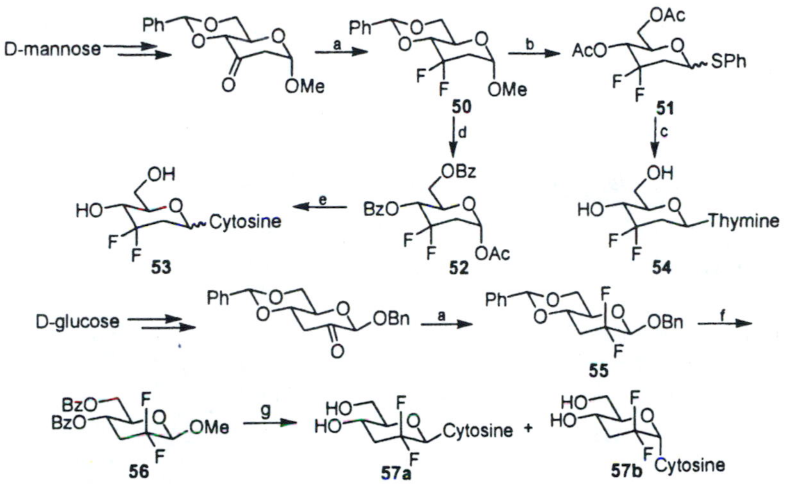

Scheme 8. *Reagent and Conditions:* (a) i. DAST, CH_2Cl_2; (b) i. PhSH, BF_3•Et_2O; ii. Ac_2O, pyridine; (c) i. NIS, TfOH, bis-(trimethylsilyl)thymine, CH_2Cl_2; ii. NH_3, MeOH; (d) i. HCl, EtOH; ii. BzCl, pyridine; iii. H_2SO_4, Ac_2O; (e) i. TMSOTf, DCE, silyl-protected cytosine; ii. NH_3, MeOH; (f) i. HCl, EtOH, ii. BzCl, pyridine; iii. H_2, Pd/C; iv. MsCl, Et_3N; (g) silylated acetylcytosine, TMSOTf, 1,1,2,2-tetrachloroethane.

Synthesis of *gem*-difluoromethylated thionucleosides

Base on the fact that 3'-thiocytidine (3TC) and Gemcitabine have been known as the highly active antitumour agents, Yoshimura et. al. (*21*) synthesized 4'-thiogemcitabine starting from D-glucose (Scheme 9). 3-*O*-benzyl-xylofuranoside **58**, derived from D-glucose, was converted to the bicyclic compound **59** in good yield via treatment with MsCl/pyridine followed by Na_2S/DMF. Acid hydrolysis and hydride reduction of the compound **59** produced 1,4-anhydro-4-thioarabinitol **60** in 90% yield. Protection of the primary alcohol of **60** with *tert*-butyldimethylsilyl group followed by oxidation gave the 2-keto derivative **61**, which was further treated with DAST to afford 2-deoxy-2,2-difluoro derivative **62**. After transformation of 3'-protecting group, Pummerer type glycosylation between resultant sulfoxide and silylated acetylcytosine, followed by deprotection, gave the 4'-thiogemcitabine **63a** and **63b.** Biological test demonstrated that *β*-4'-thiogemcitabine **63b** exhibited weak activity against CCRF-HSB-2 cell line.

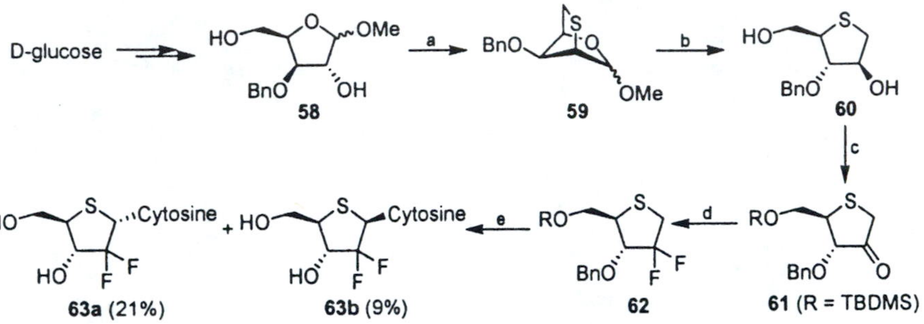

Scheme 9. *Reagent and Conditions:* (a) i. MsCl, pyridine; ii. Na_2S, DMF, 100°C; (b) i. 4 M HCl, THF; ii. $NaBH_4$, MeOH; (c) i. TDMPSCl, imidazole, DMF; ii. DMSO, Ac_2O; (d) i. DAST, CH_2Cl_2; (e) i. BCl_3, CH_2Cl_2, -78°C then MeOH, pyridine; ii. Bz_2O, Et_3N, DMAP, CH_3CN; iii. *m*-CPBA, CH_2Cl_2, -78°C; iv. silylated *N*-acetylcytosine, TMSOTf, v. TBAF, THF; vi. aq. NH_3, MeOH, then HPLC separation.

Later, Jeong and co-workers synthesized L-2'-deoxy-2',2'-difluoro-4'-thionucleosides (*22*) (Scheme 10). Their synthesis started from commercially available 1 ,2-isopropylidene-D-xylose **64**, which was converted to the thiosugar **65** in 68% overall yield over 5 steps. Reaction of the compound **65** with $Hg(OAc)_2$/AcOH followed by treatment of resultant acetate with Et_3SiH/TMSOTf and subsequent removal of benzyl protecting group with NaOMe/MeOH gave the key intermediate **66** in 73% yield. Oxidation of the compound **66** and fluorination of the resulting ketone provided *gem*-difluoromethylated derivative **67**, which was further subjected to Pummerer glycosylation and removal of protecting groups to afford L-2'-deoxy-2',2'-difluoro-4'-thionucleosides **68a** and **68b.** Biological tests shown that neither of them was effective against tumor cell lines at concentrations >100 μg/mL.

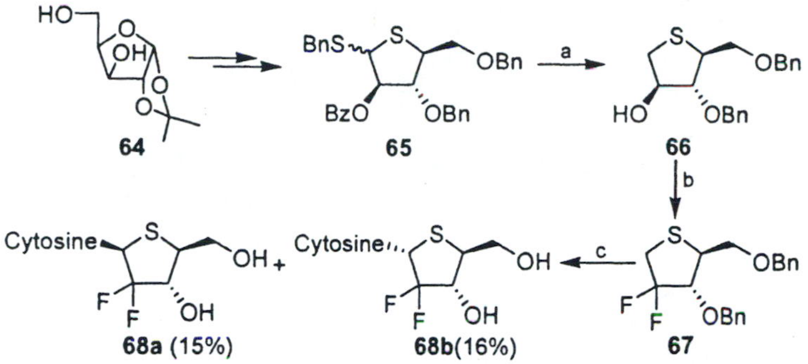

Scheme 10. *Reagent and Conditions:* (a) i. $Hg(OAc)_2$, AcOH, rt.; ii. Et_3SiH, TMSOTf, rt.; iii. NaOMe, MeOH, CH_2Cl_2, rt; (b) i. DMSO, Ac_2O, rt.; ii. DAST, CH_2Cl_2, rt.; (c) i. *m*-CPBA, CH_2Cl_2, -40°C, i i. silylated *N*-benzoylcytosine, TMSOTf, $ClCH_2CH_2Cl$, 0°C to rt.; iii. BBr_3, CH_2Cl_2, -40 °C, then NaOMe, MeOH.

Recently, the aforementioned intermediates, *gem*-difluorohomoallyl alcohol **14** and 3-deoxy-3,3-difluoro-D-arabinofuranose **16,** were also used to synthesize *gem*-fluoromethylated thionucleosides (*12,23*) (Scheme 11). Ring-opening of the compound **16** with $NaBH_4$/MeOH afforded the diol **69,** which was treated with MsCl/pyridine followed by Na_2S/DMF to provide 4'-thiofuranose **70.** Treatment of the compound **70** with BCl_3 and protection of resulting hydroxyl group with benzoyl group gave the benzoate **71**. Unexpectedly, Pummerer glycosylation of compound **71** with silylated *N*-benzoylcytosine delivered compounds **72a**, **72b**, which featured that both hydroxymethyl and base were located in the Cl' position. In addition, 2',3'-dideoxy-6',6'-difluoro-3'-thionucleoside **78** was also synthesized starting from intermediate **14** (*23*). Introduction of azide group via $S_N{}^2$ reaction of compound **14** followed by reduction of resultant azide gave the amine **73**. After protection of the amino group with Boc, dihydroxylation was carried out and the diol **74** was provided. Utilizing the similar reaction conditions of preparing 4'-thiofuranose **70,** thiofuranose **75** was provided in a straightforward fashion. Deprotection of the compound **75** and condensation of the resultant amine with 3-ethoxy-2-propenoyl isocyanate provided the compound **76**, which was subjected to ring closure with 2 M H_2SO_4 and removal of benzyol group to gave the nucleoside **77**. Compound **77** was further converted to 2',3'-dideoxy-6',6'-difluoro-3'-thionucleoside **78** using the standard reaction condition.

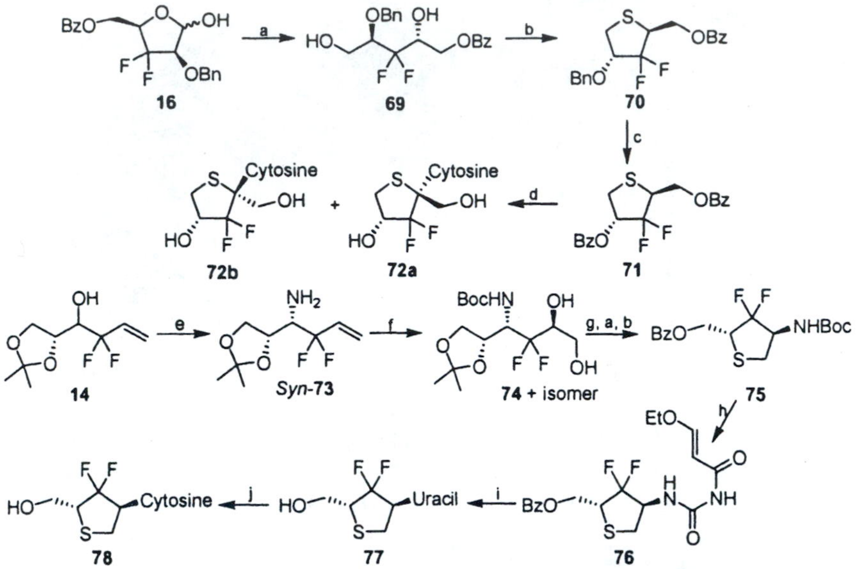

Scheme 11. *Reagent and Conditions:* (a) $NaBH_4$/MeOH; (b) i. MsCl, pyridine; ii. Na_2S, DMF; (c) i. BCl_3, MeOH; ii. Bz_2O, Et_3N, DMAP; (d). i. *m*-CPBA; ii. silylated *N*-benzoylcytosine, TMSOTf, DCE; iii. sat. NH_3/MeOH; (e) i. Tf_2O, pyridine; ii. NaN_3, DMF; iii. PPh_3, THF, iv. H_2O; (f) i. Boc_2O; ii. OsO_4, NMMO; (g) i. AcOH; ii. $NaIO_4$; (h) i. TFA; ii. 3-ethoxy-2-propenoyl isocyanate; (i) i. 2 M H_2SO_4; ii. Sat. NH_3/MeOH; (j) i. Ac_2O, DMAP; ii. TPSCl, DMAP, Et_3N; in. conc. $NH_3 \bullet H_2O$.

During the synthesis of *D*- and *L*-*β*-3'-fluoro-4'-thionucleosides, Chu and co-workers (*24*) also prepared *D*- and *L*-3', 3'-difluoro-4'-thionucleosides starting from commercially available 2-deoxy-*D*-ribose (Scheme 12). Difluorination of the ketone **79**, prepared from 2-deoxy-*D*-ribose in three steps, afforded the difluorinated intermediate **80**. Ring-opening reaction of the compound **80** with BnSH/$BF_3 \cdot Et_2O$(cat.) gave the thioacetal **81**. Treatment of the compound **81** with Tf_2O/2,6-lutidine delivered the corresponding triflate, which was cyclized to provide the thiosugar **82** in 78% yield. Transglycosylation of the compound **82** by the treatment with $Hg(OAc)_2$ in AcOH afforded the key intermediate **83** in 72% yield. Coupling the compound **83** with the various silyl-protected pyrimidine and silylated purine followed by removal of protection groups gave the corresponding *L*-3',3'-difluoro-4'-thionucleosides **84a**, **84b**. Starting from 2-deoxy-*L*-ribose and utilizing the same reaction route, *D-3*',3'-difluoro-4'-thionucleosides were also been synthesized.

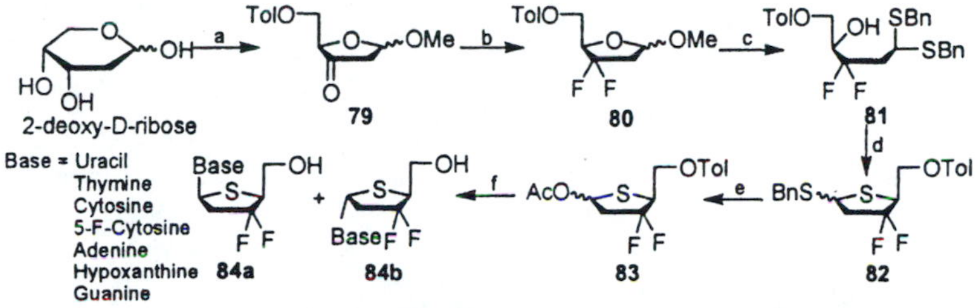

Scheme 12. *Reagent and Conditions:* (a) i. H_2SO_4, MeOH; ii. Tol-Cl, pyridine; iii. PDC, CH_2Cl_2; (b) DAST, CH_2Cl_2; (c) BnSH, $BF_3{\cdot}Et_2O$, CH_2Cl_2; (d) i. Tf_2O, 2,6-lutidine, CH_2Cl_2; ii. $BaCO_3$, TBAI, CH_3CN; (e) $Hg(OAc)_2$, Ac_2O, AcOH; (f) i. BSA, pyrimidines or purines, CH_3CN or 1,4-dioxane, TMSOTf; ii. NH_3, MeOH or 2-mercaptoethanol, NaOMe/MeOH.

Synthesis of *gem*-difluoromethylated azanucleosides

Replacement of the sugar ring oxygen of a nucleoside by nitrogen atom could cause the effects of biological significance. Besides of simple heteroatom effect of nitrogen atom (*25*), nitrogen atom could bind with exceeding high affinity and specificity to a variety of base-excision DNA repair (BER) enzymes, which may suggest a transition-state model for the glycosyl transfer reaction leading to base excision (*26*). In 2003, Tyler et. al. (*27*) firstly reported the synthesis and bioactivity of the *gem*-difluoromethylated azanucleoside 2'-deoxy-2',2'-difluoro-immucillin-H **89a** and **89b** (Scheme 13). With D-Serine as chiral pool and $BrCF_2CO_2Et$ as *gem*-difluoromethyl source, lactam **85** was afforded in a straightforward fashion. Lithiation of 9-bromo-9-deazahypoxanthine derivative **86** by bromine-lithium exchange followed by addition of the lactam **85** to the reaction mixture provided the alcohol **87**, reduction of which with $NaBH_3CN$ afforded the separable mixture of **88a** and **88b** in 1:3 ratio. Removal of benzyl group via hydrogenolysis followed by hydrolysis delivered the 2'-deoxy-2',2'-difluoro-immucillin-H **89a** and **89b**, respectively.

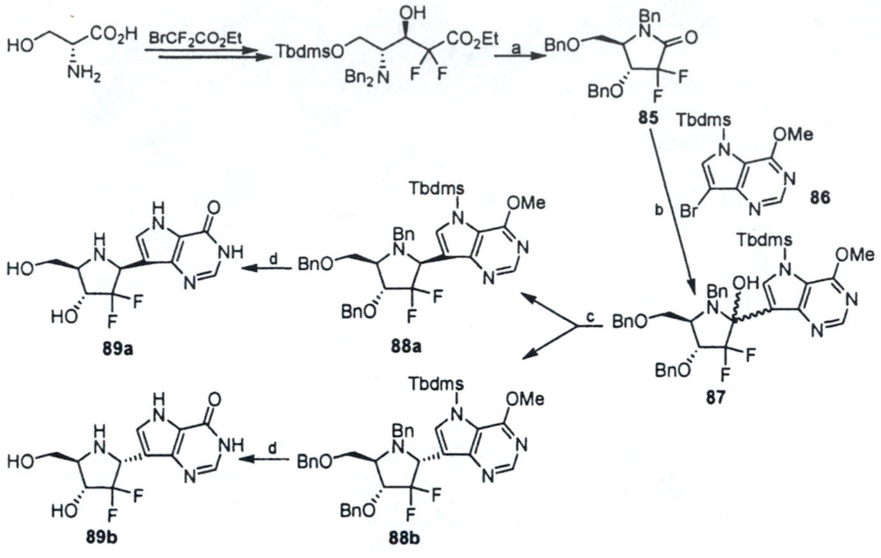

Scheme 13. *Reagent and Conditions:* (a) i. H_2, Pd/C; ii. aq. TFA; iii. NaH, BnBr; (b) BuLi, THF, -70°C; (c) $NaBH_3CN$, HOAc, MeOH; (d) i. H_2, Pd/C; ii. aq. HC1, MeOH.

Synthesis of *gem*-difluoromethylated carbocyclic nucleosides.

Due to the absence of a glycosidic linkage, carbocylic nucleosides are chemically more stable and not subjected to the phosphorylases that cleave the *N*-glycosidic linkage in conventional nucleosides. Earliest synthesis of *gem*-difluoromethylated carbocyclic nucleosides was described by Borthwick group (*28*) in 1987 (Scheme 14). Starting from the amino-triol (±)-**90,** the alcohol (±) **-91** was afforded by protection of the amino group with 2,4-dinitrophenyl followed by protection of C3',C5' hydroxyl groups as oxybis(di-isopropylsilyl) (TIPS) ether. Swern oxidation of the compound (±)-**91** gave the corresponding ketone (±)-**92**, which was treated with DAST to provide *gem*-difluoromethylated compound (±)-**93** in 12-27% yield. Removal of the dinitrophenyl-protecting group by hydrolysis and treatment of the resulting salt (±)-**94** with EtOCH=C(Me)CONCO/DBU delivered the carbocylic 1-(2'-deoxy-2',2'-difluororibofuransyl)-5-methyluracil (±)-**95** in 51% yield. Biological test shown that *gem*-difluoromethylated carbonucleoside (±)-**95** was inactive against both of HSV-1 and HSV-2.

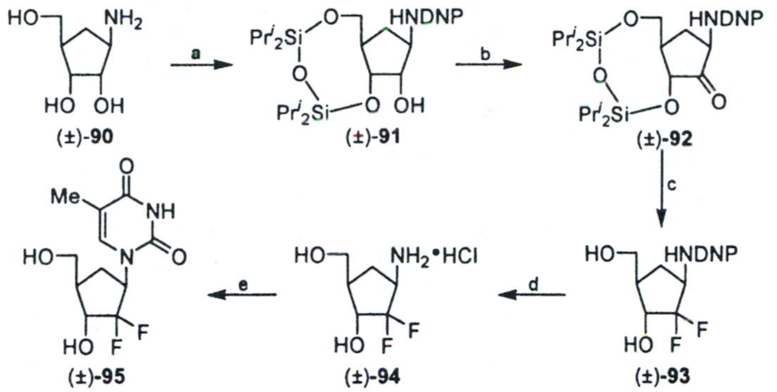

Scheme 14. *Reagent and Conditions:* (a) i. DNP-F, DMF, Na_2CO_3, rt.; ii. $O(Pr^i_2SiCl)_2$, DMF, imidazole; (b) Me_2SO, $(CF_3CO)_2O$, CH_2Cl_2, -65°C to rt, then Pr^i_2NEt; (c) DAST, CH_2Cl_2, rt., then TBAF, THF; (d) Amberlite IR 400 (OH^-), Et_2O, H_2O then 2M HCl; (e) EtOCH=C(Me)CONCO, DBU, DMF, -20°C, then 2M HCl, heat.

In 2004, synthesis of 2',3'-dioxy-6',6'-difluorocarbocyclic nucleosides was also described (*29*), which featured the construction of the carbocylic ring via ring-closing metathesis (RCM) and the incorporation of *gem*-difluoromethylene group by way of silicon-induced Reformatskii-Claisen reaction of chlorodifluoroacetic ester (Scheme 15). That is, protection of one of hydroxyl group in butane-1,4-diol **96** as benzyl ether and esterification of another hydroxyl group with ClF_2CO_2H gave the ester **97**. Silicon-induced Reformatskii-Claisen reaction of compound **97** was conducted and the resulted acid was further esterified to afford the ester **98**. Conversion of the compound **98** into the corresponding Weinreb amide **99** and treatment of the compound **99** with allylmagnesium bromide resulted in unsaturated ketone **100** in good yield. RCM of the compound **100** with second-generation Grubbs' catalyst gave the key intermediate **101** in 98% yield, which was subjected to Luche reduction to afford the separable alcohol (±)-**102a** and (±)-**102b** in a 2.9:1 *cis/trans* ratio. After hydrogenation of the compound (±)-**102a** with Pd black, treatment of the resultant product (±)-**103** with Tf_2O/Pyridine followed by substitution reaction with NaN_3 and reduction gave the cyclic amine (±)-**104**. Condensation of amine (±)-**104** with (*E*)-EtOCH=CHCONCO and removal of benzyl group via hydrogenation gave the *gem*-difluoromethylated carbocyclic nucleoside (±)**105a**. With the same synthetic route, isomer (±)-**105b** of nucleoside (±)-**105a** was also prepared from (±)-**102b**.

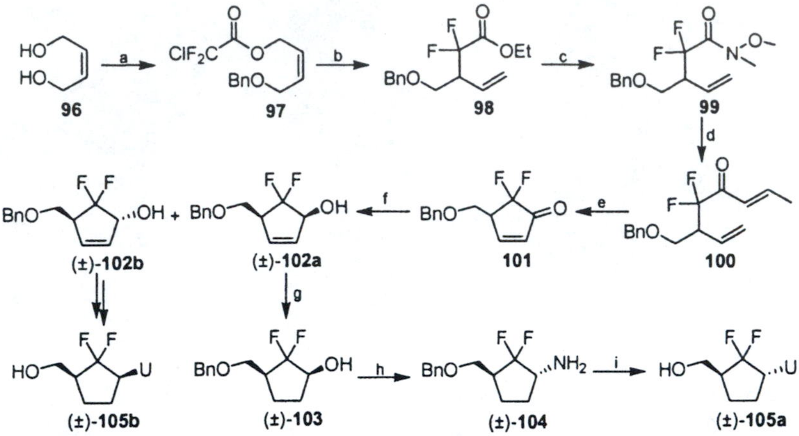

Scheme 15. *Reagent and Conditions:* (a) i. NaH, DMF, BnBr, 0°C; ii. $ClCF_2CO_2H$, cat. H_2SO_4, toluene, 140°C c; (b) i. Zinc dust, TMSCl, CH_3CN, 100°C; ii. Cat. H_2SO_4, EtOH, 40°C; (c) *N,O*-dimethylhydroxylamine, $AlMe_3$, toluene; (d) i. CH_2=CH_2CH_2MgBr, THF; ii. Et_3N, THF; (e) Second generation Grubbs' catalyst, toluene, 80°C; (f) $CeCl_3 \cdot 7H_2O$, $NaBH_4$, 0°C; (g) H_2, Pd black, benzene; (h) i. Tf_2O, pyridine, CH_2Cl_2, -40°C; ii. NaN_3, DMF; iii. H_2, Cat. Pd black, benzene; (i) i. (*E*)-EtOCH=CHCONCO, DMF, -25°C; ii. 2N H_2SO_4, reflux; iii. H_2, Cat. Pd/C.

Acknowledgement. Financial support from National Natural Science Foundation of China, Ministry of Education of China and Shanghai Municipal Scientific Committee are greatly appreciated.

References

1. For carbocyclic nucleosides reviews, see (a) Crimmins, M. T. *Tetrahedron* **1998**, *54,* 9229-9272. (b) Borthwick, A. D.; Biggadike, K. *Tetrahedron* **1992**, *48,* 571. (c) Agrofoglio, L.; Suhas, E.; Farese, A.; Condom, R.; Challand, S. R.; Earl, R. A., Guedj, R. *Tetrahedron* **1994**, *50,* 10611-10670. (d) Marquez, V. E., Lim, M. -U *Med. Res. Rev.* **1986**, *6,* 1-40. (e) Wu, Q.; Simons, C. *Synthesis* **2004**, 1533-1553. (f) Wang, J.; Froeyen, M.; Herdewijn, P. *Advances in Antiviral Drug Design* **2004,** *4,* 119-145. (g) Rodriguez, J. B.; Comin, M. J. *Mini-Rev. in Med. Chem.* **2003,** *3,* 95-114.
2. For thionucleosides reviews, see: Yokoyama, M. *Synthesis* **2000**, 1637-1655.
3. Yamashita, M.; Kato, Y.; Suzuki, K.; Reddy, P. M.; Oshikawa, T. *Abstracts of 29th Congress of Heterocyclic Chemistry* **1998**, 461.

4. For azanucleosides reviews, see: Yokoyama, M.; Momotake A. *Synthesis* **1999**, 1541-1554.
5. (a) Chu, C. K.; Baker, D. C., Eds. *Nucleosides and Nucleotides as Antitumor Agents and Antiviral Agents:* Plenum Press: New York, **1993**. (b) De Clercq, E. *J. Med. Chem.* **1995**, *38,* 2491-2517. (c) De Clercq, E. *J. Clin. Virol.* **2004**, *30,* 115-133. (d) Simons, C.; Wu, Q.; Htar, T. T. *Curr. Topics in Med. Chem.* **2005**, *5,* 1191-1203.
6. Some fluorinated nucleosides reviews, see: (a) Pankiewicz, K. W. *Carbohydr. Res.* **2000**, *327,* 87-105. (b) Pandiewicz, K. W.; Watanabe, K. A. *J. Fluorine Chem.* **1993**, *64,* 15-36. (c) Viani, F. in *Enantiocontrolled Synthesis of Fluoro-Organic Compounds;* John Wiley & Sons Ltd., Chichester, UK, **1999**, p419-449.
7. (a) Lee, D. *Clinical Lymphoma* **2003**, *4,* 152-153. (b) Noble, S.; Goa, K. L. *Drugs* **1997**, *54,* 447-472.
8. Hertel, L. W.; Kroin, J. S.; Misner, J. W.; Tustin, J. M. *J. Org. Chem.* **1988**, *53,* 2406-2409.
9. Chou, T. S.; Heath, P. C.; Patterson, L. E.; Poteet, L. M.; Lakin, R. E.; Hunt, A. H. *Synthesis* **1992**, 565-570.
10. Hertel, L. W.; Grossman, C. S.; Kroin, J. S.; Mineishib, S.; Chubbb, S.; Nowak, B.; Plunkettb, W. *Nucleosides Nucleotides* **1989**, *8,* 951-955.
11. Mabry, T. E.; Jones, C. D.; Chou, T. S.; Colacino, J. M.; Grindey, G. B.; Worzalla, J. F.; Pearce, H. L. *Nucleosides Nucleotides* **1994**, *13,* 1125-1133.
12. Zhang, X.; Xia, H.; Dong, X.; Jin, J.; Meng, W.-D.; Qing, F.-L. *J. Org. Chem.* **2003**, *68,* 9026-9033.
13. (a) Xiang, Y. J.; Kotra, L. P.; Chu, C. K. *Bioorg. Med Chem. Lett.* **1995**, *5,* 743-748. (b) Kotra, L. P.; Xiang, Y.; Newton, M. G.; Schinazi, R. F.; Cheng, Y.-C.; Chu, C. K. *J. Med Chem.* **1997**, *40,* 3635-3644.
14. Kotra, L. P.; Newton, M. G.; Chu, C. K. *Carbohydr. Res.* **1998**, *306,* 69-80.
15. Chen, G. *Faming Zhuanli Shenqing Gongkai Shuomingshu* China Patent, **2003**, 1442420.
16. Jarvi, E. T.; Sunkara, P. S.; Bowlin, T. L. *Nucleosides Nucleotides* **1989**, *8,* 1111-1114.
17. Bergstrom, D. E.; Mott, A. W.; Clercq, E. D.; Balzarini, J.; Swartling, D. J. *J. Med Chem.* **1992**, *35,* 3369-3372.
18. (a) Gumina, G.; Schinazi, R. F.; Chu, C. K. *Org. Lett.* **2001**, *3,* 4177-4180; (b) Chong, Y.; Gumina, G.; Mathew, J. S.; Schinazi, R. F.; Chu, C. K. *J. Med Chem.* **2003**, *46,* 3245-3256.
19. Zhou, W.; Gumina, G.; Chong, Y.; Wang, J.; Schinazi, R. F., Chu, C. K. *J. Med Chem.* **2004**, *47,* 3399-3408.
20. Castillón, S.; Fernández, R. *Tetrahedron* **1999**, *55,* 8497-8508.
21. (a) Yoshimura, Y.; Kitano, K.; Watanabe, M.; Satoh, H.; Sakata, S.; Miura, S.; Ashida, N.; Machida, H.; Matsuda, A. *Nucleosides Nucleotides* **1997**, *16,* 1103-1106. (b) Yoshimura, Y.; Kitano, K.; Satoh, H.; Watanabe, M.; Miura, S.; Sakata, S.; Sasaki, T.; Matsuda, A. *J. Org. Chem.* **1996**, *61,* 822-823. (c) Yoshimura, Y.; Kitano, K.; Yamada, K.; Satoh, H.; Watanabe, M.;

Miura, S.; Sakata, S.; Sasaki, T.; Matsuda, A. *J. Org. Chem.* **1997**, *62,* 3140-3152.

22. Jeong, L. S.; Moon, H. R.; Choi, Y. J.; Chun, M. W.; Kim, H. O. *J. Org. Chem.* **1998**, *63,* 4821-4825.
23. Wu, Y.-Y.; Zhang, X.; Meng, W.-D.; Qing, F.-L. *Org. Lett.* **2004**, *6,* 3941-3944.
24. Zhu, W.; Chong, Y.; Choo, H.; Mathews, J.; Schinazi, R. F.; Chu, C. K. *J. Med. Chem.* **2004**, *47,* 1631-1640.
25. Akiba, K.-Y. *Yuki Gosei Kyokai Shi* **1984**, *42,* 378.
26. Schärer, O. D.; Verdine, G. L. *J. Am. Chem. Soc.* **1995**, *117,* 10781-10782.
27. Evans, G. B.; Furneaux, R. H.; Lewandowicz, A.; Schramm, V. L.; Tyler, P. C. *J. Med. Chem.* **2003**, *46,* 3412-3423.
28. (a) Biggadike, K.; Borthwick, A. D.; Evans, D.; Exall, A. M.; Kirk, B. E.; Roberts, S. M.; Stephenson, L.; Youds, P.; Slawin, A. M. Z.; Williams, D. J. *Chem. Soc., Chem. Commun.* **1987**, 251-254. (b) Brothwick, A. D.; Evans, D. N.; Kirk, B. E.; Biggadike, K.; Exall, A. M.; Youds, P.; Roberts, M.; Knight, D. J.; Coates, J. A. V. *J. Med Chem.* **1990**, *33,* 179-186.
29. Yang, Y.-Y.; Meng, W.-D.; Qing, F.-L. *Org. Lett.* **2004**, *6,* 4257-4259.

Chapter 19

Polyfluoropyridyl Glycosyl Donors

Christopher A. Hargreaves[1], Graham Sandford[1,*], and Benjamin G. Davis[2]

[1]Department of Chemistry, University of Durham, South Road, Durham, DH1 3LE, United Kingdom
[2]Chemistry Research Laboratory, University of Oxford, Mansfield Road, Oxford OX1 3TA, United Kingdom

Synthesis of stereochemically defined oligosaccharides by a series of glycosylation processes involving the reaction between a glycosyl donor and acceptor is essential to synthetic carbohydrate chemistry and glycobiology. However, despite the importance of glycosylation chemistry and the development of much sophisticated methodology, there remains no general and stereoselective procedure for the synthesis of oligo- and polysaccharides. Families of novel glycosyl donors have been conveniently synthesized from polyfluoropyridine derivatives and, in a short series of model glycosylation reactions, have shown that they offer great possibilities for the controlled stereoselective synthesis of oligosaccharide systems. In particular, their tunable reactivity offers the promise of broad ranging activity suitable for cascade synthesis.

Carbohydrates are, of course, of tremendous importance in biological chemistry and play key roles in many natural processes including metabolism, as structural components of many biosynthetic pathways and as messengers in precise communication events.[1] For example, monosaccharidic galactolipids are important binding elements in the initial infective interaction of the outer proteins of the HIV-1 virion with the surface of leukocytes, an anchoring process that is severely disrupted simply by the alteration of only one stereocentre in the carbohydrate. [2] Such factors have recently provided increased impetus for research into the rapidly growing field of glycobiology where, for example, the concise synthesis of stereochemically defined oligosaccharides and glycocconjugates,[3-5] consisting of complex glycan units attached to a peptide or protein system is of fundamental importance. However, the synthesis of such systems for use as both mechanistic probes or as therapeutic agents presents a formidable challenge because, unlike oligopeptide and nucleic acid synthesis, there are no general automated, synthetic methodologies yet available.

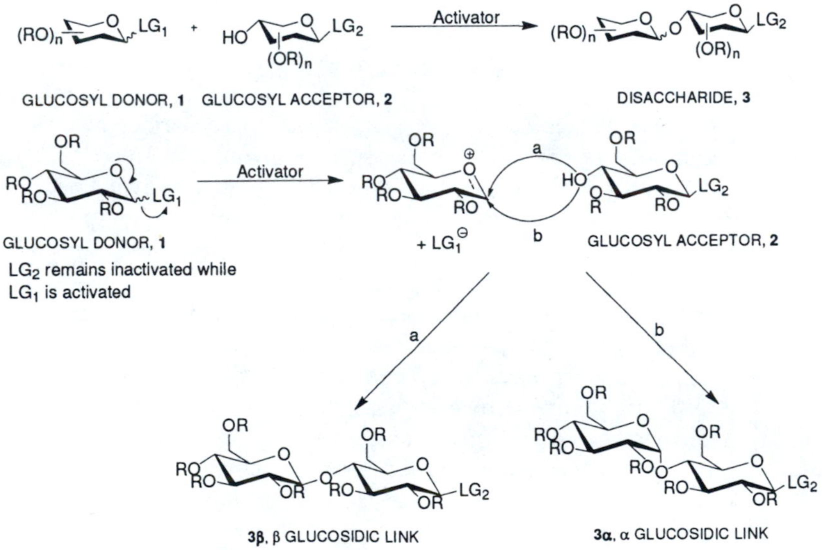

Scheme 1. Glycosylation reactions

Oligosaccharides are synthesized [6-8] by a series of key *glycosylation reactions* in which, for example, a *glycosyl donor* **1** reacts with a *glycosyl acceptor* **2** in the presence of a suitable activating system to give disaccharide **3**. (Scheme 1).

While regioselectivity may be controlled by protecting group strategies, a successful glycosylation reaction requires both chemoselectivity, that is, the desired leaving group must be displaced by the appropriate hydroxyl group to give the correct disaccharide, and stereoselectivity, attack of the nucleophilic hydroxyl group must occur specifically at either the α or β face to give a single diastereoisomer. Consequently, for the coupling of two monosaccharide residues **4** and **5** bearing single free hydroxyl groups, Scheme 2, eight possible products may, in principle, be obtained from a glycosylation reaction that has poor selectivity.

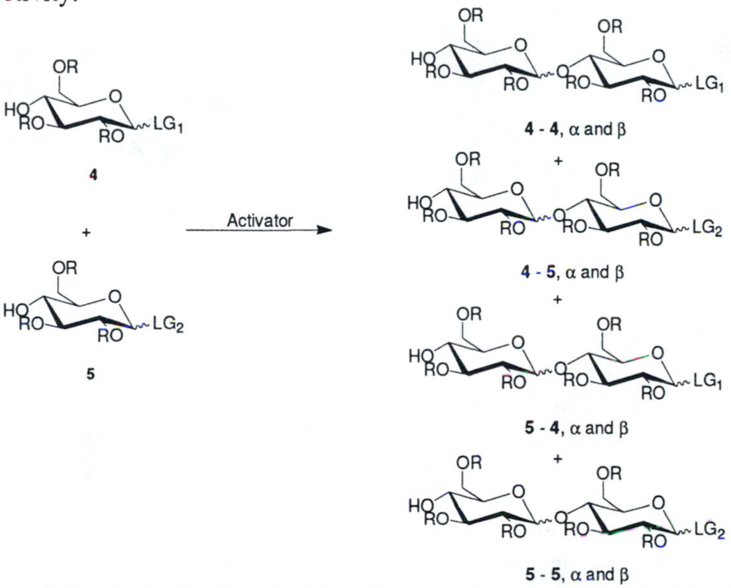

Scheme 2. Synthesis of a disaccharide with no reaction control from dual function donor-acceptor building blocks

The challenge for the synthetic chemist is, therefore, to devise strategies that overcome these difficulties and allow syntheses of simple disaccharides (e.g., α-**4-5**) from two residues. Moreover, such dual-function acceptor-donor systems have the potential, with correct selectivity, to not only allow access to di- but also oligo-saccharides through iterative reaction cascades.

Chemoselectivity can be achieved by the use of glycosyl donors that bear substituents at the anomeric position which have different leaving group abilities and, consequently, may be selectively activated towards nucleophilic attack by specific activating species. This is illustrated in Scheme 3, where **4**, bearing leaving group LG_1 which is activated specifically over LG_2 by LG_1-activator, is attacked by the hydroxyl group of the unactivated glycosyl donor **5**-LG_2, to give the required disaccharide **4-5** only. This process is termed *tuning* and the reactivity of a glycosyl donor can be adjusted through the use of different protecting groups, activators and/or anomeric leaving groups. [9,10]

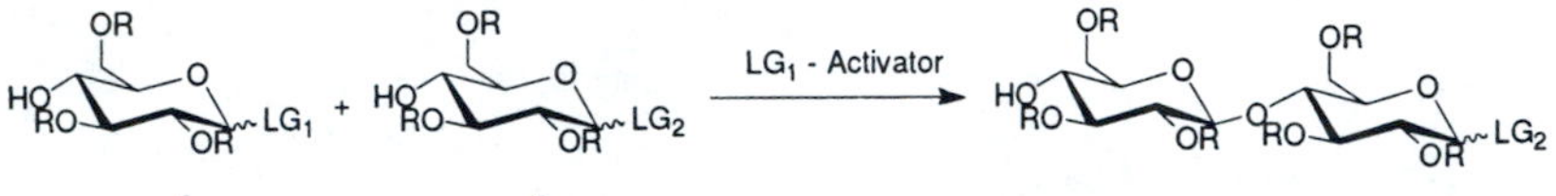

Scheme 3. Controlled chemoselective glycosylation by selective activation of LG_1

Given the importance of glycosylation reactions to glycobiology, many classes of glycosyl donors have been synthesized and their use in oligosaccharide synthesis assessed. [6-8] Whilst systems such as thioglycosides **6**, trichloroacetamidates **7**, alkenyl glycosides **8** and glycosyl fluorides **9** (Scheme 4) have been exploited in some very elegant oligosaccharide syntheses, there is still no general solution to the glycosylation problem. Typically, each donor system displays its own peculiar selectivities in specific carbohydrate systems. Furthermore, in many cases, such as the trichloroacetamidate and glycosyl fluoride systems, it is often difficult to synthesise a related family of donors which bear similar but slightly different substituents at the anomeric position that have different leaving group abilities that would make it possible for their reactivity to be subtly 'tuned' as required.

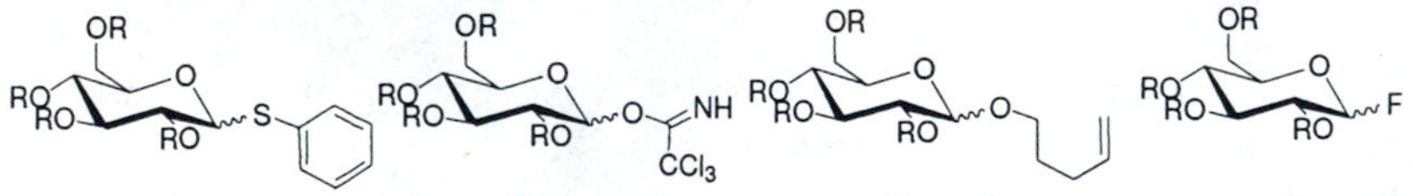

Scheme 4. Glycosyl donors in general use

Consequently, we have initiated a research programme that aims to devise synthetic approaches to families of related glycosyl donor systems, in which the leaving groups may be readily tuned, that provide a generally applicable solution to the glycosylation problem. We have pursued a glycosylation strategy whereby a range of electronically varied electron deficient pyridyl systems **10** bearing a range of functionality on the heterocyclic ring are utilized as leaving groups which may be activated towards glycosylation by a Lewis acid. (Scheme 5).

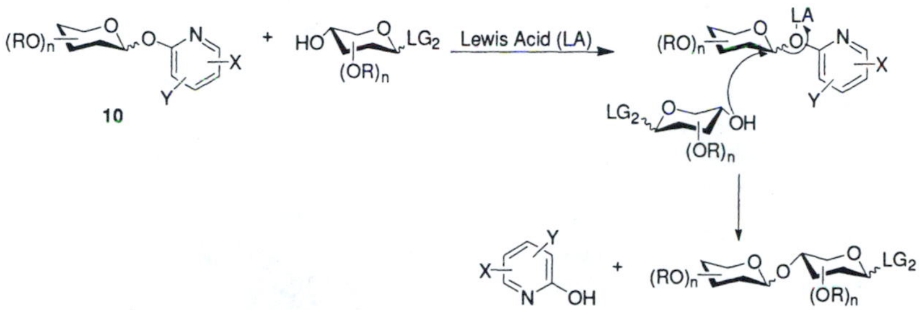

Scheme 5. Novel glycosyl donor systems

We opted to use pentafluoropyridine **11** as our polyfunctional pyridine scaffold because this system is highly susceptible towards nucleophilic attack[11] and could, in principle, be readily attached to a sugar residue and/or functionalized by a sequence of nucleophilic aromatic substitution reactions. The potential use of such systems has been noted previously.[11]

Using the strategy postulated in Scheme 6, families of related glycosyl donors could, potentially, be synthesized by reaction of pentafluoropyridine **11** with a suitable carbohydrate derivative and a series of nucleophiles to give several related classes of glycosyl donors **12**, **13** and **14**, depending simply on the order of nucleophilic substitution. Class **12** is formed by reaction of pentafluoropyridine **11** directly with a saccharide system, Class **13** by reaction of a nucleophile with PFP followed by a saccharide, and so on. Whilst donor **12** has been synthesised before,[12] the full opportunities for using highly fluorinated pyridine systems for reactivity tuning have not been exploited. For all of these glycosyl donors, leaving group ability could, we hoped, be controlled by both the substituents on the pyridine rings (Nuc_1, Nuc_2, etc.) and/or the Lewis acid activator.

Reactions of Polyfluoropyridine derivatives

The chemistry of pentafluoropyridine **11** has been developing over many years[11,13] since the its first viable synthesis[14] and is dominated by nucleophilic aromatic substitution reactions. A wide variety of nucleophiles (Nuc_1, Scheme 7) react with pentafluoropyridine to give, in

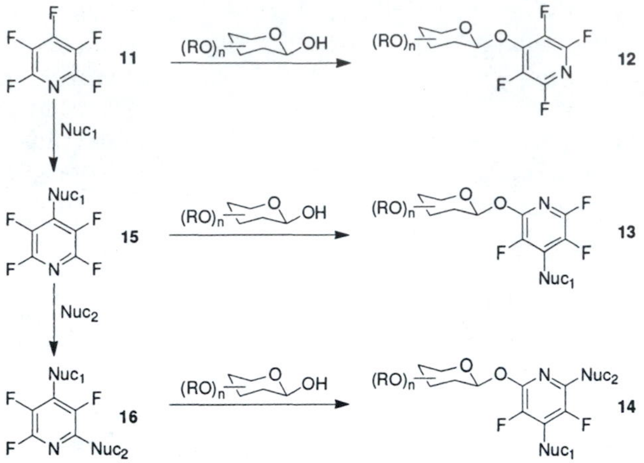

Scheme 6. *Synthetic strategy for the synthesis of related families of polyfluoropyridyl glycosyl donors*

the vast majority of cases, products **15** arising from the selective displacement of the fluorine atom located at the 4-position.

Many tetrafluorinated pyridine derivatives **15** have been synthesized by this methodology but the further reaction of such systems with subsequent nucleophiles to, potentially, give **16** have not been explored to any great extent, although some initial results have been reported. [15] In principle, reaction of a second nucleophlile (Nuc_2, Scheme 7) could give rise to three products, **16**, **17** and **18**, arising from the displacement of the *ortho* and *meta* fluorine atoms or the substituent, Nuc_1, located at the 4-position, respectively. However, the factors that determine this potential regioselectivity have not been established.

Synthesis of glycosyl donors could, therefore, potentially be achieved by reaction of a suitable carbohydrate nucleophile and pentafluoropyridine to give donors of type **12**. Suitable model reactions involving pentafluoropyridine have been recorded.[12] In contrast, however, little is known of **13** or **14**. Before embarking on the synthesis of glycosyl donor types **13** and **14** we decided to carry out model reactions between a series of tetrafluoropyridine derivatives (Nuc_1 = OC_2H_5, NH*i*Pr, CN, NO_2, Scheme 7) and sodium ethoxide to establish the regioselectivity of nucleophilic substitution on these systems which may be dependent upon the nature of the substituent located at the 4-position (Scheme 8).

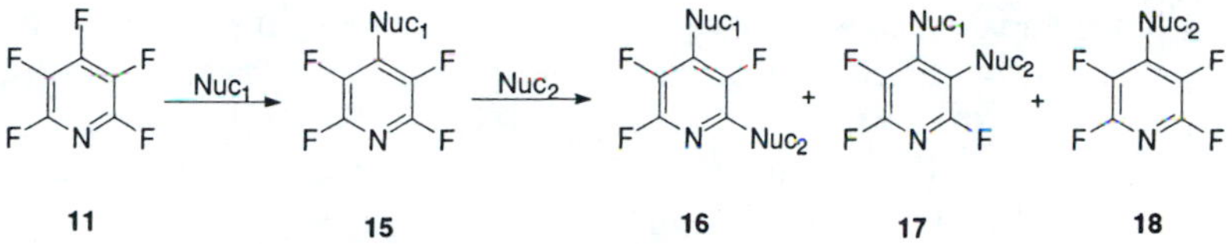

Scheme 7. Nucleophilic substitution in penta- and tetra-fluoropyridine systems

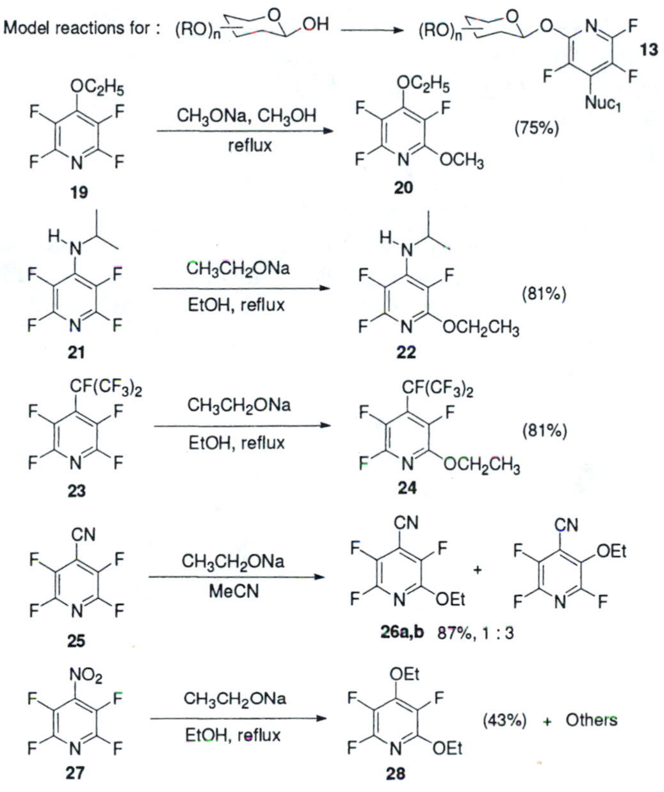

Scheme 8. Reactions of tetrafluoropyridine derivatives with alkoxides as models for preparation of donor type **13**

Reaction of 4-methoxy-tetrafluoropyridine gave a complex mixture of products arising from substitution of both the fluorine atom at position 2 and the methoxy substituent itself. In contrast, ethoxy substituents, which are poorer leaving groups than methoxy, were not displaced by sodium methoxide and the 2-methoxy-4-ethoxy pyridine system **20** was obtained in high yield from **19**. The 4-amino- and 4-perfluoroisopropyl derivatives, **22** and **24**, gave products from substitution of the *ortho* fluorine selectively. In contrast, the cyano system **25** led to a mixture of *ortho* and *meta* substituted products **26a,b** due to the activating influence of the electron withdrawing cyano substituent on the adjacent 3-position which competes effectively with the activating influence of ring nitrogen. Nitro substrate **27** gave product **28** arising from displacement of the labile nitro group only. These simple model reactions indicated that **11**, **19**, **21** and **23** could be useful substrates for the preparation of glycosyl donors whereas **25** and **27** are less suitable.

Synthesis of Polyfluoropyridyl Glycosyl Donors

Using the methodology from the model studies described above, a short series of glycosyl donors was synthesized by reaction of the tetrabenzyl glucose salt **29** and either pentafluoropyridine **11** or an appropriate tetrafluoropyridyl system such as **22**. (Scheme 9)

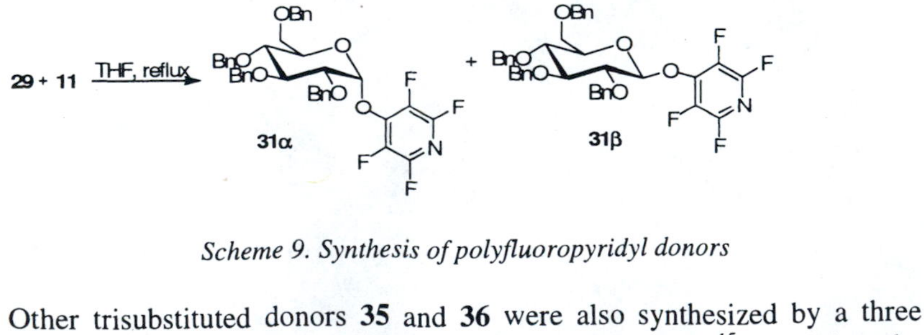

11 + NH_2 —THF, reflux→ 22, 94% —THF, reflux, 29→ 30, 51%, α : β 1 : 3

29 + 11 —THF, reflux→ 31α + 31β

Scheme 9. Synthesis of polyfluoropyridyl donors

Other trisubstituted donors **35** and **36** were also synthesized by a three stage strategy (Scheme 10) adapted from earlier work[15] involving the chemistry of perfluoro-4-isopropylpyridine, thus extending the number of possible donors that may be accessed by the general strategy outlined in Scheme 6.

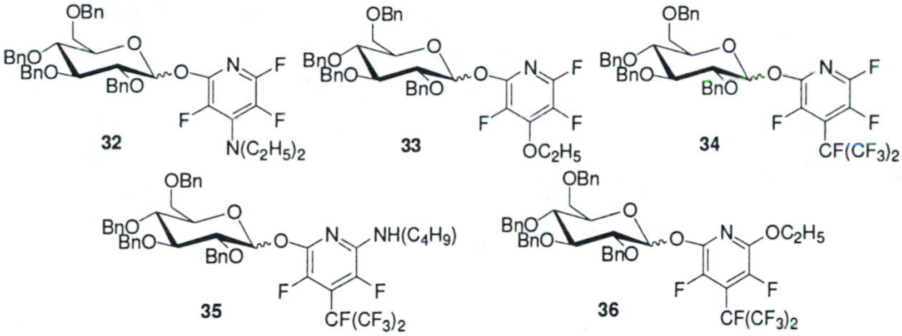

Scheme 10. Synthesis of polyfluoropyridyl donors

All donors were found to be hydrolytically stable and, in some cases, the individual anomers could be separated and purified by high pressure flash chromatography. As well as making use of the usual characterization methods, we observed that the α and β anomers showed distinct ^{19}F nmr resonances, making the identification and measurement of anomeric ratios a ready process.

Model Glycosylation reactions

We began our studies to establish the feasibility of utilizing these novel glycosyl donors by carrying out some model glycosylation reactions using cyclohexanol as the glycosyl acceptor, to mimic a sugar residue bearing a secondary hydroxyl group, and activation by a range of Lewis acids including boron trifluoride, copper (II) triflate, titanium tetrachloride and aluminium trichloride, amongst others. All reactions were carried out in acetonitrile at room temperature, monitored by ^{19}F nmr and the product disaccharide **37** was isolated in each case. Reactions are shown in Scheme 11 which gives the yields of disaccharide **37** after 24 h.

The data from Scheme 11 reveal that donor **31** is activated selectively by copper triflate whereas donors **34** and **30** are not. Titanium tetrachloride activates donors **31** and **30** but not **34** whilst all three donors are activated by boron trifluoride and aluminium trichloride at various rates. Although only initial, these results clearly demonstrate that glycosyl donor families of this type have variable reactivity that may be tuned *via* the substituents attached to the pyridine ring and the Lewis acid activator - an exciting prospect.

Furthermore, activation of anomerically pure β-glycosyl donor **31β** by copper triflate gave the corresponding disaccharide **37** as a mixture of α and β anomers in the ratio 98 : 2., indicating an almost complete stereochemical inversion of configuration at the anomeric site upon

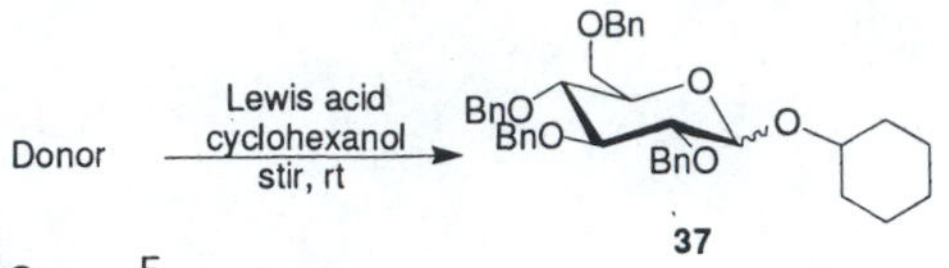

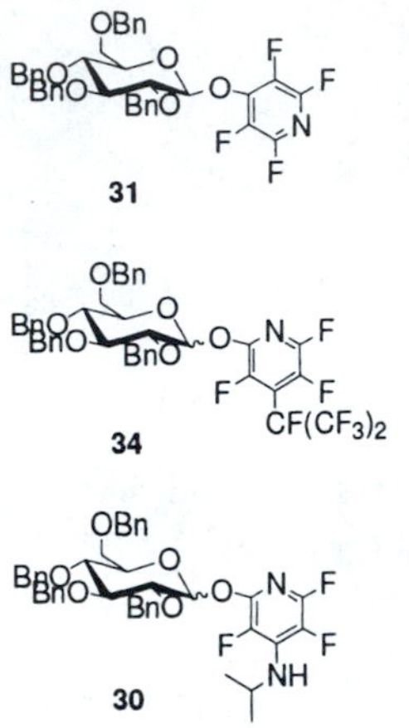

Donor	Lewis Acid	Yield after 24 h
31	$Cu(SO_3CF_3)_2$	75
34	$Cu(SO_3CF_3)_2$	0
30	$Cu(SO_3CF_3)_2$	0
31	$TiCl_4$	55
34	$TiCl_4$	0
30	$TiCl_4$	12

Scheme 11. Glycosylation reactions activated by a variety of Lewis Acids

glycosylation. (Scheme 12) This indicates that the mechanism of this particular process may be S_N2 in character which is especially unusual in a potentially coordinating solvent such as acetonitrile. In future, this may allow us to control the stereochemistry of the glycosidic bond formation step in oligosaccharide synthesis.

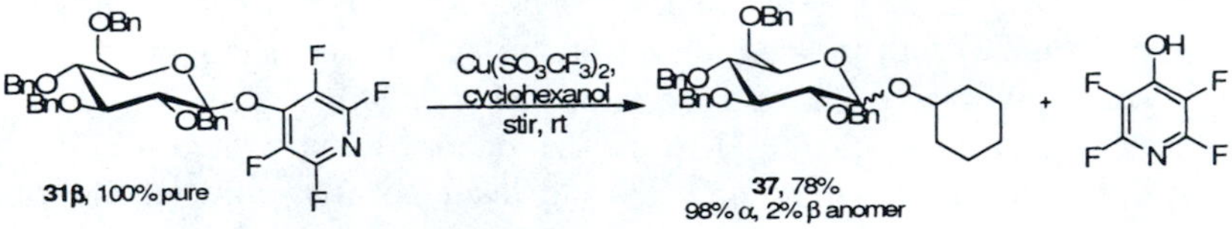

Scheme 12. Stereochemistry of the glycosylation process

The presence of fluorine atoms within the glycosyl donors allow all the glycosylation reactions to be directly monitored by ^{19}F nmr spectroscopy very accurately and conveniently. Scheme 13 shows reaction between donor **31β** with cyclohexanol initiated by copper(II) triflate over time. Resonances attributed to the donor decrease over time with a corresponding increase in the resonances arising from the pyridinol side product formed after glycosylation has occurred. This 'real time' reaction monitoring is a great benefit for convenient analysis of these processes and an advantage of using fluorinated systems.

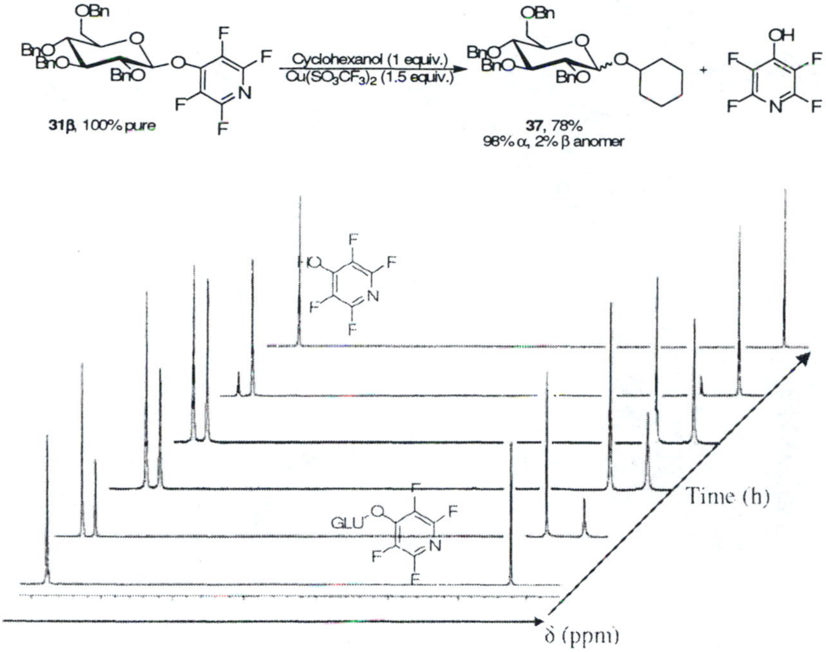

Scheme 13. Monitoring a glycosylation reaction over time by ${}^{19}F$ nmr

The possibility of measuring the conversion of glycosyl donor to products very conveniently and non-intrusively by ${}^{19}F$ nmr over time has allowed us to measure rates of reactivity of each donor system upon activation by a variety of Lewis acids, adapting a process analogous to that carried out on thioglycosde donors by Ley[10] and Wong.[16] Analysis of data concerning the conversion of glycosyl donor **33** to products using second-order kinetic analysis, in line with our previous observation that these processes may be S_N2 in character, allows the tentative determination of the initial rate constant by utilizing a plot of $1/[\text{Donor}] - 1/[\text{Donor}]_0$ versus time where the gradient is the rate of glycosylation for the acceptor upon activation by a range of Lewis acids. (Scheme 14)

For glycosylation of **33** with cyclohexanol, we see that varying the Lewis acid can have a profound effect on the rate of reaction. For example, reaction using boron trifluoride etherate as activator is 120 times faster than when titanium tetrachloride is used. Wong has suggested[16] that, for an ideal set of glycosyl donors, it would be desirable for a series of related donors to have relative glycosylation rates of between 1 and 1000. Even

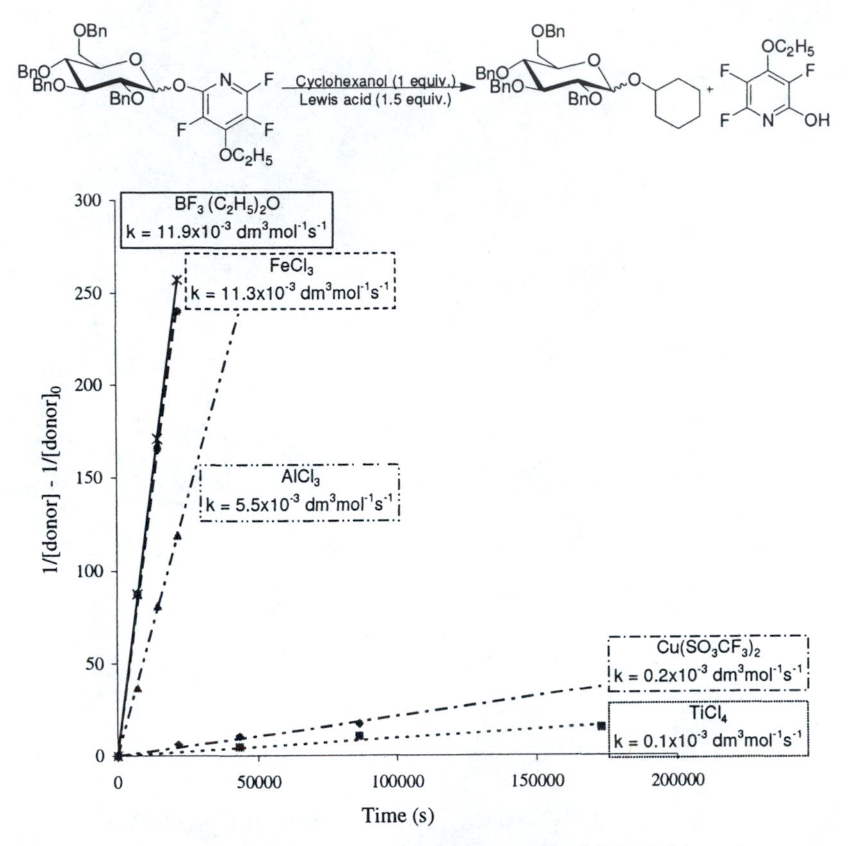

Scheme 14. Kinetic data for the reaction of **33** *with cyclohexanol*

with this single donor, a range of relative rates from 1 to 120 is already available by varying the Lewis Acid activator alone. A larger range of donors, based on fluoropyridine scaffolds, would, we believe span a much wider range.

Finally, we have used this family of glycosyl donors for the synthesis of disaccharide **38** to demonstrate the potential of using these systems in oligosaccharide synthesis.(Scheme 15)

Conclusion

We have developed general routes for the synthesis of families of related polyfluoro-pyridyl glycosyl donors from reactions of highly

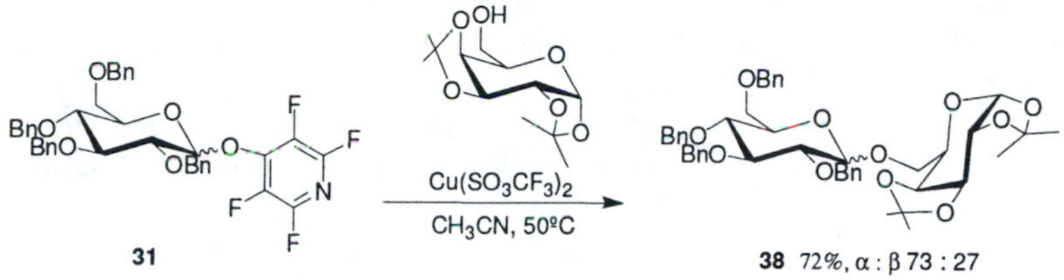

Scheme 15. Synthesis of disaccharides

fluorinated pyridine derivatives and appropriately protected monosaccharide systems using nucleophilic aromatic substitution processes. The applicability of the glycosyl donors has been demonstrated in several glycosylation reactions and we find that the activation of donors towards attack by glycosyl acceptors and the rates of glycosylation depends on both the substituents that are located on the pyridine ring and the Lewis acid activator present. The demonstrated possibility of 'tuning' the reactivity of the polyfluoro-pyridyl glycosyl donors provides a new approach to the important 'glycosylation problem'. Several early advantages of using fluorinated donors have been observed, such as analysis by ^{19}F nmr spectroscopy and, in particular, improved stereochemical control for this very difficult transformation. We hope, in future, to be able to utilise our donor systems to address key synthetic challenges in the field of glycobiology.

References

1. Davis, B. G. *Chem. Ind.* **2000**, 134.
2. Bhat, S. *Proc. Natl. Acad. Sci. USA* **1991**, *88,* 7131.
3. Davis, B. G. *J. Chem. Soc., Perkin Trans 1* **1999**, 3215.
4. Gridley, L. J.; Osborn, H. M. I. *J. Chem. Soc., Perkin Trans 1* **2000**, 1471.
5. Pratt, M. R.; Bertozzi, C. R. *Chem. Soc. Rev.* **2005**, *34*, 58.
6. Paulsen, H. *Angew. Chem. Intl. Ed. Engl.* **1982**, *21*, 155.
7. Boons, G. J. *Contemp. Org. Synth.* **1996**, *3,* 173.
8. Davis, B. G. *J. Chem. Soc., Perkin Trans 1* **2000**, 2137.
9. Mootoo, D. R.; Konradsson, P.; Udodong, U.; Fraser-Reid, B. *J. Am. Chem. Soc.* **1988**, *110*, 5583.
10. Grice, P.; Ley, S. V.; Pietruszka, J.; Osborn, M. I.; Henning, W. M.; Priepke, H. W. M.; Warriner, S. L. *Chem. Eur. J.* **1997**, *3,* 431.

11. Brooke, G. M. *J. Fluorine Chem.* **1997**, *86,* 1.
12. Huchel, U.; Schmidt, C.; Schmidt, R. R. *Eur. J. Org. Chem.* **1998**, 1353.
13. Chambers, R. D.; Sargent, C. R. *Adv. Heterocycl. Chem.* **1981**, *28,* 1.
14. Chambers, R. D.; Hutchinson, J.; Musgrave, W. K. R. *J. Chem. Soc.* **1964**, 3573.
15. Chambers, R. D.; Hassan, M. A.; Hoskin, P. R.; Kenwright, A.; Richmond, P.; Sandford, G. *J. Fluorine Chem.* **2001**, *111*, 135.
16. Zhang, Z. Y.; Rollman, I.; Ye, X. S.; Wischnat, R.; Baasov, T.; Wong, C. H. *J. Am. Chem. Soc.* **1999**, *121,* 734.

Chapter 20

New 10-Trifluoromethyl Monomers, Dimers, and Chimeras of Artemisinin from a Key Allyl Bromide Precursor

Danièle Bonnet-Delpon[1], Jean-Pierre Bégué[1], Benoit Crousse[1], Fabienne Grellepois[1], Constance Chollet[1], Fatima Chorki[1], Guillaume Magueur[1], Michèle Ourévitch[1], Nguyen Thi Ngoc Tam[1], Philippe Grellier[2], Nguyen Van Hung[3], Truong Van Nhu[4], Doan Hanh Nhan[4], and Nguyen Thi Minh Thu[4]

[1]BIOCIS UMR 8076, CNRS, Université Paris-Sud, Rue J. B. Clément, F–92296, Châtenay/Malabry, France
[2]USM 0504, Muséum National d'Histoire Naturelle, 61 rue Buffon, 75231 Paris Cedex 05, France
[3]VAST, Institute of Chemistry, 18 Hoang Quoc Viet, Cau Giay, Hanoi, Vietnam
[4]National Institute of Malaria (NIMPE), Luong The Vinh Street, Thanh, Xuan, Hanoi, Vietnam

The effect of the CF_3 group on the metabolism of bioactive molecules is well exemplified in the herein reported antimalarial artemisinin derivatives. The design, synthesis and biological activity of various CF_3-substituted artemisinin derivatives are described.

Introduction

Endoperoxides are a promising class of antimalarial drugs. Artemisinin **1**, the lead natural compound, is a sesquiterpenic lactone (*1*). Because of its poor pharmacological profile, scientists rapidly developed derivatives from dihydroartemisinin (DHA) **2** (Figure 1) with better properties, in particular better solubility in oil or water (*2*). The semisynthetic ethers or esters, such as artemether **3** (*3*), arteether **4** (*4*), artesunate **5** (*5*), and artelinate (*7*) are being widely used for therapy in Asia and in Africa (*6*). These endoperoxides have several advantages over existing antimalarial drugs. First, they clear the peripheral blood from parasites very rapidly, and second, there is no cross-resistance with other antimalarial drugs. However, they have some disadvantages, in particular a short plasmatic half-lives. Ethers and esters of DHA undergo in vivo cytochrome P450 oxidation (artemether, arteether) and/or are cleaved under mild acidic conditions (artesunate). Both processes provide active DHA, which has itself a short plasmatic half-life (*8*). As a result, a short course treatment is generally associated with a high rate of recrudescent parasitemia (*9*). An urgent objective is thus to design new artemisinin derivatives that enable prolonged efficiency with a reasonable plasmatic half-life in order to allow shorter treatments or smaller doses, while preventing the emergence of resistance.

A valuable approach is to design non-acetal artemisinin derivatives functionalized on the 16-methyl, since the presence of a polar substituent is beneficial for better water solubility properties, required for oral antimalarial treatments. Until now, there have been only a few attempts to synthesize 16-functionalized artemisinin derivatives. The total synthesis of such compounds has been described previously, as well as semi-synthesis from minor natural products also isolated from *Artemisia annua* : artemisinic acid or artemisitene, compounds less readily available than artemisinin. The more interesting approach uses artemisitene, which can also be synthesized from artemisinin (*10*). Several groups reported the preparation of 16-functionalized artemisinin derivatives through a Michael addition to artemisitene. However most of these derivatives still possess the acetal pattern at C-10, which results in a low metabolic stability toward acid conditions such as those found in stomach on oral administration (*11-15*).

From our part, we decided some years ago to combine this approach with the "low-down metabolism by fluorine substitution" concept (*16,17*). Fluorinated substituents (Rf) are known to impart greater protection to molecules from metabolic (oxidative and hydrolytic) degradation. We have shown that the introduction of a fluoroalkyl group onto a crucial position for metabolism of artemisinin derivatives, for instance on carbon C-10 (compound **6**) or on the alkoxy chain of ethers of DHA (compound **7**), resulted in excellent *in vitro* and *in vivo* activities, surpassing those of DHA and its ethers **4** and **5** (Figure 2) (*18,19*). Measurement of plasmatic half-lives of CF_3 hemiketal **6** ($T_{1/2}$ = 85.9

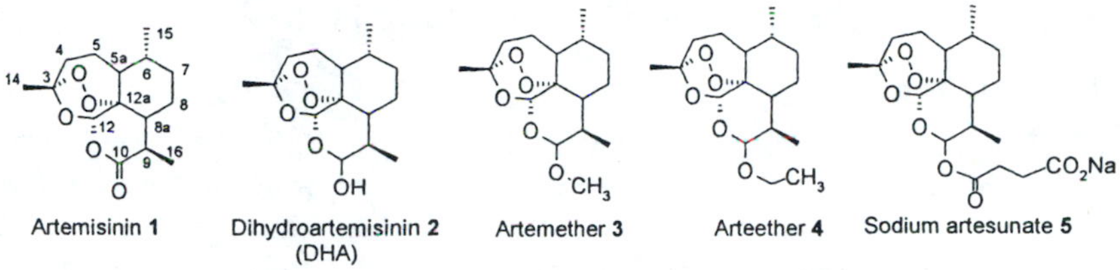

Artemisinin 1 Dihydroartemisinin 2 (DHA) Artemether 3 Arteether 4 Sodium artesunate 5

Figure 1. First generation of artemisinins

min) and DHA ($T_{1/2}$ = 26.3 min), after intravenous administration to rats, seems to indicate that the CF_3 group at C-10 effectively protects artemisinin derivatives against oxidation or hydrolytic metabolism (*20, 21*).

This chapter describes our studies on the design of orally available antimalarial non-acetal artemisinin derivatives functionalized at C-16 and protected from metabolism by a fluoroalkyl substituent at C-10.

We thought that the potent antimalarial 10-*R*-(trifluoromethyl) dihydroartemisinin **6** could be a good precursor of new fluorinated antimalarial derivatives bearing a functionality at C-16, using the following retrosynthetic scheme (Figure 3). The allyl bromide **9** was *a priori* a valuable key-intermediate, since it could also be used for the design of many other interesting artemisinin derivatives with other objectives, such as dual molecules, as exemplified in this chapter by the preparation of artemisinin-mefloquin chimera, or dimers of artemisinin which are potential anti-tumor compounds.

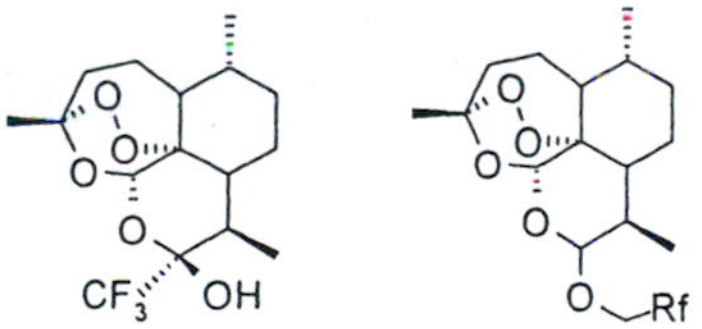

Figure 2

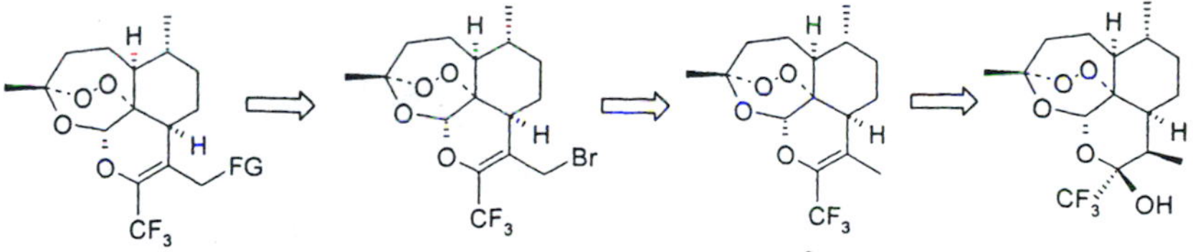

Figure 3

Preparation of the allyl bromide 9

1. Preparation of the glycal 8:

The standard methods (*22, 23*) using P_2O_5 or $BF_3;Et_2O$ to prepare the non-fluorinated parent compound **10** from DHA failed from the hemiketal **6**. This is due to the great stability of α-CF_3-substituted alcohols under acidic conditions: the electron-withdrawing character of the trifluoromethyl group disfavors a positive charge development on the α-carbon (*24*). However, the glycal **8** has been selectively prepared in good yields (76 %) by treatment of acetal **6** with thionyl chloride and pyridine in large excess (*25*). These conditions allowed to govern the reaction towards the elimination process to the detriment of the intramolecular substitution leading to the 10-chloro-10-trifluoromethyl-10-deoxoartemisinin (Figure 4).

At this stage, we checked if the "low-down metabolism by fluorine substitution" concept was validated in the case of glycal **8**. The antimalarial activity of glycal **8** has thus been evaluated and compared to that of the non-fluorinated glycal **10**. Whereas the *in vitro* activities (IC_{50}) on *P. falciparum* are similar (6 and 20 nM respectively), their *in vivo* activities are remarkably different. With glycal **10** there is no clearance of parasitemia at the end of the *i.p.* treatment (35.5 $\mu mol.kg^{-1}$ daily for four days according to the Peters' test) of mice infected with *P. berghei NK 173*, and there were no survival at day 10. With **8**, all mice survived until day 20 (*26*). This very clear validation of the effect of the CF_3 substituent at C-10 comforted us in the use of compounds **8** and **9** as precursors for the preparation of functionalized derivatives at C-16.

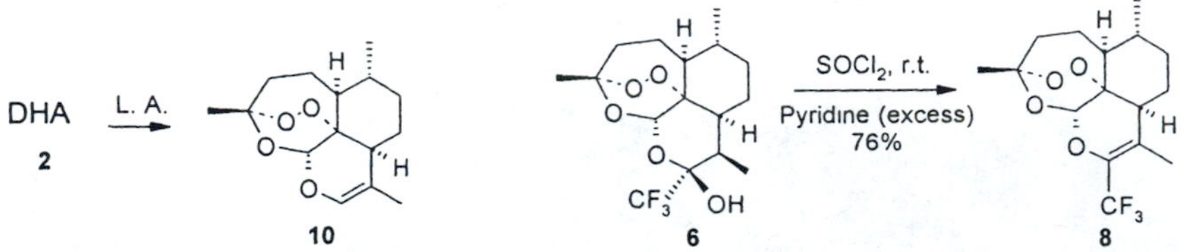

DHA **2** → (L. A.) **10**; **6** → ($SOCl_2$, r.t., Pyridine (excess), 76%) **8**

Figure 4. Preparation of 10-CF_3-10-deoxoartemisinin

2. Preparation of the allyl bromide 9:

Ionic bromination of non-fluorinated glycal **10** providing dibromides (*27, 28*) and bromohydrins (*28, 29*) was well documented. However, surprisingly the allylic radical bromination was not reported, although the bromide **11** could obviously provide a shorter route to 16-substituted derivatives than described approaches from artemisitene (*12, 30*) or from artemisinic acid (*31*).

Allylic bromination of glycals **10** and **8** was thus investigated under the usual *Wohl-Ziegler* conditions (NBS/CCl_4/reflux) in the presence of an initiator (AIBN) (*32*). Accurate conditions had to be found for the preparation of the non-fluorinated allyl bromide **11**. Although yield was high (> 70 %), **11** had to be used without isolation because of its instability (*33*). The parent 10-CF_3-16-bromo derivative **9** was obtained in good yield (72 %) from the trifluoromethyl glycal **8**, and appeared to be much more stable than bromide **11** (Figure 5). It could be purified by crystallization and stored several months at 0°C. Clearly the electron withdrawing character of the CF_3 group makes the allyl bromine less labile.

A deeper investigation of this reaction with glycal **8** revealed that the presence of the initiator (AIBN) was not necessary. Moreover the yield was improved to 90 % when the reaction was performed without AIBN. More surprisingly, the allylic bromination also occurred when the reaction was performed with Br_2 without initiator. The dibromoadduct, resulting from an electrophilic addition of Br_2 on the double bond, was never observed. This result was unexpected since i) stable dibromoadducts resulted from the treatment with Br_2 of the non fluorinated glycal **10** (*23*, *24*) and ii) dibromides **14** are also obtained from the trifluoromethyl parent enol ethers **13** (*34*). The easiness of the allylic bromination of **8** with Br_2, even in absence of AIBN, was attributed to the endoperoxide group which can itself initiate the radical reaction (36).

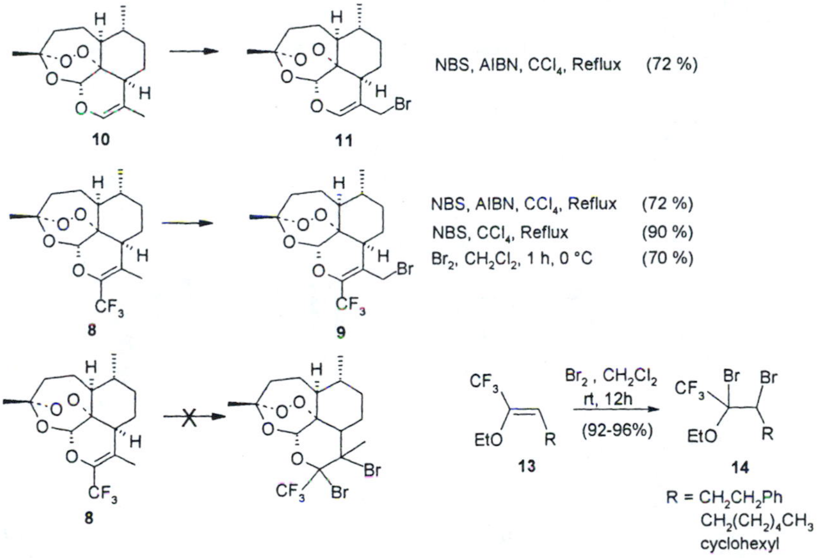

*Figure 5. Bromination reactions of glycal **10** and **8** and enol ethers **13**.*

Reactions of the allyl bromide 9 with nucleophiles

Having in hand the trifluoromethyl allyl bromide **9**, we investigated the substitution with various nucleophiles. Bromide **9** reacted selectively with amines and alcoholates, leading only to 16-substituted compounds **15** (S_N2-type products), with no trace of the allylic rearrangement product ($S_N'2$-type products) **16** (*26*, *33*). Excellent yields were obtained with secondary amines

(Table 1). With primary amines, polyalkylation could be avoided using a syringe pump and dilute conditions. Preparation of the primary amine at C-16 was more problematic. Initially we envisaged to prepare by reduction of the azide **15f**. However, amongst the few reductive conditions compatible with the endoperoxide bridge (*37*), none allowed the access to the primary amine **15g**. This latter was finally prepared using NH_3 as nucleophile, in a very large excess. Sodium alkoxides, also reacted well to afford corresponding ethers in good to excellent yields, but addition of a catalytic amount of KI was required (Table 2). With ethylene glycol, the formation of a small amount of diether (9 %) was always observed regardless of the experimental conditions. The hydrolysis of bromine **9** using KOH in DMSO provided the alcohol **15n** in a moderate yield. A better yield was obtained by performing the reaction with sodium acetate, followed by hydrolysis of the resulting 16-acetoxy **15m** (69 % overall yield) with MeOH/NaOMe. Dimethyl sodium malonate reacted with bromide **9** leading to the diester **15o** in excellent yield (90 %).

Table 1 : Reactions of bromide **9** with *N*-nucleophiles.

Nu = Amines, Alcoholates, Malonate, etc...

N-Nucleophile[a]	Equiv	Time (h)	15	Yield[b]
Morpholine	4	6	**a**	90%
Piperazine ethanol	4	6	**b**	87%
$EtNH_2$	10	4	**c**	85%
$MeNH_2$	10	3	**d**	98%
$NH_2CH_2CH_2NH_2$	10	3	**e**	95%
NaN_3[c]	1.5	1	**f**	95%
NH_3[d]		4	**g**	77%

[a] Reactions performed in THF. [b] Isolated yield. [c] Reaction performed in DMSO.
[d] Reaction performed at -15°C, large excess of NH_3 in a mixture NH_3/THF 1:1

The preparation of the allyl bromide **9** and its substitution constitute thus a very successful and useful route for an access to a large scope of 16-functionalized compounds whose solubility properties could improve oral antimalarial treatment. (*vide infra*). We were also highly tempted to reverse the SN2 selectivity of the substitution of bromide **9** in favour of the SN' substitution, in view to have an access to SN' products, the 10-trifluoromethyl artemisitene-like derivatives. Taking into account that the presence of the bulky CF_3 group with a negatively charged surface prevents from the approach of strong nucleophiles such as amines or sodium alcoholates, we anticipated that softer or

Table 2 : Reactions of bromide **9** with *O*- and *C*-nucleophiles.

Nucleophile[a]	equiv	NaH (equiv)	Solvent	Time (h)	**15**	Yield[b]
EtOH	20	3	THF	18	**h**	98%
BnOH	1.5	3	THF	18	**i**	97%
$MeOCH_2CH_2OH$	3	2	DMSO	1	**j**	96%
CH_2=CH-CH_2OH	3	2.5	DMSO	2	**k**	81%
$HOCH_2CH_2OH$	4	1.5	DMSO	2	**l**	69%
NaOAc[c]	1.5		DMF	18	**m**	82%
OH					**n**	69%
$CH_2(COOMe)_2$	1.5	1.8	THF	5	**o**	90 %

[a] in presence of KI (0.1 equiv.). [b]Isolated yield. [c]in presence of KI (0.2 equiv.).

non-charged nucleophile could be more prone to react at C-10. Furthermore a poor nucleophile would disfavor an SN2 process, while an SN1 type process should be governed by the relative stability of the C-16 primary carbenium ion and the oxonium ion.

Consequently, we investigated the solvolysis of bromide **9** in MeOH. Reaction required reflux (2 h) for a complete conversion, providing an 87:13 ratio of SN' (**16a**) and SN (**15a**) products (92 % yield) (*38*). The presence of a silver salt (triflate, acetate) did not significantly changed the 85:15 ratio of **16/15**. The reaction was extended to other alcohols: in all cases the S_N' products **16** were obtained as major products, even with the hindered isopropanol, albeit in a lower ratio (70:30) (Table 3).

Reactions are stereoselective, with only the β-alkoxy isomer obtained. Bromide **9** was selectively hydrolysed, with water alone, into the hemiketal **16n**, with no trace of alcohol **15n**. In this case, both α and β hemiketals were kinetically formed. However the β-hydroxy ketal slowly isomerised to give the more thermodynamically stable α-hydroxy isomer. It is interesting to remark that in saturated series, where the Me-9 is β, the more stable isomer is the β-hydroxy compound **6**. In contrast the repulsive interaction between Me-9 and CF_3 does not exist anymore in compounds **16**.

Most of the 16-functionalized fluoroartemisinins exhibited strong antimalarial activity *in vitro* and *in vivo*, often associated with an enhanced water solubility (*26*). At this time, the more promising candidates for an oral treatment are the amines **15a-c**, more active than sodium artesunate (*26, 39*).

Preparation of chimeras.

Taking advantage of the easy access to 16-functionalized CF_3-artemisinin derivatives via the allyl bromide **9**, we planned to synthesize two types of CF_3-artemisinin–mefloquine dual derivatives. The World Health Organization (WHO) has strongly recommended the use of artemisinin derivatives in

Table 3. Substitution of bromide **9**

9 → (ROH) → 16 + 15

(For R : H, β-CF_3, α-OH)

ROH		Ratio **16/15**	Yields % (**16** + **15**)
MeOH	**a**	87:13	92
EtOH	**p**	85:15	86
CH_2=CH-CH_2OH	**k**	85:15	80
HC≡C-CH_2OH	**q**	88:12	97
CH_3-CHOH-CH_3	**r**	70:30	91
$(CH_3)_3$-C-OH	**s**		No reaction
CF_3-CH_2OH	**t**	80:20	52 %
H_2O	**n**	100:0	90 %

combination with other antimalarial (ACT) for the treatment of malaria (*40*). The combination most prescribed in areas of highest parasite resistance is artesunate + mefloquine (*41*). We expected that a "double-drug" or chimera, in which the two active principles, a fluorinated artemisinin derivative and mefloquine, are covalently bound, would reduce the risk of drug resistance by the mutual protection of each moiety and would be easy to use. This "covalent bi-therapy" approach seems to be promising for the treatment of malaria (*42*, *43*).

We investigated the synthesis of two types of CF_3-artemisinin–mefloquine dual antimalarials (*44*). In the first case, the artemisinin moiety is linked to mefloquine by a covalent, indivisible bond. In the second case, the fluoro artemisinin and the mefloquine are bound *via* a diester linker, which should be hydrolysed by esterase(s) to allow the liberation *in vivo* of both active drugs.

The synthesis of the indivisible chimera **17** was based on the substitution of the 10-CF_3 allylic bromide **9** by the piperidinyl moiety of mefloquine (Figure 6).

The preparation of the divisible chimera **18** involved an esterification reaction between the acid **19** and the hydroxyether artemisinin derivative **15l**, the antimalarial activity of which being already proved (*26*) (Figure 7). The acid **19** was prepared by the reaction between the N-Boc mefloquine with succinic anhydride. After deprotection, the chimera **18** was isolated in good yield.

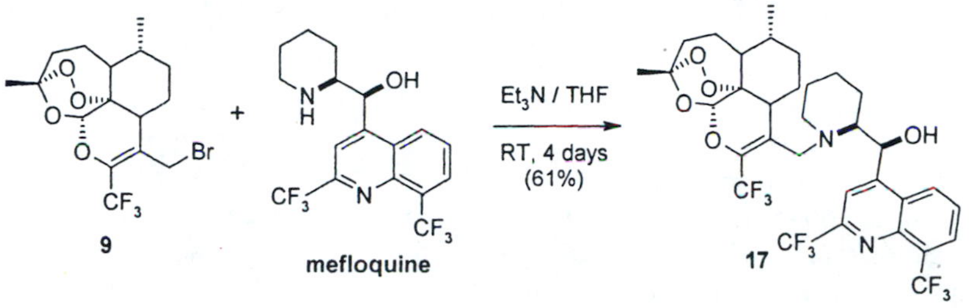

9 + mefloquine

Et_3N / THF

RT, 4 days (61%)

17

Figure 6

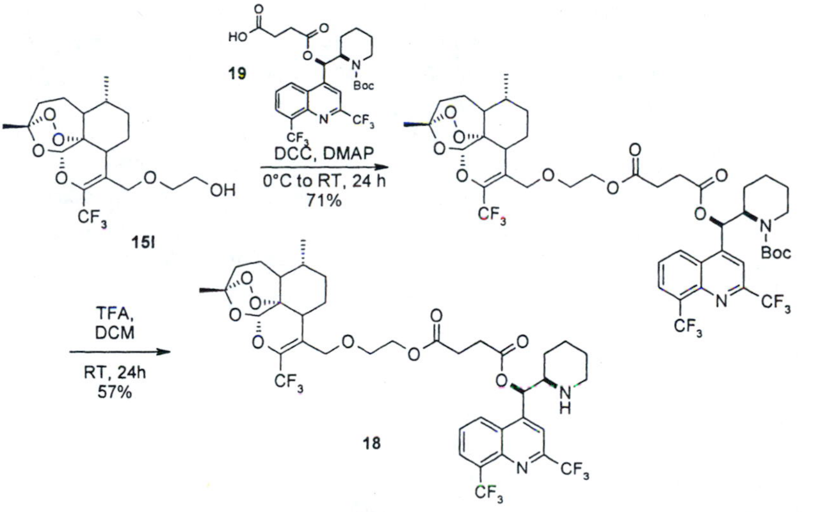

15l

19

DCC, DMAP

0°C to RT, 24 h

71%

TFA, DCM

RT, 24h

57%

18

Figure 7

In vitro, compounds **17** and **18** (as mixture of diastereomers) showed excellent efficacy (IC_{50} values ranging from 2 to 17 nm) against four different strains of *P. falciparum*, which possess different degrees of resistance to mefloquine and chloroquine. *In vivo*, chimera **18** and, to a lesser extent, chimera **17** were highly active, more efficient than the reference drug, artemether (*44*).

Preparation of dimers of fluoroartemisinin as potential antitumor compounds.

Besides their high antimalarial activities, artemisinin derivatives are also toxic against tumor cells, the most potent being their corresponding dimers.[45] Most of them are dimers at C-10 and are derived from ethers of dihydroartemisinin (*45*, *46*) or from more metabolically robust nonacetal analogues (*47*). Their preparation generally involves the addition of various kinds of linkers on the corresponding oxonium salt. A few others are dimers at C-16 and are prepared from the naturally occurring artemisitene, via a Michael reaction (*14*, *48*).

We investigated the olefin cross-metathesis reaction to prepare structurally more diversified dimers of artemisinin derivatives (*49*). The most important question raised was the unknown tolerance of the peroxide function toward ruthenium catalysts required for metathesis reaction. The endoperoxide bridge, which is essential for the biological activity, is known to be sensitive to the reductive action of various metals, not only iron involved in the mechanism of action, but also Mn, Pd, and Cu (*50*). With few other ones, such as Zn, the sensitivity of the endoperoxide depends on the reaction conditions (*50, 47c*).

CF_3-artemisinin derivatives functionalized at C-16 could be excellent precursors of these new dimers. Dimerization of the 10-CF_3-16-allylic ether **15k** was examined first. Treatment of **15k** with the ruthenium catalyst **20** (*51*) (5 or 10%), at r.t., led to the homodimer **22** with 80% conversion (Table 3). Since C-10-fluorinated artemisinin derivatives are much more stable than the non fluorinated ones, the reaction could be performed at reflux of CH_2Cl_2. Under these conditions, in the presence of only 5% of catalyst **20**, a complete conversion of the starting ether **15k** was obtained, within 1.5 h instead of 6 h.

Dimer **22** was isolated in good yield (71%) as a 90:10 mixture of *E* and *Z* isomers. The *E*/*Z* selectivity could be improved by performing the metathesis reaction with the second-generation catalyst **21**(*52*). Treatment of the allylic ether **15k** with 10% of catalyst **21** at r.t. led to the homodimer **22** (90% of conversion), obtained as a single *E* isomer (Table 4).

We next undertook the synthesis of artemisinin derivative dimers containing a free alcohol function. The previously reported allylic alcohol **15n** was oxidized with MnO_2 into the corresponding aldehyde. This latter then reacted with allyl bromide in the presence of zinc under Barbier's conditions (*53*), providing the alcohol **23**. It is worth noting that the peroxide bridge was not sensitive to Zn under these reaction conditions. Alcohol **23** was isolated in high yield (86%) as a 95:5 mixture of two diastereoisomers at C-16 (Figure 8).

Table 4. Self-cross-metathesis reaction of the allylic ether **15k** with Ru catalysts:

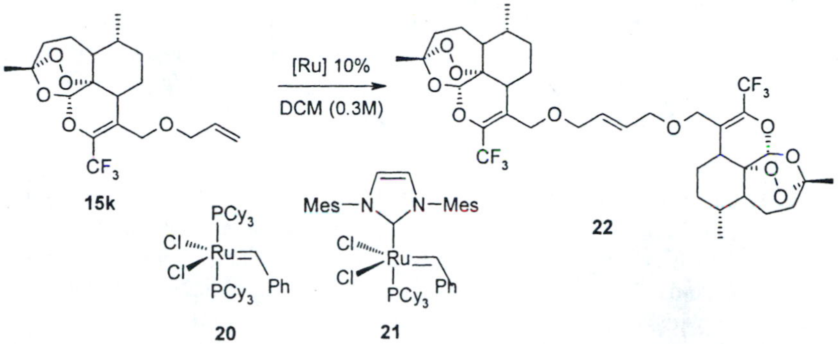

Catalyst (%)	T (°C)	Time	% Conversion **15k**	**22** yield % (% ratio *E/Z*)
20 (10)	rt	4[a]	80	(90:10)[b]
20 (5)	Reflux	1.5	100	71 (90:10)
21 (10)	rt	24	90	69 (100:0)

[a] A longer reaction time did not improve the percentage of conversion. [b] **20** was not isolated.

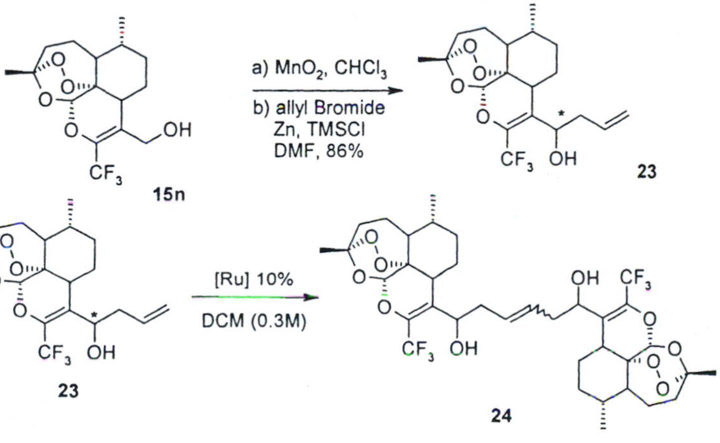

*Figure 8. Preparation and self-cross-metathesis reaction of alcohol **23**.*

With a catalytic amount (10%) of catalyst **21**, this homoallylic alcohol was converted at r.t. into the corresponding homodimer **24**, isolated in 50% yield as a mixture of *E* and *Z* isomers (>90:10 ratio).

Preliminary growth inhibitory activities were evaluated *in vitro* at the National Cancer Institute (NCI) using a diverse panel of 60 human cancer cell lines (*54*). Dimer **22** was efficient in cancer cell growth inhibition with a GI_{50} (*55*) around 10 nM in many cases, and with a selectivity against a few cancer cell lines (*e.g.*, leukaemia HL-60, nonsmall cell lung cancer NCI-H226, colon cancer COLO 205, and KM-12, CNS cancer SF-295).

In conclusion, new metabolically stable antimalarial fluorinated derivatives of artemisinin, bearing a CF_3 group at C-10 and diversified functionalities at C-16 could be easily obtained from the allyl bromide **9**. This latter is prepared in large scale from artemisinin and is a key precursor of various new derivatives. It could react with a wide range of nucleophiles (amines, alcoholates...) under basic conditions to afford C-16 substituted compounds, or, under neutral conditions, to provide functional C10-CF_3-artemisinin derivatives, in an S_N' process. Some of these novel CF_3-artemisinin derivatives exhibited an excellent antimalarial activity *in vivo* in comparison with the non fluorinated artemisinin derivatives, with better metabolic stability and solubility properties. The allyl bromide **9** also allowed the access to dual antimalarials, as exemplified with the chimera mefloquine-fluoroartemisinin. Homodimers, studied for their cytotoxic properties, could be prepared using a self cross-metathesis reaction with ruthenium catalyst, without alteration of the endoperoxide bridge.

Acknowledgments

We would like to thank the MENRT, VIH-PAL, PAL+ programs and CNRS for financial support and fellowship.

References:

1 Klayman, D. L. *Science* **1985**, *228*, 1049-1055.
2 Wu, Y. L.; Li, Y. *Med. Chem. Res.* **1995**, *5*, 599-586.
3 a) Li, Y.; Yu, P. L.; Chen, Y. X.; Li, L. Q.; Gai, Y. Z.; Wang, D. S.; Zheng, Y. P. *Acta Pharm. Sin.* 1981, *16*, 429-439. b) Wang, T.; Xu, R. *J. Tradit. Chin. Med.* **1985**, *5*, 240-242.
4 Brossi, A.; Venugopalan, B.; Gerpe, L.; Yeh, H. J. C.; Flippen- Anderson, J. L.; Buchs, P.; Luo, X. D.; Milhous, W.; Peters, W. *J. Med. Chem.* **1988**, *31*, 645-650. *TDR News*, **2000**, June (No. 62).
5 China Cooperative Research Group on Qinghaosu and Its Derivatives as Antimalarials. *J. Tradit. Chin. Med.* **1982**, *2*, 9-16.
6 Hien, T. T.; White, N. J. *Lancet* **1993**, *341*, 603- 608.
7 Lin, A. J. L.; Klayman, D. L.; Milhous, W. K. *J. Med. Chem.* **1987**, *30*, 2147-2150.

8 a) Leskovac, V.; Theoharides, A. D. *Comp. Biochem. Physiol.* **1991**, *C99*, 383-390. b) Leskovac, V.; Theoharides, A. D. *Comp. Biochem. Physiol.* **1991**, *C99*, 391-396. c) Baker, J. K.; McChesney, J. D.; Chi, H. T. *Pharm. Res.* **1993**, *10*, 662-666. d) Grace, J. M.; Aguilar, A. J.; Trotman, K. M.; Brewer, T. G. *Drug Metab. Dispos.* **1998**, *26*, 313-317. e) Navaratnam, V.; Mansor, S. M.; Sit, N. W.; Grace, J.; Li, Q. G.; Olliaro, P. *Drug Dispos.* **2000**, *39*, 255-570.

9 Meshnick, S. R.; Taylor, T. E.; Kamchonwongpaisan, S. *Microbiol. Rev.* **1996**, *60*, 301-315.

10 El-Feraly, F.S.; Ayalp, A.; Al-Yahya, M.A.; McPhail, D.R.; McPhail, A.J. *J. Nat. Prod.* **1990**, *53*, 66-71.

11 Pitayatat, S.; Tarnchompoo, B.; Thebtaranonth, Y.; Yuthavong, Y. *J. Med. Chem.* **1997**, *40*, 633-638.

12 Ma, J.; Weiss, E.; Kyle, D. E.; Ziffer, H. *Bioorg. Med. Chem. Lett.* **2000**, *10*, 1601-1603.

13 Liao, X. B.; Han, J. Y.; Li, Y. *Tetrahedron Lett.* **2000**, *42*, 2843-2845.

14 Ekthawatchai, S.; Kamchonwongpaisan, S.; Kongsaeree, P.; Tarnchompoo, B.; Thebtaranonth, Y.; Yuthavong, Y. *J. Med. Chem.* **2001**, *44*, 4688-4695.

15 Avery, M. A.; Alvim-Gaston, M.; Vroman, J. A.; Wu, B.; Ager, A.; Peters, W.; Robinson, B. L.; Charman, W. *J. Med. Chem.* **2002**, *45*, 4321-4335.

16 a) Bégué, J. P.; Bonnet-Delpon, D. *Chimie Bioorganique et Médicinale du Fluor*, EDP-Sciences/CNRS Edition, Paris, **2005**. b) Bégué, J. P.; Bonnet-Delpon, D. *ChemistryToday*, **2005**, *23,* 1-3.

17 a) Irurre, J., Jr.; Casas, J.; Ramos, I.; Messeguer, A. *Bioorg. Med. Chem.* **1993**, *1*, 219-225. b) Irurre, J., Jr.; Casals, J.; Messeguer, A. *Bioorg. Med. Chem. Lett.* **1993**, *3*, 179-182. c) Edwards, P. N. Uses of Fluorine in Chemotherapy. In *Organofluorine Chemistry: Principles and Commercial Applications*; Banks, R. E., Smart, B. E., Tatlow, J. C., Eds.; Plenum Press: New York, **1994**; pp 501-541.

18 a) Abouabdellah, A.; Bégué, J. P.; Bonnet-Delpon, D.; Gantier, J. C.; Truong Thi Thanh, N.; Truong Dinh, T. *Bioorg. Biomed. Chem. Lett.* **1996**, *6*, 2717-2720. b) Nga, T. T.; Ménage, C.; Bégué, J. P.; Bonnet-Delpon, D.; Gantier, J. C.; Pradines, B.; Doury, J. C.; Thac, T. D. *J. Med. Chem.* **1998**, *41*, 4101-4108. c) Binh, P. D.; Le Dinh Cong; Nhan, D. H.; Nhu, T. V.; Tien, N. T.; Bégué, J. P.; Bonnet-Delpon, D.; Truong Thi Thanh, Nga. *Trans. R. Soc. Trop. Med. Hyg.* **2002**, *96*, 677-683.

19 a) Pu, Y. M.; Torok, D. S.; Ziffer, H.; Pan, X.-Q.; Meshnick, S. R. *J. Med. Chem.* **1995**, *20*, 4120-4124. b) Luo, X. D.; Shen, C. C. *Med. Res. Rev.* **1987**, *7*, 29-52.

20 Pharmacokinetic studies on 4 and DHA have been performed at Monash University: Charman, W. N.; McIntosh, L.

21 Magueur, G.; Crousse, B.; Charneau, S.; Grellier, P.; Bégué, J.P.; Bonnet-Delpon, D. *J. Med. Chem.* **2004**, *47*, 2694-2699.

22 Lin, A. J.; Lee, M.; Klayman, D. L. *J. Med. Chem.* **1989**, *32*, 1249-1252.

23 Lin, A.J.; Li, L.; Klayman, D.L.; George, C.F.; Flippen-Anderson, J.L.T. *J. Med. Chem.* **1990**, *33*, 2610-2614.

24 a) Abouabdellah, A.; Aubert, C.; Bégué, J.P.; Bonnet-Delpon, D.; Lequeux, T. *J. Org. Chem.* 1991, 56, 5800. b) Abouabdellah, A.; Aubert, C.; Bégué, J.P.; Bonnet-Delpon, D.; Guilhem, J. *J. Chem. Soc. Perkin Trans. 1* **1991**, 1397-1403.
25 Grellepois, F.; Chorki, F.; Crousse, B.; Ourévitch, M.; Bonnet-Delpon, D.; Bégué, J.P. *J. Org. Chem.* **2002**, *67*, 1253-1260
26 Grellepois, F.; Chorki, F.; Ourévitch, M.; Charneau, S.; Grellier, P.; McIntosh, K. A.; Charman, W. N.; Pradines, B.; Crousse, B.;; Bégué, J. P. *J. Med. Chem.* **2004**, *47*, 1423-1433.
27 Grellepois, F.; Bonnet-Delpon, D.; Bégué, J.P. *Tetrahedron Lett.* **2001**, *42*, 2125-2127.
28 Venugopalan, B.; Bapat, C.P.; Karnik, P.J.; Lal, B.; Chatterjee, D.K.; Iyer, S.N.; Blumbach, J. *Eur. Pat. EP.* **1991**, 456, 149; *Chem. Abstr.* **1992**, *116*, 83708z.
29 Venugopalan, B.; Bapat, C.P.; Karnik, P.J. *Bioorg. Med. Chem. Lett.* **1994**, *4*, 750-752.
30 a) Ekthawatchai, S.; Kamchonwongpaisan, S.; Kongsaeree, P.; Tarnchompoo, B.; Thebtaranonth, Y.; Yuthavong, Y. *J. Med. Chem.* **2001**, *44*, 4688-4695. b) Liao, X.B.; Han, J.Y.; Li, Y. *Tetrahedron Lett.* **2000**, *42*, 2843-2845. c) Ma, J.; Katz, E.; Kyle, D.E.; Ziffer, H. *J. Med. Chem.* **2000**, *43*, 4228-4232.
31 a) Jung, M.A.; Lee, K.; Jung, H. *Tetrahedron Lett.* **2001**, *42*, 3997-4000. b) R. Haynes *Current Topic Med. Chem.* **2006**, in press.
32 Djerassi, C. *Chem. Rev.* **1948**, *43*, 271-317.
33 Grellepois, F.; Chorki, F.; Ourévitch, M.; Crousse, B.; Bonnet-Delpon, D.; Bégué, J.P. *Tetrahedron Lett.* **2002**, *43* 7837-7840.
34 Bonnet-Delpon, D.; Bouvet, D.; Ourévitch, M.; Rock, M. *Synthesis* **1998**, 288-292.
35 Liu, J.M.; Ni, M.Y.; Fan, Y.F.; Tu, Y.Y.; Wu, Z.H.; Wu, Y.L.; Zhou, W.S. *Acta. Chim. Sinica* **1979**, *37*, 129-141.
36 a) Marquet, A.; Dvolaitzky, M.; Kagan, H.B.; Mamlok, L.; Ouannes, C.; Jacques, J. *Bull. Soc. Chim. Fr.* **1961**, 1823-1831. b) Gaudry, M.; Marquet, A. *Bull. Soc. Chim. Fr.* **1969**, 4169-4178.
37 Scriven, E.F.V.; Turnbull, K. *Chem. Rev.* **1988**, *88*, 297-368.
38 Constance Chollet, Benoit Crousse, Michèle Ourévitch, Danièle Bonnet-Delpon, *J. Org. Chem.* **2006**, *71*, 3082-3085.
39 Bégué, J. P. ; Bonnet-Delpon, D. *Drugs of the Future* **2005**, *30 (3)*, 509-518.
40 WHO. Repot of a WHO Technical Consultation. World Health Organization, Geneva, April **2001**, WHO/CDS/RBM/2001.35.
41 Nosten, F.; Luxemburger, C.; Ter Kuile, F.O.; Woodrow, C.; Eh, J.P.; Chongsuphajaisiddhi, T.; White, N.J.; *J. Infect. Dis.* **1994**, *170 (4)*, 971-977.
42 a) Dechy-Cabaret, O.; Benoit-Vidal, F.; Robert, A.; Meunier, B. *ChemBio Chem* **2000**, *1*, 281-283 b) Robert, A.; Dechy-Cabaret, O.; Cazelles J.; Meunier, B. *Acc. Chem. Res.* **2 002**, *35*, 167-174; c) Dechy-Cabaret, O.; Benoit-Vical, F.; Loup, C.; Robert, A; Gornitzka, H.; Bonhoure, A.; Vial, H. Magnaval, J. F.; Seguela, J. P.; Meunier, B. *Chem. Eur. J.* **2004**, *10*, 1625-1636

43 Romeo, S.; Dell'Agli, M.; Parapini, S.; Rizzi, L.; Galli, G.; Mondani, M.; Sparatore, A.; Taramelli, D.; Bosisio, E. *Bioorg. Med. Chem. Lett.* **2004**, *14*, 2931-2934.

44 Fabienne Grellepois, Philippe Grellier, Danièle Bonnet-Delpon, Jean-Pierre Bégué. *Chem Bio Chem*. **2005,** *6*, 648-652

45 a) Woerdenbag, H. J.; Moskal, T. A.; Pras, N.; Malingré, T. M. *J. Nat. Prod.* **1993**, *56*, 849. b) Beekman, A. C.; Barentsen, A. R. W.; Woerdenbag, H. J.; Uden, W. V.; Pras, N. *J. Nat. Prod.* **1997**, *60*, 325. c) Jeyadevan, J. P.; Bray, P. G.; Chadwick, J.; Mercer, A. E.; Byrne, A.; Ward, S. A.; Park, B. K.; Williams D. P.; Cosstick, R.; Davies, J.; Higson, A. P.; Irving, E.; Posner, G. H.; O'Neill, P. M. *J. Med. Chem.* **2004**, *47*, 1290-1298. d) Posner, G. H.; McRiner, A. J.; Paik, I. H.; Sur, S.; Borstnik, K.; Xie, S.; Shapiro, T. A.; Alagbala, A.; Foster, B. *J. Med. Chem.* **2004**, *47*, 1299-1301.

46 Zheng, Q. H.; Darbie, L. G. PCT Int. Appl. WO 9701548, **1997**. d) Lai, H. E.; Singh, N. P. PCT Int. Appl. WO 003103588, **2003**.

47 a) Posner, G. H.; Ploypradith, P.; Parker, M. H.; O'Dowd, H.; Woo, S. H.; Northrop, J.; Krasavin, M.; Dolan, P.; Kensler, T. W.; Xie, S.; Shapiro, T. A. *J. Med. Chem.* **1999**, *42*, 4275-4280. b) Jung, M.; Lee, S.; Ham, J.; Lee, K.; Kim, H.; Kim, S. K. *J. Med. Chem.* **2003**, *46*, 987-994. c) Posner, G. H.; Park, I. H.; Sur, S.; McRiner, A. J.; Borstnik, K.; Xie, S.; Shapiro, T. A. *J. Med. Chem.* **2003**, *46*, 1060-1065.

48 For a review, see: Jung M.; Lee, K.; Kim, H.; Park, M. *Current Med. Chem.*, **2004**, *11*, 1265.

49 Grellepois, F.; Crousse, B.; Bonnet-Delpon, D.; Bégué, J. P. *Org. Lett.* **2005**, *7*, 5219-5222.

50 a) Liu, J. M.; Ni, M. Y.; Fan, Y. F.; Tu, Y. Y.; Wu, Z. H.; Wu, Y. L.; Zhou, W. S. *Acta Chim. Sinica* **1979**, *37*, 129. b) Robert, A.; Meunier, B. *J. Am. Chem. Soc.* **1997**, *119*, 5968-5969. c) Wu, Y. K.; Liu, H. H.; *Helv. Chim. Acta* **2003**, *86*, 3074-3080. d) Drew, M. G. B.; Metcalfe, J.; Ismail, F. M. D. *THEOCHEM* **2004**, *711*, 95-105.

51 Schwab, P.; Grubbs, R. H.; Ziller, J. W. *J. Am. Chem. Soc.* **1996**, *118*, 100-110.

52 Huang, J.; Stevens, E. D.; Nolan, S. P. *J. Am. Chem. Soc.* **1999**, *121*, 2674-2678.

53 Yamamoto, Y.; Asao, N. *Chem. Rev.* **1993**, *93*, 2207-2293.

54 Boyd, M. R.; Paull, K. D. *Drug Dev. Res.* **1995**, *64*, 91.

55 GI_{50} is the concentration of drug that inhibits percentage growth by 50%; TGI is the concentration of drug required to achieve total growth inhibition (i.e., cytostasis).

Chapter 21

Synthesis of Novel *gem*-Difluorinatedcyclopropane Hybrids: Applications for Material and Medicinal Sciences

Toshiyuki Itoh

Department of Materials Science, Faculty of Engineering, Tottori University, 4–101 Koyama Minami, Tottori 680–8552, Japan

Recent developments in application of gem-difluorocyclopropane derivatives are reviewed. They are useful tool for various types of biologically active compounds, for unique polymers, and liquid crystal compounds. Further, the results of interesting DNA cleavage property for 9-anthracenecarboxylic acid substituted gem-difluorocyclopropane derivatives that we recently discovered are reported.

Introduction

The cyclopropyl group is a structural element present in a wide range of naturally occurring biologically active compounds found in both plants and microorganisms (*1*). Since substitution of two fluorine atoms on the cyclopropane ring is expected to alter both chemical reactivity and biological activity due to the strong electron-withdrawing nature of fluorine, gem-difluorocyclopropanes are expected to display unique biological and physical properties and a great deal of attention has been paid to the chemistry and unique properties of gem-difluorocyclopropanes (*2-5*). We review here the chemistry of gem-difluorocyclopropanes that have unique biological activities or physical properties. In particular, we report the results of our attempt to use gem-difluorocyclopropane derivatives as novel DNA cleavage agents.

Boilogicall active gem-difluorocyclopropanes

Numerous types of gem-difluorocyclopropanes have been synthesized, however, examples of biologically active compounds are not so many (*6-14*). Examples of biologically active gem-difluorocyclopropanes are shown in Figure 1.

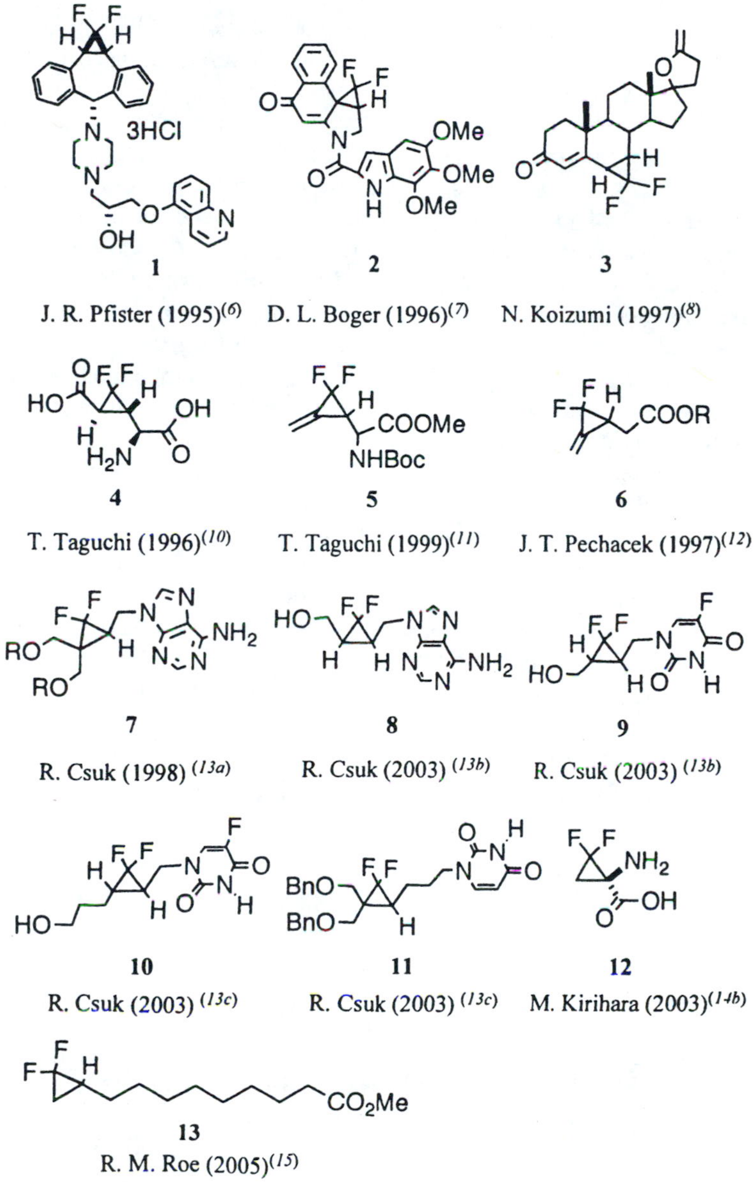

1 J. R. Pfister (1995)[6]

2 D. L. Boger (1996)[7]

3 N. Koizumi (1997)[8]

4 T. Taguchi (1996)[10]

5 T. Taguchi (1999)[11]

6 J. T. Pechacek (1997)[12]

7 R. Csuk (1998)[13a]

8 R. Csuk (2003)[13b]

9 R. Csuk (2003)[13b]

10 R. Csuk (2003)[13c]

11 R. Csuk (2003)[13c]

12 M. Kirihara (2003)[14b]

13 R. M. Roe (2005)[15]

Figure 1. Biology active gem-difluorocyclopropane compounds.

Pfister and co-workers reported that a gem-difluorocyclopropane derivative of dibensosuberabe **1** became a clinically useful modulator of P-glycoprotein-mediated multiple drug resistance (*6*). Bogers synthesized gem-difluorocyclopropane derivative **2** as a potent antitumor drug duocarmycin and showed that introduction of difluorocyclopropane substitution increased the activity drastically (*7*). Koizume accomplished the synthesis of gem-difluorocyclopropane derivative **3** (*8*).

Optically active multi-gem-difluorocyclopropanes are challenging targets for synthetic organic chemists. However, little attention has been paid to the chemistry of the synthesis of gem-difluorocyclopropane synthesis at early stage, and the first synthesis of optically active gem-difluorocyclopropanes were accomplished by the Taguchi group in 1994 (*9*). Taguchi group synthesized various types of gem-difluorocyclopropanes, such as metabotropic glutamante receptor agonist **4** (*10*) and methylene cyclopropane analogues **5** (*11*). Methylene-gem-difluorocyclopropane derivative **6** was also prepared by Pechacek and co-workers and they reported a significant inhibitory action of **6** towards Nematode induced root galling (*12*).

Csuk and co-workers reported the synthesis of various types of gem-difluorocyclopropane nucreoside, **7-11** (*13*); some compounds have been revealed to possess antitumor activity and anti-HIV activity. Kirihara established very good method for preparing optically active gem-difluorocyclopropane derivative of 1-aminocyclopropane-1-carboxylic acid **12** *via* chemo-enzymatic strategy (*14*). We developed a synthetic methodology for preparing optically active gem-difluorocyclopropanes using lipase technology (*5*). Although all of the reported compounds by Csuk's group are racemic ones, optically active ones might be prepared by the use of Kirihara's and our method.

Recently, an interesting biologically active gem-difluorocyclopropane has been reported by Roe; gem-difluorocyclopropane **13** acted as a competitive inhibitor of juvenile hormone-epoxide hydrolase inhibitor (*15*).

gem-Difluorocyclopropanes as Sources for Unique Functional Molecules

Since substitution of two fluorine atoms on the cyclopropane ring is expected to alter both chemical reactivity, it is expected that gem-difluorocyclopropane derivatives have unique physical properties. Several examples of such compounds are shown in Figure 2 (*16-19*).

Hillmyer's group developed novel polymers that have gem-difluorocyclopropane moiety **14-16** using the reaction of difluorocarben produced by the thermolysis of hexafluoropropylene oxide (*16*). Meijere and co-workers reported a unique gem-difluorocyclopropane derivative **17** in optically active form *via* chemo-enzymatic process and converted it to a novel ferroelectric liquid crystalline compound (*17*). We also accomplished the synthesis of novel optically active liquid crystal molecules **18** and **19** that possess a gem-difluorocyclopropane moiety (*18*). It should be noted that compound **19** exhibited an unidentified smectic phase below the SmA phase, while 18 showed only a SmA phase over a wide temperature range.

We recently accomplished the synthesis of pyrenyl ester of tetrakis-gem-difluorocyclopropane **20** and hoping that this may become a unique sensitive marking agent for DNA or an oligo peptide (*19*).

These examples clearly show that gem-difluorocyclopropane compounds become a useful tool for designing novel functional molecules.

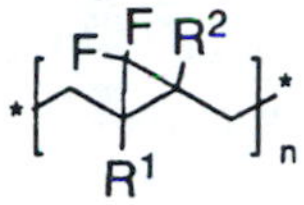

14 M. A. Hillmyer (1998)[16a]

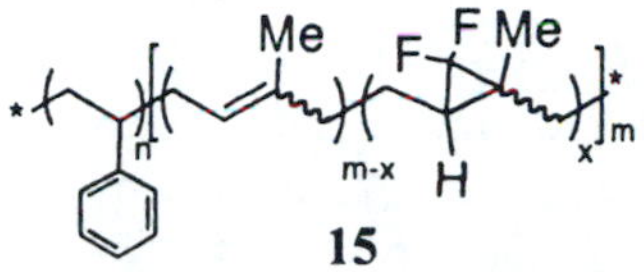

15 M. A. Hillmyer (2000)[16b]

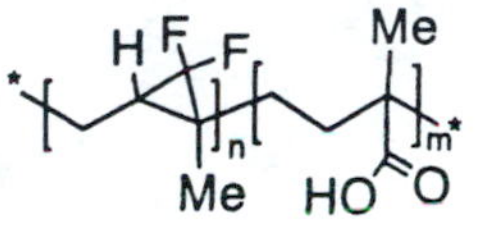

16 M. A. Hillmyer (2004)[16c]

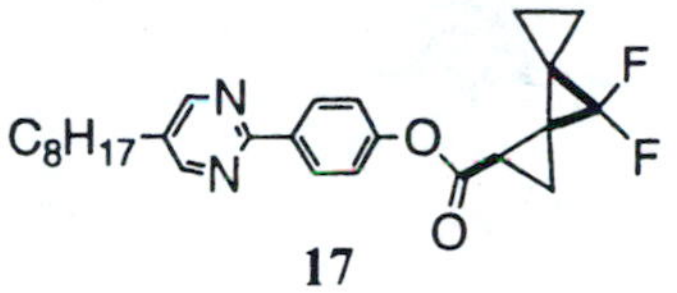

17 A de Meijere (2000)[17]

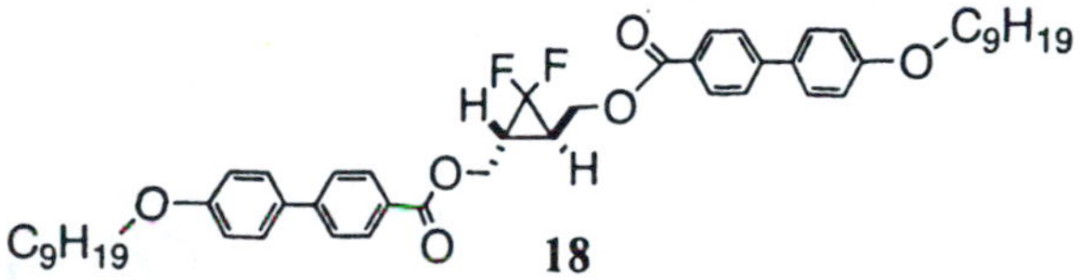

18

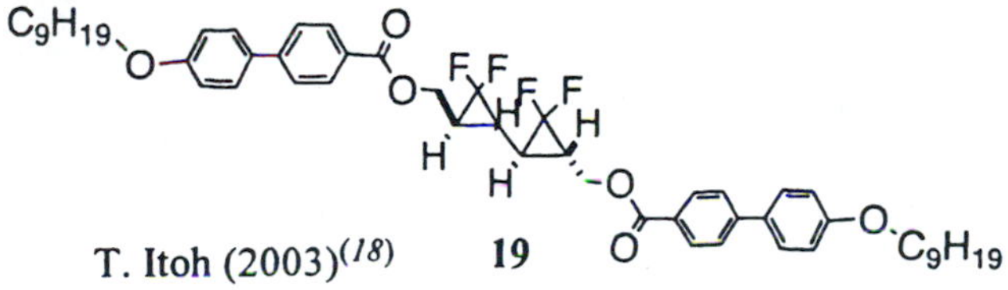

19 T. Itoh (2003)[18]

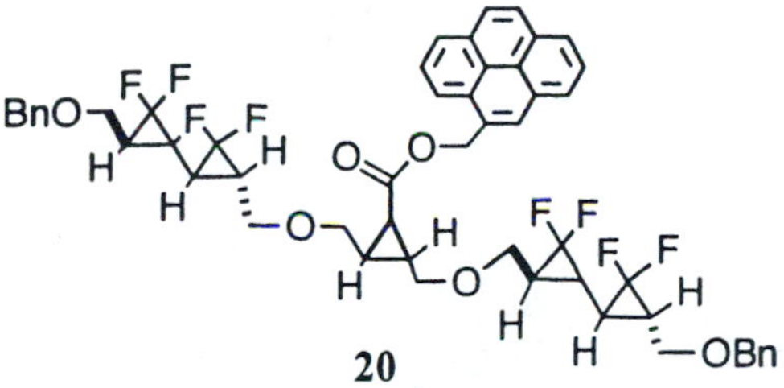

20 T. Itoh (2004)[19]

Figure 2. gem-Difluorocyclopropane compounds that show unique physical properties.

Anthracene-Cyclopropane Hybrids as DNA Cleavage Agents Switched by Photo Irradiation

During the course of our study to establish the stereochemistry of 1,3-bishydroxymethyl-2,2-difluorocyclopropane (**21**) using CD spectroscopic analysis of the corresponding 9-anthracencarboxylic diester **22**, we found the interesting facts that the esterification of **21** with 9-anthracencarboxylic acid chloride must be carried out in a flask with a blackout shield, and that the produced diester **22** was so unstable in dichloromethane solution under sunlight that it formed unidentified complex compounds by opening the cyclopropane ring (5f). We also found that bis-gem-difluorocyclopropane **23**(5c) and tetrakis-gem-difluorocyclopropane **24**(5c) were also rapidly decomposed by irradiation of sunlight in dichloromethane solution (Figure 3).

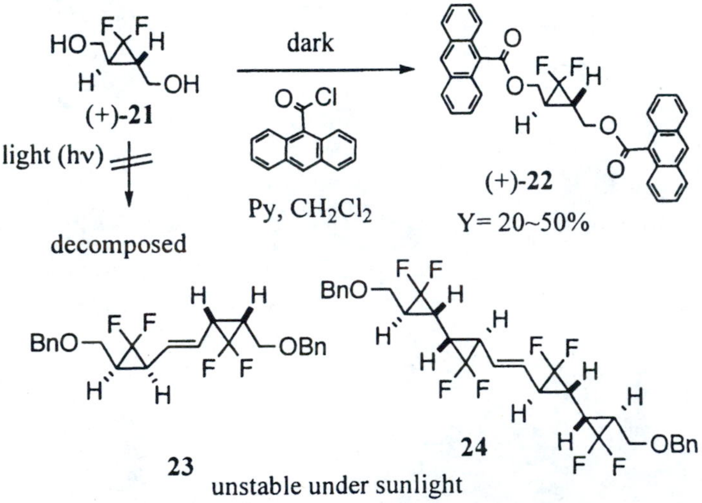

(+)-**21** —dark; Py, CH_2Cl_2→ (+)-**22** Y= 20~50%

light (hν) → decomposed

23 **24** unstable under sunlight

Figure 3. Unstable nature of gem-difluorocyclopropane under sun light.

Interest has been growing about DNA damage induced by various types of natural and unnatural compounds, in particular, the search for organic compounds that possess DNA cleavage property switched by photo irradiation (*20*). Anthryl or anthraquinone derivatives have been reported to be potent DNA cleavage agents: Schuster and co-workers reported anthraquinonecarboxylic amide caused GG selective cleavage of duplex DNA (*21*), and Kumar and co-workers reported anthracene substituted alkylamine derivatives showed potent DNA cleavage properties (*22*). Further, Toshima reported a quinoxaline-carbohydrate hybrid to be a GG-selective DNA cleaving agent with DNA cleaving and binding abilities which are dependent on the structure of the sugar moiety (*23*).

Form these results, we anticipated that our anthracene-gem-difluorocyclopropane compounds might display DNA cleavage activity by photo irradiation, because it was reported that decomposition of a gem-difluorocyclopropane ring proceeded via a radical intermediate (*24*), and it has been suggested that DNA damage was caused by radical species (*25*) (Figure 4). Hence, we prepared optically active gem-difluorocyclopropane derivatives **25-27** from (*S,S*)-**21** or (*R,R*)-**21** (Figure 5*)* and tested their DNA cleavage activities (*26*) (Figure 6).

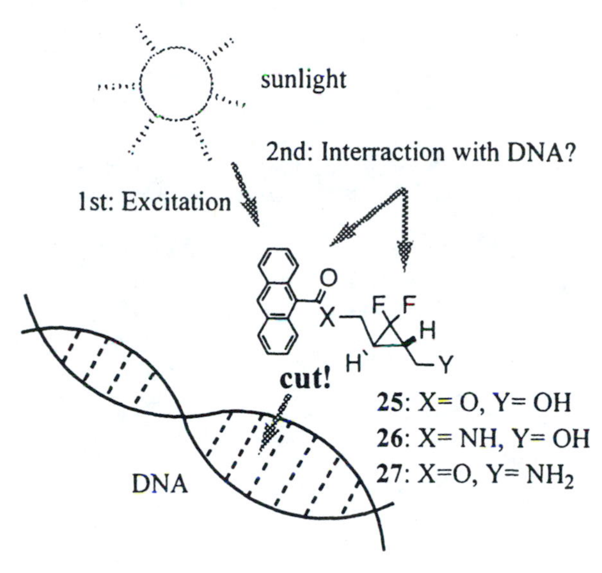

Figure 4. Working hypothesis of DNA cleavage by gem-difluorocyclopropane.

As shown in Figure 6, we found that (*S,S*)-**22** showed a weak DNA cleavage activity: 23% of supercoiled ϕX174 DNA (Form I) was converted to the relaxed form (Form II) by photo irradiation in the presence of 250 μM of (*S,S*)-**22** (Figure 6, lane II), though very weak DNA cleavage was also obtained for ethyl 9-anthracenecarboxylare (lane I). A slight increased activity was found for (*S,S*)-**25** (lane III), and, to our delight, very strong activity was observed for gem-difluorocyclopropane **26** (lane IV and V); DNA was cleaved by the photo irradiation and supercoiled ϕX174 DNA (Form I) was converted to the relaxed form (Form II) almost completely in the presence of 100 μM of (*S,S*)-**26**; this concentration corresponded to 1.3 equivalent versus the DNA base pair (Figure 6, lane V). However, no significant difference was found in the activity between enantiomers of (*R,R*)-**26** and (*S,S*)-**26** at a higher concentration of 250 μM. Further, nonfluorinated cyclopropane derivative (*S,S*)-**28** also caused potent DNA cleaving (lane VI).

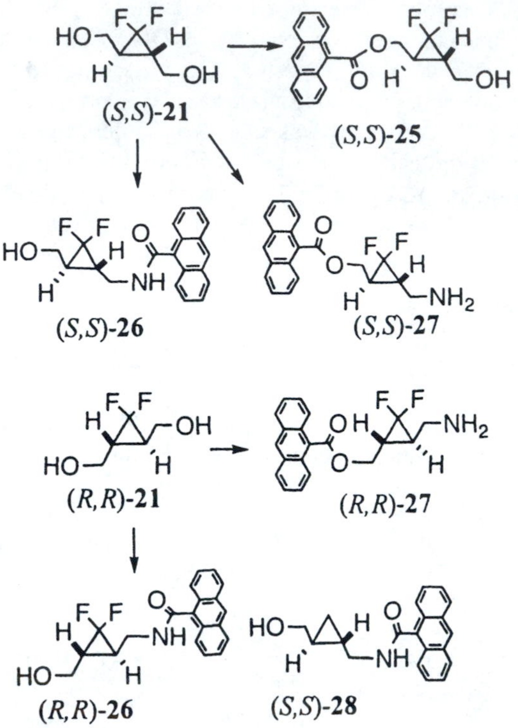

Figure 5. gem-Difluorocyclopropane derivatives.

We expected that enhanced DNA cleavage activity might be obtained by introduction of the terminal amino group in our compounds due to increased interaction of the anionic part of DNA. We prepared difluorocyclopropane analogues (*S,S*)-**27** and (*R,R*)-**27**, which possess the terminal amino group and tested DNA cleavage activity (Figure 7). The DNA cleaving test gave very interesting results: a clear contrast of the reactivity was found between (*R,R*)-**27** and (*S,S*)-**27**. Supercoiled ϕX174 DNA was converted to the corresponding relaxed form when photo irradiated in the presence of 85 μM of (*R,R*)-**27**, while 1,200 μM was required to show the same activity for the (*S,S*)-**27**; (*R,R*)-**27** showed more than 10 times stronger activity compare to the corresponding (S,S)-isomer.

DNA cleavage activity might be significantly influenced in the configurational fashion if it were caused mainly by intercalative or minor groove binding of the anthracenyl group with the DNA. Therefore, these results seem to suggest that the present DNA cleavage might be caused mainly by intercalative or minor groove binding of the anthracenyl group with the DNA.

The cyclopropane group may play a very important role in determining the binding fashion with DNA. We now hypothesize that the terminal amino group

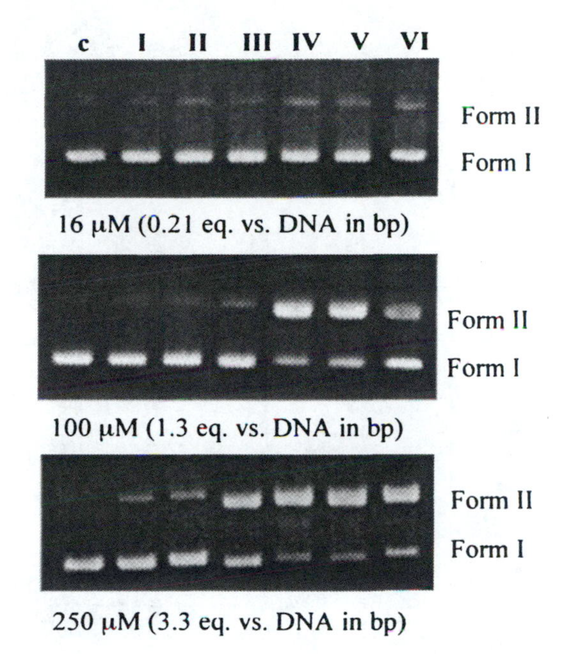

Figure 6. Photocleavage of supercoiled ϕX174 plasmid DNA by cyclopropane derivatives. lane ***c:*** *control,* ***I:*** *ethyl 9-anthracenecarboxylate* ***II:*** *(S,S)-**22**,* ***III:*** *(S, S)-**25**,* ***IV:*** *(R,R)-**26**,* ***V.*** *(S,S)-**26**,* ***VI:*** *(S,S)-**28** Reaction buffer: 20 mM sodium phosphate pH7.0 (20 % DMSO). Samples were irradiated at 25°C for 45 min at a distance of 6 cm using a Xe lamp with a polystyrene filter (365 nm). Electrophoresis (0.8% agarose gel, TAE buffer, 100 V, 40 min). Gel was stained by EtBr and visualized by UV-B lamp (transilluminator).*

may bind to the phosphate group of DNA, and then intercalative or minor groove binding of the anthracenyl group with the DNA sequence takes place. This is the first demonstration that chirality of a small cyclopropane derivative could control the DNA cleavage ability of anthracene hybrid compounds. We thus developed a useful tool molecule for designing a novel DNA cleavage agent, though the origin of the DNA cleavage property is not still unclear. Our cyclopropane compound **21** have two hydroxyl groups, thus making it easy to modify the structure by introducing DNA sequence recognition groups such as oligonucleotides and may be able to design an artificial restriction enzyme based on cyclopropanes.

Summary and Outlook

The substitution of two fluorine atoms on the cyclopropane ring can alter both its chemical reactivity and biological activity due to the strong electron-withdrawing nature of the fluorine, and this makes it possible to create new

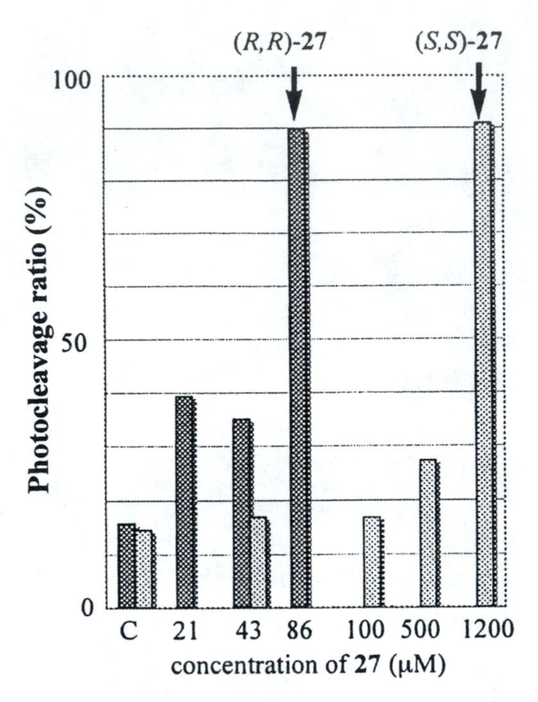

*Figure 7. Photocleavage of supercoiled ϕX174 plasmid DNA by 9-anthracenecarboxylate (R,R)-**27** and (S,S)-**27**. Reaction buffer: 20 mM sodium phosphate pH 7.0 (20% DMSO). Samples were irradiated at 25 °C for 40 min at a distance of 3 cm using a Xe lamp with a polystyrene filter (365 nm).*

molecules that exhibit unique biological activities or functionalities. We discovered that 9-anthracenecarboxylic acid substituted cyclopropane derivatives have strong DNA cleavage property switched by photo irradiation. It is quite noteworthy that a clear contrast in the activity was observed between the enantiomers; chirality of the cyclopropane ring controlled the binding fashion of the attached anthracene group with DNA sequence. It is my hope that the information in this chapter will be helpful in designing novel biologically active compounds or unique functional molecules based on gem-difluorocyclopropane moiety.

Acknowledgments

The author is grateful to Dr. Keiko Ninomiya for performing DNA cleavage test of our anthracene-gem-difluorcyclopropane hybrids. The author is also grateful to Professor Masahiko Sisido of Okayama University for helpful discussions throughout this work. The author records here his acknowledgments to his students, in particular, Miss Nanane Ishida and Mr. Kuhihiko Tanimoto who made great contribution to complete this study.

References

1. For reviews see: (a) "Cyclopropanes and Related Rings", *Chem. Rev.* **2003**, *103*, 931-1648. (b) Tozer, M. J.; Herpin, T. F. *Tetrahedron,* **1996,** *52,* 8619-8683.
2. For reviews, see. (a) Welch, J. T. *Tetrahedron,* **1987**, *43,* 3123-3197. (b) Resnati, G. *Tetrahedron,* **1993**, *49*, 9385-9445. (c) "Enantiocontrolled Synthesis of Fluororganic Compounds: Stereochemical Challenges and Biomedicinal Targets", Ed. Soloshonok, V. A., Wiley, Chichester, UK (1999).
3. (a) Dolbier, Jr. W. R.; Battiste, M. A. *Chem. Rev.* **2003,** *103,* 1071-1098. (b) Fedory'nski, M. *Chem. Rev.* **2003,** *103,* 1099-1132.
4. Methods for preparing gem-difluorocyclopropanes, see. (a) Sargeant, P. B. *J. Org. Chem.* **1970,** *35,* 678-682. (b) Seyferth, D.; Hopper, S. P. *J. Org Chem.* **1972,** *37,* 4070-4075. (c) Schlosser, M.; Chau, L-V.; Bojana, S. *Helv. Chim. Acta,* **1975,** *58,* 2575-2585. (d) Schlosser, M.; Spahi'c; B.; Chau, L-V. *Helv. Chim. Acta,* **1975,** *58,* 2586-2604. (e) Schlosser, M.; Bessard, Y. *Tetrahedron,* **1990,** *46,* 5222-5229. (f) Morikawa, T.; Uejima. M.; Kobayashi, Y. *Chem. Lett.* **1988,** 1407-1710. (g) Bessard, Y.; Müller, U.; Schlosser, M. *Tetrahedron,* **1990,** *46,* 5213-5221. (h) Schlosser, M.; Bessard, Y. *Tetrahedron,* **1990,** *46,* 5222-5229. (i) Tian, F.; Battiste, M. A.; Dolbier, Jr. W. R. *Org. Lett.* **1999,** *1,* 193-195. (j) Tian, F.; Kruger, V.; Bautista, O.; Duan, J-X.; Li, A-R.; Dolbier, Jr. W. R.; Chen, Q-Y. *Org. Lett.* **2000,** *2,* 563-564.
5. Our examples for preparing optically active gem-difluorocyclopropanes. (a) Itoh, T.; Mitsukura, K.; Furutani, M. *Chem. Lett.* **1998**, 903-904. (b) Mitsukura, K.; Korekiyo, S.; Itoh, T. *Tetrahedron Lett.* **1999**, *40,* 5739-5742. (c) Itoh, T.; Mitsukura, K.; Ishida, N.; Uneyama, K. *Org. Lett.* **2000**, *2,* 1431-1434. (d) Itoh, T. *J. Synth. Org. Chem. Jpn.* **2000,** *58,* 316-326. (e) Itoh, T.; Ishida, N.; Mitsukura, K.; Uneyama, K. *J. Fluorine Chem.* **2001**, *112,* 63-69. (f) Itoh, T.; Ishida, N.; Mitsukura, K.; Hayase, S.; Ohashi, K. *J. Fluorine Chem.* **2004**, *125,* 775-783.
6. (a) J. R. Pfister, F. Makra, A. V. Muehldorf, H. Wu, J. T. Nelson, P. Cheung, N. A. Bruno, S. M. Casey, N. Zutshi and D. L. Slate, *Biorganic Med. Chem. Lett.* **1995**, *5*, 2473-2476. (b) Barnett, Charles J.; Huff, Bret; Kobierski, Michael E.; Letourneau, Michael; Wilson, Thomas M. *J. Org. Chem.* **2004**, *69,* 7653-7660.
7. Boger, D. L.; Jenkins, T. J. *J. Am. Chem. Soc.* **1996**, *118,* 8860-8870.
8. Koizumi, N.; Takrgawa, S.; Mieda, M.; Shibata, K. *Chem. Pharm. Bull.* **1996**, *44,* 2162-2164.
9. Taguchi, T.; Shibuya, A.; Sasaki, H.; Endo, J-i.; Morikawa, T. ; Shiro, M. *Tetrahedron: Asymmetry,* **1994**, *5,* 1423-1426.
10. (a) Taguchi, T.; Shibuya, A.; Sasaki, H.; Endo, J-i.; Morikawa, T. ; Shiro, M. *Tetrahedron: Asymmetry,* **1994**, *5,* 1423-1426. (b) Taguchi, T.; Kurishita, M.; Shibuya, A.; Aso, K. *Tetrahedron,* **1997**, *53,* 9497-9508.
11. (a) Shibuya, A.; Kurishita, M.; Ago, C.; Taguchi, T. *Tetrahedron,* **1996**, *52,* 271-278. (b) Taguchi, T.; Kurishita, M.; Shibuya, A.; Aso, K. *Tetrahedron,* **1997**, *53,* 9497-9508. (c) Shibuya, A.; Sato, A.; Taguchi, T. *Bioorganic & Medicinal Chem. Lett.* **1998**, *8,* 1979-1984. (d) Shibuya, A.; Okada, M.; Nakamura, Y.; Kibashi, M.; Horikawa, H.; Taguchi, T. *Tetrahedron,* **1999**, *55,* 10325-10340.
12. Pechacek, J. T.; Bargar, T. M.; Sabol, M. R. *Bioorg. Med. Chem. Lett.* **1997**, *7,* 2665-2668.

13. (a) Csuk, R.; Eversmann, L. *Tetrahedron,* **1998**, *54,* 6445-6459. (b) Csuk, R.; Thiede, G. *Tetrahedron,* **1999**, *55,* 739-750. (c) Csuk, R.; Eversmann, L. *Zeitschrift fuer Naturforschung, B: Chemical Sciences,* **2003**, *58,* 997-1004. (d) Csuk, R.; Thiede, G. *Zeitschrift fuer Naturforschung, B: Chemical Sciences,* **2003**, *58,* 853-860. (e) Csuk, R.; Eversmann, L. *Zeitschrift fuer Naturforschung, B: Chemical Sciences,* **2003**, *58,* 1176-1186.
14. (a) Kirihara, M.; Takuwa, T.; Kawasaki, M.; Kakuda, H.; Hirokami, S-i.; Takahata, H. *Chem. Lett.* **1999**, 405-406. (b) Kirihara, M.; Kawasaki, M.; Takuwa, T.; Kakuda, H.; Wakikawa, T.; Takeuchi, Y.; Kirk, K. L. *Tetrahedron. Asymm.* **2003**, *14,* 1753-1761.
15. Roe, R. M.; Kallapur, V.; Linderman, R. J.; Viviani, F. *Pesticide Biochemistry and Physiology,* **2005**, 83(2-3), 140-154.
16. (a) Ren, Y.; Lodge, T. P.; Hillmyer, M. A. *J. Am. Chem. Soc.* **1998**, *120,* 6830-6831. (b) Ren, Y.; Lodge, T. P.; Hillmyer, M. A. *Macromolecules,* **2000**, *33,* 866-876. (c) Reisinger, J. J.; Hillmyer, M. A. *Prog. Polym. Sci.* **2002**, *27,* 971-1005. (d) Hillmyer, M. A.; Schmuhl, N. W.; Lodge, T. P. *Macromol. Symp.* **2004**, *215,* 51-56.
17. Miyazawa, K.; Yufit, D. S.; Howard, J. A. K.; de Meijere, A. *Eur. J. Org. Chem.* **2000**, 4109-4117.
18. Itoh, T.; Ishida, N.; Ohashi, M.; Asep, R.; Nohira, H. *Chem. Lett.* **2003**, *32,* 494-495.
19. Itoh, T. *ACS Symposium Series #911,* Ed, V. A. Soloshonok, Oxford University Press/ American Chemical Society: Washington DC, Chapter 25, pp.430-439 (2005).
20. For reviews, see: (a) Ros, T. D.; Spalluto G.; Boutorine, A. S.; Bensasson, R. V.; Prato, M. *Current Pharmaceutical Design,* **2001**, *7,* 1781-1821. (b) Armitage, B. *Chem. Rev.* **1998**, *98,* 1171-1200.
21. (a) Ly, D.; Kan, Y.; Armitage, B.; Schuster, G. B. *J. Am. Chem. Soc.* **1996**, *118,* 8747-8748. (b) Gasper, S. M.; Armitage, B.; Shi, X.; Hu, G. G.; Yu, C.; Schuster, G. B.; Williams, L. D. *J. Am. Chem. Soc.* **1998**, *120,* 12402-12409.
22. Kumar, C. V.; Punzalan, E. H. A.; Tan, W. B. *Tetrahedron,* **2000**, *56,* 7027-7040.
23. (a) Toshima, K.; Takano, R.; Maeda, Y.; Suzuki, M.; Asai, A.; Matsumura, S. *Angew. Chem. Int. Ed* **1999**, *38,* 3733-3735. (b) Toshima, K.; Maeda, Y.; Ouchi, H.; Asai, A.; Matsumura, S. *Bioorganic & Medicinal Chem. Lett.* **2000**, *10,* 2163-2165. (c) Toshima, K.; Takai, S.; Maeda, Y.; Takano, R.; Matsumura, S. *Angew. Chem. Int. Ed.* **2000**, *39,* 3656-3658. (d) Toshima, K.; Takano, R.; Ozawa, T.; Matsumura, S. *Chem. Commun.* **2002**, 212-213.
24. Tian, F.; Bartberger, M. D.; Dolbier, W. R. Jr. *J. Org. Chem.* **1999**, *64,* 540-545.
25. (a) Boger, D. L.; Patane, M. A.; Zhou, J. *J. Am. Chem. Soc.* **1994**, *116,* 8544. (b) Helissey, P.; Bailly, C.; Vishwakarma, J. N.; Auclair, C.; Waring, M. J.; Giorgi-Renault, S. *Anticancer Drug Des.* **1996**, *11,* 527. (c) Amishiro, N.; Nagamura, S.; Murakata, C.; Okamoto, A.; Kobayashi, E.; Asada, M.; Gomi, K.; Tamaoki, T.; Okabe, M.; Yamaguchi, N.; Yamaguchi, K.; Saito, H. *Bioorganic & Medicinal Chem.* **2000**, *8,* 381-391.
26. Ninomiya, K.; Tanimoto, K.; Ishida, N.; Horiii, D.; Sisido, M.; Itoh, T. *J. Fluorine Chem.* **2006**, *127,* in press.

Chapter 22

The Synthesis of an Antiviral Fluorinated Purine Nucleoside: 3′-α-Fluoro-2′,3′-dideoxyguanosine

Kunisuke Izawa[1], Takayoshi Torii[1], Tomoyuki Onishi[1], and Tokumi Maruyama[2]

[1]AminoScience Laboratories, Ajinomoto Company, Inc., 1–1 Suzuki-cho, Kawasaki-ku, Kawasaki, Japan 210–8681

[2]Faculty of Pharmaceutical Sciences at Kagawa Campus, Tokushima Bunri University, Shido 1314–1, Sanuki-city, Kagawa, Japan 769–2193

In this review, the synthesis of FddG, a potential HIV reverse transcriptase inhibitor is described focusing on regio- and stereoselective fluorination at the C3’α-position of the nucleoside. These results, which include a synthetic strategy employing retentive fluorination, may provide a new approach towards a variety of C3’α -substituted nucleosides.

Introduction

Acquired immunodeficiency syndrome (AIDS) and the viral infectious diseases, which are represented in hepatocellular carcinoma by hepatitis B and C viruses, have become serious medical problems worldwide. Human immunodeficiency virus (HIV), the pathogenic virus that causes AIDS, belongs to the retrovirus. This type of virus utilizes a reverse transcriptase (RT) for synthesizing the proviral DNA from the RNA of HIV genome to enable sustained infection of HIV Thus, the RT has borne important function for the multiplication of the HIV, therefore, it has become the target enzyme to develop the novel anti-HIV drugs. As a result of computational structure activity relationship studies, many nucleoside derivatives appeared in development candidacy. In the course of these researches, a variety of nucleosides possessing a fluorine atom in the sugar moiety were synthesized, and their activities were duly evaluated. A fluorine atom is similar to a hydrogen atom in size but possesses very different electronic properties which not only reinforce activity but also improve bioavailability; in addition, the presence of a fluorine atom may bring about increased metabolic stability. The substitution of a fluorine atom at the C3'α - or C2'β-position of a nucleoside was shown in a number of cases to confer significant anti-HIV activity. Conversely, those compounds with the opposite configuration exhibited lower activity (*1*). A typical example of the former is Lodenosine (FddA, **1**) (*2*), and of the latter Alovudine (FLT, **2**) (*3*). Recently, FddG **3** which has the same sugar moiety as FLT **2** (Figure 1) was taken into clinical development having attracted much attention for both its anti-HIV activity (*4*) and its anti-HBV activity (*5*). As regards the synthesis of FddG **3**, the development of an industrial-scale process is somewhat problematic due to the peculiar physical properties of guanine base and no suitable manufacturing method has been reported. In this chapter, we review the recent research on the synthesis of FddG **3** including our own latest results.

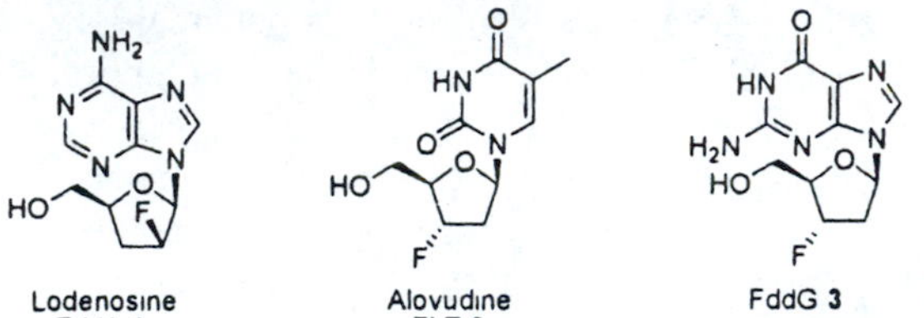

Lodenosine
FddA **1**

Alovudine
FLT **2**

FddG **3**

Figure 1. The nucleosides derivatives bearing fluorine in the sugar moiety

Results and Discussion

Synthesis of FddG from fluorinated sugars

A convenient synthetic approach to FddG **3** is to couple the fluorinated sugar with the purine base. There have been many reports of the preparation of 3α-fluoro-2,3-dideoxyribose derivatives from natural monosaccharides such as D-xylose **4** (*6*) and 2-deoxy-D-ribose **9** (*7*), since the fluorinated sugars are key intermediates for FLT **2**. The first synthesis of such fluorinated sugars was reported by Fleet *et al.* in 1987 for the preparation of FLT **2** (*6a*). The key fluorination step was performed via triflate formation followed by a fluorination reaction with tetrabutylammonium fluoride to give the fluorinated sugar **8** (Scheme 1).

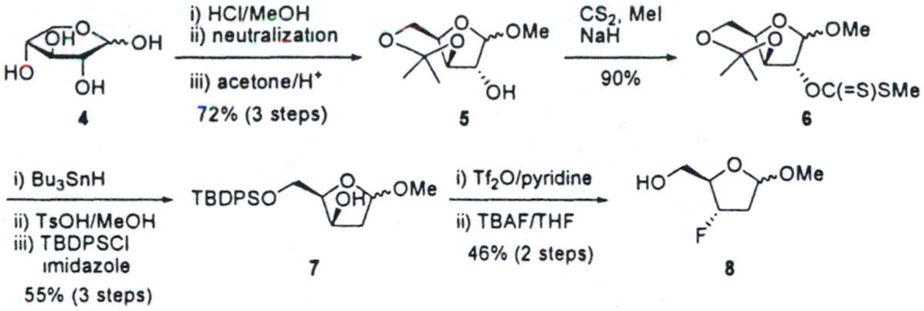

Scheme 1

A more efficient method was reported by Saischek *et al.* in 1991 (Scheme 2) (*7a*). They also used the same fluorination conditions as those reported by Fleet *et al.*, although the yield was slightly improved probably because of the changed protecting group.

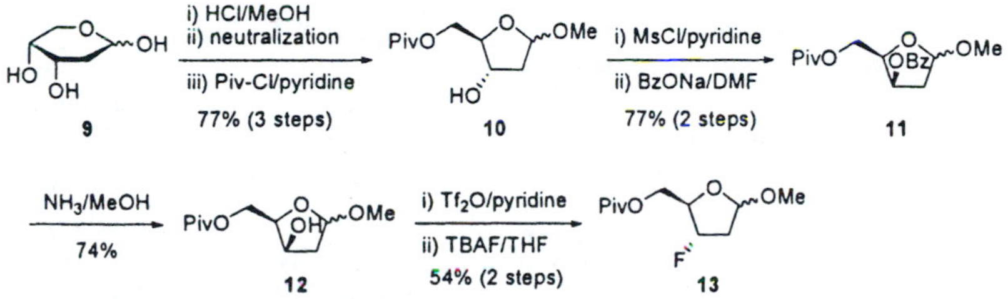

Scheme 2

Thus, 2-deoxy-D-ribose **9** was converted to compound **12** in 6 steps involving the inversion of configuration of the OH group at C3-position. The fluorination of compound **12** gave the desired fluorinated sugar **13** in 54% yield. One big advantage of this method is that it obviates the need to carry out the unfavorable deoxygenation using a toxic tin compound required for the synthesis from D-xylose **4** above. Alternatively, a similar fluorinated sugar can also been synthesized by the reaction of the 2-deoxy-D-ribose **9** derivative with diethylaminosulfur trifluoride (DAST) (Scheme 3) (*7c*). It is worthy of note that this method can utilize 2-deoxy-D-ribose **9** as a starting material without the inversion of the configuration of the C3-OH group. In this reaction, fluorination takes place with retention of the configuration at the C3-position probably due to the neighboring participation of the substituted benzoate group at the C5-position. Interestingly, the fluorination of β-methyl glycoside **14** afforded the fluorinated sugar **16** in a moderate yield. However, α-isomer **15** gave the fluorination product **17** in very low yield probably because of steric hindrance.

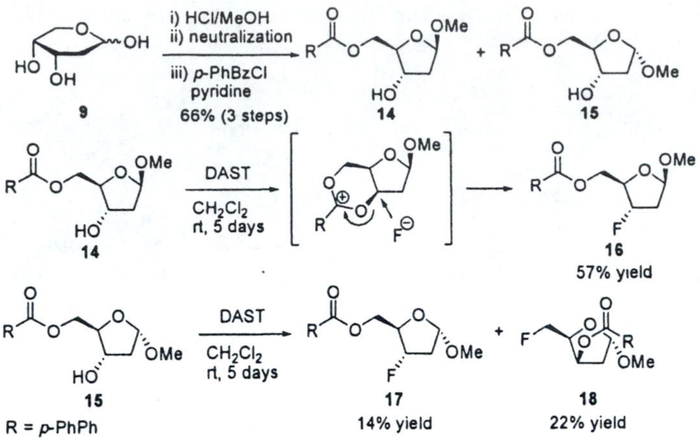

Scheme 3

There are several other reported procedures to obtain 3α-fluoro-dideoxyribose (*8*).

The coupling reaction of the fluorinated sugar with silylated 2-amino-6-chloropurine (2-ACP) was reported by Chu *et al.* using Vorbrüggen methodology, although they only reported the reaction of the corresponding L-sugar. The coupling products were obtained in 47% yield as a 1:1 mixture of α- and β-isomers. The coupling products **20** and **21** were hydrolyzed with 2-mercaptoethanol to give a mixture of guanine derivatives **22** and **23**. After isomeric separation, L-FddG **22** was obtained in 31% yield (Scheme 4) (*9*). A drawback in this reaction is that chromatographic separation of the α-isomer **23** is necessary to obtain the pure product.

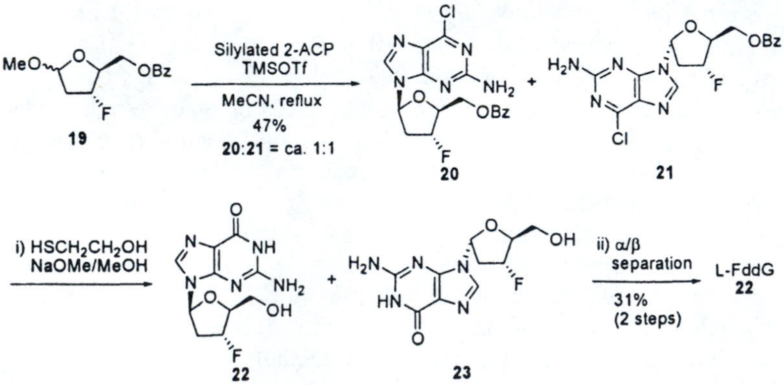

Scheme 4

Recently, Komatsu *et al.* reported an enzymatic β-selective glycosylation of the sugar moiety with several nucleic bases, in which the formation of the α-isomer was completely avoided during the coupling reaction (*10a*). They successfully applied this technology to synthesize FddG **3** (*10b,c*). The fluorinated sugar **17**, prepared from 2-deoxy-D-ribose **9** by the Abdel-Bary *et al.* method (*7c*), was first converted to 1-Cl-sugar by acetylation and subsequent treatment with HCl. After the reaction with phosphorous acid in the presence of tributylamine, the sugar phosphates **24** and **25** were obtained as cyclohexylamine salts in 59% yield in a 4:1 mixture of α/β isomers. Finally, the deprotection of the C5-OH group followed by enzymatic glycosylation with guanine in the presence of $CaCl_2$ afforded FddG **3** in 63% yield (Scheme 5).

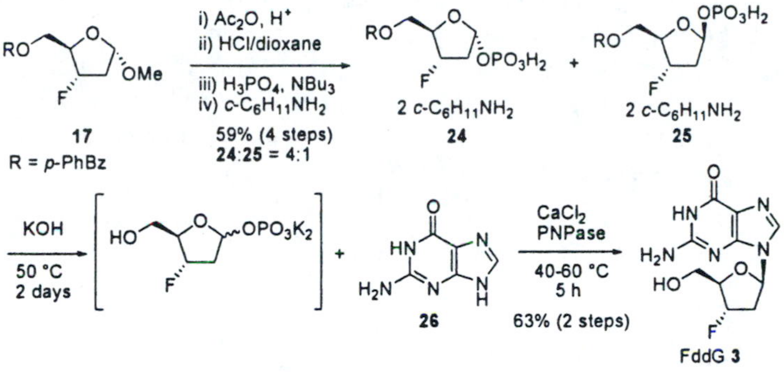

Scheme 5

Synthesis of FddG from pyrimidine nucleosides

An efficient synthetic method for FLT was reported by Etzold *et al.* O2-C3'-Anhydrothymidine **28** (*11*) prepared from thymidine **27**, was treated with HF in the presence of AlF_3 to give FLT **2** after the deprotection of C5'-OH group (Scheme 6) (*3a,b*).

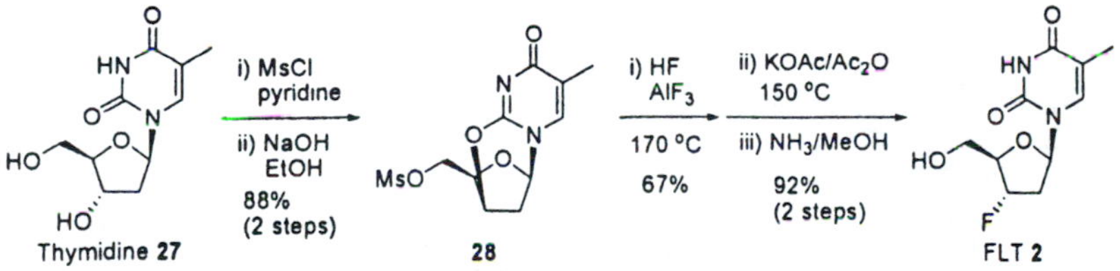

Scheme 6

Saizewa *et al.* succeeded in obtaining FddG **3** via the chemical transglycosylation of Ac-FLT **29** (*12*) with *N*-palmitoyl tris(trimethylsilyl)-guanine **30** in the presence of a Lewis acid (Scheme 7) (*13*). A drawback to this process is the low transglycosylation yield.

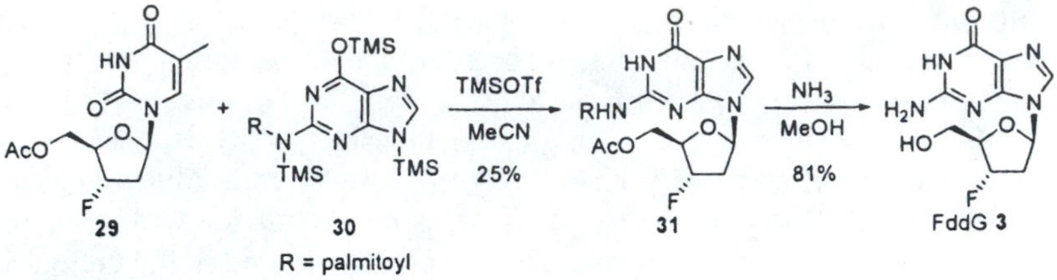

Scheme 7

Burns *et al.* reported significant improvements to the transglycosylation yield by means of the use of an enzyme catalyst (Scheme 8) (*14*). They used FddU **32** as a starting material which was easily prepared by the same procedure described above in the synthesis of FLT **2** (*15*). The enzymatic glycosylation of FddU **32** with 2-amino-6-cyclobutoxypurine **33** (*16*) gave the coupling product **34** in 71% yield. Although the authors did not disclose an appropriate method to convert compound **34** to FddG **3**, if such a method could be developed, this approach might indeed prove convenient.

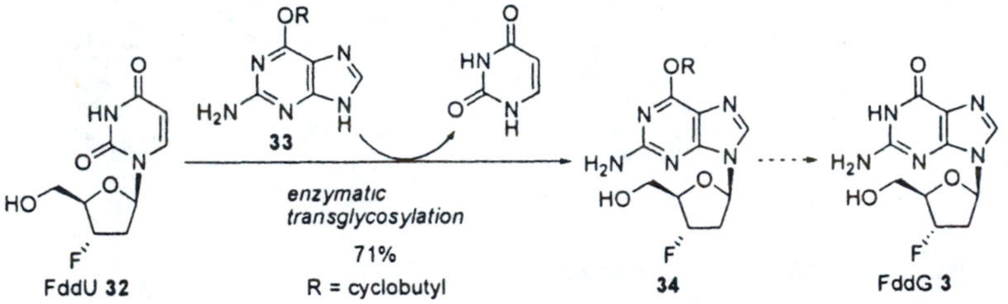

Scheme 8

Synthesis of FddG from guanosine

There have been several reports of the synthesis of FddG **3** directly from guanosine **35**. This method is the most straightforward because it avoids glycosylation, which normally gives a mixture of α/β isomers thereby necessitating a subsequent separation step.

Synthesis of FddG with the inversion of the configuration at the C3'-position of guanosine

The first synthesis of FddG **3** from guanosine **35** was reported by Herdewijn *et al.* in 1988 (*17*). The key intermediate **37** for the fluorination at the C3'-position was prepared according to the method reported by Hansske *et al.* (*18*). After protection of the C5'-OH group with benzoic anhydride and triethylamine, compound **37** was subjected to fluorination using DAST followed by deprotection to give FddG **3** (Scheme 9). The fluorination took place with the inversion of the configuration at the C3'-position contrary to the case of 2-deoxy-o-ribose derivative **14** (Scheme 3) *(7c)*. Although the fluorination yield was rather low at ^{350}o, this method may be regarded as one of the most convenient and efficient approaches to the synthesis of FddG **3.**

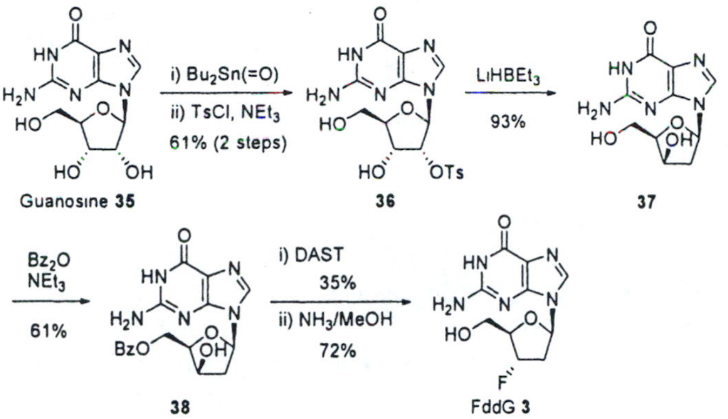

Scheme 9

Previously, we succeeded in significantly improving the fluorination yield of 6-chloropurine-3'-deoxyriboside with DAST during the synthesis of FddA **1** (*2b, 19*). We speculated that the introduction of a chlorine atom at the 6-position of the purine base might have a beneficial influence - even on the reactivity of the C3'-position of the riboside. The starting 2-amino-6-chloropurine riboside **39** is easily prepared from guanosine **35** (*20*) and the chlorine atom once introduced at the 6-position can be readily transformed to guanine base by hydrolysis after modification of the sugar moiety. As we expected, we were able to confirm that the fluorination yield with DAST increased to 60% from 35% in the case of guanine. The fluorination product **43** was then subjected to deprotection of the trityl group at the C5'-position followed by treatment with 2-mercaptoethanol under basic hydrolytic conditions to give FddG **3** in 67% yield (Scheme 10) (*21*).

Synthesis of FddG with bromine rearrangement during fluorination

We previously reported that the bromohydrin derivative **47** easily prepared from adenosine **45** (*22,23*) gave the C3'α-fiuorinated compound **48** in 75% yield

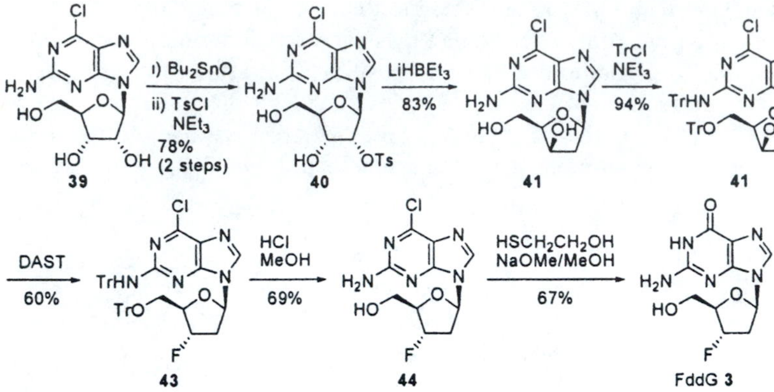

Scheme 10

under fluorination conditions using morpholinosulfur trifluoride (MOST). The reaction can be explained in terms of the bromine atom rearranging from the C3'β- to the C2'β-position via a bromonium ion intermediate to which the fluoride anion attacks from the α-side of the C3'-position. The C3' α-fluorinated product **48** is converted to 3' α-FddA **50** by radical debromination and deprotection (Scheme 11) (*24*).

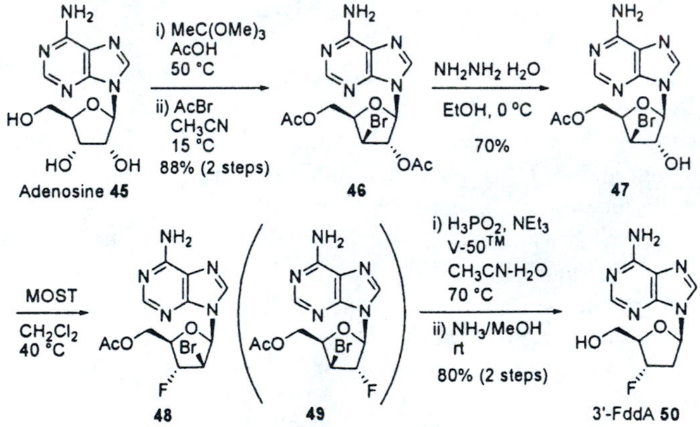

Scheme 11

We initially assumed that the fluorination reaction might proceed via a unique intramolecular rearrangement of bromine from the C3'β- to C2'β-position simultaneously with C3'α-fiuorination (Figure 2) (*23a*).

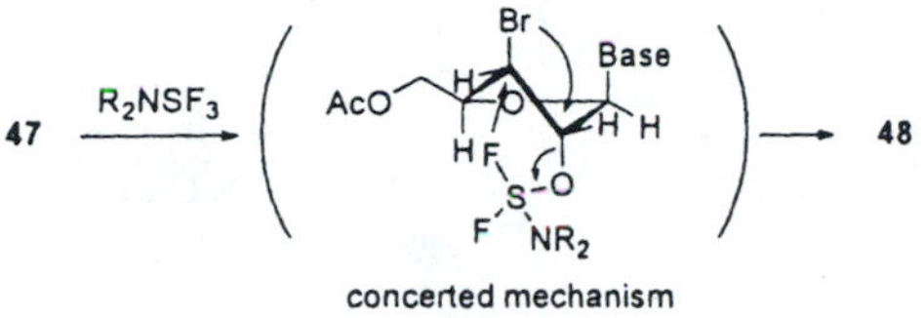

47 $\xrightarrow{R_2NSF_3}$ (concerted mechanism) → 48

Figure 2

However, we later recognized that the fluorination also gave small quantities of the regioisomer **49** (*25*). Therefore, we carried out the same fluorination reaction using the regioisomer **51** of the bromohydrin **47** prepared from adenosine **45** (Scheme 12) (*24c*).

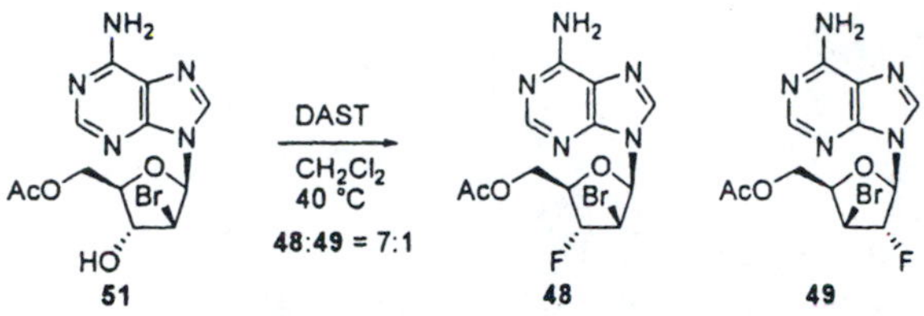

51 → 48 + 49 (DAST, CH_2Cl_2, 40 °C; 48:49 = 7:1)

Scheme 12

As a result, we obtained the same isomeric ratio (7:1) of the product suggesting that the reaction might proceed via the bromonium ion intermediate as shown in Figure 3. This putative mechanism is strongly supported by the formation of a C2'α-isomer **49** as a byproduct.

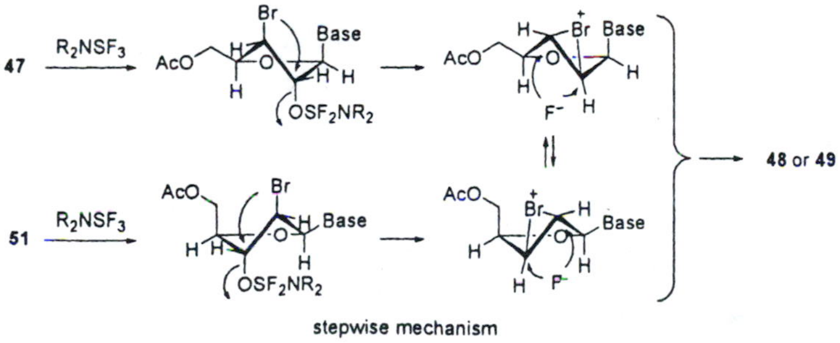

stepwise mechanism

Figure 3

Next, we turned our attention to the synthesis of FddG **3** by applying the same methodology described above. Thus, the bromohydrin **53** prepared from guanosine **35** (*22c*) was allowed to react with DAST to give the desired fluorination product **54** in 59% yield along with the regioisomer **55** in 21% yield (Scheme 13) (*25*). After separation of the isomer **55**, the fluoronucleoside **54** was transformed into FddG **3** by the same method.

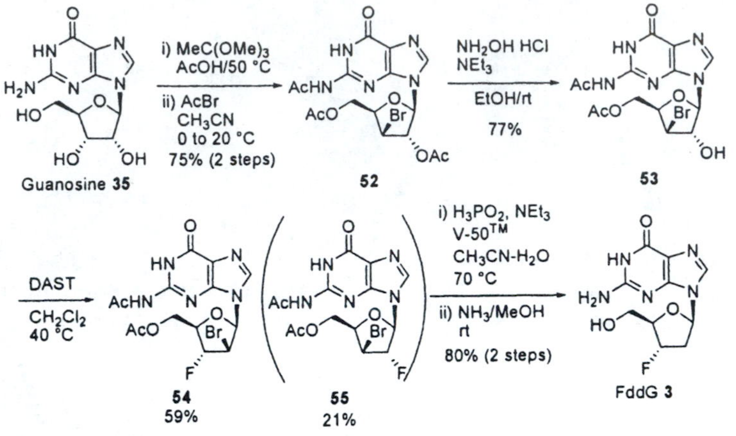

Scheme 13

While the ratio of the C3'α-F-isomer **48** to the C2'α-F-isomer **49** was 7.0:1 in the case of the adenosine derivative **47**, the bromohydrin **53** prepared from guanosine **35** gave the products in a ratio of 2.8:1 with lower selectivity. In the case of bromohydrin **56**, which is the *N*-deacetylated compound of **53**, the ratio of fluorination was somewhat improved to 4.0:1 (Scheme 14) (*25*). Variations in the ratio were nucleic base dependent between adenine, *N*-acetylguanine and guanine. Although the reason for this is not clear, we may speculate it is related to the electronic properties of each respective base.

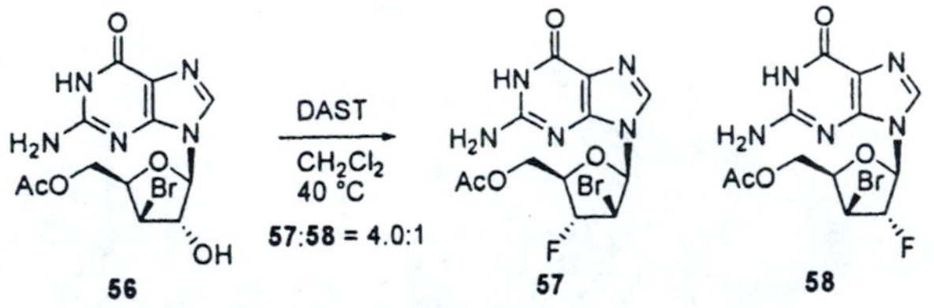

Scheme 14

We further investigated the fluorination conditions with a view to improving the yield of the C3'α-product **54** (Table 1). We were interested to discover that the fluorination of compound **53** with SF_4 furnished the C3'α-product **54** in 79% yield in a ratio of 7.8:1 (Run 3) (*26*). This suggests that the diethylamino group of DAST may have a negative influence on the regioselectivity of the reaction. On the contrary, when we attempted the fluorination of compound **53** via the mesylate with spray dried (sd) KF and NEt_3/3HF, we could not detect the desired product at all. In the case of sdKF, a major product was the furan derivative **59** (Run 4), and the starting material was recovered in the case of

$NEt_3/3HF$ (Run 5). To our surprise, we found that the reaction of compound **53** with perfluorobutanesulfonyl fluoride (NfF) in the presence of NEt_3 produced the 2',3'-didehydro-2',3'-dideoxyguanosine derivative **60** in 62% yield (Run 6) (*27*).

Table 1. Fluorination of compound 53 under a variety of conditions

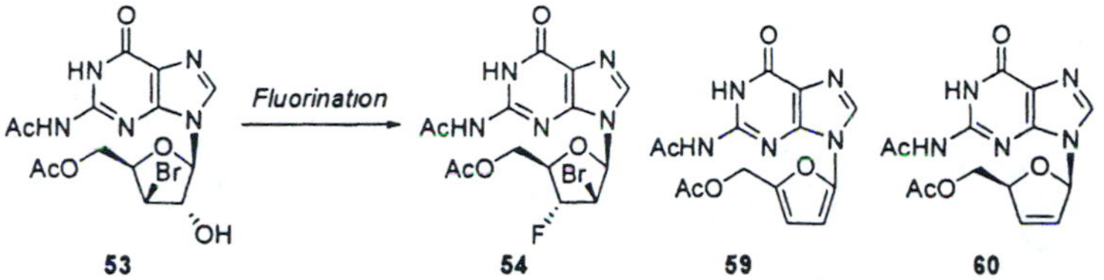

run	reagent	solvent	temp	result
1	DAST	CH_2Cl_2	40 °C	**54** 59% yield (**54**:**55** = 2.8:1)
2	MOST	CH_2Cl_2	40 °C	**54** 58% yield (**54**:**55** = 2.7:1)
3	SF_4	CH_2Cl_2	rt	**54** 79% yield (**54**:**55** = 7.8:1)
4	MsCl/py, then sdKF	DMF	80 °C	major product : **59**
5	MsCl/py, then 3HF NEt_3	MeCN	60 °C	No Reaction
6	NfF, NEt_3	MeCN	60 °C	**60** 62% yield

Synthesis of FddG via retentive fluorination at the C3'α-position

Concluding that the fluorination with bromine rearrangement described above might proceed via a bromonium ion intermediate, we suspected that a similar reaction may take place via a sulfonium ion intermediate. Thus, we chose the readily available 8,2'-anhydro-8-mercaptoguanosine **61** (*28*) as a starting material, anticipating that the fluorination might proceed with participation of sulfur atom, which facilitate the attack of fluoride ion to C3'α- rather than C2'α-position because of possible steric requirement (Scheme 15).

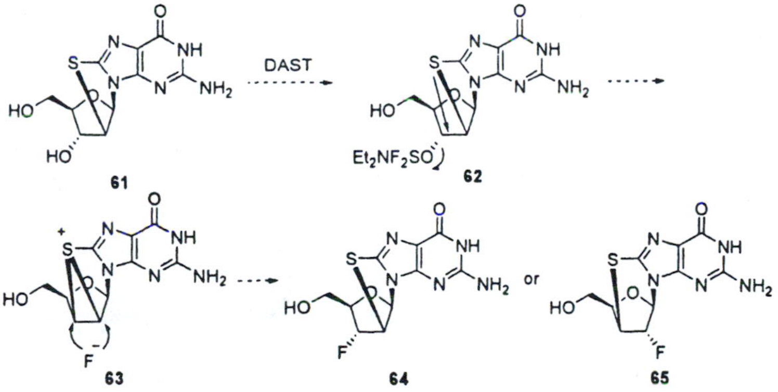

Scheme 15

8,2'-Anhydro-8-mercaptoguanosine **61** was first converted to the N2,O5'-diacetyl derivative **66** which was subjected to fluorination. As anticipated, we obtained the C3'α-fluorinated product **67** in 47% yield without any formation of the C2'α-fiuorinated product **68** when we used DAST (Scheme 16) (*29*). In the case of the fluorination with NfF, however, we found that elimination predominated.

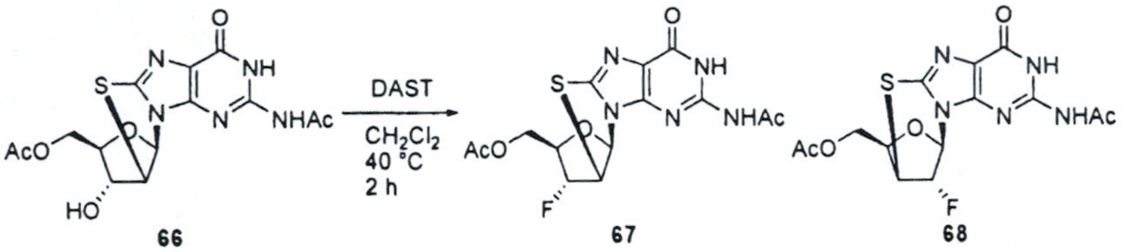

66 **67** 47% yield **68** Not obtained

Scheme 16

Encouraged by these results, we further investigated the fluorination of the ditrityl protected compound **69** using several fluorinating reagents (Table 2). Although the reaction with DAST gave a complex mixture (Run 1), the desired C3'α-product **70** was obtained in 63% yield using NfF/NEt_3 along with the elimination product **71** (Run 2). The ratio of **70** to **71** was 4:1 in this case. Next, we attempted optimization of the reaction conditions. When we used diisopropylethylamine (DIPEA) as a base, elimination was completely suppressed, though the yield of the C3'α-product **70** was not improved. The best yield 91% was obtained, when we used an excess amount of NfT and DIPEA (Run 7) (*29,30*).

Table 2. Fluorination of compound 69

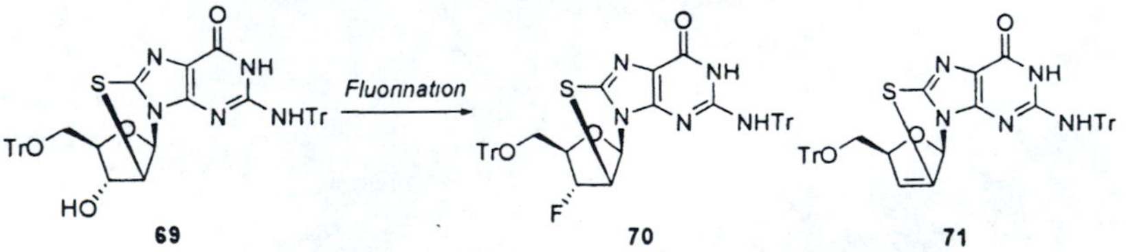

69 **70** **71**

run	F reagent (eq)	base (eq)	solvent	temp.	results
1	DAST (5)	none	CH_2Cl_2	40 °C	complex mixture
2	NfF (4)	NEt_3 (4)	AcOEt	65 °C	**70** 63% yield(**70 71** = 4 0.1)
3	NfF (4)	$i$$Pr_2$NEt (4)	AcOEt	65 °C	**70** 67 area%(**70.71** >50 1)
4	NfF (4)	DBU (4)	AcOEt	65 °C	**70** 98% yield
5	OctF (4)	$i$$Pr_2$NEt (4)	AcOEt	85 °C	**70** 17 area% (**70 71** >50.1)
6	NfF (10)	$i$$Pr_2$NEt (10)	AcOEt	65 °C	**70** 90 area% (**70 71** = 41 1)
7	NfF (12)	$i$$Pr_2$NEt (12)	AcOEt	65 °C	**70** 91% yield (**70.71** >50.1)

The fluorination product **70** was then treated with acetic acid followed by reductive desulfurization with Raney Ni in aq. NaOH affording ddG **72** but not FddG **3** (Scheme 17).

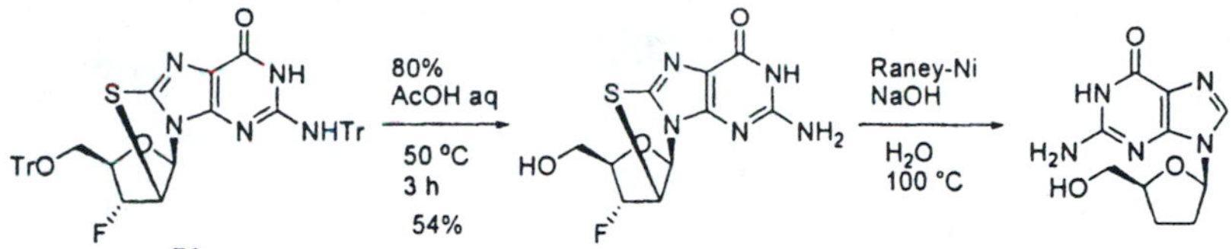

80% AcOH aq, 50 °C, 3 h, 54%; Raney-Ni, NaOH, H_2O, 100 °C

70 → **64** → **72**

Scheme 17

Accordingly, we tried to achieve the desulfurization of compound **70** in various solvent systems prior to deprotection with the acid. The results are listed in Table 3. The reaction of **70** with Raney Ni in EtOH gave a small amount of the desired Tr_2-FddG **73** along with Tr_2-ddG **74** as a major product in a ratio of **73**:**74** = 1:4 (Run 1). By elevating the reaction temperature using 1-BuOH, the ratio of **73**:**74** was improved slightly to 1:3 (Run 2). To our delight, when we used toluene as a solvent, the preferential formation of Tr_2-FddG **73** vs. Tr_2-ddG **74** was observed. Eventually, we discovered that the desulfurization of compound **70** with Raney Ni in toluene without any additives gave the best ratio of **73**:**74** in 6:1 (Run 5); the desired product **73** was isolated in 61% yield after column chromatography.

Table 3. Desulfurization of ditritylated compound 70

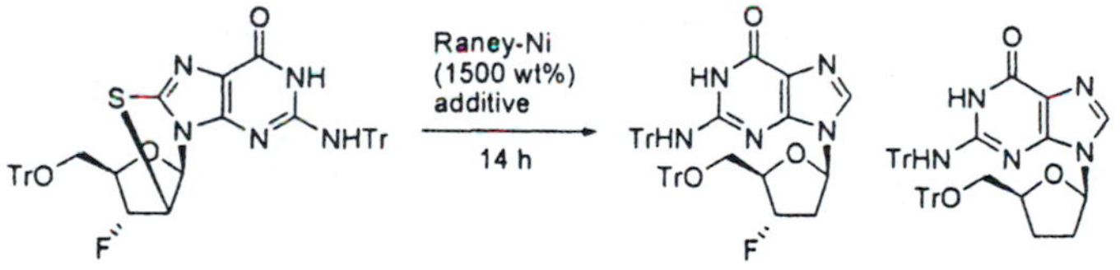

Raney-Ni (1500 wt%), additive, 14 h

70 → **73** + **74**

run	solvent	additive (eq)	temp	results
1	EtOH	NEt_3 (4)	80 °C	**73**:**74** = 1:4
2	1-BuOH	NEt_3 (4)	100 °C	**73**:**74** = 1.3
3	toluene	NEt_3 (4)	90 °C	**73**:**74** = 1.7:1
4	1-BuOH	none	100 °C	**73**:**74** = 1:1.5
5	toluene	none	90 °C	**73** 61% yield, **73**:**74** = 6:1

The Tr_2-FddG **73** thus obtained was deprotected under acidic conditions to give FddG **3** in 69% yield (Scheme 18). This synthetic method using retentive fluorination at the C3'-position has the advantage of providing FddG **3** in high yields with a safer fluorination agent, NfF (*30*). However, there is a drawback for industrial scale synthesis in the need to perform desulfurization using an excess amount of Raney Ni.

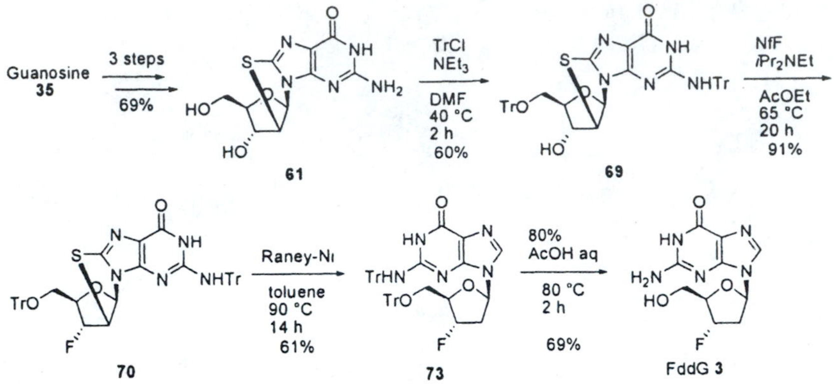

Scheme 18

Conclusion

This review has examined the synthetic methods for FddG **3** focusing in particular on fluorination methods. Initial approaches using a nucleoside as the starting material required SF_4 reagents such as DAST or MOST for the fluorination since other agents afforded mainly elimination products. However, SF_4 reagents are not desirable for industrial scale synthesis due to their poor availability and inherent toxicity. In order to overcome this problem, we developed a new nucleoside fluorination method utilizing neighboring participation. Applying this methodology, we succeeded in carrying out the fluorination of a guanosine derivative at the C3'-position in good yield using readily available NfF. This may provide a novel stereoselective method for introducing a fluorine atom into the sugar moiety of nucleosides.

References

1. Jeong, L. S.; Lim, B. B.; Marquez, V. E. *Carbohydr Res.* **1994**, *262*, 103-114.
2. (a) Izawa, K.; Takamatsu, S.; Katayama, S.; Hirose, N.; Kozai, S.; Maruyama, T. *Nucleosides, Nucleotides & Nucleic Acids* **2003**, *22*, 507-517. (b) Herdewijn, P.; Pauwels, R.; Baba, M.; Balzarini, J.; De Clercq, E. *J. Med. Chem.* **1987**, *30,* 2131-2137. (c) Wysocki, R. J.; Siddiqui, M. A.; Barchi, J. J.; Driscoll, J. S.; Marquez, V. E. *Synthesis* **1991**, 1005-1008. (d) Siddiqui, M. A.; Driscoll, J. S.; Marquez, V. E. *Tetrahedron Lett.* **1988**, *39,* 1657-1660. (e) Jin, F.; Wang, D.; Confalone, P. N.; Pierce, M. E.; Wang, Z.; Xu, G.; Choudhury, A.; Nguyen. D. *Tetrahedron Lett.* **2001**, *42*, 4787-4789.

3. (a) Etzold, G.; Hintsche, G. E. R.; Kowollik, G.; Langen, P. *Tetrahedron* **1971**, *27,* 2463-2472. (b) Kowollik, G.; Langen, P. *Nucleic Acid Chemistry;* Wiley: New York, 1978; Part 1, pp 299-302. (c) Green, K.; Blum, D. M. *Tetrahedron Lett.* **1991**, *32,* 2091-2094. (d) Herdewijn, P.; Balzarini, J.; De Clercq. E.; Pauwels, R.; Baba, M.; Broder, S.; Vanderhaeghe, H. *J. Med. Chem.* **1987**, *30,* 1270-1278. (e) Fleet, G. W. J.; Son, J. C.; Derome, A. E. *Tetrahedron* **1988**, *44,* 625-636.
4. (a) Balzarini, J.; Baba, M.; Pauwels, R.; Herdewijn, P.; Wood, S. G.; Robins, M. J.; De Clercq, E. *Mol. Pharmacol.* **1988**, *33,* 243-249. (b) Hartmann, H.; Vogt, M. W.; Durno, A. G.; Hirsch, M. S.; Hunsmann, G.; Eckstein, F. *AIDS Res. Hum. Retroviruses* **1988**, *4,* 457-466.
5. (a) Hafkemeyer, P.; Keppler-Hafkemeyer, A.; al Haya, M. A.; Von Janta-Lipinski, M.; Matthes, E.; Lehmann, C.: Offensperger, W.-B.; Offensperger, S.; Gerok, W.; Blum, H. E. *Antimicrob. Agents Chemother.* **1996**, *40,* 792-794. (b) Schröder, I.; Holmgren, B.; Öberg, M.; Löfgren, B. *Antivir. Res.* **1998**, *37,* 57-66.
6. (a) Fleet. G. W. J.; Son, J. C. *Tetrahedron Lett.* **1987**, *28,* 3615-3618. (b) Fleet, G. W. J.; Son, J. C.; Derome, A. E. *Tetrahedron* **1988**, *44,*. 625-636.
7. (a) Saischek, G.; Fuchs, F.; Dax, K.: Billiani, G. EP patent 450585, 1991. (b) Motawia, M. S.; Pedersen, E. B. *Liebigs Ann. Chem.* **1990**, 1137-1139. (c) Abdel-Bary, H. M.; El-Barbary, A. A.; Khodair, A. I.; Abdel Megied, A. E.; Perdersen, E. B.; Nielsen, C. *Bull. Soc. Chim. Fr.* **1995**, *132,* 149-155. (d) Guo, Z.-W.; Huang, B.-G.; Xiao, W.-J.; Hui, Y.-Z.; Lang, S. A. *Chinese J. Chem.* **1995**, *13,* 363-367.
8. (a) Morizawa, Y.; Asai, T.; Yasuda, A.; Uchida, K. JP patent 0129390, 1989. (b) Mikhailopulo, I. A.; Pricota, T. I.; Sivets, G. G.; Altona, C. *J. Org. Chem.* **2003**, *68,* 5897-5908.
9. Chun, B. K.; Schinazi, R. F.; Cheng, Y.-C.; Chu, C. K. *Carbohydr. Res.* **2000**, *328,* 49-59.
10. (a) Komatsu, H.; Awano, H.; Tanikawa, H.; Itou, K.; Ikeda, I. *Nucleosides, Nucleotides & Nucleic Acids* **2001**, *20,* 1291-1293. (b) Komatsu, H.; Araki, T. *Tetrahedron Lett.* **2003**, *44,* 2899-2901. (c) Komatsu, H.; Ikeda, I.; Araki, T.; Kamachi, H. JP patent 200355392, 2003.
11. (a) Michelson, A. M.; Todd, A. R. *J. Chem. Soc.* **1955**, 816-823. (b) Horwitz, J. P.; Chua, J.; Urbanski, J. A.; Noel, M. *J. Org. Chem.* **1963**, *28,* 942-944. (c) Fox, J. J.; Miller, N. C. *J. Org. Chem.* **1963**, *28,* 936-941. (d) Balagopala, M. I.; Ollapally, A. P.; Lee, H. J. *Nucleosides & Nucleotides* **1996**, *15,* 899-906.
12. Huang, J.-T.; Chen, F-C.; Wang, L.; Kim, M-H.; Warshaw, J. A.; Armstrong, D.; Zhu, Q.-Y.; Chou, T.-C.; Watanabe, K. A.; Matulic-Adamic, J.; Su, T.-L.; Fox, J. J.; Polsky, B.; Baron, P. A.; Gold, J. W. M.; Hardy, W. D.; Zuckerman, E. *J. Med. Chem.* **1991**, *34,* 1640-1646.
13. Saizewa, G. W.; Kowollik, G.; Langen, P.; Mikhailopulo, I. A.; Kvasjuk, E. I. DD patent 209197, 1984.

14. (a) Burns, C. L.; Koszalka, G. W.; Krenitsky, T. A.; Daluge, S. M. WO patent 9313778, 1993. (b) Burns, C. L.: Koszalka, G. W.; Krenitsky, T. A.; Daluge. S. M. US patent 5637574, 1997.
15. (a) Kowollik, G.; Etzold, G.; Von Janta-Lipinski, M.; Gaertner, K.; Langen, P. *J. Prakt. Chem.* **1973**, *315,* 895-900. (b) Etzold, G.; Kowollik, G.; Von Janta-Lipinski, M.; Gaertner, K.; Langen, P. DD patent 103241, 1974.
16. Rideout, J. L.; Freeman, G. A.; Short, S. A.; Almond, M. R.; Collins, J. L. EP patent 421739, 1991.
17. (a) Herdewijn, P.; Balzarini, J.; Baba, M.; Pauwels, R.; Van Aerschot, A.: Janssen, G.; De Clercq, E. *J. Med. Chem..* **1988**, *31,* 2040-2048. (b) Marchand, A.; Mathé, C.; Imbach, J.-L.; Gosselin, G. *Nucleosides, Nucleotides & Nucleic Acids* **2000**, *19,* 205-217.
18. Hansske, F.; Robins, M. J. *J. Am. Chem. Soc.* **1983**, *105,* 6736-6737.
19. Takamatsu, S.; Maruyama, T.; Katayama, S.; Hirose, N.; Naito, M.; Izawa, K. *J. Org. Chem.* **2001**, 66, 7469-7477.
20. (a) Nair, V.; Sells, T. B. *Synlett* **1991**, 753-754. (b) Robins, M. J.; Uznanski, B. *Can. J. Chem.* **1981**, *59,* 2061-2067. (c) Roncaglia, D. I.; Schmidt, A. M.; Iglesias, L. E.; Iribarren, A. M. *Biotechnol. Lett.* **2001**, *23,* 1439-1443.
21. Torii, T.; Maruyama, T.; Demizu, Y.; Onishi, T.; Izawa, K.; Neyts, J.; De Clercq, E. *Nucleosides, Nucleotides & Nucleic Acids* **2006**, *25,* in press.
22. (a) Norman. D. G.; Reese, C. B. *Synthesis* **1983**, 304-305. (b) Talekar, R. R.; Coe, P. L.; Walker, R. T. *Synthesis* **1993**, 303-306. (c) Shiragami. H.; Amino, Y.; Honda, Y.; Arai, M.; Tanaka, Y.; Iwagami, H.; Yukawa, T.; Izawa, K. *Nucleosides & Nucleotides* **1996**, *15,* 31-45. (d) Kondo, K.; Adachi, T.; Inoue. I. *J. Org. Chem.* **1977**, *42,* 3967-3968.
23. Shiragami, H.; Tanaka, Y.: Uchida, Y.; Iwagami, H.; Izawa, K.; Yukawa, T. *Nucleosides & Nucleotides* **1992**, *11,* 391-400.
24. (a) Takamatsu, S.; Katayama, S.; Naito, M.; Yamashita, K.; Ineyama, T.; Izawa, K. *Nucleosides, Nucleotides & Nucleic Acids* **2003**, *22,* 711-713. (b) Takamatsu, S.; Naito, M.; Yamashita, K.; Ineyama, T.; Izawa, K. JP patent 2001122891, 2001. (c) Katayama, S.; Takamatsu, S.; Naito, M.; Tanji, S.; Ineyama, T.; Izawa, K. *J. Fluor. Chem.* **2006**, *127,* in press.
25. Torii, T.; Onishi, T.; Tanji, S.; Izawa, K. *Nucleosides, Nucleotides & Nucleic Acids* **2005**, *24,* 1051-1054.
26. Ishii, A.: Otsuka, T.; Kume, K.; Kuriyarna, Y.; Torii, T.; Onishi, T.; Izawa. K. JP patent 2006022009, 2006.
27. Torii, T.; Onishi, T.; Izawa, K. FP patent 1550665, 2005.
28. (a) Ogilvie, K. K.; Slotin, F.; Westmore, J. B.; Lin, D. *Can. J. Chem.* **1972**, 50, 1100-1104. (b) Kaneko, M.; Kimura, M.; Shimizu, B. *Chem. Pharm. Bull.* **1972**, *20,* 635-637. (c) Ogilvie, K. K.; Slotin, L.; Westmore, J. B.; Lin. D. *Can. J. Chem.* **1972**, *50,* 2249-2253.
29. Torii, T.; Onishi, T.; Izawa, K. JP patent 2006052182, 2006.
30. Torii, T.; Onishi, T.; Izawa, K.; Maruyama, T. To be submitted.

Chapter 23

Solvent-Peptide Interactions in Fluoroalcohol–Water Mixtures

C. Chatterjee, G. Hovagimyan, and J. T. Gerig*

Department of Chemistry and Biochemistry, University of California at Santa Barbara, Santa Barbara, CA 93106

Mixtures of trifluoroethanol and water were observed to stabilize certain conformations of peptides and proteins more than 40 years ago. Other fluorinated alcohols are often more potent in this regard. We describe recent results of intermolecular NOE studies of the interaction of hexafluoroisopropanol and other fluoroalcohols with small peptides, including melittin, analogs of angiotensin containing fluorophenylalanine and "trp-cage", a 20-residue peptide designed by N. H. Anderson and co-workers. In many cases, the results provide evidence of strongly preferential interactions of peptides with the fluoroalcohol component of a solvent mixture.

Background

The peptide hormone angiotensin II plays a central role in the processes that lead to coronary heart disease and hypertension because it is linked to the regulation of blood volume, electrolyte balance and vasoconstriction. A precursor to the octapeptide, angiotensiogen, is produced by the liver. The enzyme renin catalyzes production of an inactive decapeptide that is the substrate for angiotensin-converting enzyme (ACE) which removes two amino acids to produce the physiologically active hormone. Angiotensin II interacts with several classes of G-protein-coupled receptors. Inhibitors of ACE such as Captopril are important drugs for control of hypertension, as are materials that block interaction of angiotensin II with its receptors. Figure 1 summarizes these biochemical events.

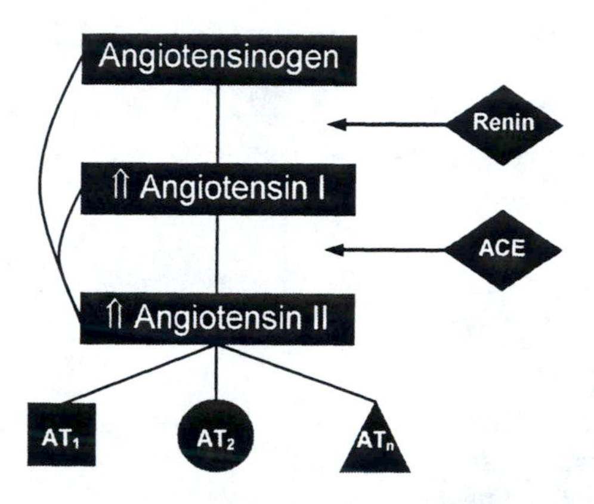

Figure 1*. Schematic of the renin-angiotensin system (ACE, angiotensin-converting enzyme; AT1, AT2, ATn, various angiotensin II receptors). The figure is based on the website www.hypertensiononline.org.*

Much effort has been devoted over the last 30 years to attempts to define the conformation or conformations of angiotensin II in solution, in the hope that knowledge of the preferred structure of the octapeptide would guide design of small molecules that would inhibit ACE or be effective in blocking interaction of the hormone with its receptors. Physical studies, including circular dichroism, Raman spectroscopy and NMR spectroscopy, have led to a variety of suggested structures. A recent NMR study of the peptide in aqueous solution at 4°C has supported the conclusion that angiotensin II under these conditions exists predominantly as a U-shaped molecule (1). Similar experiments with angiotensin II in phospholipid micelles led to a similar structure (2).

Fluorine-containing analogs of angiotensin II have been prepared to probe the interaction of the hormone with its receptor and to explore enhancement of the bioavailability of drugs. Vine and co-workers showed that 4-fluorophenylalanine at position 8 gave a compound equally as potent as the native hormone (3). The same group confirmed an earlier observation by Bumpus and Smedley that 4-fluorophenylalanine at position 4, the tyrosine position, produced an inactive compound that is an antagonist of native angiotensin II. Replacing valine-5 by hexafluorovaline afforded a peptide that was at least as active as the native material but was more resistant to proteolytic degradation (4). Replacement of phenylalanine-8 with pentafluorophenylalanine led to reduced contractile activity and antagonism of the effects of the native material (5).

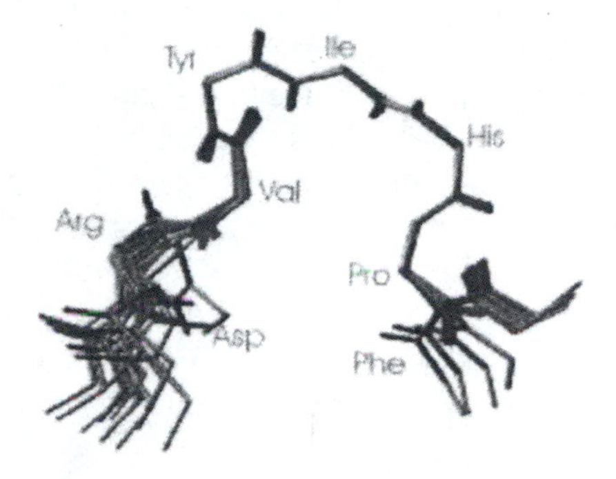

Figure 2*. Solution conformation of angiotensin II proposed by Tzakos,* et al. *(1).*

Several methods are available for introducing fluorine "labels" into peptide and protein structures. One of the reasons for doing this is that fluorine NMR spectroscopic studies of a system are thereby enabled (6). However, it is presently not possible to predict how the presence of fluorine will alter local or global structure and dynamics of a biomolecule.

Goals of Our Work

Our goal when starting this work was comparison of the conformations and conformational dynamics of analogs of angiotensin II in which position 4 is occupied by phenylalanine, tyrosine or 4-fluorophenylalanine. That is, we hoped to compare compounds that differed only by the presence of a hydrogen atom, a fluorine atom, or a hydroxyl group at the *para* position of the residue occupying a portion of the octapeptide that appears to be critical for biological activity.

$$ASP_1 - ARG_2 - VAL_3 - X_4 - VAL_5 - HIS_6 - PRO_7 - PHE_8$$

X =

OH F H

Bumpus and collaborators have shown that adding fluorinated alcohols to aqueous solutions of angiotensin II analogs apparently has the effect of stabilizing certain conformations (7). We planned to use alcohols such as trifluoroethanol or hexafluoroisopropanol as cosolvents, hoping that this would produce systems in which a single peptide conformation was dominant. Conformational studies of peptides by NMR rely on the establishment of constraints that must be satisfied by the deduced structure; the reliability of the structure is dependent on the number and quality of the experimental constraints. Typically, the constraints used are short-range internuclear distances that are estimated by intramolecular nuclear Overhauser effects (NOEs). In an effort to increase the number of conformation-determining constraints available we planned to explore the use of intermolecular NOEs. As is almost always the case in research, things have not developed exactly along the path that was initially envisioned and all that can be done for the present is to describe some observations made as we progress toward the original goals.

Intermolecular Overhauser Effects

The nuclear Overhauser effect (NOE) is the change in the observed intensity of a NMR signal that results when the magnetization associated with another spin is perturbed (8). The NOE is a consequence of nuclear spin-nuclear spin dipole-dipole interactions that contribute to relaxation.

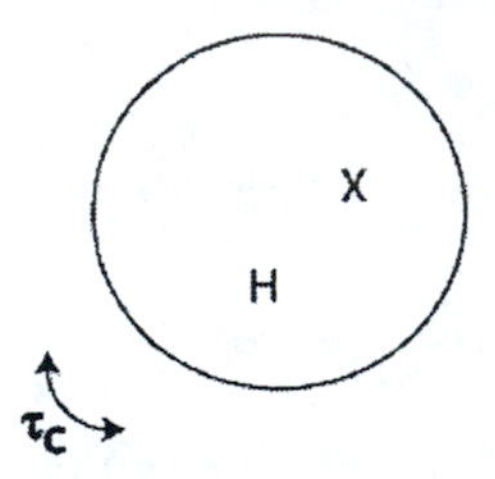

Figure 3. *Intramolecular dipole-dipole relaxation in a molecule containing spins H and X.*

Intramolecular NOEs arise when two interacting spin dipoles, say a proton H and some other spin X, both assumed to be spin 1/2 nuclei here, are contained in the same molecule (Figure 3). It is common to imagine that spin H is at the center of a sphere of radius $\mathbf{r}_{HX}$, the distance between the two spins. Nucleus X "skates" on the surface of the sphere as the molecule tumbles in solution, changing the orientation of the H-X internuclear vector. The re-orientation of the molecule is described in terms of a rotational correlation time τ_C which is approximately the time required for the molecule to rotate in any direction through an angle of about one radian (9). The intramolecular NOE resulting from the H—X interaction depends on τ_C and $\mathbf{r}_{HX}$.

When the interacting spins H and X are on different molecules, their dipolar interaction is modulated by mutual diffusion of the two molecules. The simplest description of the interaction in this situation assumes that the spins are contained within two spherical molecules that have radii $\mathbf{r}_H$ and $\mathbf{r}_X$, respectively (Figure 4). The closest the two spins can approach each other is $\mathbf{r}_H + \mathbf{r}_X = \mathbf{a}$. The time dependence of the interaction of the spheres is assumed to be described by the mutual diffusion coefficient ($\mathbf{D} = \mathbf{D}_H + \mathbf{D}_X$) where $\mathbf{D}_H$ and $\mathbf{D}_X$ are the corresponding translational diffusion coefficients. A correlation time (τ) for the interaction of H and X can be taken as the time required to diffuse the distance **a** with

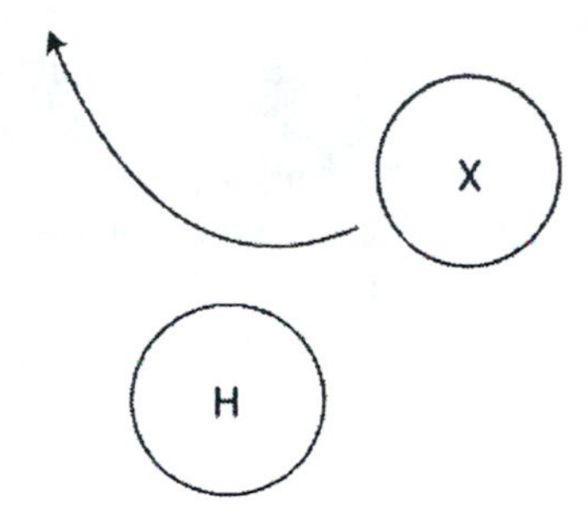

Figure 4. *Dipolar interaction between spins H and X when the spins are in different molecules. Typically, H is contained in a solute molecule and X is part of the solvent molecules.*

$\tau = \mathbf{a}^2/\mathbf{D}$ (10-12). The cross relaxation rate σ_{HX} for the intermolecular interaction of the spins H and X when the signal for spin H is observed is given by

$$\sigma_{HX} = \frac{3\gamma_H^2\gamma_X^2 \boldsymbol{h} \boldsymbol{N}_X}{10\pi \boldsymbol{Da}}\left(6J_2\left(\omega_H + \omega_X\right) - J_2\left(\omega_H - \omega_X\right)\right)$$

where $J_2(\omega)$ is a spectral density function dependent on the magnetic field used for the experiment and the correlation time τ, and $\mathbf{N}_X$ is the concentration of molecules containing spin X (10).

The equation for σ_{HX} is based on the assumption that solvent molecules can approach the sphere representing the solute equivalently from all directions. Real solute molecules have a shape and a solvent molecule will make approaches to the solute from different directions that are non-equivalent due to the shape of the solute molecule. (For the present, we continue to regard the solvent molecule as a sphere with spin X located at its center.) Some solvent approach paths will allow the solute hydrogen and solvent molecule to interact at their van der Waals contact distance. Other approaches will involve interactions at distances longer than this because of steric interference. To take into account the shape of the solute molecule as it interacts with solvent molecules, we used an empirical method that assumes that the contribution to σ_{HX} of a single H-solvent spin interaction over a small element of the solute-solvent contact surface can be computed using the standard equations. Summing the contributions associated with all surface elements is assumed to give the aggregate NOE for a solute hydrogen.

Reliability of Calculated Intermolecular NOEs

We have done several tests of the reliability of the computation method described for estimating intermolecular cross relaxation rates. One test system consisted of 1,3-di-t-butyl benzene dissolved in tetramethylsilane (TMS). This solute was chosen because the steric bulk of the t-butyl groups makes solvent approaches to proton H2 difficult. TMS was used as the solvent; this molecule is nearly spherical and we believed that this non-polar molecule would likely not exhibit strong interactions with the solute.

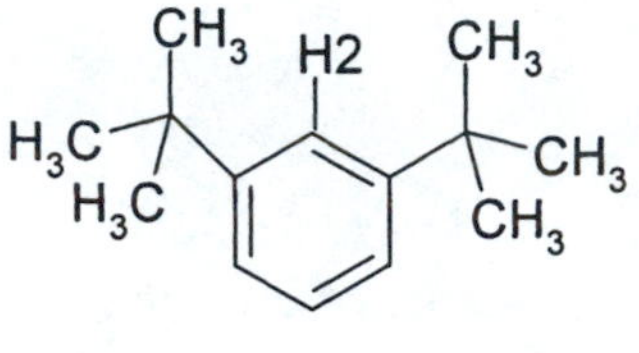

1,3-di-t-butylbenzene

Cross relaxation rates were compared to those calculated employing the method outlined and experimental values of the diffusion coefficients for solute and solvent in the samples examined. We found reasonably good agreement between the experimental cross relaxation terms and those predicted from theory (13). In particular, the "screening" of proton H2 by the adjacent t-butyl groups seemed to be captured correctly by the calculations.

Another test was done with the cyclic dipeptide alanyglycyldiketopiperazine dissolved in water (14). Solute proton-water proton NOEs were determined. At low pH, where the complications from exchange of the N-H protons of the solute with the solvent protons are relatively minor, there is again good agreement between the observed and calculated intermolecular cross relaxation terms (Table I). The samples examined contained traces of trimethylsilylpropionic acid (TSP) to provide a reference signal; solvent-methyl proton cross relaxation of this reference species was also in good agreement with σ_{HH} computed by our method.

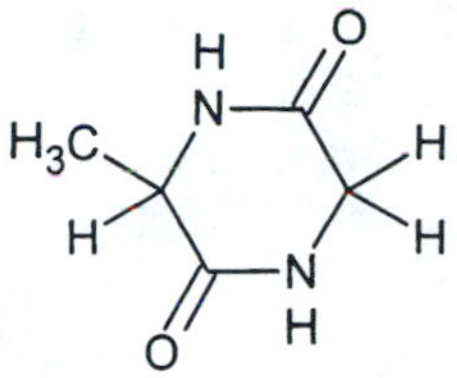

alanylglycyldiketopiperazine

Table I. Solvent-alanylglycyldiketopiperazine intermolecular cross relaxation rates in water (25°)[a]

Solute proton	σ_{HH}^{NOE} *x 10^3 s^{-1}*	σ_{HH}^{ROE} *x 10^3 s^{-1}*
Ala NH	-73. (-67., 20.)	nd
Ala CαH	29. (19., 19.)	22. (25.)
Ala CH_3	24. (20., 20.)	27. (27.)
Gly NH	-100. (-110., 21.)	nd
Gly CH_2	29. (20., 20.)	26. (26.)
TSP	19. (20.)	23. (26.)

[a] The first number given in parentheses is the calculated initial cross relaxation rate. Its value depends on intermolecular dipolar interactions with water as well as exchange with protons of water. The second number is the calculated contribution of intermolecular dipolar interactions only. When only a single number is present in parentheses it is the calculated contribution of the intermolecular dipolar interactions. The calculated rotating frame cross relaxation rates includes only the effects of intermolecular dipolar interactions with water. The rate constants for solvent exchange of alanyl and glycyl N-H protons were 0.96 and 0.14 s^{-1}, respectively (14).

Our original notion was that a solvent-peptide intermolecular NOE could provide structural constraints because its magnitude should depend on the distance of closest approach and thus reveal spins of a molecule that are "exposed" to the solvent, as opposed to buried within a structure. Calculated cross relaxation terms confirm that side chain protons of a small peptide are generally exposed to solvent to about the same extent in all conformations. The big differences are in the solvent-backbone proton NOEs. In helical conformations peptide N-H and Cα-H protons are protected from interactions with solvent spins and should exhibit reduced cross relaxation interactions solvent spins.

Interactions of Peptides with Fluoroalcohols

Alanylglycyldiketopiperazine. Our first indications of "trouble" ahead came when we determined the proton-*fluorine* cross relaxation rates for the cyclic dipeptide alanylglycyldiketopiperazine dissolved in aqueous solutions of a variety of fluorinated alcohols (15). In every solvent, measured cross relaxation rates exceeded those predicted, sometimes by a substantial amount (Table II).

Berger and his coworkers in Leipzig reported similar results about the same time (16, 17). They attributed the enhanced proton-fluorine cross relaxation rates to selective solvation. In effect, the local concentrations of fluoroalcohol near some of the spins of the peptide were assumed to be higher than the concentration of the alcohol in the bulk of the solution. Their results were consonant with the observation that trifluoroethanol can accumulate on the surface as well as penetrate to the interior of protein molecules (18).

Table II. $^{1}H\{^{19}F\}$ Intermolecular Cross Relaxation Rates of Alanylglycyldiketopiperazine Protons in Fluoroalcohols[a]

$\sigma_{HF} \times 10^3\ s^{-1}$		$\sigma_{HF} \times 10^3\ s^{-1}$		$\sigma_{HF} \times 10^3\ s^{-1}$		$\sigma_{HF} \times 10^3\ s^{-1}$	
50% TFE		*50% HFA*		*35% HFIP*		*80% PTFB*	
Obs.	*Calc.*	*Obs.*	*Calc.*	*Obs.*	*Calc.*	*Obs.*	*Calc.*
12.	5.1	20.	4.7	13.	4.6	49.	9.0

[a] TFE, trifluoroethanol; HFA, hexafluoroacetone hydrate; HFIP, hexafluoroisopropanol; PTFB, perfluoro-t-butanol

Angiotensin analogs. Figure 5 shows the proton NMR spectrum of phe-4,val-5 angiotensin II in 40% trifluoroethanol/water at 10° as well as a peptide proton-solvent fluorine NOE spectrum. There are strong, positive but variable values of σ_{HF} for the Phe, Val, Pro, His side chains. The observed cross relaxation rates for some side chain protons are close to the values expected (~3 x $10^{-3}\ s^{-1}$) on the basis of the bulk concentration of the trifluoroethanol and the diffusion constants of the peptide and solvent components. Other side chain cross relaxation rates are 3-4 times larger than expected, a result that appears to be consistent with the notion that the local concentration of the fluoroalcohol is larger than the bulk concentration, particularly near the histidine and phenylalanine side chains. Solvent-solute cross relaxation rates for the peptide N-H or the Cα-H protons are close to zero in this system. That is, there are at best weak contacts between the solvent fluorines and the backbone protons of the octapeptide, a result that indicates the peptide is predominately in a helical conformation.

We have also determined the cross relaxation rates for water protons of the solvent with the solute. These are generally smaller than the predicted values by up to a factor of 2, an observation consistent with the presence of high local concentrations of fluoroalcohol.

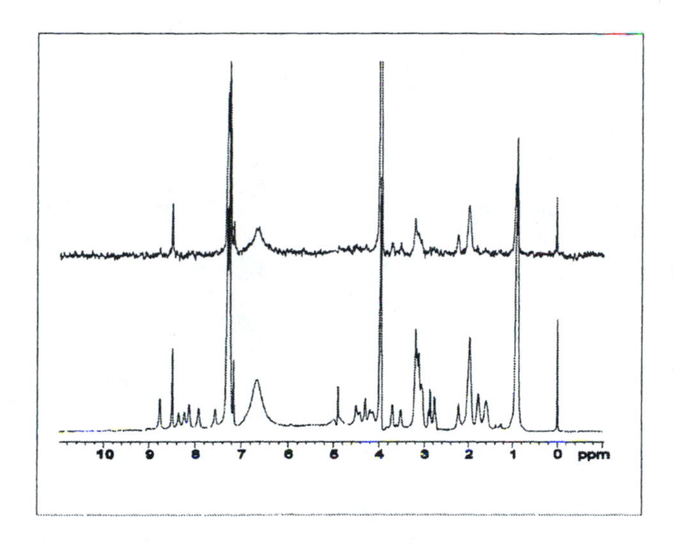

Figure 5. *Lower spectrum: 1D proton NMR spectrum of 4-phe,5-val angiotensin II in 40% trifluoroethanol/water at 10° and 500 MHz. Upper spectrum:* $^{1}H\{^{19}F\}$ *intermolecular NOE spectrum obtained by inversion of the fluorine line of the solvent. The mixing time for the NOE spectrum was 500 ms.*

Melittin. We have done similar experiments with melittin, a 26-residue peptide from bee venom (19). The structure of melittin in 35% hexafluoroisopropanol is a bent helix (Figure 6). The break in helical structure takes place from amino acids gly-12 to leu-16.

Intermolecular NOEs were determined between the hexafluoroisopropanol fluorine spins and the solute protons. There are some overlaps in the proton spectrum, so the determination of the σ_{HF} was not always clean as we would like. However, most side chains show positive solvent NOEs that are close to those predicted by simply assuming diffusive encounters of the peptide and the solvent molecules. Similarly, many N-H protons have NOEs that are close to those predicted (Figure 7). However, the backbone N-H protons in the helix-break region show large effects that are in the opposite direction (negative) from those predicted.

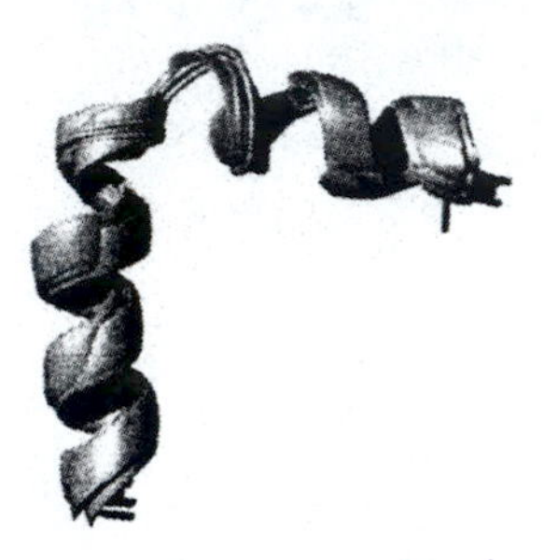

Figure 6. *Representation of the conformation of melittin in 35% hexafluoroisopropanol/water. The drawing represents an overlay of the ten "best-fit" structures determined by analysis of intramolecular $^1H\{^1H\}$ NOEs (19).*

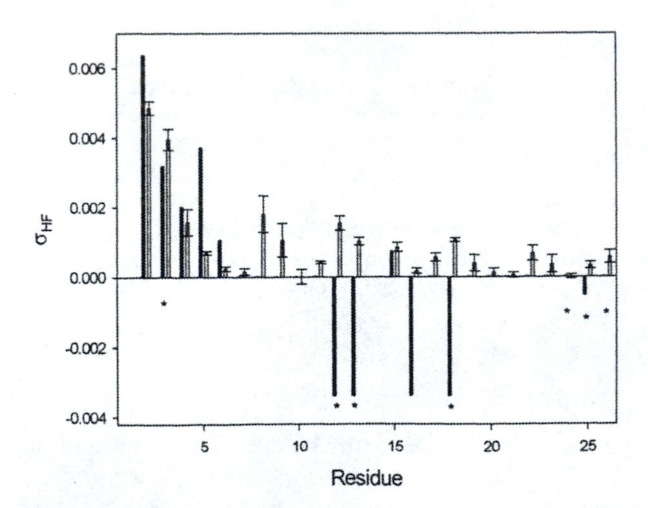

Figure 7*. Comparisons of observed (dark bars) and calculated (light bars) values of σ_{HF} of the peptide backbone N-H signals in melittin in 35% HFIP/water. Asterisks ndicate ambiguous experimental results. The signals for gly-12, leu-13 and ser-18 are overlapped; one or more of the experimental NOEs for any of these could be zero. The signals for gly-3 and arg-24 are overlapped; based on the calculations it has been assumed that the observed NOE signal arises from gly-3. The signal for gln-25 could be influenced by the nearby signals of gly-12, leu-13 and ser-18.*

Our published interpretation of these observations included the notion that the peptide was being selectively solvated by hexafluoroisopropanol molecules and that, in some parts of the peptide, the alcohol interactions were long lived (19). Some of our conclusions about melittin dissolved in water/hexafluoroisopropanol mixtures have been supported by recently published molecular dynamics simulations (20). The simulations show the tendency for alcohol molecules to accumulate near the peptide surface, presumably due to hydrophobic interactions. In disagreement with our observations, in the simulations fluoroalcohol molecules tended to accumulate more around the helices than near the bend region. The authors did not specifically consider the lifetimes of solvent-peptide interactions—this information would be important in calculating the strength of dipolar interactions arising in the selectively solvated peptide.

Trp-cage ("tc5b"). We are presently involved in looking at the interactions of fluoroalcohols with the 20-residue peptide "trp-cage" ("tc5b") reported by Anderson's group several years ago (21). The peptide was developed as part of a design project and was shown to take up the conformation illustrated in Figure 8 in both water and trifluoroethanol/water. The peptide has subsequently become a popular subject for molecular dynamics and other theoretical studies (22-24). The peptide conformation in hexafluoroisopropanol appears to be similar to that shown. We were attracted to this system because the fold shows regions of helical, turn and beta structure and we thought it might be an appropriate system to determine if there was any specificity in fluoroalcohol-peptide interactions due to secondary structure.

***Figure 8**. Conformation of the trp-cage protein as determined by NMR methods. The rendering is based on the first structure given in the Protein Data Bank file (accession code 1L2Y).*

Figure 9 shows solvent fluorine-peptide proton intermolecular NOEs in the low field portion of the proton spectrum at 45°. Most of the proton signals are missing in the NOE spectrum, with only protons of the tyrosine-3 and tryptophan-6 side chains showing $^{1}H\{^{19}F\}$ NOEs. There are relatively few signals in the upfield region as well, all of them associated with amino acid side chains. At this temperature, the cross relaxation rates (σ_{HF}) for the peaks observed range are positive, ranging from about 2 x 10^{-3} s^{-1} to 10 x 10^{-3} s^{-1}. These cross relaxation rates are 2-4 times larger than those expected based on the diffusion constants of the solvent and peptide, a result that is reminiscent of the results described earlier for alanylglycyldiketopiperazine and the angiotensin analogs.

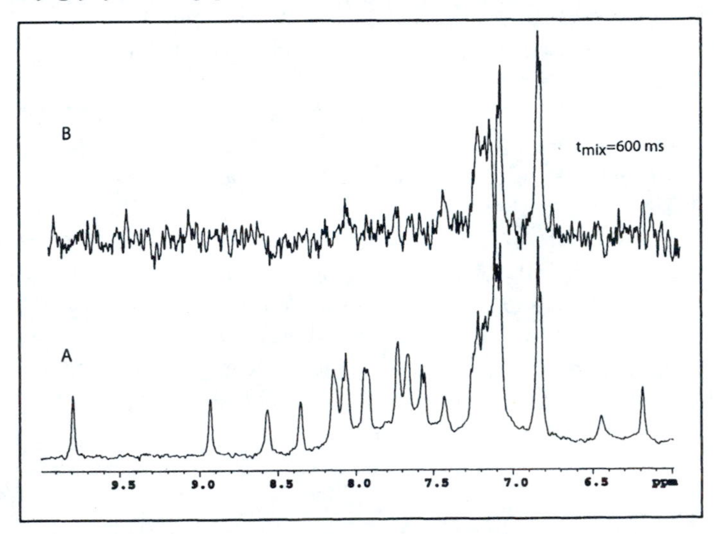

Figure 9. *$^{1}H\{^{19}F\}$ Intermolecular NOEs between the trp-cage peptide and the solvent fluorines of 30% hexafluoroisopropanol at 500 MHz (Spectrum B). A control 1D proton spectrum is shown in trace A. The sample temperature was 45°.*

When the temperature of the trp-cage solution in 30% HFIP is lowered to 5° more $^{1}H\{^{19}F\}$ NOEs are observed in both the upfield and low field portions of the proton NMR spectrum (Figure 10), but now the cross relaxation rates characteristic of these NOEs are negative (-2 x 10^{-3} s^{-1} to -6 x 10^{-3} s^{-1}) while the NOEs are predicted to be positive with values 2 to 4 x 10^{-3} s^{-1}.

A possibly over-facile interpretation of these observations is that the fluoroalcohol selectively solvates the peptide and that at low temperature the peptide-fluoroalcohol interactions become sufficiently long-lived that the NOEs arise from a dynamic situation that is more like an intramolecular interaction and less like a collisional-intermolecular interaction.

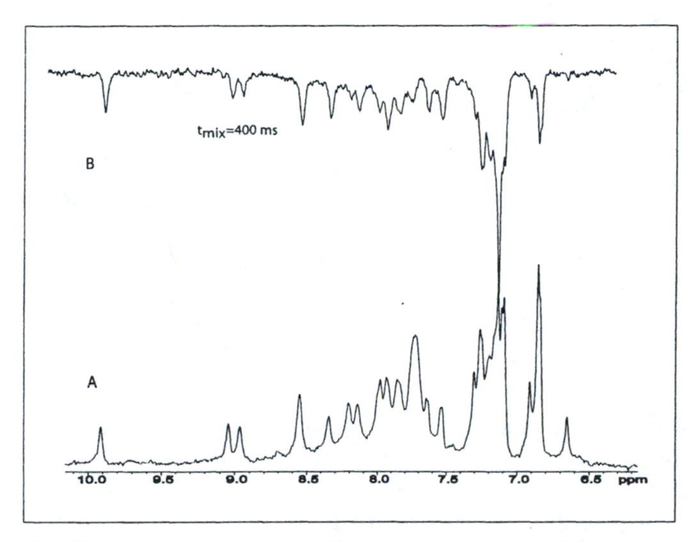

Figure 10. $^1H\{^{19}F\}$ Intermolecular NOEs between the trp-cage peptide and the solvent fluorines of 30% hexafluoroisopropanol at 500 MHz (Spectrum B). A control 1D proton spectrum is shown in trace A. The sample temperature was 5°.

Conclusions

Intermolecular $^1H\{^{19}F\}$ NOE experiments suggest that certain analogs of angiotensin II in fluoroalcohol/water mixtures exist in conformations that protect the backbone N-H and Cα-H protons from interactions with solvent molecules. Intermolecular NOE experiments indicate that in water/fluoro-alcohol mixtures small peptides can preferentially accumulate fluoroalcohol solvent molecules near their surface. In some situations, the time scale for interactions of fluoroalcohol molecules with peptides appears to be that characteristic of simple bimolecular collisions. In other instances, the formation of relatively long-lived peptide-alcohol complexes seems to take place. There are indications that the strength of fluoroalcohol interactions depend on the nature of amino acid side chains and possibly local secondary structure. Efforts to understand the considerations that define the nature of fluoroalcohol-peptide interactions will have to continue before it will be possible to make *a priori* predictions of these nature of these interactions in a particularly system.

Acknowledgments

This work was supported by the National Science Foundation (CHE-0408415) and, in its early stages, by the Petroleum Research Fund of the American Chemical Society (PRF 36776-AC). We thank these sponsors.

References

1. Tzakos, A.G., Bonvin, A. M. J. J., Troganis, A., Cordopatis, P., Amzel, M. L., Geronthanasis, I. P., *Eur. J. Biochem.*, **2003,** 270, 849-860.
2 Carpenter, K.A., Wilkes, B.C. and Schiller, P.W., *Eur. J. Biochem.*, **1998**, 251, 448-453.
3. Vine, W.H., Bruecker, D. A., Needleman, P., Marshall, G. R., *Biochemistry*, **1973**, 12, 1630-1635.
4. Hsieh, K., P. Needleman, and G. R. Marshall, *J. Med. Chem.*, **1987**, 30, 1097-1100.
5. Bovy, P.E., Getman, D. P., Matsoukas, J. M., Moore, G. J., *Biochim. Biophys. Acta*, **1991,** 1079, 23-39.
6. Gerig, J. T., *Prog. Nucl. Magn. Reson. Spectrosc.*, **1994**, 26, 293-370.
7. Greff, D., Fermandjian, S., Fromageot, P., Khosla, M. C., Smeby, R. R., Bumpus, F. M., *Eur. J. Biochem.*, **1976**, 287-305.
8. Neuhaus, D., and Williamson, M. P., *The Nuclear Overhauser Effect in Structural and Conformational Analysis*, 2nd ed., Wiley: New York, 2000.
9. Freeman, R., *Spin Choreography*, University Science Books: Sausalito, CA, 1997.
10. Ayant, Y., Belorizky, E., Fries, P., Rosset, J., *J. Phys. France*, **1977**, 38, 325-337.
11. Cowan, B. *Nuclear Magnetic Resonance and Relaxation*, Cambridge: Cambridge, 1997.
12. Skrynnikov, N. R., Khazanovoch, T. N., and Sanctuary, B. C., *Mol. Phys.*, **1997**, 91, 977-992.
13. Gerig, J. T., *J. Org. Chem.*, **2003**, 68, 53244-5248.
14. Hovagimyan, K. G. and Gerig, J. T., *J. Phys. Chem B*, **2005**, 109, 24142-24151.
15. Gerig, J. T. and Strickler, M. A., *Biopolymers*, **2002**, 64, 227-235.
16. Diaz, M. D. and Berger, S., *Magn. Reson. Chem.*, **2001**, 39, 369-373.
17. Diaz, M. D., Fioroni, M., Burger, K., Berger, S., *Chem. Eur. J.*, **2002**, 8, 1663-1669.
18. Kumar, S., Modig, K., and Halle, B., *Biochemistry*, **2003**, 42, 13708-13716.
19. Gerig, J.T., *Biophys. J.*, **2004,** 86, 3166-3175.
20. Roccatano, D., Fioroni, M., Zacharias, M., Columbo, G., *Prot. Sci*, **2005**, 14, 2582-2589.
21. Neidigh, J. W., Fesinmeye, R. M., and Anderson, N. H., *Nature Struct. Biol.*, **2002**, 9, 425-430.
22. Snow, C. D., Zagrovic, B., and Pande, V. J. *J. Am Chem. Soc.*, **2002**, 124, 14548-14549.
23. Ding, F., Buldyrev, S.V., and Dokholyan, N.V. *Biophys. J.*, **2005**, 88, 147-155.
24. Iavarone, A. T., and Parks, J. H., *J. Am. Chem. Soc.*, **2005**, 127, 8606-8607.

Chapter 24

Fluorinated Methionines as Probes in Biological Chemistry

John F. Honek

Department of Chemistry, University of Waterloo, Waterloo, Ontario N2L 3G1, Canada

Fluorinated analogs of the amino acid methionine have been synthesized and utilized as replacements for this amino acid in a variety of proteins. Application of ^{19}F nuclear magnetic resonance spectroscopy to these fluorinated proteins have increased our knowledge of protein structure and ligand binding events. As well, characterization of these fluorinated proteins using enzyme or bioactivity assays has indicated that little perturbation of the normal activity of the protein occurs upon incorporation of these analogs into proteins.

Introduction

The amino acid methionine (**1**) plays important roles as one of the twenty standard amino acids found in proteins (*1*) as well as being the precursor to *S*-adenosyl-*L*-methionine, a key molecule which can serve as the source of amino, propylamino, methyl, adenosyl and ribose carbons in biochemistry (*2, 3, 4, 5*). In the specific case of its role in proteins, there are a number of proteins that specifically require methionine in their active sites. Yet it is the third-least abundant amino acid found in proteins (*1*). The additional flexibility and polarizability of the unbranched thioether side chain of methionine compared to other "hydrophobic" amino acids such as leucine, valine, and isoleucine endows methionine with distinctive properties useful in biomolecular recognition events. For example, methionine has enhanced flexibility around its χ_3 torsion angle and

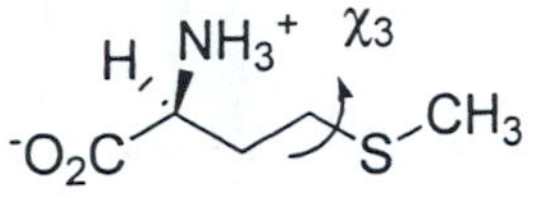

H, NH_3^+ χ_3

^-O_2C S CH_3

1

results in methionine having "hydrophobic pliability" quientessential to interaction processes such as in sequence-independent recognition of non-polar protein surfaces (*6*). Examples of these types of interactions can be seen in the biochemistry of the calcium-regulated protein calmodulin which utilizes its several methionine residues in concert to bind to a number of different peptides unrelated in primary sequence (*7, 8, 9*). Other examples of methionine contributions to molecular recognition can be found for troponin C (*10*) and collagen/gelatin binding by the protein fibronectin (*11*). Other proteins have been found to contain methionine rich regions (M-regions) and these include signal recognition particle 54 (SRP54) and PcoC, a methionine-rich copper resistance protein found in *Escherichia coli* which may use these methionine-rich regions for protein-protein interactions (*12, 13, 14*). The thioether functionality of methionine can act as a ligand to a variety of metal centers including iron centers such as is found in various *c* cytochromes (*15*) and to copper centers such as is found in plastocyanin (*16*), several azurins (*17, 18*), nitrite reductase (*19*), ascorbate oxidase (*20*) and dopamine β-hydroxylase (*21*). Control of the metal reactivity and reduction-oxidation potential of the metal can be accomplished in this way.

In order to probe the biochemistry of methionine residues in proteins, a subtle, sterically non-demanding set of substitutions would be critical. Although replacements for methionine have been made utilizing norleucine and seleno- and telluromethionine among other analogs (*22, 23, 24*), it was felt that a stepwise alteration in the side chain chemistry would be an important contribution to this area. In addition, the inclusion of a spectroscopic probe into the side chain would be ideal. Based on these and several other considerations, the application of fluorinated methionine (*S*-fluoromethyl-*L*-homocysteine) analogs to these biochemical systems seemed very appropriate. Monofluoromethionine (MFM), difluoromethionine (DFM), and trifluoromethionine (TFM) were therefore considered as suitable probes for these studies (Figure 1). The small steric size of the fluorine atom (van der

Waals radius of fluorine is 1.47 Å compared to 1.2 Å for hydrogen) should allow for minimal perturbation of the protein structure once incorporated (*25*). Since an initial approach would be to attempt to incorporate these analogs into proteins using intact bacterial cells, it was hoped that the presence of fluorine in these analogs would allow for "normal" cellular processing including ligation to the methionyl-tRNA and subsequent incorporation of these analogs into biosynthesized proteins.

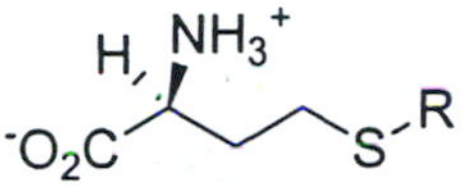

$R = CH_2F$ MFM
$= CHF_2$ DFM
$= CF_3$ TFM

Figure 1. Chemical structures of the various fluorinated methionine analogs described in this article.

Synthetic Approaches to Fluorinated Methionines

The synthesis of racemic *D,L*-TFM had been reported previously in 1957 by Dannley and Taborsky (overall yield of 11% in 5 chemical steps and utilizing $(CF_3S)_2Hg$ as the source of the trifluoromethylthio group) (*26*); however, there had been only indirect evidence of TFM incorporation into TCA-insoluble protein fractions from *S. cerevisiae* (*27*) . No detailed characterization of a TFM-containing protein had been reported.

Since MFM, DFM and TFM analogs were considered to be an ideal set of biochemical probes to explore the biological effects of stepwise introduction of fluorine into methionine, facile synthetic procedures for DFM and TFM were developed and attempts were made to prepare MFM. A number of synthetic procedures were explored to prepare protected MFM. The most successful involved the reaction of diethylaminosulfur trifluoride with protected methionine sulfoxide (*28*). Although protected MFM was indeed prepared, attempts to deprotect the monofluorosulfide were unsuccessful with evidence for hydrolysis of the monofluorosulfide moiety to homocysteine, fluoride and formaldehyde (Figure 2). A report by Janzen and co-workers had also appeared and made use of XeF_2 to prepare protected MFM but no attempts to isolate MFM itself were reported (*29*).

The DFM and TFM analogs were readily prepared (Figure 3) by basic ring opening of racemic N-acetylhomocysteine thiolactone to form the thiol(ate) which could then be alkylated with either chlorodifluoromethane or trifluoromethyl iodide followed by enzymatic resolution using porcine acylase to produce optically pure *L*-DFM or *L*-TFM respectively (*28*). *L*-DFM has also been independently synthesized by Tsushima and co-workers in a very facile manner by Na/NH_3 reduction of *L*-homocystine followed by reaction with $CHClF_2$ (*30*). This method is also applicable to the synthesis of *L*-TFM (unpublished results) although *L*-homocystine is a more expensive starting material than racemic N-acetylhomocysteine thiolactone.

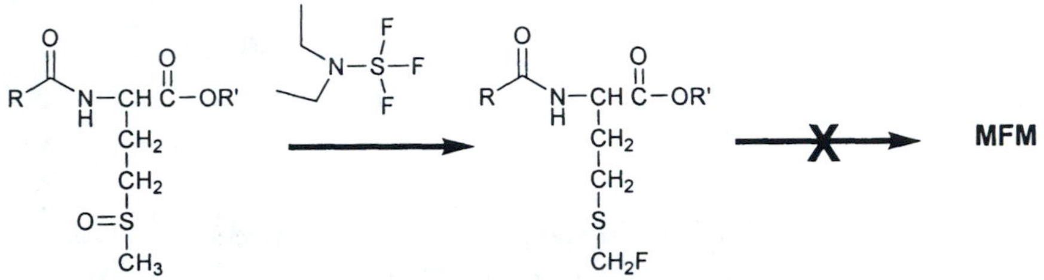

Figure 2. Overview of attempts to synthesize monofluoromethionine (MFM).

Properties of Fluorinated Methionines

An evaluation of the effects fluorination might have on the steric, electronic and conformational properties of methionine is necessary to ascertain the nature of contributions that fluorination might make on protein structure and function. As such the largest perturbation to protein structure would be found in the CF_3^- group. The trifluoromethyl group has been suggested to be similar in size to an isopropyl group (*25*) and *ab initio* calculations at the RHF/6-31G* level indicate van der Waals' volumes of 71.2, 80.9, and 84.4 $Å^3$ for CH_3SCH_3, CH_3SCHF_2, and CH_3SCF_3, respectively (*31*). Electronic properties of the thiomethyl function in methionine are also seen to change upon fluorination. Recent calculations at the B3LYP/6-31+G(d,p)//B3LYP/6-31+G(d,p) levels have determined the Mulliken charges on the sulfur atom to be 0.079 e, 0.139 e and 0.212 e for CH_3SCH_3, CH_3SCHF_2 and CH_3SCF_3 respectively. We have previously noted the reduced nucleophilicity of the sulfur in trifluoromethionine compared to the sulfur in methionine (*32, 33*). Potential energy scans for methyl 1-propyl sulfide and trifluoromethyl 1-propyl sulfide at the B3LYP/6-31+G(d,p) level are shown in Figure 4 (*34*). Rigid potential energy scans for these two compounds as models for the respective side chains of methionine and TFM indicate that there is a higher energy minima for TFM that occurs at angles of ~ 105° and ~255° (at ~ 1 kcal above the global minium) but at ~ 80° and ~ 280° for the methyl 1-propyl sulfide (at ~ 0.7 kcal above its global minimum). Relaxed potential energy scans reduce these differences and place the secondary minima at ~80° and ~ 280° for both sulfide model compounds.

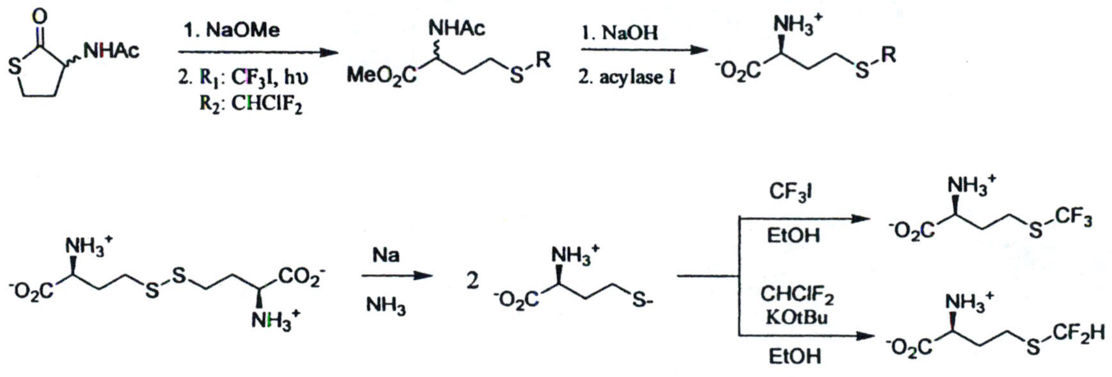

Figure 3. Synthetic approaches to L-DFM and L-TFM analogs.

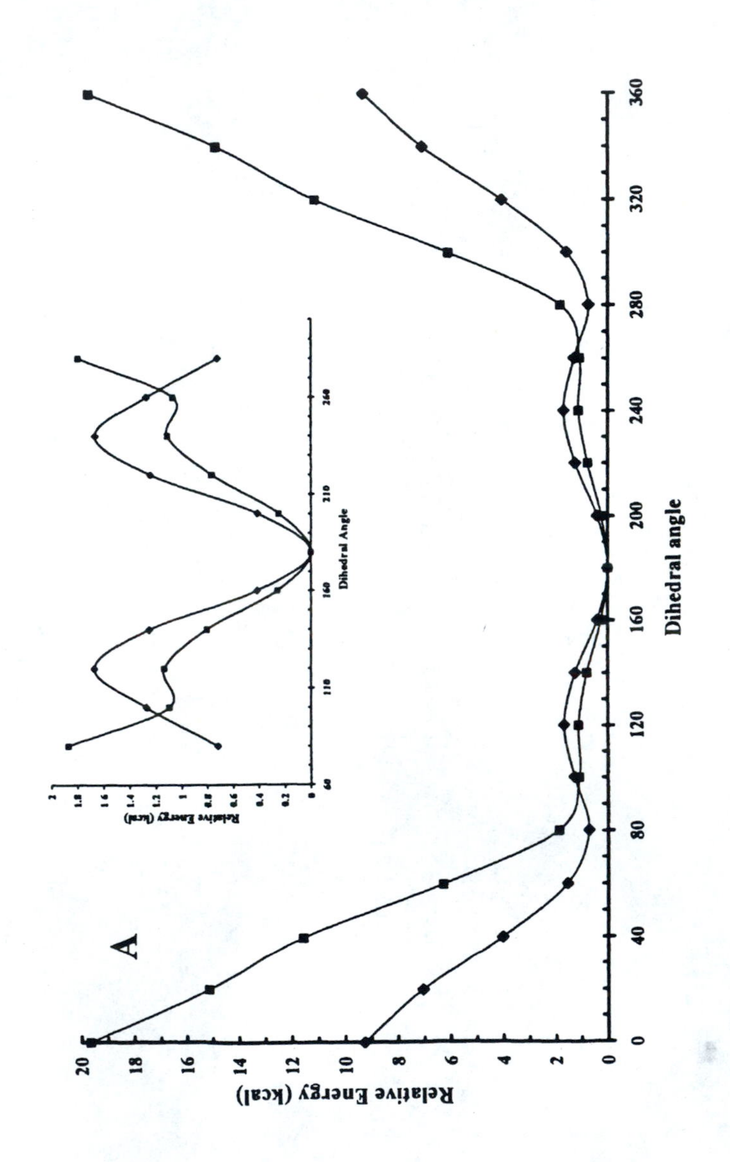

A
Relative Energy (kcal)
Dihedral angle
0
2
4
6
8
10
12
14
16
18
20
0
40
80
120
160
200
240
280
320
360
Relative Energy (kcal)
Dihedral Angle

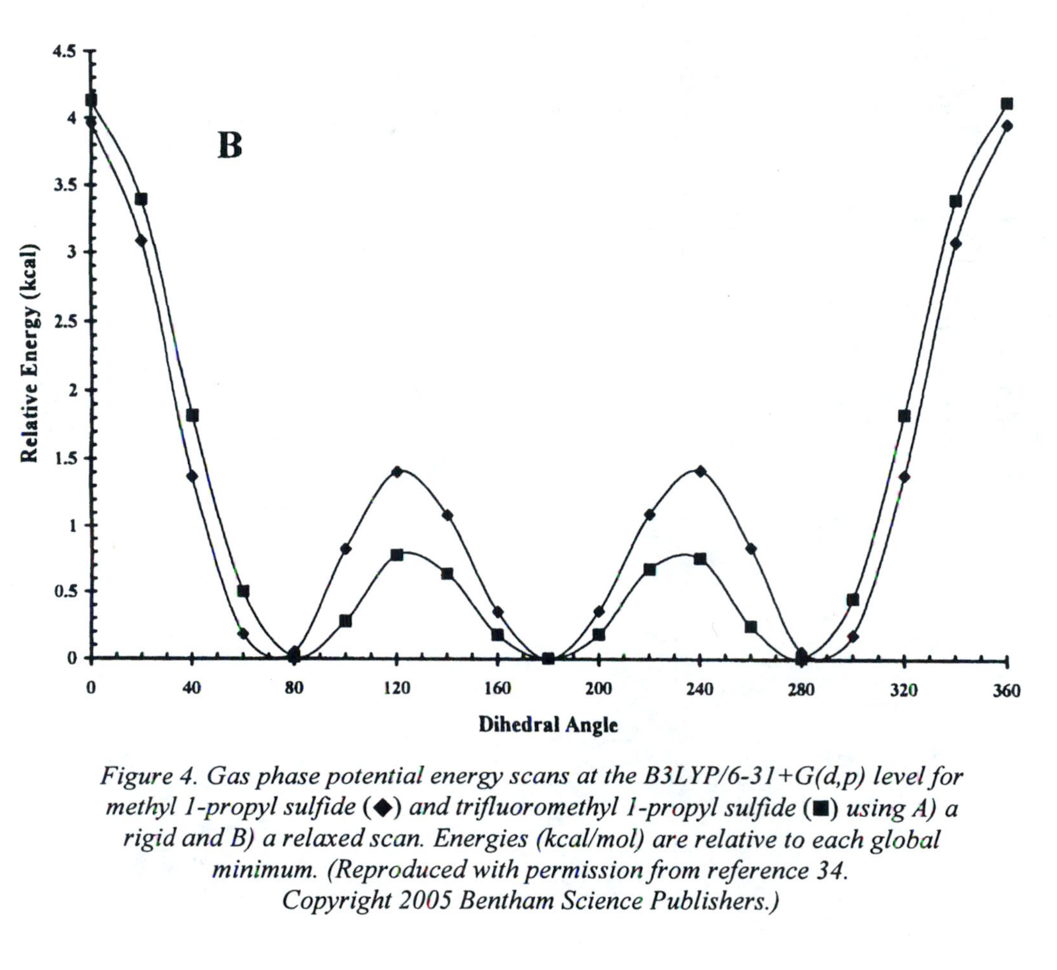

Figure 4. Gas phase potential energy scans at the B3LYP/6-31+G(d,p) level for methyl 1-propyl sulfide (◆) and trifluoromethyl 1-propyl sulfide (■) using A) a rigid and B) a relaxed scan. Energies (kcal/mol) are relative to each global minimum. (Reproduced with permission from reference 34. Copyright 2005 Bentham Science Publishers.)

A first application of these fluorinated analogs was to evaluate their effects on a simpler system such as a bioactive peptide which contained a critical methionine residue for activity. The chemotactic tripeptide N-formylMet-Leu-Phe was studied with DFM and TFM replacing the key methionine residue (*31*). This methionine residue had been shown to be a sensitive residue to alter and almost all analogs at this position have resulted in substantially lower chemotactic activity to human leukocytes. Of interest is that the DFM and TFM containing peptides were found to enhance the chemotactic activity of the peptide to human neutrophiles and indicated that not only were these fluorinated analogs capable of replacing methionine residues in critical biomolecules but that they could endow the resulting peptides with enhanced bioactivity. Further studies of bioactive peptides containing DFM and TFM are certainly warranted.

Bioincorporation into Proteins

Since the natural abundance of the ^{19}F isotope in fluorine compounds is 100%, its sensitivity to nuclear magnetic resonance (NMR) detection has been determined to be 83% that of ^{1}H and there is high sensitivity of ^{19}F shifts to surrounding chemical environments, ^{19}F NMR can be an extremely valuable technique for investigating protein structure and dynamics (*35, 36, 37*). Hence it was of interest to investigate the ability of DFM and TFM to be incorporated into proteins and if successful, to study their ^{19}F NMR characteristics.

The lysozyme from bacteriophage lambda (LaL) is a late gene product and is critical to the propagation of this phage from *Escherichia coli (38)*. It is involved in the hydrolysis of bacterial peptidoglycan to produce 1,6-anhydromuropeptides, and as such is mechanistically different from the more familiar hen egg white lysozyme. This monomeric protein (158 amino acids; M_r = 17,825 Da) contains methionine at positions 1, 14 and 107. Induction of LaL expression in the presence of the fluorinated methionines resulted in incorporation of DFM and TFM into the enzyme (*32*). The modified enzymes were found to be catalytically active. In the case of TFM incorporation, small amounts of methionine itself also were required due to cellular toxicity of TFM in its absence. Incorporation levels for TFM ranged from production of 2 mg/L of 70 % TFM-labeled LaL to 15 mg/L of 31 % TFM-labeled LaL. The resulting ensemble of TFM-labeled LaL (such as one TFM in either positions 1, 14 or 107, as well as dilabeled LaL containing TFM at either positions 1/14 or 1/107 or 14/107 as well as the completely trilabeled TFM-LaL) exhibited ^{19}F spectra as shown in Figure 5 (*32*). The spectrum is interesting in that instead of the expected three resonances resulting from each of the three positions substituted with TFM, the spectrum of the ensemble of proteins exhibited four separate resonances centered at – 39.32, -39.82, -39.99 and -40.11 ppm with the two resonances at positions -39.99 ppm and -40.11 ppm varying in intensity depending upon the extent of labeling (low incorporation producing a higher intensity for the ^{19}F resonance at -39.99 ppm and higher incorporation levels enhancing the resonance at -40.11 ppm). Assignment of the ^{19}F resonances were made using site directed mutagenesis. Application of a gadolinium(III) ethylenediaminetetraacetic acid (Gd(III)EDTA) paramagnetic line broadening agent indicated solvent exposure for TFM at positions 1 and 107 but not position 14 as well as the fact that TFM at position 107 could produce two distinct resonances depending on whether Met or TFM was present in another position of the protein (Figure 5) (*39*). X-ray structure determination of LaL indcated that indeed Met14 is a buried core amino acid (*40*). The structure also offered insight into the unusual two independent resonances for position 107. Met 14 is a core amino acid and is tightly surrounded by a number of side chains from other parts of LaL, including residue side chains from an alpha helix that is in contact with Met107. It appears that when Met is present at position 14, TFM at position 107 exhibits a resonance at -39.99 ppm, but if TFM is also present at position 14, TFM at position 107 exhibits a resonance at - 40.11 ppm. Hence the CF_3– group can result in minor structural perturbation which is detected by the fluorine nucleus at position 107, indicating the extreme sensitivity of ^{19}F to its environment. Nevertheless, the fluorinated enzyme is still catalytically active against *Escherichia coli* cells.

Investigation of the incorporation of DFM into LaL was also of interest. The smaller CHF_2^- group was accommodated better by the bacterial expression systems such that overproduction of labeled LaL with DFM was quantitative and expression levels were identical to wild type protein (> 20 mg/L) (*41*). The proton coupled ^{19}F NMR spectrum of the homogeneous $(DFM)_3$-LaL exhibited resonances centered at – 91.2 ppm (doublet), - 92.5 ppm (broader doublet) and two sets of quartets centered at – 94.2 ppm and – 95.5 ppm. Resonances at – 91.2 ppm and – 92.5 ppm each integrated to one DFM whereas the entire set of two quartets integrated to one DFM. Site-directed mutagenesis and paramagnetic line broadening experiments indicated resonances centered at – 91.2 ppm and – 92.5 ppm originated from DFM at positions 1 and 107 respectively and the resonances centered at – 94.2 ppm and – 95.5 ppm originated from DFM at position 14 (and confirmed by two dimensional ^{19}F-^{19}F-COSY NMR experiments). In order to explain this complexity, it is important to note that the fluorine atoms in L-DFM itself are diastereotopic due to the chiral center in the amino acid. This diastereotopicity results in the two fluorines being chemically shift inequivalent. In the the tightly packed core of the protein, DFM14 side chain rotation is slowed and causes the chemical shift inequivalence between the two fluorines in the CHF_2^- moiety to increase. The two quartets are the result of the two fluorines having separate chemical shifts as well as ^{19}F-^{19}F coupling and proton coupling (^{1}H-^{19}F) from the hydrogen on the CHF_2^- moiety.

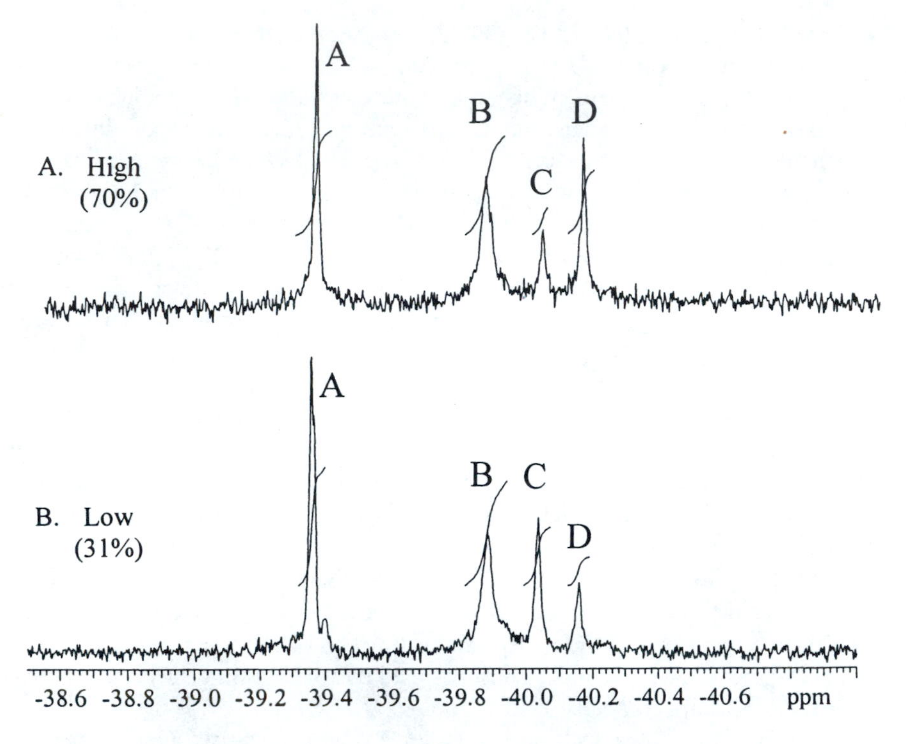

Figure 5. ^{19}F NMR spectra (376.3 MHz) of high- and low-level incorporated TFM-labeled lysozymes. (Reproduced from reference 32. Copyright 1997 American Chemical Society.)

This effect can be readily exemplified for the amino acid itself when the ^{19}F spectrum of L-DFM is determined at low temperatures (-80 °C) (*41*). Hence DFM appears to be an excellent ^{19}F probe for studying the rotational dynamics of methionine residues in proteins. The labeled enzyme has identical catalytic activity to the wild type enzyme indicating little structural change to the enzyme on DFM incorporation. Other proteins have also been successfully studied with TFM or DFM and include calmodulin (*42*), leucine-isoleucine-valine (LIV) binding protein (*43*), a mutant of the green fluorescent protein (GFP) (*44*) and the alkaline protease from *Pseudomonas aeruginosa (45).*

Direct Interactions of DFM and TFM with Cells and Enzymes

DFM and TFM are also interesting due to their direct interactions with various cells and enzymes. For example, it has been reported that TFM is cytotoxic to cells (*27, 32, 46*) and that TFM can be decomposed to toxic difluorothiophosgene by the enzyme γ-cystathioninase (*47*) and methionine γlyase (*48, 49*), the latter being of interest as a potential antimicrobial against *Trichomonas vaginalis.* DFM has also been reported to have activity against aphids (*30*). With respect to the above research on incorporation into proteins, it was of interest to determine the nature of the differential ease with which DFM was incorporated into proteins compared to TFM. Detailed crystallographic studies were taken with these and other methionine analogs in order to obtain atomistic level information on molecular recognition with the fluorinated analogs of methionine. A critical step in the efficiency of incorporation of a methionine analog into a protein is its ability to be charged onto the tRNA for methionine so that it can be used in ribosomal protein biosynthesis. The enzyme Met-tRNA synthetase catalyzes the reaction of ATP and methionine as shown in Figure 6.

X-ray structures of the *E. coli* methionyl-tRNA synthetase were obtained in the presence of DFM and TFM and compared to that obtained for normal methionine (*50*). In the case of normal methionine, the substrate binding site is fabricated when the enzyme binds methionine which induces the rearrangement of several aromatic residues including Tyr15, Trp253 and Phe300 which move toward the liganded amino acid. These rotations are propagated away from the active site, up to Phe304 and Trp229. These conformational changes create a set of new stacking interactions between Phe300 and Trp253 as well as between Phe304 and Trp229 and thereby contribute to a reduction of the binding pocket volume. This set of interactions also occurs when DFM binds to the methionyl-tRNA synthetase. However in the case of TFM binding the Trp229, Phe300 and Phe304 side chains remain in the open state characteristic of the free enzyme. Trp253 does move towards TFM but does not reach the exact position encountered in the enzyme-methionine or the enzyme-DFM complexes since steric hindrance by one fluorine atom of TFM occurs. In parallel, Tyr15 flips towards the NH group of TFM, but like Trp253, the movement of this residue is incomplete. The reduced affinity of TFM to the enzyme compared to normal methionine or even DFM is seen by the K_d values of 220 μM, 0.6 μM and 38 μM respectively for complexes of the enzyme in the presenceof adenosine and pyrophosphate (*50*). However, the binding affinities of the Met and TFM analogs of the reactive adenylate intermediate (Figure 7) have K_i values of 0.25 nM and 2.4 nM respectively (*34*). This indicates that although TFM does not bind to the enzyme as well as methionine, the addition of adenosine (or ATP) would allow an increased binding interaction once it was coupled to the adenyl group.

Once the methionine residue is attached to its appropriate tRNA (in the case of Met there are two tRNA, an initiator tRNA and an elongator tRNA primed by the same methionyl-tRNA synthetase), bacteria formylate the amino group of the methionine which is linked to the initiator tRNA (*51*). This formyltransferase utilizes N-10-formyltetrahydrofolate to produce the formylated methionyl-tRNA. This charged tRNA is now able to serve to initiate protein biosynthesis. Once a

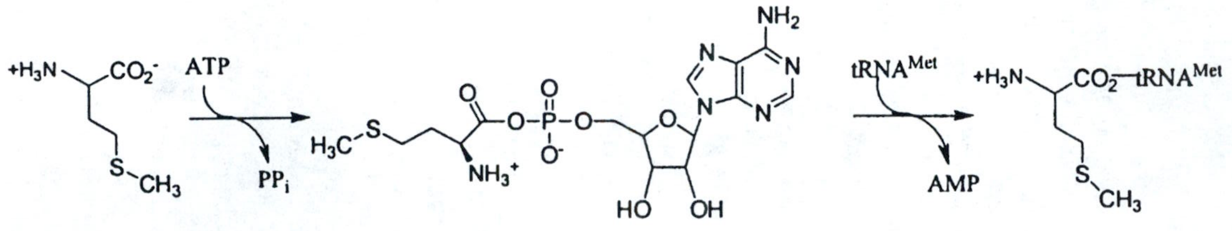

Figure 6. The chemical reaction catalyzed by methionyl-tRNA synthetase.

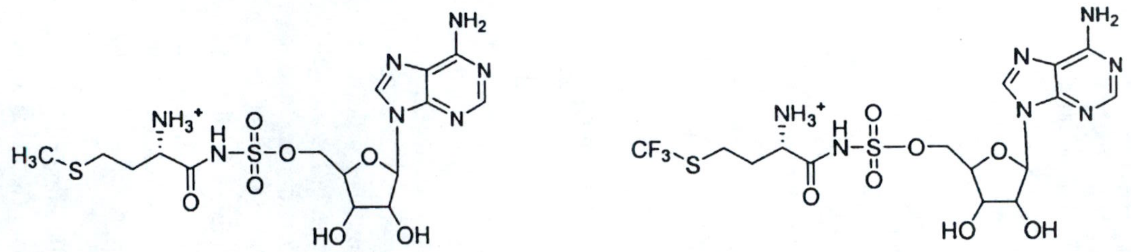

Figure 7. Chemical structures of the methionyl- and trifluoromethionyl-sulfonamide analogs of the acyl adenylate intermediate formed in the methionyl-tRNA synthetase reaction.

protein is biosynthesized, the formyl group may be removed by an intracellular deformylase. The resulting protein may also undergo a subsequent reaction, that of excision of the N-terminal methionine. The enzyme, methionine aminopeptidase, is selective for the methionine residue and will hydrolytically remove the N-terminal methionine depending upon the nature of the amino acid at position 2 which needs to be small and uncharged. It has been found that deformylation and methionine removal are essential for prokaryotic viability and hence these enzymes have become targets for recent drug development. It was of interest to determine how the fluorinated methionines themselves would bind to the *E. coli* methionine aminopeptidase. Co-crystallization of TFM and other methionine analogs with the *E. coli* methionine aminopeptidase were undertaken in collaboration with Professor Matthews' group (*52*). TFM was found to bind to the methionine aminopeptidase utilizing the same cavity to accommodate the methionyl side chain. Very little perturbation seems to occur to accommodate the bulkier fluorinated side chain. An interesting question is whether the fluorination of methionine affects its processing from proteins that contain it by the methionine aminopeptidase. Recently we have incorporated DFM into the alkaline protease from *P. aeruginosa* (*45*). The DNA construct for this protein produces an N-terminal methionine as well as the active site methionine important in this class of protease. DFM incorporation produces a labeled protein that is virtually identical to the unlabeled protease with respect to catalytic efficiency as well as thermal stability. However it was noted that there appears to be reduced processing of the DFM N-terminal residue from the protein compared to the unfluorinated protease by methionine aminopeptidase. It has been noted in studies of incorporation of TFM into a mutant of the green fluorescent protein that the TFM residue is not removed from the expressed protein (*44*). Both these studies indicate that fluorination of the N-terminal methionine reduces or may even prevent excision from a protein by the enzyme methionine aminopeptidase. Further studies are warranted in this regard.

Conclusions

It is evident that fluorinated methionine analogs can be incorporated into peptides and proteins and serve as useful surrogates of methionine and as ^{19}F NMR spectroscopic probes of protein structure and side chain dynamics as well as of ligand binding. In addition, these analogs have allowed us to obtain detailed atomistic information on the interaction of fluorinated molecules and enzymes. They have allowed us to increase our knowledge of methionine binding site cavities and protein accommodation of nonnatural side chain functionalities in general. Future work will need to explore DFM and TFM interactions with metal centers such as inorganic complexes as well as metalloenzyme centers. As well, the effect that fluorination has upon methionine processing from proteins is another area worthy of further exploration, as well as their application to the further study of methionine-rich regions in proteins.

Acknowledgements

The author's gratitude goes to former and current members of his laboratory who have made not only wonderful contributions to our understanding of carbon-sulfur chemistry and biochemistry but to contributing to an outstanding scientific environment. The author gratefully acknowledges his past and present collaborators mentioned in the references cited in this article. Funding from NSERC (Canada) and the University of Waterloo are also gratefully acknowleged.

References

1. Voet, D. and Voet, J. G., *Biochemistry*. 3rd ed. 2004: John Wiley. 1591.
2. Fontecave, M., Atta, M. and Mulliez, E. *Trends Biochem Sci*. **2004**, *29*, 243-249.
3. Ikeguchi, Y., Bewley, M. C. and Pegg, A. E. *J Biochem (Tokyo)*. **2006**, *139*, 1-9.
4. Schubert, H. L., Blumenthal, R. M. and Cheng, X. *Trends Biochem Sci*. **2003**, *28*, 329-335.
5. Iwata-Reuyl, D. *Bioorg Chem*. **2003**, *31*, 24-43.
6. Gellman, S. H. *Biochemistry*. **1991**, *30*, 6633-6636.
7. Meador, W. E., Means, A. R. and Quiocho, F. A. *Science*. **1992**, *257*, 1251-1255.
8. Ikura, M., Clore, G. M., Gronenborn, A. M., Zhu, G., Klee, C. B. and Bax, A. *Science*. **1992**, *256*, 632-638.
9. Zhang, M. and Vogel, H. J. *J Mol Biol*. **1994**, *239*, 545-554.
10. Lin, X., Krudy, G. A., Howarth, J., Brito, R. M., Rosevear, P. R. and Putkey, J. A. *Biochemistry*. **1994**, *33*, 14434-14442.
11. Miles, A. M. and Smith, R. L. *Biochemistry*. **1993**, *32*, 8168-8178.
12. Gowda, K., Clemons, W. M., Jr., Zwieb, C. and Black, S. D. *Protein Sci*. **1999**, *8*, 1144-1151.

13. Jiang, J., Nadas, I. A., Kim, M. A. and Franz, K. J. *Inorg Chem*. **2005**, *44*, 9787-9794.
14. Wernimont, A. K., Huffman, D. L., Finney, L. A., Demeler, B., O'Halloran, T. V. and Rosenzweig, A. C. *J Biol Inorg Chem*. **2003**, *8*, 185-194.
15. Ubbink, M., Campos, A. P., Teixeira, M., Hunt, N. I., Hill, H. A. and Canters, G. W. *Biochemistry*. **1994**, *33*, 10051-10059.
16. Collyer, C. A., Guss, J. M., Sugimura, Y., Yoshizaki, F. and Freeman, H. C. *J Mol Biol*. **1990**, *211*, 617-632.
17. Murphy, L. M., Strange, R. W., Karlsson, B. G., Lundberg, L. G., Pascher, T., Reinhammar, B. and Hasnain, S. S. *Biochemistry*. **1993**, *32*, 1965-1975.
18. Salgado, J., Jimenez, H. R., Moratal, J. M., Kroes, S., Warmerdam, G. C. and Canters, G. W. *Biochemistry*. **1996**, *35*, 1810-1819.
19. Basumallick, L., Szilagyi, R. K., Zhao, Y., Shapleigh, J. P., Scholes, C. P. and Solomon, E. I. *J Am Chem Soc*. **2003**, *125*, 14784-14792.
20. Messerschmidt, A., Ladenstein, R., Huber, R., Bolognesi, M., Avigliano, L., Petruzzelli, R., Rossi, A. and Finazzi-Agro, A. *J Mol Biol*. **1992**, *224*, 179-205.
21. Blackburn, N. J., Hasnain, S. S., Pettingill, T. M. and Strange, R. W. *J Biol Chem*. **1991**, *266*, 23120-23127.
22. Yuan, T. and Vogel, H. J. *Protein Sci*. **1999**, *8*, 113-121.
23. Gassner, N. C., Baase, W. A., Hausrath, A. C. and Matthews, B. W. *J Mol Biol*. **1999**, *294*, 17-20.
24. Budisa, N., Karnbrock, W., Steinbacher, S., Humm, A., Prade, L., Neuefeind, T., Moroder, L. and Huber, R. *J Mol Biol*. **1997**, *270*, 616-623.
25. O'Hagan, D. and Rzepa, H. S. *J Chem Soc Chem Commun*. **1997**, 645-652.
26. Dannley, R. L. and Taborsky, R. G. *J Org Chem*. **1957**, *22*, 1275-1276.
27. Colombani, F., Cherest, H. and de Robichon-Szulmajster, H. *J Bacteriol*. **1975**, *122*, 375-384.
28. Houston, M. E. and Honek, J. F. *J Chem Soc Chem Commun*. **1989**, 761-762.
29. Janzen, J. F., Wang, P. M. C. and Lemire, A. E. *J Fluorine Chem*. **1983**, *22*, 557-559.
30. Tsushima, T., Ishihara, S. and Fujita, Y. *Tetrahedron Lett*. **1990**, *31*, 3017-3018.
31. Houston, M. E., Harvath, L. and Honek, J. F. *Bioorg Med Chem Lett*. **1997**, *7*, 3007-3012.
32. Duewel, H., Daub, E., Robinson, V. and Honek, J. F. *Biochemistry*. **1997**, *36*, 3404-3416.
33. Duewel, H. S. and Honek, J. F. *J Protein Chem*. **1998**, *17*, 337-350.

34. Vaughan, M. D., Sampson, P. B., Daub, E. and Honek, J. F. *Med Chem.* **2005**, *1*, 227-237.
35. Gerig, J. T. *Prog NMR Spectrosc.* **1994**, *26*, 293-370.
36. Danielson, M. A. and Falke, J. J. *Annu Rev Biophys Biomol Struct.* **1996**, *25*, 163-195.
37. Frieden, C., Hoeltzli, S. D. and Bann, J. G. *Methods Enzymol.* **2004**, *380*, 400-415.
38. Duewel, H. S., Daub, E. and Honek, J. F. *Biochim Biophys Acta.* **1995**, *1247*, 149-158.
39. Duewel, H. S., Daub, E., Robinson, V. and Honek, J. F. *Biochemistry.* **2001**, *40*, 13167-13176.
40. Leung, A. K., Duewel, H. S., Honek, J. F. and Berghuis, A. M. *Biochemistry.* **2001**, *40*, 5665-5673.
41. Vaughan, M. D., Cleve, P., Robinson, V., Duewel, H. S. and Honek, J. F. *J Am Chem Soc.* **1999**, *121*, 8475-8478.
42. McIntyre, D. D., Yuan, T. and Vogel, H. J. *Prog Biophys Mol Biol.* **1997**, *65*, P-A1-27.
43. Salopek-Sondi, B., Vaughan, M. D., Skeels, M. C., Honek, J. F. and Luck, L. A. *J Biomol Struct Dyn.* **2003**, *21*, 235-246.
44. Budisa, N., Pipitone, O., Siwanowicz, I., Rubini, M., Pal, P. P., Holak, T. A. and Gelmi, M. L. *Chem Biodivers.* **2004**, *1*, 1465-1475.
45. Walasek, P. and Honek, J. F. *BMC Biochem.* **2005**, *6*, 21.
46. Singer, R. A., Johnston, G. C. and Bedard, D. *Proc Natl Acad Sci U S A.* **1978**, *75*, 6083-6087.
47. Alston, T. A. and Bright, H. J. *Biochem Pharmacol.* **1983**, *32*, 947-950.
48. Yoshimura, M., Nakano, Y. and Koga, T. *Biochem Biophys Res Commun.* **2002**, *292*, 964-968.
49. Coombs, G. H. and Mottram, J. C. *Antimicrob Agents Chemother.* **2001**, *45*, 1743-1745.
50. Crepin, T., Schmitt, E., Mechulam, Y., Sampson, P. B., Vaughan, M. D., Honek, J. F. and Blanquet, S. *J Mol Biol.* **2003**, *332*, 59-72.
51. Vaughan, M. D., Sampson, P. B. and Honek, J. F. *Curr Med Chem.* **2002**, *9*, 385-409.
52. Lowther, W. T., Zhang, Y., Sampson, P. B., Honek, J. F. and Matthews, B. W. *Biochemistry.* **1999**, *38*, 14810-14819.

Chapter 25

Synthesis and Conformational Analysis of Fluorine-Containing Oligopeptides

Takashi Yamazaki[1], Takamasa Kitamoto[2], and Shunsuke Marubayashi[2]

[1]Strategic Research Initiative for Future Nano-Science and Technology, Institute of Symbiotic Science and Technology, Tokyo University of Agriculture and Technology, 2–24–16, Nakamachi, Koganei 184–8588, Japan
[2]Department of Applied Chemistry, Graduate School of Engineering, Tokyo University of Agriculture and Technology, 2–24–16, Nakamachi, Koganei 184–8588, Japan

Preparation of hexapeptide Tyr-D-Ser-Gly-Phe-Leu-Thr possessing an interesting μ-selective opioid activity as well as its F_3-analog at the Thr terminal methyl group was carried out to investigate the effect of highly electronegative fluorine atoms toward the conformation of the molecules on the basis of information obtained from various types of ^{1}H NMR techniques.

Modification of organic molecules with fluorine is one of the major strategies for enhancement or improvement of the original biological activities (*1*). As already documented (*2*), this atom is known to be the second smallest and the most electronegative element of all atoms. So, introduction of fluorine atoms into organic molecules imposes only the slightest steric perturbation, but, at the same time, causes possible and significant repulsive or attractive interaction with neighboring electronically negative or positive functional groups, respectively. Such effects including the hydrogen bond (HB) forming ability should result in conformational changes which sometimes lead to a serious decrease in or even loss of the inherent biological activities.

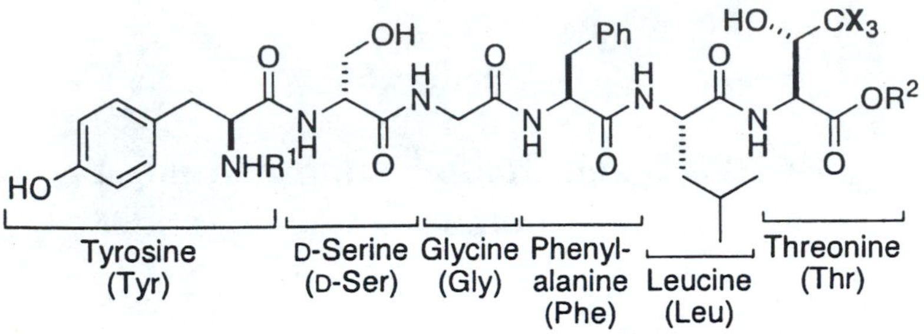

Compound	X	R^1	R^2
1a	H	H	H
2a	H	Boc	Bn
1b	F	H	H
2b	F	Boc	Bn

Figure 1. Structure of the Target F_0- and F_3-Hexapeptides

Our strong interest in the three dimensional shapes of organic molecules prompted us to initiate a new research project for clarification of the relationship between conformations of molecules and fluorine substitution. For this purpose, the hexapeptide Tyr-D-Ser-Gly-Phe-Leu-Thr **1a** which was reported to possess interesting μ-selective opioid activity (*3*) was selected as our target material. The introduction of three fluorine atoms at the methyl moiety of the *C*-terminal Thr was proposed. Comparison between the F_3-hexapeptide **2b** and native **2a** with *N*- as well as *C*-protective groups unambiguously demonstrated clear conformational alteration on the basis of extensive ^{1}H NMR spectroscopy analyses.

Preparation of the Target Peptides 1 and 2

In Scheme 1 is shown the preparation scheme for Tyr-D-Ser-Gly-Phe-Leu-Thr **1a** and its fluorinated analog **1b** with three fluorine atoms at the Me group of the *N*-terminal Thr. This peptide formation was mainly carried out by the solution phase method using WSC (*4*) and HOBt (*4*) in the presence of triethylamine as the standard condensation conditions. Among various types of condensation required for two amino acids, preparation of dipeptide, Boc-Tyr-D-Ser-OBn, was problematic and furnished the product only in moderate yield (Entry 1 in Table 1). Change of the solvents with higher polarity (entries 2 and 3), base (entry 4), or the condensation reagent to DPPA (*4*) (entries 5 to 7)

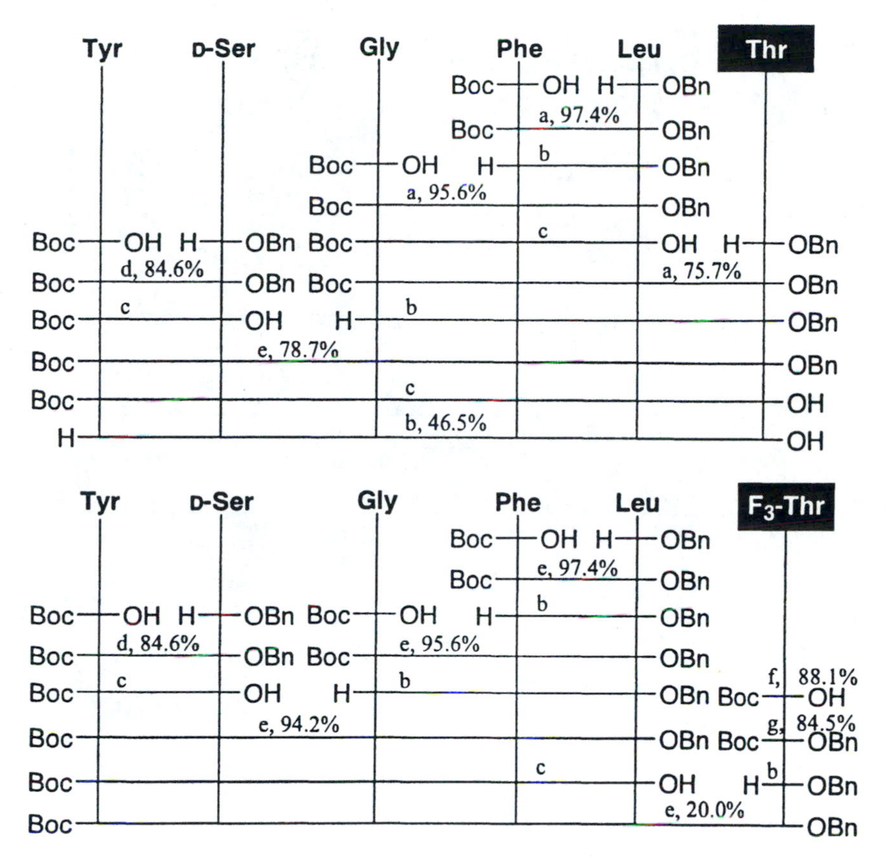

a) WSC·HCl, HOBt·H_2O, Et_3N/CH_2Cl_2, overnight, rt. b) CF_3CO_2H/CH_2Cl_2, 1~2 h, 0 °C. c) H_2, Pd/C/MeOH, 6 h ~ overnight, rt. d) CDMT, NMM (2 equiv)/AcOEt, overnight, rt. e) WSC·HCl, HOBt·H_2O, Et_3N/DMF, overnight, rt. f) $(Boc)_2O$, Na_2CO_3/1,4-dioxane, H_2O, overnight, rt. g) BnBr, $NaHCO_3$/DMF, overnight, rt.

*Scheme1. Preparation of F_0- and F_3-Hexapeptides **1a** and **1b***

affected the chemical yields slightly. After some investigation, it was found that CDMT was an appropriate reagent (*4*) as reported previously by Kaminski (*5*). This substance worked in a proper manner in the presence of 2.2 equiv of NMM for condensation, DMT-MM (*6*). NMM played an activating role with (*4*) where an equimolar amount of base was utilized for the facile conversion of CDMT at ambient temperature to the "true" reagent responsible carboxylic acids and the appropriate amines then attacked to the intermediary ester (Scheme 1). As shown in entry 8, in spite of reaching a similar yield as in entry 1, utilization of AcOEt as the solvent demonstrated significant increase the efficiency of the reaction to attain 84.6% isolated yield in the construction of the desired

Table 1. Condensation Conditions of Boc-Tyr and D-Ser-OBn

Boc-Tyr-OH + TsOH·H-D-Ser-OBn —(rt, overnight)→ Boc-Tyr-D-Ser-OBn

Entry	Solvent	Condensation Reagent[1)]	Base[1)]	(equiv)	Isolated Yield (%)
1	CH_2Cl_2	WSC·HCl	Et_3N	1.1	42.5
2	THF	WSC·HCl	Et_3N	1.1	36.6
3	DMF	WSC·HCl	Et_3N	1.1	49.1
4	DMF	WSC·HCl	*i*-Pr_2NEt	1.1	27.2
5	DMF	DPPA	Et_3N	1.3	18.0
6	DMF	DPPA	Et_3N	2.0	53.0
7	DMF	DPPA	*i*-Pr_2NEt	2.2	58.8
8	THF	CDMT	NMM	2.2	48.5
9	AcOEt	CDMT	NMM	2.2	84.6

1) For abbreviation names, see ref 4.

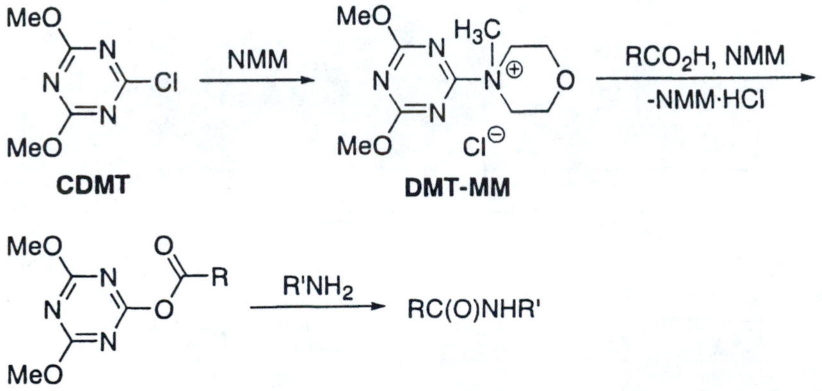

Scheme 2. Reaction of CDMT and DMT-MM

dipeptide Boc-Tyr-D-Ser-OBn with the both hydroxy groups in Tyr and D-Ser unprotected (entry 9). Another advantage worthy of mention is that this process does not require an inert atmosphere.

Trifluorinated Thr was prepared following the procedure of Soloshonok and coworkers (*7*). *N*-benzylated Pro **3** was condensed with *o*-aminobenzophenone to yield the amide **4** which was then utilized for the formation of the nickel-Gly complex **5** (Scheme 3). Due to the original stereogenic center in Pro, the *N*-protecting benzyl group efficiently blocks the top side of this square-planar Ni complex **5**, which allowing the diastereoselective *si* face approach of electrophiles after conversion of **5** to the corresponding enolate. Moreover, in

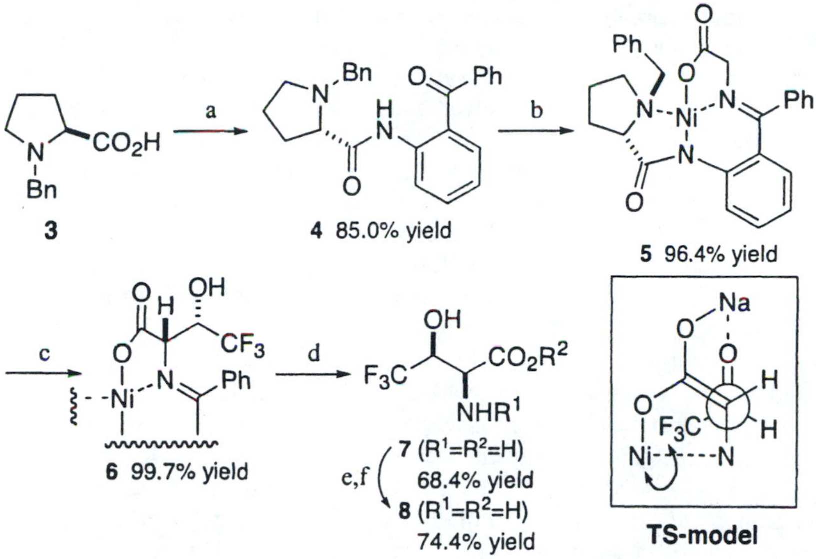

3

4 85.0% yield

5 96.4% yield

6 99.7% yield

7 (R^1=R^2=H) 68.4% yield

8 (R^1=R^2=H) 74.4% yield

TS-model

a) *o*-aminobenzophenone, $SOCl_2/CH_2Cl_2$, overnight, rt. b) glycine, $Ni(NO_3)_2 \cdot 6H_2O$, KOH/MeOH, 1 h, 60 °C. c) CF_3CHO, MeONa/MeOH, 15 min, 60 °C. d) HCl aq./MeOH, 15 min, reflux. e) Boc_2O, Na_2CO_3/1,4-dioxane, H_2O, overnight, rt. f) BnBr, $NaHCO_3$/ DMF, overnight, rt.

Scheme 3. Preparation of F_3-Thr 7

spite of the sterically more congested environment as opposed to the case when the opposite carbonyl face of CF_3CHO was used (thus, relationship between H and CF_3 in **TS-model** is opposite), possible interaction of fluorine with the nickel atom renders the electrophilic trifluoroacetaldehyde accessible to the *si* face of the carbonyl moiety. Accordingly, the adduct **6** is constructed with the desired correct stereochemistry as the major product. Requisite F_3-Thr **7** was obtained in 68.4% yield by the simple acidic hydrolysis of the complex **6** and in 55.9% overall yield from the compound **3** in four steps.

The diastereoselectivity of F_3-Thr **7** was determined to be >96.0% de: the ^{19}F NMR of **7** afforded three sets of doublet peaks due to contamination by a small impurity even after chromatographic purification. We used the integration values of the largest and the second largest peaks to guarantee the least level of selectivity. For the determination of enantio-selectivity, **7** was converted to a diastereomeric mixture of the corresponding α-phenethyl amides giving five peaks by ^{19}F NMR. This fact led to the conclusion that the selectivity was at least 91.7% ee. **7** was then subjected to the usual Boc protection condition, followed by the base-mediated benzylation of the carboxylic acid (*8*) to successfully convert it into the Boc-protected benzyl ester **8**.

To obtain the target hexapeptides **1a** and **1b,** one of the most efficient routes would be the combination of the common intermediate *N*-terminal pentapeptide and Thr or F_3-Thr. However, contrary to our expectation, this sequence using both amino acids gave rise to lower chemical yields, 20~40%, along with epimerization of the products which rendered the isolation difficult. After investigation of the order of the condensation while neglecting this efficiency, non-fluorinated **2a** was successfully obtained as shown in Scheme 1. The usual deprotection procedure converted **2a** into **1a** in moderate yield. In the case of F_3-Thr after reaction, Boc removal was found to be difficult. Condensation with Boc-Gly-Phe-Leu-OH or even with Boc-Leu-OH disclosed formation of undesirable materials by TLC analysis, apparently proving epimerization possibly at the α-position of F_3-Thr. At this position, the proton would be activated by the strongly electron-withdrawing CF_3 moiety. For this reason, along with the fact that we fortunately managed to purify the crude stereoisomeric mixture, condensation of the *N*-terminal pentapeptide and F_3-Thr yielded **2b** in 20% isolated yield. The final step was deprotection of the Boc and benzyl groups from **2b**, but in spite of our extensive investigation of the reaction conditions, we still did not find out acceptable routes for cleavage at the both terminii. These facts prompted us to compare conformational preference of **2a** and **2b** by ^{1}H NMR.

Spectroscopic Comparison of 2a and 2b

As described above, with the desired target hexapeptides Boc-Tyr-D-Ser-Gly-Phe-Leu-Thr-OBn **2a** and -F_3-Thr-OBn **2b** in hand, one-dimensional ^{1}H NMR (*9*) and then two-dimensional COSY and NOESY spectra were required for unambiguous assignment of the resonances. In Figure 2 are shown the NOE correlation diagrams for the both materials in DMSO-d_6. The NOE is expressed in three levels, strong, medium, and weak, corresponding to the different threshold of the NOE.

For the non-fluorinated **2a**, the most remarkable characteristics is the long range correlation between the OH groups of the *N*-terminal Tyr and the *C*-terminal Thr. Taking the pKa values of these OH groups into account (*10*), it is quite apparent that the more acidic Tyr OH moiety should serve as the HB donor and Thr OH as the acceptor. Strong NOEs were observed between the Phe-Leu and Leu-Thr residues with the CH_α(i)-NH(i+1) type (*11*) correlation recognized as supportive of a β-sheet structure. NOEs between Tyr-D-Ser and D-Ser-Gly were medium, and those of Gly-Phe weak. It would be reasonable to consider that the shape of **2a** deviated from planarity at the part where weaker NOE cross peaks were observed as a reflection of longer distances between two protons in question. In connection with the above mentioned HB, this information suggests that this molecule possesses a somewhat helical- or spherical-type shape, which allows other residues like Tyr-NH and Leu-CH_3 to come closer to one and another.

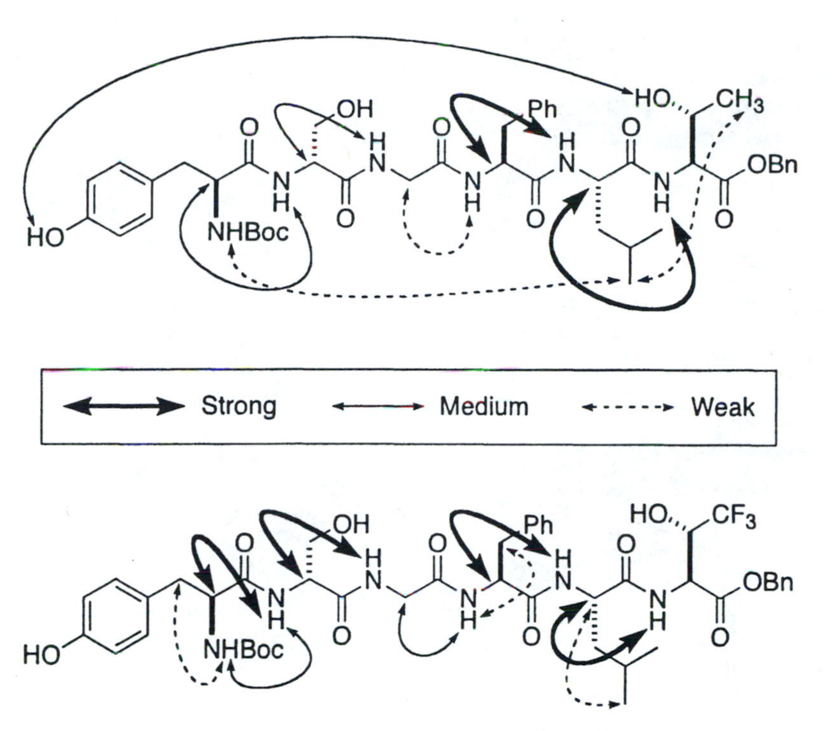

*Figure 2. NOESY Spectra Correlation of the F_0- and F_3-Hexapeptides **2a** and **2b***

We have tried the similar NOESY experiments using three different solvent system such as $CDCl_3$:DMSO-d_6=4:1, 2:3, and 0:1 for clarification of the influence of solvent polarity on hexapeptide conformations. In the least polar solvent, $CDCl_3$:DMSO-d_6=4:1, strong long range correlations were found between Tyr-O*H* and Thr-O*H* as well as D-Ser-O*H*, and Leu-N*H* and Tyr aromatic protons as well as benzyl methylene protons of the *C*-terminal ester. On the other hand, the characteristic NOEs between the CH_α(i) and N*H*(i+1) protons were almost extinguished except for the Leu-Thr positions. Although the former Tyr-O*H*-related HB still existed in the solvent system of $CDCl_3$:DMSO-d_6=2:3, HB based on Leu-N*H* extinguished. HB with Tyr-O*H* and Thr-O*H* was consistently observed in all cases clearly reflecting their inherently intense interaction. In general, the greater the solvent polarity, the higher the possibility of the extended shape as a consequence of interruption of HB by coordination of polar solvents. This fact allowed us to readily understand the strength of the HB between Tyr-O*H* and Thr-O*H*.

Interestingly, on assumption of the conformation of **2b** there is a total disappearance of the long range NOE correlation between Tyr-O*H* and Thr-O*H*, the dominant factor for control of the conformation of **2a**. This phenomenon would be readily understood when we take decrease of the F_3-Thr-*O*H basicity into account due to the neighboring strongly electron-withdrawing CF_3 moiety.

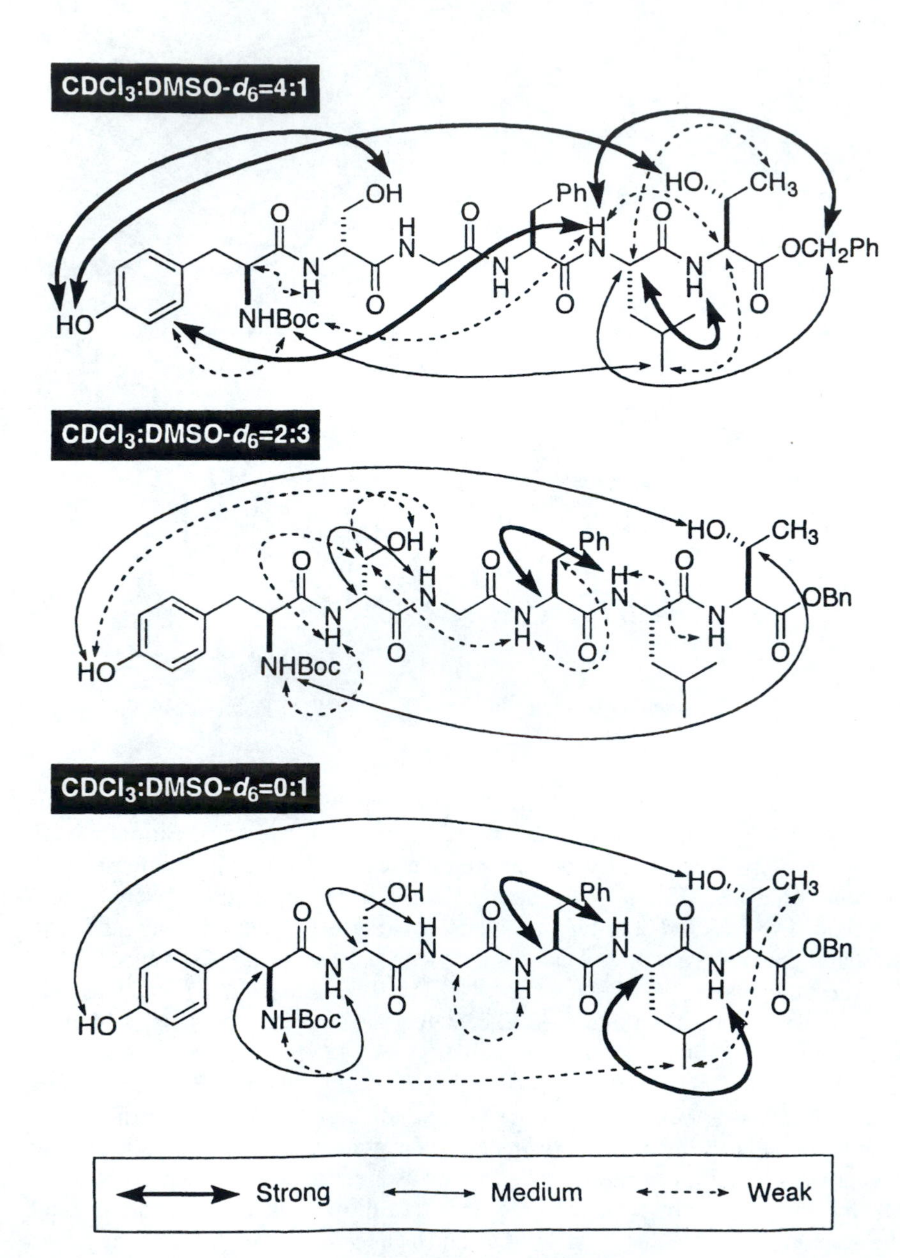

*Figure 3. NOE Correlation of **2a** in Different Solvent Polarity*

Such variation in physical properties would be at least roughly estimated from the pKa difference between ethanol and 2,2,2-triflouoroethanol of ΔpKa=3.1 (*10*): thus, entry of three fluorine atoms should effectively weaken the HB accepting ability of the alcoholic oxygen in F_3-Thr and, as a consequence, the long range correlation was completely vanished. As a result, more intense four CH_α(i) and N*H*(i+1) relationships out of the five were observed as the strong NOE correlation with the other medium, indicating that the F_3-hexapeptide **2b** possessed the extended conformation with the β-sheet-like flat shape of the hexapeptide backbone.

Investigation was also performed for **2a** and **2b** to identify which amide protons participated in HB formation. These interactions would be indicated by the chemical shift measurement with continuous variation of the NMR solvent polarity by mixing less polar $CDCl_3$ to more polar DMSO-d_6 while maintaining the total solvent volume and sample concentration constant. Free amide protons should experience grater effects of the polarity while protons included in HB are usually insensitive to such alteration.

In Table 2 and 3 were shown the chemical shift data of amide as well as alcoholic protons in five types of combination of $CDCl_3$ and DMSO-d_6 in a range of 0 to 80% of the former component. Comparison of $\Delta\delta_{0\text{-}40}$ values of relatively polar regions in both compounds clearly manifested chemical shifts of Gly-N*H*, Phe-NH, D-Ser-N*H*, and D-Ser-O*H* that were affected only slightly with the chemical shift difference of less than 0.1 ppm (50.0 Hz). The F_3-Thr-O*H* proton was added to this list when the hexapeptides **2b** were the substrate. Approximate 2 ppm lower field shift of this O*H* group with respect to the Thr-O*H* moiety unambiguously supported the above discussion of the less basic nature of alcoholic oxygen of F_3-Thr-O*H*. This more acidic proton shifts to the lower magnetic field.

Table 2. Polarity Dependence of Selected Protons in 2a

	$CDCl_3$ in DMSO-d_6-$CDCl_3$ (%)					$\Delta\delta_{0\text{-}40}$ [a]
	0	16	40	64	80	(ppm)
Leu-N*H*	8.197	8.204	8.074	7.835	7.632	-0.123
Gly-N*H*	8.069	8.086	8.109	8.183	8.208	**0.040**
Phe-N*H*	7.998	8.005	7.920	7.587	7.658	**-0.078**
Thr-N*H*	7.962	7.954	7.762	7.918	7.352	-0.200
D-Ser-N*H*	7.940	7.954	7.920	7.240	7.826	**-0.020**
Tyr-N*H*	6.876	6.874	6.729	6.676	6.400	-0.147
Tyr-O*H*	9.136	9.133	9.008	8.956	8.869	-0.128
Thr-OH	5.018		4.898		4.607	-0.120
D-Ser-OH	4.989		5.007		5.080	**0.018**

a) $\Delta\delta_{0\text{-}40}$ is the chemical shift difference between the δ values in $CDCl_3$-DMSO-d_6=0:100 and in 40:60.

Table 3. Polarity Dependence of Selected Protons in 2b

	$CDCl_3$ in DMSO-d_6-$CDCl_3$ (%)					$\Delta\delta_{0\text{-}40}$ [a]
	0	16	40	64	80	(ppm)
Leu-N*H*	8.266	8.164	8.076	7.960	7.706	-0.190
Gly-N*H*	8.121	8.128	8.130	8.192	8.150	**0.009**
Phe-N*H*	8.006	8.036	8.0 14	7.892	7.692	**-0.052**
F_3-Thr-N*H*	8.450	8.343	8.260	8.014	7.827	-0.190
D-Ser-N*H*	7.941	7.924	7.924	7.856	7.629	**-0.017**
Tyr-N*H*	6.898	6.828	6.78 1	6.676	6.248	-0.117
Tyr-O*H*	9.147	9.133	9.043	8.938	8.831	-0.104
F_3-Thr-O*H*	6.996	6.947	6.904	6.804	6.7 18	-0.092
D-Ser-OH	5.047	5.058	5.090	5.086	4.917	0.043

a) $\Delta\delta_{0\text{-}40}$ is the chemical shift difference between the δ values in $CDCl_3$-DMSO-d_6=0:100 and in 40:60.

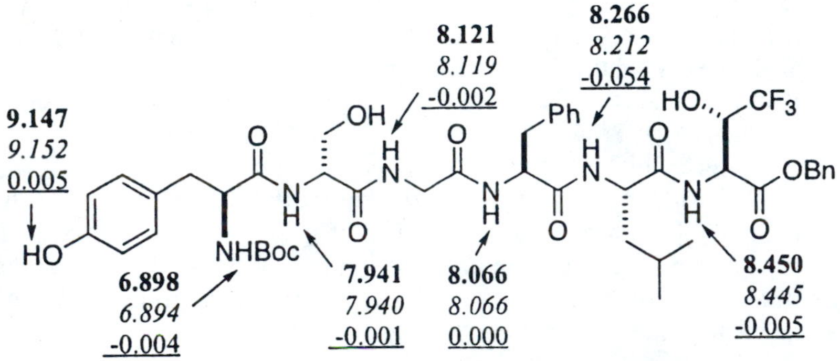

Chemical shift in **20 mM** and *0.1 mM* DMSO-d_6 solutions and their difference in ppm.

Figure 4. Concentration Dependence of Selected Protons in F_3-Hexapeptides ***2b***

As described above, the repeated NOE correlations between CH_α(i) and N*H*(i+l) in the case of both **2a** and **2b** anticipated their β-sheet-type linear conformation. Moreover, it was spectroscopically established that these amide protons were not included in any intermolecular interaction as they showed quite small deviation of up to 0.005 ppm (2.5 Hz) in many cases after 200 fold dilution of the NMR sample (Figure 4). This NMR information led us to the unambiguous conformational conclusion that the hexapeptide **2b** possessed an extended structure, and in the case of nonfluorinated **2a**, the left half portion of the molecule demonstrating weaker NOEs suggested molecular deformation to allow an energetically favorable contact between Tyr-O*H* and Thr-O*H* at both ends.

Conclusion

Substitution of hydrogen atoms for fluorine is one of the most familiar strategies for attainment of better biological activities. We have successfully demonstrated in this article that such processes can affect the original three dimensional shape of compounds. These conformational changes would further lead to possible reduction of favorable contacts with enzyme active sites and hence an undesirable decrease in biological activity, illustrating the importance of the appropriate design of molecules, *i.e.* the site and number of fluorine atoms to be introduced.

References

1. (a) Ramachandran, P. V., Ed.; *Asymmetric Fluoroorganic Chemistry;* ACS Symposium Series 746; American Chemical Society: Washington, DC, 2000. (b) Soloshonok, V. A., Ed.; *Enantiocontrolled Synthesis of Fluoro-Organic Compounds;* John Wiley & Sons: New York, 1999.
2. (a) Smart, B. E., *J. Fluorine Chem.* **2001**, *109,* 3. (b) Hiyama, T.; *Organofluorine Compounds. Chemistry and Applications;* Springer-Verlag: Berlin, 2000. (c) Kitazume, T., Yamazaki, T.; *Experimental Methods in Organic Fluorine Chemistry;* Kodansha, Gordon and Breach: Tokyo, 1998. (d) Smart, B. E. In *Organofluorine Chemistry. Principles and Commercial Applications;* Banks, R. E., Smart, B. E., Tatlow, J. C., Eds.; Plenum Press: New York, 1994, pp 57.
3. Roques, B. P. *J. Pept. Sci.* **2001**, *7,* 63.
4. WSC: Water soluble carbodiimide hydrochloride or l-ethyl-3-(3-dimethyl-aminopropyl)carbodiimide hydrochloride. HOBt: 1-Hydroxybenzotriazole. DPPA: Diphenylphosphine azide. CDMT: 2-Chloro-4,6-dimethoxy-1,3,5-triazine. NMM: N-Methylmorpholine.
5. Kaminski, Z. J. *Synthesis* **1987**, 917.
6. (a) Kunishima, M.; Kawachi, C.; Morita, J.; Terao, K.; Iwasaki, F.; Tani, S. *Tetrahedron* **1999**, *55,* 13159. (b) Kunishima, M.; Kawachi, C.; Hioki, K.; Terao, K.; Tani, S. *Tetrahedron* **2001**, *57,* 1551.
7. (a) Soloshonok, V. A.; Kukhar', V. P.; Galushko, S. V.; Svistunova, N. Y.; Avilov, D. V.; Kuz'mina, N. A.; Raevski, N. I.; Struchkov, Y. T.; Pysarevsky, A. P.; Belokon, Y. N. *J. Chem. Soc. Perkin Trans. 1,* **1993**, 3143. (b) Soloshonok, V. A.; Avilov, D. V.; Kukhar', V. P. *Tetrahedron: Asym.* **1995**, *6,* 1741. (c) Belokon, Y. N.; Tararov, V. I.; Maleev, V. I.; Savel'eva, T. F.; Ryzhov, M. G. *Tetrahedron: Asym.* **1998**, *9,* 4249.
8. Xing, X.; Fichera, A.; Kumar, K. *J. Org. Chem.* **2002**, *67,* 1722.
9. All ^{1}H NMR spectra were obtained by JNM-LA500 (JEOL, 500 MHz).
10. pKa values of phenol, ethanol, and 2,2,2-trifluoroethanol are reported to be 9.99, 15.5, and 12.37 at 25 °C, respectively. See, CRC *Handbook of Chemistry and Physics, 82nd Edition;* Lide, D. R., Ed.; CRC Press: New York, 2001.
11. CH_α(i) and N*H*(i+l) mean a carbonyl α proton and an amide proton of the *i*th and *i+1*th amino residues from the *N*-terminal, respectively.

Chapter 26

Fluorinated Inhibitors of Matrix Metalloproteinases

Fiorenza Viani, Luca Bruché, Gabriele Candiani, Florent Huguenot, Monika Jagodzinska, Raffaella Maffezzoni, Nathalie Moussier, Monica Sani, Roberta Sinisi, and Matteo Zanda*

C.N.R.–Instituto di Chimica del Riconoscimento Molecolare, and Dipartimento di Chimica, Materiali ed Ingegneria Chimica "G. Natta" del Politecnico di Milano, via Mancinelli 7, I–20131 Milano, Italy

The synthesis of potent inhibitors of matrix metalloproteinases (MMPs) bearing trifluoromethyl or difluoromethyl groups in purely aliphatic positions, and the effect of the incorporation of the fluoroalkyl functions on the inhibitory potency is presented.

Incorporation of fluorine into organic molecules is an effective strategy for improving and modifying the biological activity (*1*). In particular, the trifluoromethyl (Tfm) group occupies a prominent position in medicinal chemistry as a substituent of peculiar properties. It is in fact highly hydrophobic, electron-rich, sterically demanding, moreover it can provide high *in vivo* stability, and features a good mimicry with several naturally occurring residues such as methyl, isopropyl, isobutyl, phenyl, etc. (*2*).

Matrix metalloproteinases (MMPs) are zinc (II)-dependent proteolytic enzymes involved in the degradation of the extracellular matrix (*3*). More than 25 human MMPs have been identified so far. Loss in the regulation of their activity can result in the pathological destruction of connective tissue, a process associated with a number of severe diseases, such as cancer and arthritis. The inhibition of various MMPs has been envisaged as a strategy for the therapeutic intervention against such pathologies. To date, however, a number of drawbacks have hampered the successful exploitation of MMPs as pharmacological targets. In particular, the toxicity demonstrated by many MMPs' inhibitors in clinical trials has been ascribed to nonspecific inhibition.

Recently (*4*) we described the synthesis of the Tfm-malic hydroxamate **1** (Fig. 1). Unfortunately, the effect of the replacement of the α-CH_3 group of the highly potent parent molecule **A** (*5*) with a CF_3 on the inhibition of MMP-3 and -9 was detrimental, and a drop of about 5 orders of magnitude of inhibitory potency was observed.

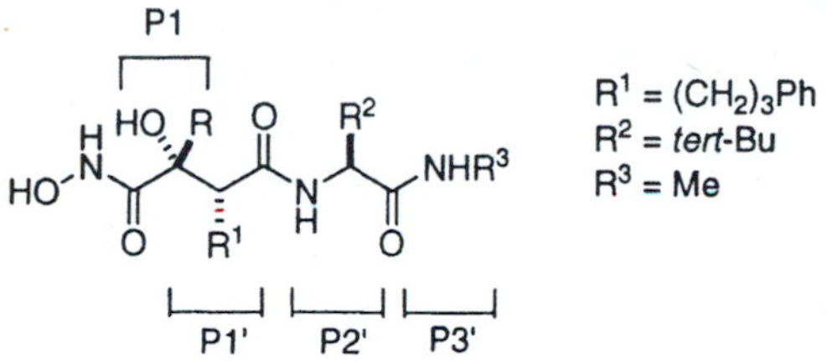

Figure 1. Low potency Tfm-malic hydroxamate inhibitor of MMP-2 and -9.

Two concurrent reasons for the reduced activity of the fluorinated inhibitors were postulated: (a) reduced coordinating strength of the neighbouring hydroxamate group by the electron-withdrawing Tfm, and (b) the need of the fluorinated molecule to adopt within the binding site a conformation which does not coincide with its minimum-energy conformation in solution.

In order to probe the importance of the first effect (reduced coordinating strength), and assess the compatibility of a Tfm group in α-position to the hydroxamic function, we decided to investigate the effect of a Tfm group positioned as R^1 substituent in structures **2** (Fig. 2), analogs of molecules **B** (Fig. 2) which were recently reported by Becker et al. as potent inhibitors of MMP-2, MMP-9 and MMP-1 3 (*6*). Remarkably, these molecules exhibited limited inhibition of MMP-1, an enzyme thought to be responsible for the musculoskeletal side effect observed clinically with the broad-spectrum MMP inhibitor marimastat (*7*). Whereas a large number of different alkyl and alkylaryl residues where well tolerated as nitrogen substituents R^2, only inhibitors **B** bearing R^1 = H, CH_3 or Ph were described.

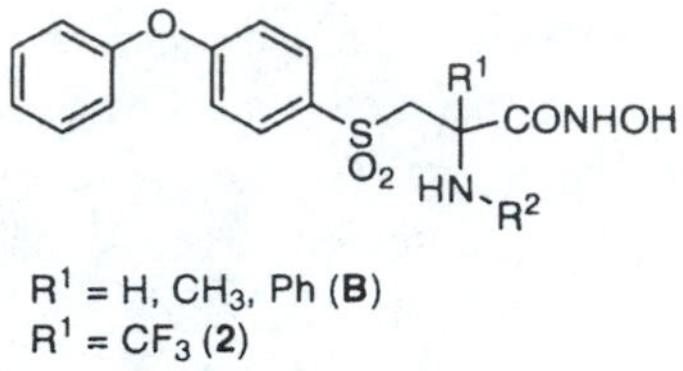

R^1 = H, CH_3, Ph (**B**)
R^1 = CF_3 (**2**)

Figure 2. α-Amino hydroxamic MMPs inhibitor.

In the synthesis of the hydroxamic acid **2a**, having a free quaternary amino group (Figure 3), the intermediate sulfone **4** was synthesized by Pd-catalyzed reaction of phenol with the *p*-Br derivative **3** (*8*). Lithiation of **4**, followed by nucleophilic addition to the *N*-Cbz imine of trifluoropyruvate **5** (*9*) afforded the α-Tfm α-amino acid derivative **6** in fair yields. Basic hydrolysis of the ester function delivered the carboxylic acid **7**, that was submitted to condensation with *O*-Bn-hydroxylamine affording the hydroxamate **8**, which was hydrogenolyzed to the target molecule **2a**.

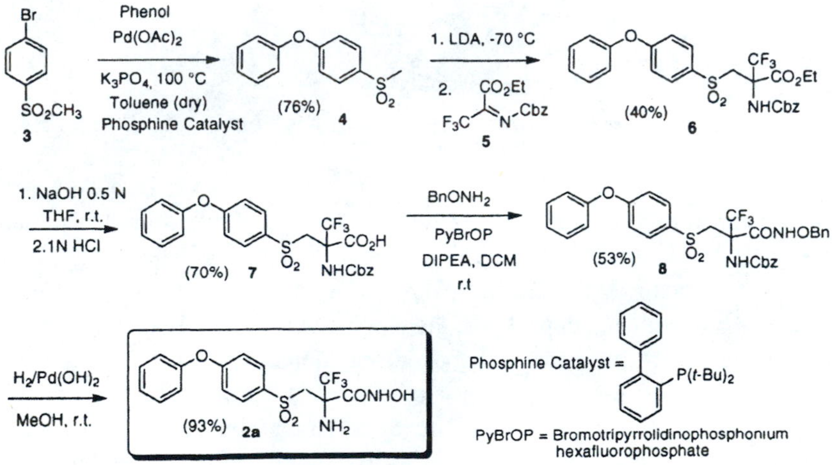

*Figure 3. Synthesis of **2a**.*

The *N*-alkylated analogs **2b-d** (Fig. 4) were prepared according to a modified procedure. To this end, the sulfenyl diaryl ether **10** was prepared from phenol and **9** using an Ullmann-type reaction (*10*), then oxidized to the sulfoxide **11**. Lithiation and Mannich-type reaction with **5** afforded a nearly equimolar mixture of sulfoxide diastereomers **12**, which were deoxygenated to the racemic sulfide **13** according to the Oae/Drabowicz protocol (*11*). *N*-Alkylation occurred in good to excellent yields affording the corresponding sulfides **14b-d.** Due to the presence of the sulfide function, which could interfere with a Pd-catalyzed hydrogenolysis, the Cbz group was cleaved with HBr (*12*), affording the secondary amines **15b-d** in nearly quantitative yields. Ester saponification could be performed smoothly, affording the carboxylic acids **16b-d** in good to excellent yields. Coupling of **16b-d** with *O*-Bn-hydroxylamine afforded the sulfenyl hydroxamates **17b-d,** which were oxidized to sulfones **18b-d.** The target hydroxamic acids **2b-d** were obtained in fair yields by hydrogenolysis with the Pearlman catalyst.

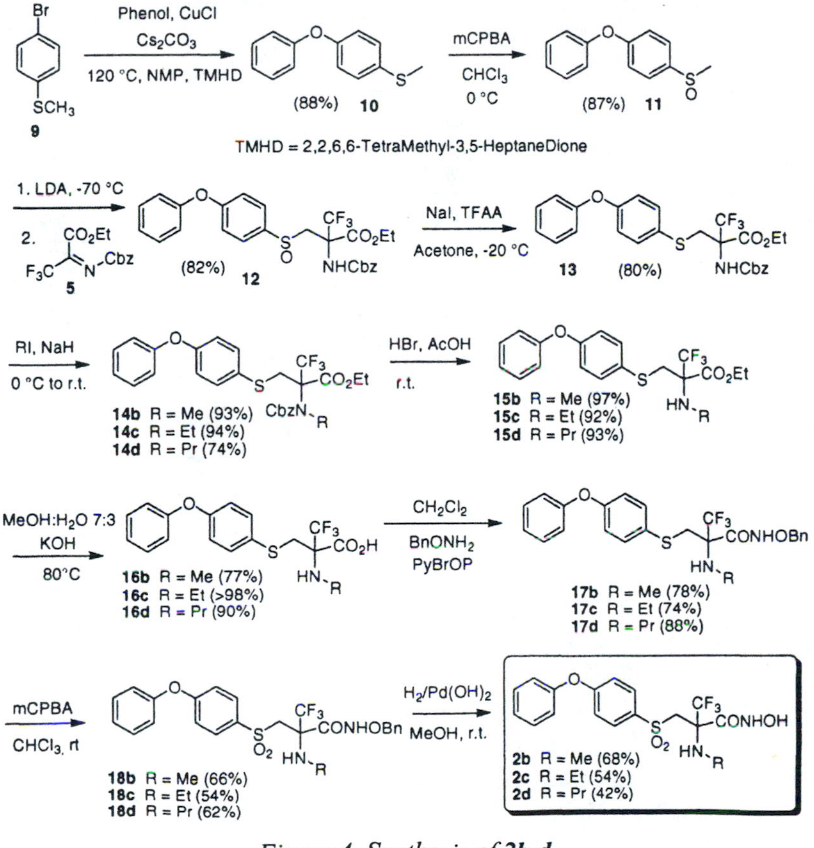

TMHD = 2,2,6,6-TetraMethyl-3,5-HeptaneDione

*Figure 4. Synthesis of **2b-d**.*

Inhibition tests on **2a-d** were performed on the catalytic domains of MMP-1, MMP-3 and MMP-9.

Table 1. Effect of the compounds 2 on different MMPs' proteolytic activity.

Compound	IC$_{50}$/MMP-3 (nM)	IC$_{50}$/MMP-9 (nM)	IC$_5$(/MMP-1 (nM)
2a	14	1	> 5000
2b	32	Ca. 20	n.a.
2c	28	63[a]	n.a.
2d	53	59	n.a.

n.a. = not available; [a] The IC$_{50}$ value for this compound was erroneously reported to be 1982 nM (see Ref. 7).

The primary α-amino hydroxamate **2a** is the most potent compound (Table 1), but it is worth noting that all of them are nanomolar inhibitors of MMP-3 and MMP-9. Even more importantly, **2a** showed excellent selectivity versus MMP-1 (>5000-fold). These results show that a Tfm group can be successfully used as a substituent in MMPs inhibitors, and is very well tolerated by the enzymes. This also suggests that the main reason for the weak potency of compounds **A** is likely to be due to a conformational change induced by the Tfm group, that lowers the affinity to the MMPs active site.

Interestingly, the X-ray crystallographic structure of the complex of **2a** with the truncated catalytic domain of MMP-9 showed that the (*R*)-enantiomer preferentially binds, whereas the Tfm group does not make significant interactions with the active site residues of the protease, resulting essentially water exposed (*13*).

Next we decided to study the effect of a fluoroalkyl group installed in a position far away from the hydroxamic function, in order to better understand the effect of the unique stereoelectronic features of fluoralkyl functions in a purely aliphatic position on MMPs inhibitory potency (*14*). For this purpose we chose as a model system a structurally very simple class of hydroxamate inhibitors bearing an arylsulfone moiety in γ-position, such as **C** (Fig. 5) which showed nanomolar inhibitory potency against MMP-2, 3 and 13 (*15,16*)

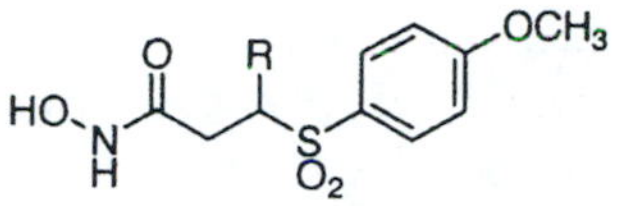
OCH3
HO
N
H
O
R
S
O2

R = H, alkyl, cycloalkyl, etc. **(C)**
R = CF_3 **(19a)**
R = CHF_2 **(19b)**

Figure 5. β-Sulphonyl hydroxamic MMPs inhibitors.

In molecules **C**, the R side chain was found to be critical not only for potency, but it could also dramatically influence the enzyme selectivity profile of the inhibition. Compounds **C** bearing large hydrophobic groups R (such as alkyl, cycloalkyl and arylalkyl groups) showed low nanomolar, and even subnanomolar affinity for MMP-2, 3 and 13, and excellent selectivity with respect to MMP- 1.

4,4,4-Trifluorocrotonic acid **20** (Fig. 6) was used as starting material for the synthesis of **19a.** The thia-Michael addition of **21** to **20** occurred in reasonable yields, delivering the carboxylic acid **22**. Coupling with *O*-Bn hydroxylamine delivered the *O*-Bn hydroxamate **23** in satisfactory yield, and the subsequent oxidation to the sulfone **24** took place in nearly quantitative yield. Hydrogenolysis with the Pearlman's catalyst afforded the target racemic hydroxamic acid **19a** in good overall yields.

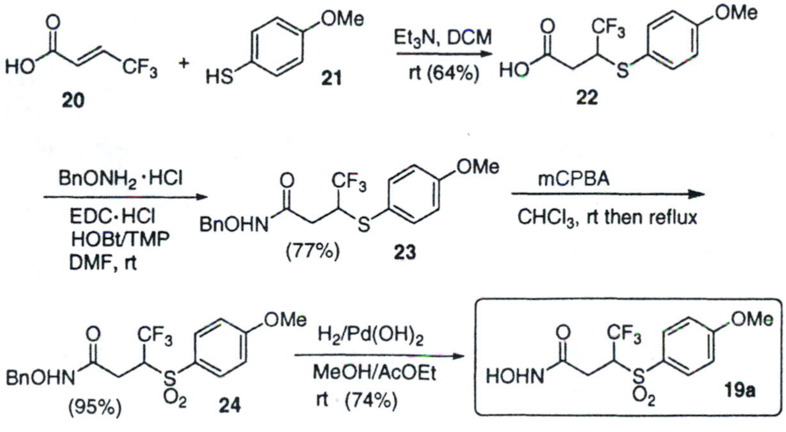
HO
CF3
20
+
HS
OMe
21
Et3N, DCM
rt (64%)
HO
CF3
S
OMe
22
BnONH2·HCl
EDC·HCl
HOBt/TMP
DMF, rt
BnOHN
CF3
S
OMe
(77%)
23
mCPBA
CHCl3, rt then reflux
BnOHN
CF3
S
O2
OMe
(95%)
24
H2/Pd(OH)2
MeOH/AcOEt
rt (74%)
HOHN
CF3
S
O2
OMe
19a

*Figure 6. Synthesis of racemic **19a**.*

Analogous reaction sequence from 4,4-difluorocrotonic acid delivered the difiuoro-hydroxamic derivative **19b** (Fig. 5).

Next, we turned our attention to the synthesis of (*R*) and (*S*)-**19a** (Fig. 8). Both an oxazolidin-2-one (*17*) **25** (Fig. 7) and an oxazolidine-2-thione (*18*) **26** were used as chiral auxiliaries, since they are known to feature a remarkably different chemistry at the cleavage stage (the oxazolidin-2-one is more chemically robust but more resistant to exocyclic cleavage).

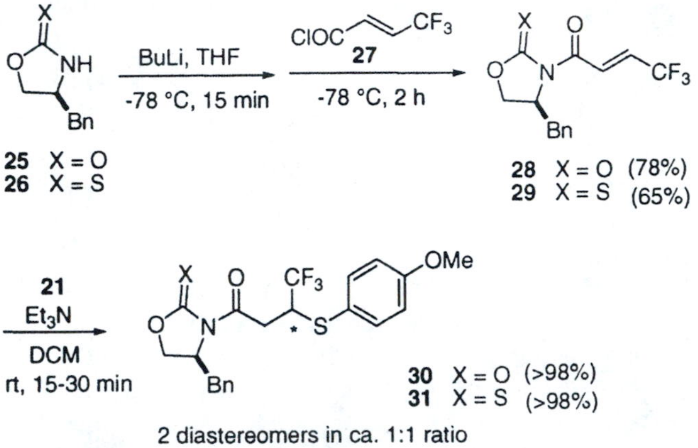

Figure 7. First part of the synthesis of stereochemically pure ***19a.***

Acylation of lithiated **25** and **26** with 4,4,4-trifluorocrotonoyl chloride **27** afforded the Michael acceptors **28** and **29**, respectively. Thia-Michael addition of thiophenol **21** delivered in quantitative yields the oxazolidin-2-one **30** and the oxazolidine-2-thione **31**, in both cases as nearly equimolar mixtures of diastereomers, which could be obtained in pure form by flash chromatography. No attempts were made to improve the stereocontrol, as both the epimeric forms were needed for biological evaluation.

The diastereomer (*S,S*)-**30** was subjected to exocyclic cleavage of the oxazolidin-2-one auxiliary, affording the carboxylic acid (*S*)-**22** (Fig. 8) in fair yields. Coupling with *O*-Bn hydroxylamine occurred efficiently, and the resulting hydroxamate (*S*)-**23** was quantitatively oxidized to the sulfone (*S*)-**24**. The enantiomerically pure hydroxamic acid (+)-(*S*)-**19a** was obtained in good yields by hydrogenolysis. An identical synthetic sequence was performed from (*S,R*)-**30** to provide (—)-(*R*)-**19a**.

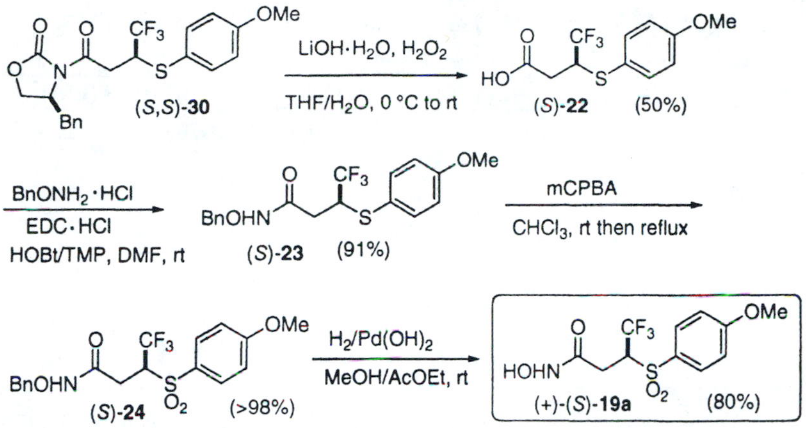

Figure 8. Completion of the synthesis of stereochemically pure ***19a*** *via oxazolidin-2-one auxiliary.*

In order to improve the synthetic process, and particularly the auxiliary cleavage step, we explored the oxazolidine-2-thione intermediate (*S,S*)-**31** (Fig. 9). Satisfactorily, we found that the hydroxamate (*S*)-**23** could be obtained directly, in reasonable yields, by reaction with *O*-Bn-hydroxylamine, that could be then converted into the final hydroxamate (*S*)-**19a** through the identical sequence described above. Analogously, the epimer (*S,R*)-**31** was converted in comparable yields into (*R*)-**23**, and then into the enantiomer (—)-(*R*)-**19a**.

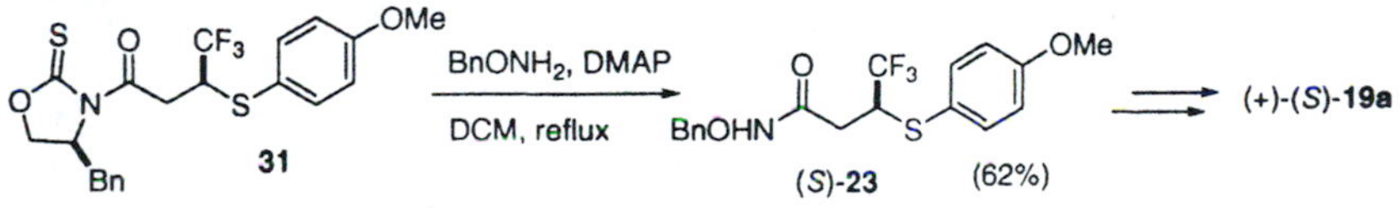

Figure 9. Completion of the synthesis of stereochemicallv pure ***19a*** *via oxazolidin-2-thione auxiliary.*

The inhibition tests on the catalytic domains of MMP-3 and MMP-9 are reported in Table 2.

Table 2. Effect of the compounds 19 on different MMPs' proteolytic activity.

Compound	IC_{50}/MMP-3 (nM)	IC_{50}/MMP-9 (nM)
(*R*) - **19a**	3.2	209.3
(*S*) - **19a**	4.3	51.9
Racemic **19a**	8.4	515.9
Racemic **19b**	1.7	6.3

Interestingly, the pure enantiomers of **19a** and the racemic compound showed nearly identical inhibitory potency, with a single digit nanomolar IC_{50} on MMP-3. Moderate selectivity was observed with respect to MMP-9, in particular (*R*)-**19a** that was ca. 65-fold selective. It is therefore apparent that the inhibitory potency of **19a** is nearly independent of its stereochemistry. This could be the result of an easy interchange of position of the Tfm and sulfone moieties in two different enzyme pockets, most likely S_1' and S_2' Alternatively, one could hypothesize that enantiopure **19a** have undergone racemization at some stage during the inhibition tests. The difluoro compound **19b** was even more potent, showing an impressive inhibitory activity on both MMP-3 and -9, and therefore lower selectivity. These findings also suggest that judicious introduction of other fluoroalkyl groups (such as CF_2Cl, C_2F_5, etc.), onto the backbone of protease inhibitors could represent a successful strategy in order to improve and modulate the inhibitory activity.

These results on Tfm-hydroxamates **2a-d** and **19a** show that (*a*) a Tfm group can be successfully used as a substituent in protease inhibitors (*19*) and is very well tolerated by the enzymes, (*b*) an electronwithdrawing Tfm group in α-position to the hydroxamic function brings about little effect on the zinc chelating capacity of the latter. Interestingly, preliminary results on **19b** show that also the CHF_2 group is a promising substituent in MMPs inhibitors. On the other hand, previous results from our group showing that replacement of an α-methyl by a Tfm group in another structural class of hydroxamate inhibitors of MMPs, namely **A** (see Fig. 1), was responsible for a dramatic loss of inhibitory potency, also show that the final outcome of Tfm incorporation is strongly dependent on the whole structure of the inhibitor, and the effect of the Tfm group on the entire structural conformation is a main factor in determining the biological activity of the molecule.

Acknowledgement. We thank the European Commission (IHP Network grant "FLUOR MMPI" HPRN-CT-2002-00181), MIUR (Cofin 2004, Project

"Polipeptidi Bioattivi e Nanostrutturati), Politecnico di Milano, and C.N.R. for economic support.

References

1. Ojima, I.; McCarthy, J. R.; Welch, J. T. *Biomedical Frontiers of Fluorine Chemistry,* Eds. ACS Books, American Chemical Society, Washington, D.C., 1996.
2. (a) Banks, R. E.; Tatlow, J. C.; Smart, B. E. *Organofluorine Chemist: Principles and Commercial Applications;* Plenum Press: New York, 1994. For a recent example of Tfm group replacing an isobutyl residue on a peptidomimetic side chain in potent Plasmepsin II inhibitors: (b) Binkert, C.; Frigerio, M.; Jones, A.; Meyer, S.; Pesenti, C.; Prade, L.; Viani, F.; Zanda, M. *ChemBioChem* **2006**, *7*, 181-186.
3. (a) Coussens, L. M.; Fingleton, B.; Matrisian, L. M. *Science* **2002**, *295*, 2387-2392. (b) Whittaker, M.; Floyd, C. D.; Brown, P.; Gearing, A. J. H. *Chem. Rev.* **1999**, *99*, 2735-2776. (c) Bode, W.; Huber, R. *Biochim. Biophys. Acta* **2000**, *1477,* 241-252. (d) Giavazzi, R.; Taraboletti, G. *Crit. Rev. Oncol. Hematol.* **2001**, *37,* 53-60.
4. (a) Sani, M.; Belotti, D.; Giavazzi, R.; Panzeri, W.; Volonterio, A.; Zanda, *Tetrahedron Lett.* **2004**, *45,* 1611–1615. See also: (b) Zanda, M. *New J. Chem.* **2004**, *28,* 1401–1411.
5. (a) Jacobson, I. C.; Reddy, P. G.; Wasserman, Z. R.; Hardman, K. D.; Covington, M. B.; Arner, E. C.; Copeland, R. A.; Decicco, C. P.; Magolda, R. L. *Bioorg. Med. Chem. Lett.* **1998**, *8,* 837-842. (b) Jacobson, I. C.; Reddy, G. P. *Tetrahedron Lett.* **1996**, *37,* 8263-8266.
6. Sinisi, R.; Sani, M.; Candiani, G.; Parente, R.; Pecker, F.; Bellosta, S.; Zanda, M. *Tetrahedron Lett.* **2005**, *46,* 6515-6518.
7. (a) Becker, D. P.; DeCrescenzo, G.; Freskos, J.; Getman, D. P.; Hockerman, S. L.; Li, M.; Mehta, P.; Munie, G. E.; Swearingen, C. *Bioorg. Med. Chem. Lett.* **2001**, *11,* 2723-2725. (b) Hockerman, S. L.; Becker, D. P.; Bedell, L. J.; DeCrescenzo, G.; Freskos, J. N.; Getman, D. P.; Heintz, R. M.; Li, M. H.; Mischke, B. V.; Villamil, C. I.; Barta, T. E. U.S. Patent 6 583 299, 2003; Chem. Abstr. **2000**, *134,* 29702.
8. Aranyos, A.; Old, D. W.; Kiyomori, A.; Wolfe, J. P.; Sadighi, J. P.; Buchwald, S. L. *J. Am. Chem. Soc.* **1999**, *121,* 4369-4378.
9. Bravo, P.; Capelli, S.; Meille, S. V.; Viani, F.; Zanda, M.; Kukhar, V. P.; Soloshonok, V. A. *Tetrahedron: Asymmetry* **1994**, *5,* 2009-2018.

10. Buck, E.; Song, Z. J.; Tschaen, D.; Dormer, P. G.; Volante, R. P.; Reider, P. *Org. Lett.* **2002**, *4*, 1623-1626.
11. Drabowicz, J.; Oae, S. *Synthesis* **1977**, 404-405.
12. Cowart, M.; Kowaluk, E. A.; Daanen, J. F.; Kohlhaas, K. L.; Alexander, K. M.; Wagenaar, F. L.; Kerwin, J. F., Jr. *J. Med. Chem.* **1998**, *41,* 2636-2642.
13. Bode, W.; Tochowicz, A.; Zanda, M. unpublished results.
14. Sani, M.; Candiani, G.; Pecker, F.; Zanda M. *Tetrahedron Lett.* **2005**, *46,* 2393-2396.
15. (a) Groneberg, R. D.; Burns, C. J.; Morrissette, M. M.; Ullrich, J. W.; Morris, R. L.; Darnbrough, S.; Djuric, S. W.; Condon, S. M.; McGeehan, G. M.; Labaudiniere, R.; Neuenschwander, K.; Scotese, A. C.; Kline, J. A. *J. Med. Chem.* **1999**, *42,* 541-544. (b) Salvino, J. M.; Mathew, R.; Kiesow, T.; Narensingh, R.; Mason, H. J.; Dodd, A.; Groneberg, R.; Burns, C. J.; McGeehan, G.; Kline, J.; Orton, E.; Tang, S.-Y.; Morrissette, M.; Labaudiniere, R. *Bioorg. Med. Chem. Lett.* **2000**, *10,* 1637-1640.
16. Freskos, J. N.; Mischke, B. V.; DeCrescenzo, G. A.; Heintz, R.; Getman, D. P.; Howard, S. C.; Kishore, N. N.; McDonald, J. J.; Munie, G. E.; Rangwala, S.; Swearingen, C. A.; Voliva, C.; Welsch, D. J. *Bioorg. Med. Chem. Lett.* **1999**, *9*, 943-948.
17. Evans, D. A.; Gage, J. R.; Leighton, J. L.; Kim, A. S. *J. Org. Chem.* **1992**, *57,* 1961-1963 and references therein.
18. (a) Crimmins, M. T.; King, B. W.; Tabet, E. A.; Chaudhary, K. *J. Org. Chem.* **2001**, *66,* 894-902. (b) Crimmins, M. T.; McDougall, P. J. *Org. Lett.* **2003**, *5,* 591-594 and references therein. (c) Palomo, C.; Oiarbide, M.; García, J. M. *Chem. Eur. J.* **2002**, *8,* 37-44.
19. Interestingly, the trifluoroethylamine group is emerging as a very effective replacement of the peptide bond in proteases (for example Cathepsins) inhibitors, providing increased potency and excellent metabolic stability: (a) Volonterio, A.; Bravo, P.; Zanda, M. *Org. Lett.* **2000**, *2,* 1827-1830. (b) Molteni, M.; Volonterio, A.; Zanda, M. *Org. Lett.* **2003**, *5,* 3887-3890. (c) Black, W. C.; Bayly, C. I.; Davis, D. E.; Desmarais, S.; Falgueyret, J.-P.; Léger, S.; Li, C. S.; Massé, F.; McKay, D. J.; Palmer, J. T.; Percival, M. D.; Robichaud, J.; Tsou, N.; Zamboni, R. *Bioorg. Med. Chem. Lett.* **2005**, *15,* 4741-4744. (d) Li, C. S.; Deschenes, D.; Desmarais, S.; Falgueyret, J.-P.; Gauthier, J. Y.; Kimmel, D. B.; Léger, S.; Massé, F.; McGrath, M. E.; McKay, D. J.; Percival, M. D.; Riendeau, D.; Rodan, S. B.; Thérien, M.; Truong, V.-L.; Wesolowski, G.; Zamboni, R.; Black, W. C. *Bioorg. Med. Chem. Lett.* **2006**, *16,* 1985-1989.

Chapter 27

Using Fluorinated Amino Acids for Structure Analysis of Membrane-Active Peptides by Solid-State ^{19}F-NMR

Parvesh Wadhwani[1], Pierre Tremouilhac[1], Erik Strandberg[1], Sergii Afonin[1], Stephan Grage[1], Marco Ieronimo[2], Marina Berditsch[2], and Anne S. Ulrich[1,2,*]

[1]Institute for Biological Interfaces, Forschungszentrum Karlsruhe, Hermann-von-Helmholtz-Platz 1, 76344 Eggenstein, Leopoldshafen, Germany
[2]Institute of Organic Chemistry, University of Karlsruhe, Fritz-Haber-Weg 6, 76131 Karlsruhe, Germany

Several different membrane-active peptides were labeled with a variety of fluorinated amino acids for structure analysis by solid state ^{19}F NMR. Namely, 4-F-Phg/4-CF_3-Phg, 3,3,3-F_3-Ala/3-F-Ala, and 2-CF_3-Ala were used to replace a single amino acid such as Ile/Leu, Ala, and Aib, respectively, without significantly perturbing the peptide conformation or function. These NMR reporter groups can be analyzed to calculate the structure and mobility of the peptide in the lipid bilayer. This review focuses on synthetic challenges with ^{19}F-labeled amino acids, such as racemization and fluorine elimination, and recent results on various antimicrobial and fusogenic peptides in model membranes will be summarized.

Introduction

Fluorine, being one of the most reactive elements, is hardly found in biology. In the recent past, however, the advantageous use of organic fluorine has become increasingly visible.[1] From antitumor drugs and antidepressants to pesticides, the incorporation of a single ^{19}F atom or a CF_3-group has been shown to improve the properties and bioavailability of such compounds.[2] The small size of ^{19}F, its high electronegativity and the strong C-F bond strength obviously induce certain changes in the molecular properties. When used as a single label for ^{19}F-NMR, however, the potentially perturbing effects are often outweighed by the exquisite sensitivity and strong dipolar couplings of this isotope. Suitable labeling strategies are not only advantageous for NMR in solution, but especially so in the solid state, where the pronounced anisotropy of the ^{19}F parameters can be directly exploited. These applications have stimulated research efforts specifically focused on the use of fluorinated amino acids in structural biology. The synthesis of these labeled amino acids has been thoroughly researched and is well documented.[3-5] The present chapter will therefore concentrate on the incorporation of such fluorinated amino acids into various membrane-active peptides to elucidate their structure and dynamics in model membranes by solid state ^{19}F NMR.

Advantages of ^{19}F as an NMR Label

Like most other conventional NMR nuclei (^{1}H, ^{13}C, and ^{15}N), also ^{19}F has a spin of ½ that is conveniently detected and analyzed by NMR. This fluorine isotope is 100% abundant, shows no natural background, and possesses a high gyromagnetic ratio. Its isotropic chemical shift (CS) and the anisotropy tensor (CSA) are very informative *per se*, being highly sensitive to changes in the local environment due to the lone pair of electrons. For this reason ^{19}F NMR in solution is often used to monitor chemical reactions. One can readily detect when the environment of a fluorinated sites changes as the product is generated, or when a protein undergoes a conformational change following an external trigger like ligand binding. Another useful parameter is the strong dipolar coupling (DD) of ^{19}F, which enables the determination of ^{19}F-^{19}F internuclear distances up to 18 Å.[6] Designated hardware, such as Teflon-free probes with separate ^{1}H and ^{19}F channels, efficient decoupling and good filters, is required for ^{19}F NMR at high field.[7]

Solid State ^{19}F NMR Structure Analysis

Solid state NMR is an ideal technique to investigate the structure and dynamics of peptides embedded in membranes.[8,9] It requires the biological samples to be labeled either selectively or uniformly with NMR-active isotopes, for which traditionally ^{2}H, ^{13}C and ^{15}N have been employed.[10] The same kind of structural information can also be obtained - but with much higher sensitivity - if the samples are selectively labeled with a single fluorinated amino acid at a strategically well-defined position, provided that it does not perturb the conformation or activity of the peptide. The most informative structural parameter that can be obtained from the orientation-dependent CSA of a single label is the angle θ of the labeled group with respect to the static magnetic field. Alternatively, if the molecule contains two ^{19}F-labels, their internuclear distance can be extracted from the homonuclear dipolar coupling DD. Furthermore, the anisotropy of the dipolar coupling within a trifluoromethyl-group can reveal the alignment θ of the CF_3-axis (Figure 1). To interpret such structural information, the tensor elements of the CSA and DD have to be known, and the label has to

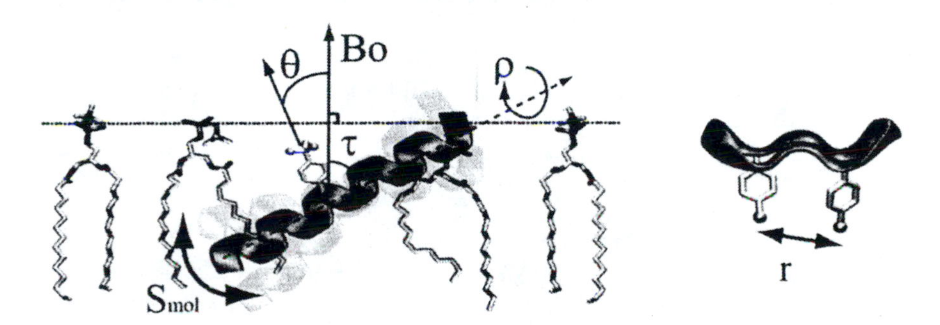

Figure 1. Solid state ^{19}F NMR analysis of single fluorinated amino acids can yield the alignment of a peptide in the lipid membrane. The NMR parameters CSA and DD yield local orientational constraints (θ) and distances (r), respectively. [θ is the angle of the ^{19}F-label with respect to external magnetic field direction B_0, and r is the internuclear distance between two ^{19}F-labels.] The data of several such labels are combined and translated into the orientation of the entire peptide in the membrane, by means of a ρ-τ analysis. [τ is the tilt of the peptide long axis, ρ is its azimuthal rotation, and S_{mol} describes any wobble that leads to a partial motional averaging of the NMR parameters.]

be placed into an unambiguous position with respect to the well defined peptide backbone. With these conditions fulfilled, the three-dimensional alignment of any regular secondary structure element (such as an α-helix or a β-strand) in the membrane can be readily determined. Its orientation and partial motional averaging can be described by three parameters, namely the tilt of the peptide axis (τ), its azimuthal rotation angle (ρ), and its wobble in terms of a molecular order parameter (S_{mol}) (Figure 1). A number of local orientational constraints θ of several individual labels have to be collected in order to construct the overall picture of the entire peptide, as described in detail elsewhere.[8,11-14] This information can be supplemented by measuring distance constraints 'r' between two ^{19}F-labels, which are useful to verify the intramolecular conformation or detect intermolecular interactions. One possibility to separate the required DD from the CSA is to use the Carr-Purcell-Meiboom-Gill (CPMG) sequence which yields a pure dipolar spectrum.[15,16]

As a general solid state ^{19}F-NMR strategy, we have synthesized a series of peptide analogues containing a ^{19}F-labeled group in one out of several different positions (or a symmetric pair in the case of gramicidin S). A list of the different peptides studied so far is summarized in Table 1. The integrity of the labeled

Table 1. Peptide sequences labeled with ^{19}F-amino acids

Entry	*Peptide*	*^{19}F-label*	*Peptide Sequence*[1]
1	B18	4-F-Phg	**LGLL**LRH**L**RHHSN**LLANI**
2	GS	4-F-Phg	*Cyclo*-[P**V**O**L**-^{D}F]$_2$
3		4-CF_3-Phg	*Cyclo*-[P**V**O**L**-^{D}F]$_2$
4	PGLa	4-F-Phg	GMASKAG**A**I**A**GK**I**AKVALKAL-amide
5		4-CF_3-Phg	GMASKAGA**IA**GK**IA**KVALKAL-amide
6		F_3-Ala	GMASK**A**G**A**I**A**GKI**A**KVALKAL-amide
7	K3	3-F-Ala	KIAGKIAKI**A**GKI**A**KIAGKIA-amide
8		4-CF_3-Phg	KIAGKIAK**IAG**K**I**AKIAGKIA-amide

[1] position singly labeled using ^{19}F amino acids are shown in boldface.

peptides with respect to secondary structure is usually checked by circular dichroism (CD) spectroscopy, and the functional activity is examined using appropriate biological assays. For antimicrobial peptides like PGLa, gramicidin S and K3, the antimicrobial and hemolytic activities are tested, while a lipid mixing assay is used to monitor the activity of the fusogenic peptide B18. Only those labeled peptides that are structurally and functionally unperturbed (which is usually the case when the labeled positions are carefully chosen) are then utilized for NMR structure analysis.

NMR Sample Preparation

Static solid state NMR measurements are carried out using either macroscopically oriented membrane samples or non-oriented multilamellar lipid vesicles (Figure 2). Oriented membranes are readily prepared on small glass slides [13,15,17-20] or on polymer films.[21] Typically, the desired ratio of peptide and lipid is solubilized in a chloroform-methanol mixture, which is then spread over the glass plates. After the solvent is removed under vacuum, ~20 plates are stacked and placed in a hydration chamber with controlled humidity for 12-48 h. The membranes orient themselves spontaneously, with each plate carrying about

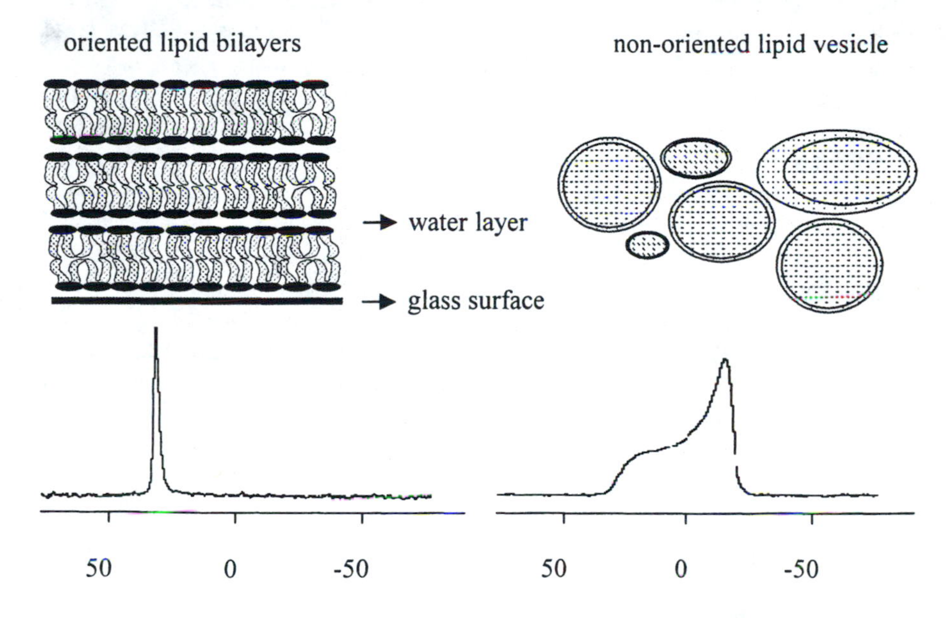

Figure 2. Typical ^{31}P NMR spectra of macroscopically oriented lipid bilayers (left) and a non-oriented multilamellar vesicle sample (right).

3000-5000 aligned lipid bilayers. The stack is sealed into polyethylene foil to prevent dehydration, and a thin container with a 1M NaF solution is placed on top during the NMR experiment for chemical shift referencing.[22] The quality of membrane alignment is measured by solid state ^{31}P-NMR (Figure 2). Non-oriented samples are prepared in similar way, except that the lipid-peptide solution is not spread onto the glass plates, but water (1/1 v/w) is added directly after removing the organic solvent. The hydrated suspension is vortexed thoroughly, followed by 5-10 freeze-thaw cycles.

Fluorine Contamination

Although the ^{19}F nucleus has no natural abundance background in a biological sample, contamination of NMR samples with unwanted fluorine can lead to spectral artifacts. Typical contaminations originate from fluorinated solvents to which the sample has been exposed prior to the NMR measurement. A common one is trifluoroacetic acid (TFA), which is used in every deprotection cycle in Boc-based solid phase peptide synthesis (SPPS), or at the final stage of Fmoc-based SPPS. It is also used as an ion-pairing agent in RP-HPLC purification of peptides. Therefore, without special precautions usually a large amount of TFA is found to be associated to a peptide (typically 1 TFA molecule per amino acid) when reconstituted in an NMR sample. This can give rise to a strong signal during ^{19}F NMR measurements in the region overlapping with that of ^{19}F-labeled peptide. It is difficult to remove TFA by solvent extraction, repeated lyophilization, or ion exchange. In our hands the most successful route is to perform the final HPLC purification in the presence of HCl on a designated column, and by keeping all TFA-exposed equipment such as other columns, pipettes and solvent bottles out of the lab.[23] Further solvents that can interfere with ^{19}F NMR include trifluoroethanol and hexafluoroisopropanol, which are often recommended for reconstituting membrane-active peptides. Even a nominally ^{19}F-free NMR probe often contains Teflon in various capacitors as an intrinsic manufacturing material, which can also give rise to a severe background signal.

Incorporation of ^{19}F-Labeled Amino Acids into Peptides

Fluorine-labeling of peptides is readily achieved by incorporating a suitable fluorinated amino acid in the synthesis. ^{19}F-containing amino acids are generally classified in three categories, namely fluorinated α-amino acids, fluorinated β-amino acids, and fluorinated cyclic amino acids, which are well described in literature.[3,24,25]

Many different ^{19}F-labeled amino acids have been used to study peptides and proteins by ^{19}F-NMR, and the most relevant ones for solid state NMR are summarized in Figure 3. Most of these are commercially available or can be synthesized, though it may not be trivial to obtain the pure L-enantiomer. Unlike this specific list, most previous, qualitative ^{19}F NMR studies have employed

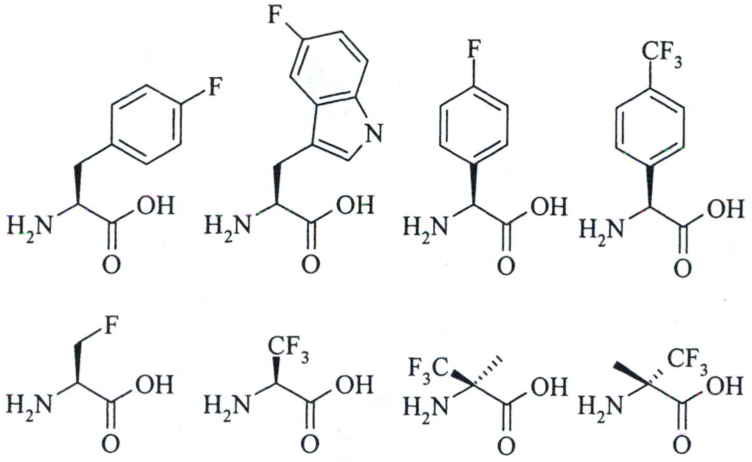

Figure 3. 4-F-Phe, 5-F-Trp, 4-F-Phg, 4-CF_3-Phg, (top row), 3-F-Ala, 3,3,3-F_3-Ala, 2-CF_3-Ala (R-form) , 2-CF_3-Ala (S-form) (bottom row).

derivatives of natural amino acids such Trp, Phe, Tyr, Leu or Val, which cause minimum structural perturbation and are often compatible with biosynthetic labeling protocols.[26] Our quantitative NMR strategy, however, relies on deducing the peptide alignment directly from the ^{19}F-labels as outlined in Figure 1. This approach requires that the labels are rigidly attached to the peptide framework such that they reflect the orientation and dynamics of the relevant backbone segment. Therefore, flexible side chains as in Phe, Trp or Leu are not ideal. Instead, the analogues of phenylglycine (4-F-Phg, 4-CF_3-Phg), alanine (3-F-Ala, 3,3,3-F_3-Ala) and aminoisobutyric acid (2-CF_3-Ala) provide the necessary rigid connection between the ^{19}F-label and the peptide backbone. Our discussion below is therefore mostly concerned with these derivatives, besides Trp.

Since the non-natural amino acids have to be incorporated by chemical synthesis, the usual SPPS protocols are limited to a length of about 60 residues, although larger proteins can be prepared via alternative methods such as chemical ligation. Here we will focus on the incorporation of ^{19}F-labeled amino acid into small membrane-active peptides using Fmoc-SPPS.[27] The synthesis of these fluorinated amino acids has been reviewed elsewhere.[3,5]

Fluorotryptophan Labeled Peptides

In membrane-spanning peptides and proteins tryptophan is often found to reside at the phosphate headgroup region where it is believed to interact with the phosphate groups.[28] These interactions are believed to be important for the activity of some ion-channels forming peptides like gramicidin A.[17,29] These phenomena have been thoroughly investigated by solid state ^{19}F-NMR of 5-fluorotryptophan, and 6-fluorotryptophan has been used for monitoring conformational changes by fluorescence spectroscopy.[30] Although slight structure perturbations are expected due to the electronegative effects of fluorine substituent, they are generally accepted in view of the advantages that ^{19}F-labeled tryptophan offers. These include easy incorporation into not only small synthetic peptides but also larger recombinant proteins, and the high NMR sensitivity of the ^{19}F nucleus combined with its response to changes in the local environment. Peptides and proteins that have been labeled with fluorinated Trp include the ion channel forming peptide gramicidin A, the immunoglobulin binding domain B1 of streptococcal protein G, and a Leu specific binding protein.[17,29,31]

Fluorophenylglycine Labeled Peptides

Phenylglycine (Phg) is a shorter homologue of the naturally occurring amino acid phenylalanine. It is found in various natural compounds like antibiotics [32] and streptogramin B[33], and has been used in artificial sweeteners and a hepatitis C virus serine protease inhibitor.[34] We have used 4-F-Phg and 4-CF_3-Phg to label several peptides as in Table 1.[23] The advantage of these fluorinated phenylglycines over Phe or Trp is that they carry the label as a straight extension on the peptide backbone. Using 4-F-Phg we have determined the membrane-bound structure of orientation of the peptides B18 and gramicidin S. One disadvantage of 4-F-Phg, however, is that it still possesses rotational freedom of the phenyl ring around the C_α-C_β bond. When the CSA of this amino acid is used to measure orientational constraints on a peptide, the torsion angle of the phenyl ring has to be known. It was found that the preferred conformation depends significantly of the secondary structure and on the neighboring amino acids. This disadvantage is overcome by using instead the trifluoromethyl-group as in 4-CF_3-Phg, which rotates fast about the CF_3-axis. In this case, the anisotropy of the axially symmetric dipolar coupling can be readily analyzed, instead of the asymmetric CSA tensor of a single ^{19}F-substituent.

As a general synthetic problem, Phg esters formed during peptide synthesis are highly sensitive to epimerization.[23,35] The presence of the electronegative ^{19}F-substituent in 4-F-Phg and 4-CF_3-Phg further enhances this susceptibility towards racemization under the basic conditions employed in standard Fmoc SPPS protocols, thereby resulting in the formation of epimeric peptides.[23] Since racemization can hardly be avoided, we tend to use a racemic mixture of 4-F-DL-Phg or of 4-CF_3-DL-Phg for peptide synthesis. Fortunately, all epimeric peptides encountered so far could be successfully resolved by HPLC even on a preparative scale. To distinguish and assign the two products, the purified peptides were acid-hydrolyzed to give the constituent amino acids, which were derivatized in-situ using Marfey's reagent.[23,36] In the case of 4-CF_3-Phg, where the pure enantiomers are not commercially available for a unique assignment, an enzymatic route was followed.[37] In brief, the free 4-CF_3-DL-Phg was acetylated and subsequently enantioselectively hydrolyzed using porcine kidney acylase to yield the free 4-CF_3-L-Phg. This was followed by the usual derivatization using Marfey's reagent and HPLC characterization. In certain cases where the HPLC separation of epimeric peptides is too complicated or if multiple CF_3-groups are to be placed within a single peptide, a base-free procedure using DIPCDI/HOBt is recommended to prevent excessive racemization. However, for ^{19}F-NMR analysis the side-products containing a fluorinated D-Phg actually turn out to be an advantage, as these peptide epimers can be analyzed to yield additional orientational constraints.[13,23]

Fluoroalanine Labeled Peptide

Various fluorinated alanines can be incorporated into peptides (see Figure 3), namely 3-F-Ala, 3,3,3-F_3-Ala, and 2-CF_3-Ala. These derivatives are expected to be relatively unperturbing when substituted for native Ala. Incorporation of 3-F-Ala under the basic conditions of Fmoc-SPPS promoted HF elimination, and produced dehydroalanine as illustrated in Figure 4. This problem can be overcome by using 2-^{2}H-3-F-Ala and performing the coupling reaction at low temperature (-10°C) to reduce the rate of ^{2}HF elimination according to the kinetic isotope effect. Working carefully under base free conditions as with DIPCDI/HOBt, it was even possible to obtain 3-F-Ala labeled peptides without any HF elimination (unpublished results). Regarding the suitability of the CH_2F-

group as an NMR label, it has been shown to undergo axial averaging on the NMR timescale.[38-40] However, a quantitative analysis in terms of angular constraints cannot be reliably carried out unless the CSA tensor is fully characterized within the molecular frame.[12,38] Given the advantages of a CF_3-group and the convenient analysis of its anisotropic dipolar coupling, the other two amino acids 3,3,3-F_3-Ala and 2-CF_3-Ala are preferred as labels. 3,3,3,-F_3-Ala was incorporated into PGLa (Table 1), without excessive HF-elimination (unpublished results). 2-CF_3-Ala is a fluorinated derivative of aminoisobutyric

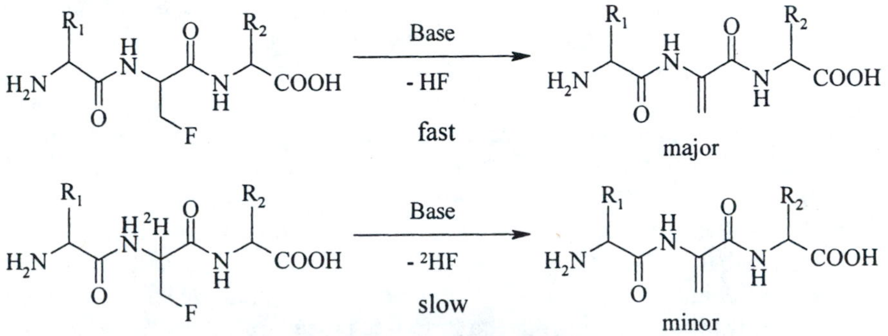

Figure 4: Base-assisted HF-elimination from 3-F-Ala, and its suppression in the α- deuterated analogue.

acid (Aib) and has been used to substitute this characteristic amino acid in a peptaibol.[25] Since there are two methyl-substitutents on C_α, HF elimination is not a problem in synthesis. However, due to the bulky and electron-withdrawing CF_3-group, extremely long times are needed to couple the subsequent amino acid. This problem was overcome by synthesizing tri-peptides, in which 2-CF_3-Ala was flanked by the respective amino acids of the desired peptide sequence.[25,41] Furthermore, having used a racemic R-/S mixture of 2-CF_3-Ala in the synthesis, HPLC separation of the full-length epimeric peptides was not trivial. Again, by preparing tri-peptides as building blocks, it was possible to separate these on a preparative scale by flash chromatography. To assign the epimers, various enzymatic routes are available.[42] Alternatively, the stereochemistry of 2-CF_3-Ala in the purified peptides could be assigned on the basis of NOE cross-peak intensities of the methyl groups in CF_3-Ala.

Solid State NMR Results on Membrane-Active Peptides

Fusogenic Peptide B18

In most fusogenic proteins, which merge lipid membranes with one another, a short peptide sequence is responsible for this activity and can function even in isolation. The 18 amino acid sequence B18 is derived from the sea urchin protein "bindin", which mediates the fusion between the sperm and oocyte membranes during fertilization. This peptide has been labeled at nine different positions one by one with 4-F-Phg (Table 1)[43] to resolve its structure in the membrane-bound state.[23,44,45] All labeled peptides, even those containing the D-enantiomer, retained their fusogenic activity, thus accommodating the stiff side chain of 4-F-Phg in the functionally relevant conformation.[43,44] Solution state ^{1}H-NMR had shown that B18 assumes a helix-turn-helix structure, hence the alignment of both helical segments was determined from the CSA of 4-F-Phg in macroscopically oriented membranes (DMPC/DMPG, 80/20). The respective tilt angles show that the N-terminal helix penetrates obliquely into the lipid bilayer at $\tau \approx 54°$, leaving the C-terminal helix flat on the membrane surface with a tilt angle of 90°.[20]

Antimicrobial Peptides Gramicidin S, PGLa and K3

Most antimicrobial peptides are cationic with an overall amphiphilic structure, and they supposedly kill bacteria by permeabilizing their membrane. Gramicidin S from *Bacillus brevis* has a symmetric cyclic β-sheet structure, and it offers some pharmaceutical potential as an antibiotic provided that its therapeutic index could be improved.[46] The peptide was labeled in a symmetric manner (Table 1) by replacing a pair of Leu (or Val) with either 4-F-Phg or 4-CF_3-Phg.[47,48] The ^{19}F-labeled peptides exhibited antimicrobial activities comparable to the native one.[23,47] The advantage of the doubly labeled peptides is that the two labels allowed to measure the internuclear ^{19}F-^{19}F distance as an additional constraint in structure elucidation. The respective distances obtained from the two substituted Leu positions (6Å) and Val (8Å) agree well with the conformation expected from the wild type peptide.[19] Orientational constraints from 4-F-Phg substituted gramicidin S revealed that at low peptide concentration (peptide/lipid ratio of 1:80) the molecule is very mobile ($S_{mol} \approx 0.3$) and resides on the membrane surface with the hydrophobic side chains facing the lipid bilayer, as expected.[19] At high peptide concentration (1:20), however, gramicidin

S can flip upright in DMPC membranes by almost 90°, presumably forming an oligmeric pore by lateral self-assembly.[11] Under those conditions it does not wobble any more (S_{mol}≈1.0), even though the putative oligomer undergoes long-axial rotation around the membrane normal. A recent crystallographic study confirmed such hydrogen-bonded assembly for another gramicidin S analogue.[49] By ^{19}F-NMR it was found that the change in membrane alignment depends on the nature of the lipid bilayer, as the immersion was favored in short-chain lipids but prevented in long-chain lipids and at high cholesterol concentration.[9,16] Finally, by substituting gramicidin S symmetrically with two 4-CF_3-Phg labels and using the CPMG experiment, we were able to detect not only the strong dipolar coupling within a single CF_3-group but also a weak coupling between the two CF_3-groups.[48]

Another antimicrobial peptide extensively investigated by ^{19}F NMR is PGLa. This α-helical amphiphilic peptide from *Xenopus laevis* was synthesized with either 4-F-Phg or 4-CF_3-Phg substituted for a single Ile or Ala (Table 1).[13,14,23] All labeled peptides were intact, as evident from CD spectroscopy and antimicrobial assays, except for one Ala position substituted with 4-CF_3-Phg in the polar region of the α-helix.[23] The ^{19}F-NMR analysis of PGLa was possible over a wide range of peptide-to-lipid ratios down to 1:3000, which would not have been accessible with conventional, less sensitive isotopes. The orientational constraints showed that at low concentration (1:200) the α-helix lies flat on the membrane surface (τ≈98°), as previously reported by ^{15}N NMR.[13,14,50] With increasing peptide concentration (1:50) the peptide was found to tilt obliquely into the membrane (τ≈125°), presumably forming antiparallel homo-dimers. This novel kind of re-alignment was confirmed with non-perturbing ^{2}H and ^{15}N NMR labels.[51]

Based on the sequence of PGLa, the peptide K3 was designed with improved antimicrobial activity.[52] This peptide was labeled with 3-F-Ala (Table 1) together with ^{13}C-Ala for REDOR analysis by Schaefer et al.[38,53] As observed for gramicidin S and PGLa, also K3 was found to change its tilt angle presumably due to self-association.[38] A comprehensive REDOR analysis was carried out to confirm the helical conformation of K3, to define the intermolecular contacts to neighboring peptides, and to characterize its interactions with phospholipid head groups. These results suggested a parallel head-to-head structure of K3 dimers.[38,39,53] K3 was also labeled with 4-CF_3-Phg (Table 1) for a direct orientational analysis, and all labeled peptides retained their antimicrobial activity (unpublished results).

Conclusions

Given the advances in chemical labeling strategies, solid state ^{19}F NMR investigations on membrane-associated peptides have matured and become routine over the recent years. 4-CF_3-Phg and 3,3,3-F_3-Ala are ideal labels to measure orientational constraints, while 4-F-Phg and 3-F-Ala are most suitable for measuring intramolecular (or intermolecular) distances. In most cases, the substitution of a single amino acid by a ^{19}F-labeled side chain of similar size did not perturb the peptide conformation or its biological activity. The main advantage of ^{19}F-labels lies in their high NMR sensitivity, requiring only small amounts of sample and measurement time, and allowing the study of low peptide concentrations that remain inaccessible to conventional isotope labels.

Acknowledgements

We thank all members of the BioNMR group who have contributed to this work, in particular Ulrich H.N. Dürr, Ralf W. Glaser, Carsten Sachse and Jesus Salgado. Special thanks also go to our colleagues Jacob Schaefer, Orsoyla Toke, Ron McElhaney, Les Kondejewski, Tim Cross, Junfeng Wang, Tetsuo Asakura, and Beate Koksch. We acknowledge financial support by the DFG [SFB 197 (TPB13), SFB 604 (TP B6)] and the Center for Functional Nanostructures.

References

1. (a) Bohm, H. J.; Banner, D.; Bendels, S.; Kansy, M.; Kuhn, B.; Muller, K.; Obst-Sander, U.; Stahl, M. *Chembiochem* **2004**, *5*, 637- 643. (b) Harper, D. B.; Ohagan, D. *Nat. Prod. Rep.* **1994**, *11*, 123-133.
2. (a) Ismail, F. M. D. *J. Fluorine Chem.* **2002**, *118*, 27-33. (b) Margiotta, N.; Papadia, P.; Lazzaro, F.; Crucianelli, M.; De Angelis, F.; Pisano, C.; Vesci, L.; Natile, G. *J. Med. Chem.* **2005**, *48*, 7821-7828. (c) Wolf, W.; Presant, C. A.; Waluch, V. *Adv. Drug Deliver. Rev.* **2000**, *41*, 55-74.
3. (a) *Fluorine-Containing Amino-Acids*; Kukhar, V. P.; Soloshonok, V. A., Eds.; John Wiley & Sons: Chichester, 1995. (b) Qiu, X. L.; Meng, W. D.; Qing, F. L. *Tetrahedron* **2004**, *60*, 6711-6745.
4. (a) Haufe, G.; Kroger, S. *Amino Acids* **1996**, *11*, 409-424. (b) Tolman, V. *Amino Acids* **1996**, *11*, 15-36. (c) Dave, R.; Badet, B.; Meffre, P. *Amino Acids* **2003**, *24*, 245-261. (d) Kelly, N. M.; Sutherland, A.; Willis, C. L. *Nat. Prod. Rep.* **1997**, *14*, 205-219.
5. Sutherland, A.; Willis, C. L. *Nat. Prod. Rep.* **2000**, *17*, 621-631.

6. (a) Gilchrist, M. L., Jr.; Monde, K.; Tomita, Y.; Iwashita, T.; Nakanishi, K.; McDermott, A. E. *J. Magn. Reson.* **2001**, *152*, 1-6. (b) Smith, S. O.; Aschheim, K.; Groesbeek, M. *Quart. Rev. Biophys.* **1996**, *29*, 395-449.
7. Doty, F. D.; Entzminger, G., Jr.; Hauck, C. D. *J. Magn. Reson.* **1999**, *140*, 17-31.
8. Strandberg, E.; Ulrich, A. S. *Concepts Magn. Reson.* **2004**, *23A*, 89-120.
9. Ulrich, A. S. *Prog. NMR Spectr.* **2005**, *46*, 1-21.
10. (a) Opella, S. J.; Marassi, F. M. *Chem. Rev.* **2004**, *104*, 3587-3606. (b) Castellani, F.; van Rossum, B.; Diehl, A.; Schubert, M.; Rehbein, K.; Oschkinat, H. *Nature* **2002**, *420*, 98-102.
11. Afonin, S.; Dürr, U. H. N.; Wadhwani, P.; Salgado, J.; Ulrich, A. S. *submitted* **2006**.
12. *Solid-State NMR on Biological Systems*; Ulrich, A. S.; Ramamoorthy, A., Eds.; Wiley, **2004**; Vol. 42.
13. Glaser, R. W.; Sachse, C.; Dürr, U. H. N.; Wadhwani, P.; Ulrich, A. S. *J. Magn. Reson.* **2004**, *168*, 153-163.
14. Afonin, S.; Glaser, R. W.; Sachse, C.; Wadhwani, P.; Ulrich, A. S. **2005**, *in preparation.*
15. Grage, S. L.; Ulrich, A. S. *J. Magn. Reson.* **1999**, *138*, 98-106.
16. Ulrich, A. S.; Wadhwani, P.; Dürr, U. H. N.; Afonin, S.; Glaser, R. W.; Sachse, C.; Tremouilhac, P.; Berditchevskaia, M. In *NMR Spectroscopy of Biological Solids*; Ramamoorthy, A., Ed.; Marcel Dekker: **2006**.
17. Grage, S. L.; Wang, J.; Cross, T. A.; Ulrich, A. S. *Biophys. J.* **2002**, *83*, 3336-3350.
18. Grage, S. L.; Ulrich, A. S. *J. Magn. Reson.* **2000**, *146*, 81-88.
19. Salgado, J.; Grage, S. L.; Kondejewski, L. H.; Hodges, R. S.; McElhaney, R. N.; Ulrich, A. S. *J. Biomol. NMR* **2001**, *21*, 191-208.
20. Afonin, S.; Dürr, U. H. N.; Glaser, R. W.; Ulrich, A. S. *Mag. Reson. Chem.* **2004**, *42*, 195-203.
21. Augé, S.; Mazarguil, H.; Tropis, M.; Milon, A. *J. Mag. Res.* **1997**, *124*, 455-458.
22. (a) Glaser, R. W.; Ulrich, A. S. *J. Mag. Res.* **2003**, *164*, 104-114. (b) Ulrich, R.; Glaser, R. W.; Ulrich, A. S. *J. Mag. Res.* **2003**, *164*, 115-127.
23. Afonin, S.; Glaser, R. W.; Berditchevskaia, M.; Wadhwani, P.; Gührs, K. H.; Möllmann, U.; Perner, A.; Ulrich, A. S. *Chembiochem* **2003**, *4*, 1151-1163.
24. Jaeckel, C.; Koksch, B. *Eur. J. Org. Chem.* **2005**, 4483-4503.
25. (a) Koksch, B.; Sewald, N.; Burger, K.; Jakubke, H. D. *Amino Acids* **1996**, *11*, 425-434. (b) Koksch, B.; Sewald, N.; Jakubke, H. D.; Burger, K. In *ACS Symposium Series* **1996**, *639*, 42-58.
26. (a) Gerig, J. T. *Prog. NMR Spec.* **1994**, *26*, 293-370. (b) Danielson, M. A.; Falke, J. J. *Annu. Rev. Biophys. Biomol. Struct.* **1996**, *25*, 163-195. (c) Sun,

S.-Y.; Pratt, E. A.; Ho, C. In *Biomedical Frontiers of Fluorine Chemistry; ACS Symposium Series*; Ojima, I.; McCarthy, J. M.; Welch, J. T., Eds.; American Chemical Society, **1996**, 296-310.

27. Fields, G. B.; Noble, R. L. *Int. J. Pept. Protein Res.* **1990**, *35*, 161-214.
28. (a) Landolt-Marticorena, C.; Williams, K. A.; Deber, C. M.; Reithmeier, R. A. *J. Mol. Biol.* **1993**, *229*, 602-608. (b) Arkin, I. T.; Brunger, A. T. *Biochim. Biophys. Acta* **1998**, *1429*, 113-128.
29. (a) Ketchem, R. R.; Roux, B.; Cross, T. A. *Structure* **1997**, *5*, 1655-1169. (b) Hu, W.; Lee, K. C.; Cross, T. A. *Biochem.* **1993**, *32*, 7035-7047. (c) Koeppe, R. E., II; Killian, J. A.; Greathouse, D. V. *Biophys. J.* **1994**, *66*, 14-24. (d) Koeppe, R. E., 2nd; Sun, H.; van der Wel, P. C.; Scherer, E. M.; Pulay, P.; Greathouse, D. V. *J. Am. Chem. Soc.* **2003**, *125*, 12268-12276.
30. Lotte, K.; Plessow, R.; Brockhinke, A. *Photochem. Photobiol. Sci.* **2004**, *3*, 348-359.
31. (a) Salopek-Sondi, B.; Luck, L. A. *Protein Eng.* **2002**, *15*, 855-859. (b) Luck, L. A.; Vance, J. E.; O'Connell, T. M.; London, R. E. *J. Biomol. NMR* **1996**, *7*, 261-272.
32. (a) Vernier, J. M.; Hegedus, L. S.; Miller, D. B. *J. Org. Chem.* **1992**, *57*, 6914-6920.
33. Mahlert, C.; Sieber, S. A.; Gruenewald, J.; Marahiel, M. A. *J. Am. Chem. Soc.* **2005**, *127*, 9571-9580.
34. Chen, K. X.; Njoroge, F. G.; Pichardo, J.; Prongay, A.; Butkiewicz, N.; Yao, N.; Madison, V.; Girijavallabhan, V. *J. Med. Chem.* **2006**, *49*, 567-574.
35. Anteunis, M. J. O.; Vanderauwera, C. *Int. J. Pept. Protein Res.* **1988**, *31*, 301-310.
36. Marfey, P. *Carlsberg Res. Commun.* **1984**, *49*, 591-596.
37. (a) Morgan, J.; Pinhey, J. T.; Sherry, C. J. *J. Chem. Soc. Perkin Trans. 1* **1997**, 613-619. (b) Ulijn, R. V.; Bisek, N.; Flitsch, S. L. *Org. Biomol. Chem.* **2003**, *1*, 621-622.
38. Toke, O.; O'Connor, R. D.; Weldeghiorghis, T. K.; Maloy, W. L.; Glaser, R. W.; Ulrich, A. S.; Schaefer, J. *Biophys. J.* **2004**, *87*, 675-687.
39. Holl, S. M.; Marshall, G. R.; Beusen, D. D.; Kociolec, K.; Redlinski, A. S.; Leplawy, M. T.; McKay, R. A.; Schaefer, J. *J. Am. Chem. Soc.* **1992**, *114*, 4830-4833.
40. Goetz, J. M.; Wu, J. H.; Lee, A. F.; Schaefer, J. *Solid State NMR* **1998**, *12*, 87-95.
41. Thust, S.; Koksch, B. *J. Org. Chem.* **2003**, *68*, 2290-2296.
42. (a) Keller, J. W.; Hamilton, B. J. *Tetrahedron Lett.* **1986**, *27*, 1249-1250. (b) Kamphuis, J.; Boesten, W. H.; Broxterman, Q. B.; Hermes, H. F.; van Balken, J. A.; Meijer, E. M.; Schoemaker, H. E. *Adv. Biochem. Eng. Biotechnol.* **1990**, *42*, 133-186. (c) Koksch, B.; Quaedflieg, P. J. L. M.;

Michel, T.; Burger, K.; Broxterman, Q. B.; Schoemaker, H. E. *Tetrahedron: Asymm.* **2004**, *15*, 1401-1407.

43. Ulrich, A. S.; Otter, M.; Glabe, C.; Hoekstra, D. *J. Biol. Chem.* **1998**, *273*, 16748-16755.
44. Glaser, R. W.; Grüne, M.; Wandelt, C.; Ulrich, A. S. *Biochem.* **1999**, *38*, 2560-2569. (b) Ulrich, A. S.; Tichelaar, W.; Förster, G.; Zschörnig, O.; Weinkauf, S.; Meyer, H. W. *Biophys. J.* **1999**, *77*, 829-841.
45. (a) Binder, H.; Arnold, K.; Ulrich, A. S.; Zschörnig, O. *Biochim. Biophys. Acta* **2000**, *1468*, 345-358. (b) Grage, S. L.; Afonin, S.; Grüne, M.; Ulrich, A. S. *Chem. Phys. Lipids* **2004**, *132*, 65-77.
46. (a) McInnes, C.; Kondejewski, L. H.; Hodges, R. S.; Sykes, B. D. *J. Biol. Chem.* **2000**, *275*, 14287-14294. (b) Bradshaw, J. *BioDrugs* **2003**, *17*, 233-240. (c) Lee, D. L.; Powers, J. P.; Pflegerl, K.; Vasil, M. L.; Hancock, R. E.; Hodges, R. S. *J. Pept. Res.* **2004**, *63*, 69-84.
47. Wadhwani, P.; Afonin, S.; Ieromino, M.; Buerck, J.; Ulrich, A. S. *J. Org. Chem.* **2006**, *71*, 55-61.
48. Grage, S.; Suleymanova, A. V.; Afonin, S.; Wadhwani, P.; Ulrich, A. S. *submitted* **2006**.
49. Grotenbreg, G. M.; Timmer, M. S.; Llamas-Saiz, A. L.; Verdoes, M.; van der Marel, G. A.; van Raaij, M. J.; Overkleeft, H. S.; Overhand, M. *J. Am. Chem. Soc.* **2004**, *126*, 3444-3446.
50. Bechinger, B.; Zasloff, M.; Opella, S. J. *Biophys. J.* **1998**, *74*, 981-987.
51. (a) Strandberg, E.; Wadhwani, P.; Tremouilhac, P.; Durr, U. H. N.; Ulrich, A. S. *Biophys. J.* **2006**, *90*, 1676-1686. (b) Tremouilhac, P.; Strandberg, E.; Wadhwani, P.; Ulrich, A. S. *Biochim. Biophys. Acta* **2006**, *in press*.
52. (a) Maloy, W. L.; Kari, U. P. *Biopolymers* **1995**, *37*, 105-122. (b) Hirsh, D. J.; Hammer, J.; Maloy, W. L.; Blazyk, J.; Schaefer, J. *Biochem.* **1996**, *35*, 12733-12741.
53. Toke, O.; Maloy, W. L.; Kim, S. J.; Blazyk, J.; Schaefer, J. *Biophys. J.* **2004**, *87*, 662-674.

Chapter 28

Asymmetric Synthesis of β-Trifluoromethylated β-Amino Carbonyl Compounds Based on the 1,2-Addition to Trifluoroacetaldehyde SAMP- or RAMP-Hydrazones

Kazumasa Funabiki[1], Masashi Nagamori[1], Masaki Matsui[1], Gerhard Raabe[2], and Dieter Enders[2]

[1]Department of Materials Science and Technology, Faculty of Engineering, Gifu University, 1–1, Yanagido, Gifu 501–1193, Japan
[2]Institut für Organische Chemie, Rheinisch-Westfälische Technische Hochschule, Professor-Pirlet-Strasse 1, 52074 Aachen, Germany

An efficient asymmetric synthesis of α-trifluoromethyl substituted primary amines *via* nucleophilic 1,2-addition of alkyl-, phenyl- and allyl-lithium reagents to trifluoroacetaldehyde SAMP- or RAMP-hydrazones, followed by benzoylation and SmI_2-promoted nitrogen-nitrogen single bond cleavage is described. Asymmetric synthesis of trifluoromethylated β-amino aldehydes, carboxylic acids, and esters as well as γ-aminoalcohol *via* the oxidation of the obtained chiral α-trifluoromethylated homoallylamines is also described.

Introduction

Because of their importance in pharmaceutical research based on the special electronic properties of the trifluoromethyl group, much attention has been addressed to the development of the efficient and enantioselective synthesis of α-trifluoromethylated amines *(1)* as well as α- or β-trifluoromethylated amino acid derivatives *(2,3)*.

The 1,2-addition of organometallic reagents to CN double bonds is one of the most efficient routes to α-branched amines *(4)*. Among them, the asymmetric 1,2-addition reaction using commercially available (*S*)- or (*R*)-1-amino-2-(methoxymethyl)pyrrolidine (SAMP or RAMP, respectively) as a chiral auxiliary provides a promising synthetic route to enantioenriched amines *(5)*.

In this chapter, the asymmetric synthesis of α-trifluoromethylated amines based on the nucleophilic 1,2-addition of alkyl-, phenyl-, and allyl-lithium reagents to trifluoroacetaldehyde SAMP- or RAMP-hydrazones as well as the β-trifluoromethylated β-amino acids *via* the oxidation of the obtained α-trifluoromethylated homoallylamine derivative, as described in in Figure 1.

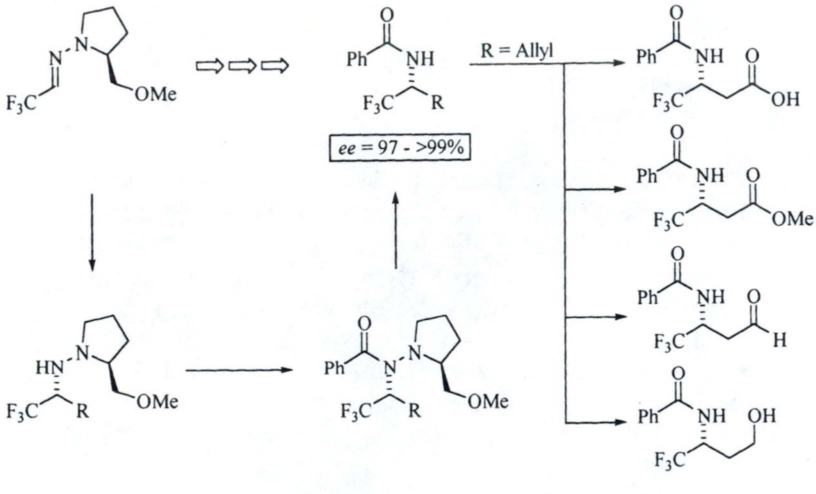

Figure 1. Asymmetric Synthesis of β-Trifluoromethylated β-Amino Carbonyl Compounds Based on the 1,2-Addition to Trifluoroacetaldehyde SAMP-or RAMP-Hydrazones

Preparation of of Trifluoroacetaldehyde Hydrazones *(6)*

Trifluoroacetaldehyde SAMP- or RAMP-hydrazones **2** or **2'** were readily obtained in 83% and 66% yields from commercially available trifluoroacetaldehyde ethyl hemiacetal **1** and SAMP or RAMP, respectively, in the presence of a catalytic amount of *p*-toluenesulfonic acid (*p*-TsOH) in benzene. Trifluoroacetaldehyde morpholinohydrazone **3** was prepared in the same manner in 58% yield (Figure 2).

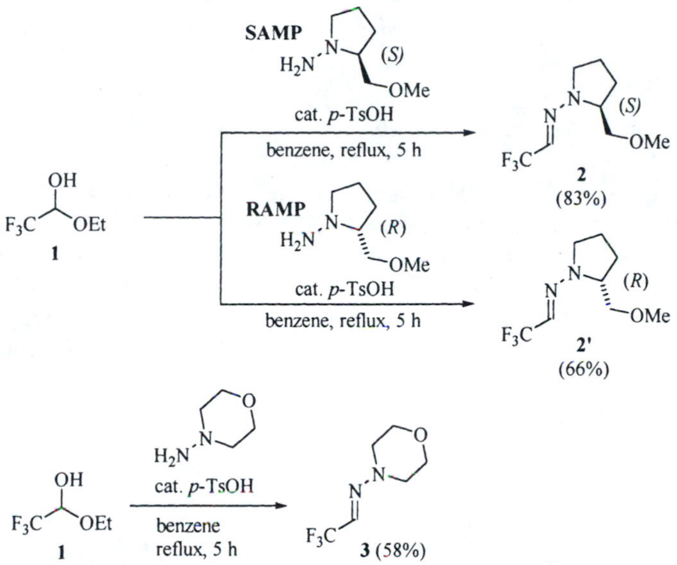

SAMP
H_2N
(*S*)
OMe
cat. *p*-TsOH
benzene, reflux, 5 h
OH
F_3C OEt
1
RAMP
H_2N
(*R*)
OMe
cat. *p*-TsOH
benzene, reflux, 5 h
N
(*S*)
OMe
F_3C
2
(83%)
N
(*R*)
OMe
F_3C
2'
(66%)
O
H_2N
cat. *p*-TsOH
OH
F_3C OEt
1
benzene
reflux, 5 h
O
N
F_3C
3 (58%)

Figure 2. Preparation of Trifluoroacetaldehyde SAMP-, RAMP-, and Morphirino Hydrazones

Trifluoroacetaldehyde SAMP-hydrazone is much more stable than the corresponding imines and this would be the reason why the hydrazone can be purified by flash column chromatography.

Very recently, Brigaud reported the preparation of other chiral trifluoroacetaldehyde hydrazone, as shown in Figure 3. In this reaction, the use of acidic catalyst produced the desired chiral hydrazone **5** along with the 1,3,4-oxadiazine derivative **6**, because of the presence of a hydroxyl group. Neutral conditions without acidic catalyst afforded only the hydrazone **5** in quantitative yield *(7)*.

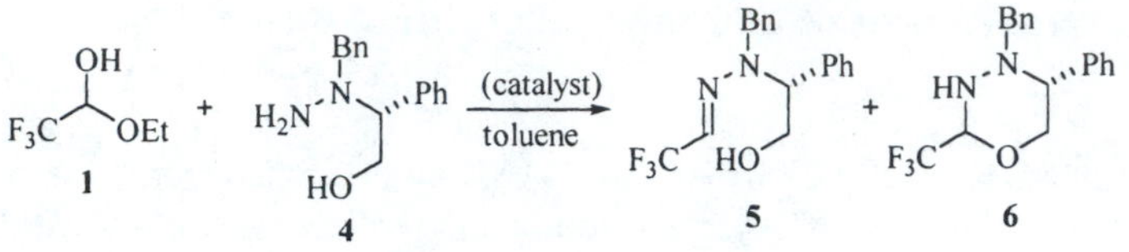

Figure 3. *Preparation of Other Chiral Trifluoroacetaldehyde Hydrazone*

1,2-Addition of Alkyllithums to Trifluoroacetaldehyde SAMP- or RAMP-hydrazones

Optimization of the Reaction Conditions of 1,2-Addition *(6)*

First, we examined the screening of the reaction conditions of nucleophilic 1,2-addition to trifluoroacetaldehyde SAMP hydrazone **2** with commercially available *n*-buthyllithium (*n*-BuLi). When 3 equiv of *n*-BuLi were added slowly to an Et_2O solution of **2** at –78 °C and the reaction mixture was gradually warmed up to room temperature, a low yield (36%) of the product **7a** was obtained together with a complex mixture, probably due to the low stability of trifluoromethylated lithium hydrazide (Figure 4, Table 1, entry 1).

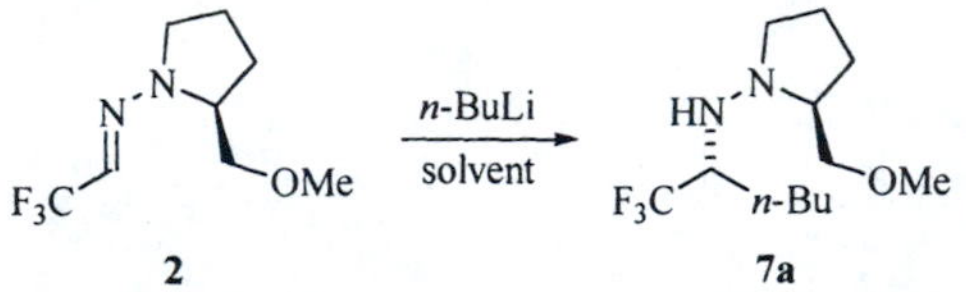

Figure 4.

Treatment of **2** with 1.5 equiv of *n*-BuLi in Et_2O at low temperature (–78 °C) for a shorter reaction time (1 h) gave trifluoromethylated hydrazine **7a** in 63% yield (entry 2). The use of 3 equiv of *n*-BuLi gave a higher yield (entry 3). Employing THF as a reaction solvent resulted in unsatisfactory yields of **7a** with decrease in diastereoselectivities (entries 4 and 5). The major diastereoisomer was readily separated by flash column chromatography.

Table 1. Screening of the Reaction Conditions of Nucleophilic 1,2-Addition of *n*-BuLi

Entry[a]	*n-BuLi*	*Solvent*	*Yield (%)*[b]	*De (%)*[c]
1[d]	3	Et_2O	36	>96 (>98)
2	1.5	Et_2O	63	>96 (>98)
3	3	Et_2O	79	>96 (>98)
4	1.5	THF	43 (37)	82 (>98)
5[c]	3	THF	45 (39)	82 (>98)

[a] The reactions were carried out with trifluoroacetaldehyde SAMP-hydrazone **2** at –78 °C for 1 h. [b] Yields of isolated pure products. Values in parentheses are for the major diastereomer. [c] Measured by ^{19}F NMR before isolation. Values in parentheses after column chromatography. [d] After *n*-BuLi was added at –78 °C, the reaction mixture was warmed up to room temperature overnight.

1,2-Addition of Various Alkyllithums and Phenyllithium *(6)*

The results of the reaction of **2** with various alkyllithiums as well as phenyllithium (PhLi) are summarized in Table 2 (Figure 5).

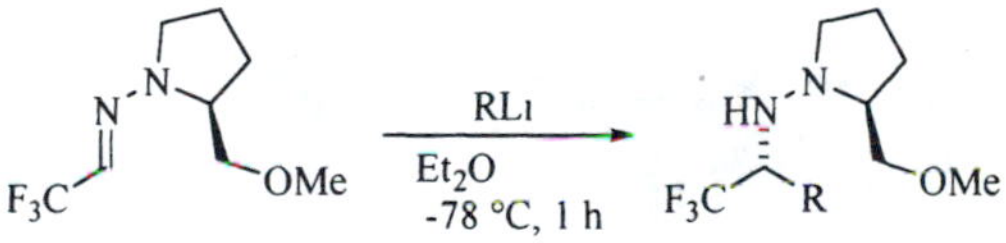

Figure 5.

Commercially available alkyllithiums, such as *n*-BuLi and *n*-hexyllithium, reacted well in the nucleophilic 1,2-addition to give the corresponding trifluoromethylated hydrazines **7a,a',d** in good yields with excellent diastereoselectivities (entries 1, 2, and 5). Ethyllithium and *n*-propyllithium, easily prepared from *t*-butyllithium (*t*-BuLi) and the corresponding iodoalkane *(8)*, also reacted with hydrazone **2** to provide the corresponding hydrazines **7c** and **d** in 48 and 65% yields, respectively (entries 3 and 4). Treatment of **2** with *t*-BuLi gave a moderate yield of **7e** in moderate *de* (entry 6). However, the diastereomer could be readily separated by column chromatography affording diastereomerically pure **7e**. When hydrazone **2** was treated with PhLi under the same conditions, 15% of the product **2f** was obtained in 86% *de* along with a complex mixture of by-products (entry 7). Unfortunately, even in the presence

Table 2. Reaction of Trifluoroacetaldehyde SAMP- or RAMP-Hydrazones

Entry[a]	*R*	*Product*	*Yield (%)*[b]	*De (%)*[c]
1	*n*-Bu	**7a**	79	>96 (>98)
2[d]	*n*-Bu	**7a'**	74	>96 (>98)
3	Et[e]	**7b**	48	>96 (>98)
4	*n*-Pr[e]	**7c**	65	>96 (>98)
5	*n*-Hex	**7d**	68	>96 (>98)
6	*t*-Bu	**7e**	58 (50)	72 (>98)
7	Ph	**7f**	15[f]	86 (88)

[a] The reactions were carried out with trifluoroacetaldehyde SAMP-hydrazone **2** and RLi (3 equiv) in Et_2O at –78 °C for 1 h. [b] Yields of isolated products. Values in parentheses are for the major diastereomer. [c] Measured by ^{19}F NMR before isolation. Values in parentheses after column chromatography. [d] Trifluoroacetaldehyde RAMP-hydrazone **2'** was used instead of SAMP-hydrazone **2**. [e] Prepared from *t*-BuLi and the corresponding iodoalkane. [f] There were many unidentified by-products in the ^{19}F NMR of the crude reaction mixture.

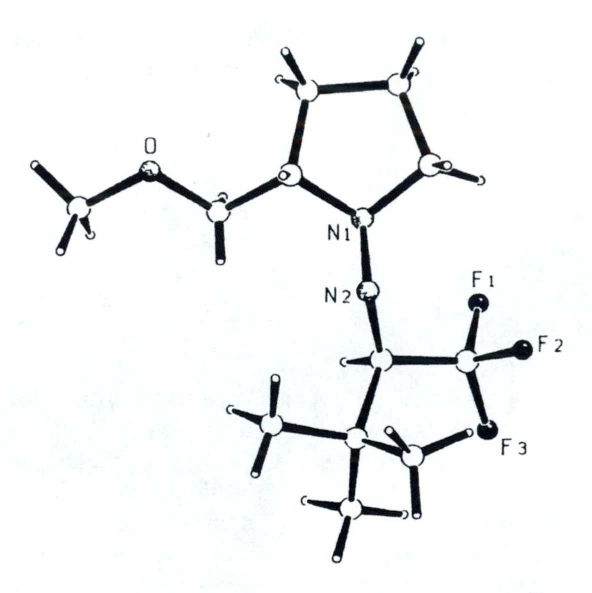

Figure 6.

of the trifluoromethyl group, the reaction of **2** with 3 equiv of methyllithium (MeLi) in Et_2O or toluene at –78 °C did not proceed efficiently giving only a small amount of the product together with recovery of **2** (58-75%). Raising the reaction temperature from –78 °C to room temperature in analogy to fluorine-free hydrazones as well as using methylmagunesium iodide at –20 °C or methycerium chloride ($MeCeCl_2$) at –78 °C in place of MeLi did not improve the reaction.

The absolute configuration of the stereogenic center generated by the 1,2-addition using SAMP was established unambiguously as *R* by X-ray crystallography of **7e** (Figure 6).

1,2-Addition of Prepared Allyllithium *(9)*

For the screening of the allylation reaction of trifluoroacetaldehyde hydrazones, the reaction between trifluoroacetaldehyde morpholinohydrazone **3** and tetraallyltin was first examined. The results are summarized in Table 3 (Figure 7).

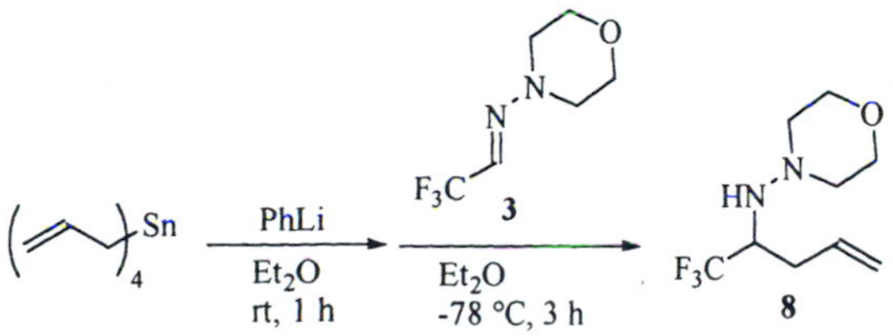

Sn 4 PhLi Et_2O rt, 1 h; F_3C **3** Et_2O -78 °C, 3 h; F_3C HN N O **8**

Figure 7.

Treatment of tetraallyltin (3 equiv) with PhLi (12 equiv) at room temperature, followed by addition of trifluoroacetaldehyde morpholino-hydrazone **3** at –78 °C, gave the corresponding trifluoroacetaldehyde morpholinohydrazine **8** in 62% yield (entry 4). The reaction of **3** with 3 equiv each of tetraallyltin and PhLi was extremely sluggish, providing only a trace amount of **8** (entry 1). Employing tetraallyltin (1-2 equiv) and PhLi (4-8 equiv) produced **8** in 21-47% yields along with the recovery of **3** (entries 2 and 3). The use of allyltriphenyltin (3 equiv) in place of tetraallyltin with PhLi (3 equiv) was also ineffective, giving **8** in 20% yield together with 15% of **3**. $Yb(OTf)_3$-catalyzed (5 mol %) allylation with 0.3 equiv of tetraallyltin in acetonitrile at room temperature did not occur at all, and **3** was recovered in quantitative yield *(10)*. With other allylic nucleophiles e.g., allylmagnesium bromide (3 equiv, –78 °C to rt, overnight), only a trace amount of **8** was formed with 67% recovery of **3**.

Table 3. Screening of the Reaction Conditions of Allylation of Trifluoroacetaldehyde Morpholinohydrazone

Entry[a]	*$(CH_2{=}CHCH_2)_4Sn$ (equiv)*	*PhLi (equiv)*	*Yield (%)*[b]
1	3	3	7 (38)
2	1	4	21 (35)
3	2	8	47
4	3	12	62

[a] The reactions were carried out with trifluoroacetaldehyde morpholinohydrazone **3**, $(CH_2{=}CHCH_2)_4Sn$, and PhLi at −78 °C for 1 h. [b] Yields of isolated pure products. Values in parentheses are for the recovery of starting hydrazone.

Next, in order to synthesize diastereo- and enantiomerically pure hydazines, the stereoselective allylation of trifluoroacetaldehyde SAMP- or RAMP-hydrazone **2** or **2'** was carried out (Figure 8) *(11)*. When SAMP-hydrazone (*S*)-**2** was treated with a mixture, obtained by the reaction between 3 equiv of tetraallyltin and 12 equiv of PhLi at room temperature, in Et_2O at −78 °C for 4 h, the allylated hydrazine **7g** was produced with high diastereoselectivty (93 : 7), and diastereomerically pure **7g** was easily obtained by flash column chromatography in 80% yield. RAMP-hydrazone **2'** also nicely underwent the

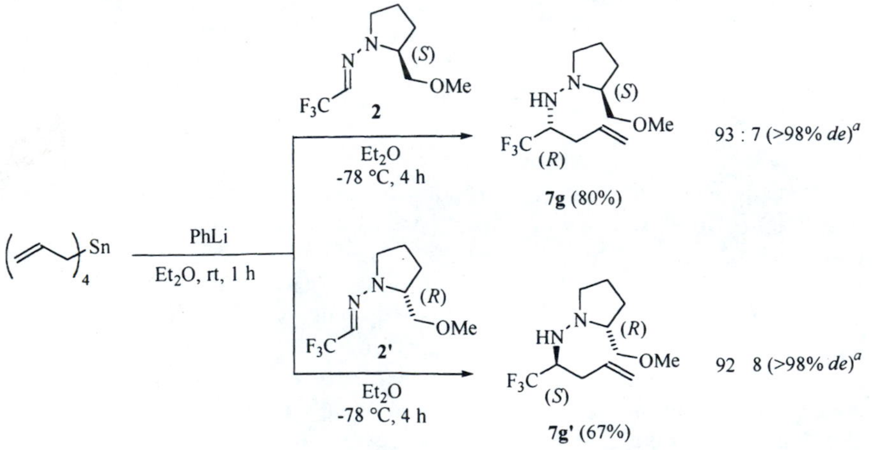

[a] Determined by ^{19}F NMR before isolation Values in parentheses after column chromatography.

Figure 8. Reaction of Trifluoroacetaldehyde SAMP- or RAMP-Hydrazones with Prepared Allyllithium

allylation reaction to give **7g'** (>98% *de*) in 67% yield after column chromatography.

Asymmetric Synthesis of α-Trifluoromethylated Amines

Benzoylation of Trifluoroacetaldehyde SAMP-Hydrazines *(6,9)*

Benzoylation of trifluoroacetaldehyde SAMP- or RAMP-hydrazines **7**, having *n*-butyl, *n*-propyl, *n*-hexyl, allyl, or phenyl groups, was carried out with an excess amount of triethylamine (10 equiv) and benzoyl chloride (10 equiv) in the presence of a catalytic amount of 4-dimethylaminopyridine (DMAP) at room temperature, to give the corresponding hydrazides in good to excellent yields (Figure 9).

However, the reaction of trifluoroacetaldehyde SAMP-hydrazine **7** with a *t*-butyl group did not proceed at all under this reaction conditions, and the starting hydrazine **7** was recovered quantitatively. It was found that lithium hydrazide, easily prepared from trifluoroacetaldehyde SAMP-hydrazine **7e** with 1.2 equiv of *n*-BuLi at –78 °C, smoothly reacted with 1.2 equiv of benzoyl chloride in a shorter reaction time, to produce **9e** in 66% yield (Figure 9).

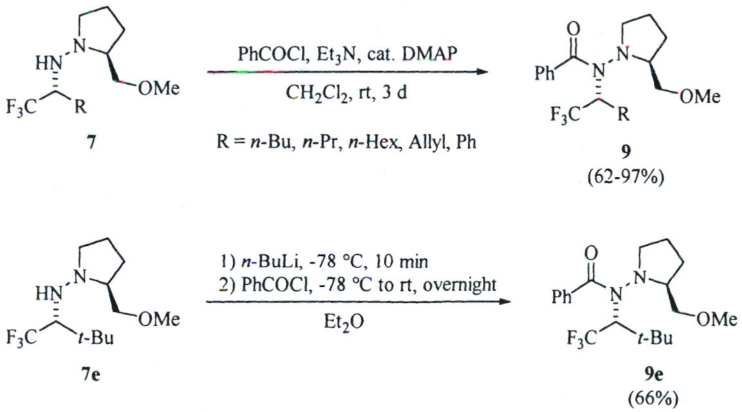

HN, N, OMe, F_3C, R — **7** — PhCOCl, Et_3N, cat. DMAP; CH_2Cl_2, rt, 3 d; R = *n*-Bu, *n*-Pr, *n*-Hex, Allyl, Ph → Ph, O, N, N, OMe, F_3C, R — **9** (62-97%)

HN, N, OMe, F_3C, *t*-Bu — **7e** — 1) *n*-BuLi, -78 °C, 10 min; 2) PhCOCl, -78 °C to rt, overnight; Et_2O → Ph, O, N, N, OMe, F_3C, *t*-Bu — **9e** (66%)

Figure 9. Benzoylation of Trifluoroacetaldehyde SAMP- or RAMP-Hydrazines with Benzoyl Chloride and Et_3N

SmI_2-Induced Nitrogen-Nitrogen Single Bond Cleavage *(6,9)*

Three equiv of SmI_2 *(12)* in the presence of 1,3-dimethyltetrahydro-2(1*H*)-pyrimidinone (DMPU) in THF at room temperature for 30 min afforded the (*R*)-*N*-benzoyl α-trifluoromethylated amines **10** without detectable epimerization or racemization (Figure 10, Table 3).

Various SAMP- or RAMP-hydrazides **9** participated successfully in the reaction to provide the corresponding amides **10** in good to excellent yields with excellent *ee* (up to >99%). The reaction in MeOH gave a trace amount of the

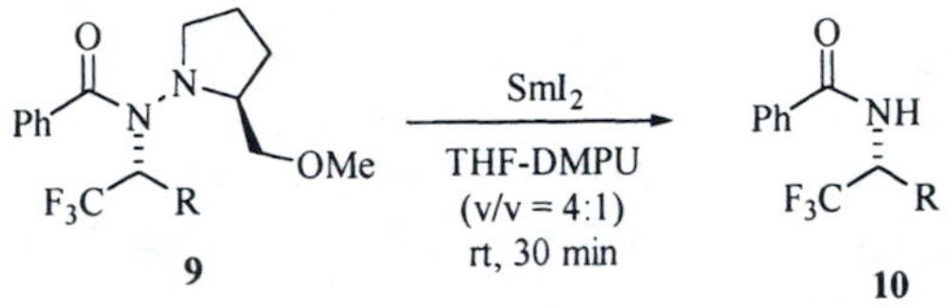

Figure 10. SmI_2-Induced Cleavage of the N-N Single Bond of Trifluoromethylated SAMP- or RAMP-Hydrazides

Table 3. SmI_2-Induced Cleavage of the *N-N* Single Bond of Trifluoromethylated SAMP- or RAMP-Hydrazides

Entry[a]	*R*	*Product*	*Yield (%)*[b]	*Ee (%)*[c]
1	*n*-Bu	**10a**	87	98
2[d]	*n*-Bu	**10a**	4 (90)	-
3[e]	*n*-Bu	**10a'**	95	97
4	*n*-Pr	**10c**	83	98
5	*n*-Hex	**10d**	97	98
6	*t*-Bu	**10e**	71	>99
7	Ph	**10f**	73	88
8	Allyl	**10g**	75	>99

[a] The reactions were carried out with hydrazide **9** and SmI_2 (3 equiv) in the presence of DMPU in THF at room temperature for 30 min. [b] Yields of isolated products. Values in parentheses are for the recovery of **9**. [c] Measured by HPLC analysis using a chiral stationary phase column (DAICEL OD or (S,S)-Welk-O, 1-heptane/2-propanol = 9/1 or 95/5). [d] MeOH was used as a solvent in the absence of DMPU. [e]RAMP-hydrazide **9a'** was used.

product **10**, together with recovery of the starting hydrazide (90%). This method using SmI_2 also effective for cleavage of nitrogen-nitrogen single bond of other chiral SAMP hydrazides, such as heteroaromatic substituted SAMP hydrazide *(13)* and *N*-SAMP 3-substituted 3,4-dihydro-2*H*-isoquinolin-1-one *(14)*.

The benzoyl group is essential for the cleavage of nitrogen-nitrogen single bond of **9**. Thus, treatment of SAMP-hydrazine **7g** with SmI_2 (3 equiv) under the same conditions gave only a trace amount of the corresponding amine, and hydrazine **7g** was recovered in 84% yield (Figure 11).

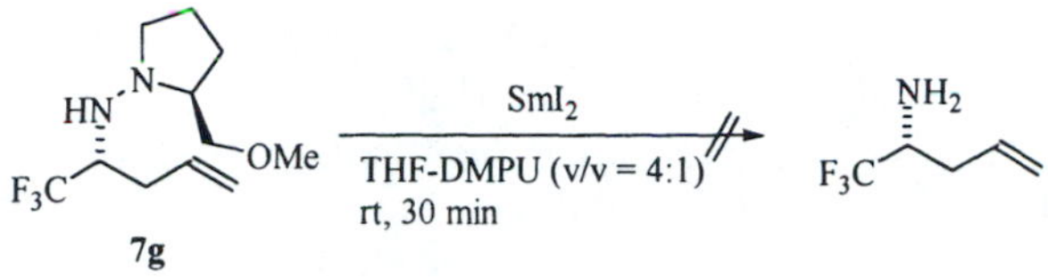

HN N OMe F$_3$C **7g** — SmI_2, THF-DMPU (v/v = 4:1), rt, 30 min —//→ NH_2 F$_3$C

Figure 11. SmI_2-Induced Cleavage of the N-N Single Bond of Trifluoromethylated SAMP- or RAMP-Hydrazides

Very recently, Friestad has examined the relationship between the protecting groups of the hydrazines and SmI_2 *(15)*. Consquently, it was found that the use of a trifluoroacetyl group as a protecting group resulted in the effective cleavage of nitrogen-nitrogen single bond of the hydrazides *via* the chelation of fluorine atom(s) and Sm metal.

Asymmetric Synthesis of β-Trifluoromethylated β-Amino Carbonyl Compounds via Oxidation of α-Trifluoromethylated Homoallyl Amines *(16)*

As shown in Figure 12, the ozonolysis of (*R*)-*N*-benzoyl-α-trifluoromethylated homoallylamine **10g** thus prepared in the presence of 2.5 M NaOH (6 equiv) methanol solution in dichloromethane at –78 °C smoothly proceeded to produce the corresponding (*R*)-β-*N*-benzoylamino-β-trifluoromethylated carboxylic acid methyl ester **11** in 69% yield with slight racemization *(17)*. Other enantiomer **10g'** also participated well in the ozonolysis to produce the **11'** in 49% yield with 96% *ee*. The strong electronegativity of the trifluoromethyl group seems to have an effect on the epimerization of the starting hydrazone or the product under the basic conditions.

Direct synthesis of acid **12'** using the combination of 1 equiv of potassium permanganate ($KMnO_4$) and 10 equiv of sodium periodinate ($NaIO_4$) has been carried out. The oxidation proceeded smoothly at room temperature to give the corresponding β-benzoylamino-β-trifluoro-methylated carboxylic acid **12'** in 65% yield. Its enantiomer ratio could be determined by esterification using chlorotrimethylsilane in methanol leading to methyl ester **11'** *(18)*, as 94% *ee*.

Finally, the carbon-carbon double bond of (*R*)-*N*-benzoyl-α-trifluoromethylated homoallylamine **10g** was successfully cleaved with a catalytic amount (1 mol%) of osmium tetroxide (OsO_4) using $NaIO_4$ (4 equiv) as the stoichiometric oxidant, to produce the corresponding β-benzoylamino-β-trifluoromethyl aldehyde **13** in 92% yield with >99% *ee*. The aldehyde **13** is stable enough to easily separate by normal flash chromatography. The *N*-benzoyl group will play a crucial role in the stability of the aldehyde like the *p*-tolylsulphinyl *(19)*, *N-t*-butoxycarbonyl (Boc) *(20)*, or carboxybenzyloxy (Cbz)*(21)* groups on the reported fluorine-free aldehydes.

The reduction of the aldehyde **13** with 0.5 equiv of sodium borohydride ($NaBH_4$) in EtOH at room temperature for 4 h produced the quantitative yield of the corresponding enantiopure trifluoromethylated γ-amino alcohol **14** without epimerization (Figure 13).

Conclusions

In summary, we have succeeded in the highly enantioselective synthesis of α-trifluoromethylated amines through the 1,2-addition of various organolithium species to trifluoroacetaldehyde SAMP- or RAMP-hydrazones and subsequent SmI_2-promoted cleavage of the nitrogen-nitrogen single bond. Furthermore, we have demonstrated that asymmetric synthesis of trifluoromethylated β-amino aldehyde, carboxylic acid, and ester as well as γ-aminoalcohol *via* the oxidation of chiral α-trifluoromethylated homoallylamine obtained.

Acknowledgements

K.F. is grateful to the Alexander von Humboldt Foundation for a postdoctoral fellowship (2000-2001). We thank Degussa AG, BASF AG, Bayer AG, and Aventis for the donation of chemicals. We also thank the Central Glass Co., Ltd., for the gift of trifluoroacetaldehyde ethyl hemiacetal and hydrate. This work has been supported by the Saijiro Endo Foundation. We also thank Dr. J. Runsink for measuring the ^{19}F NMR spectra. We thank Professors Hiroki

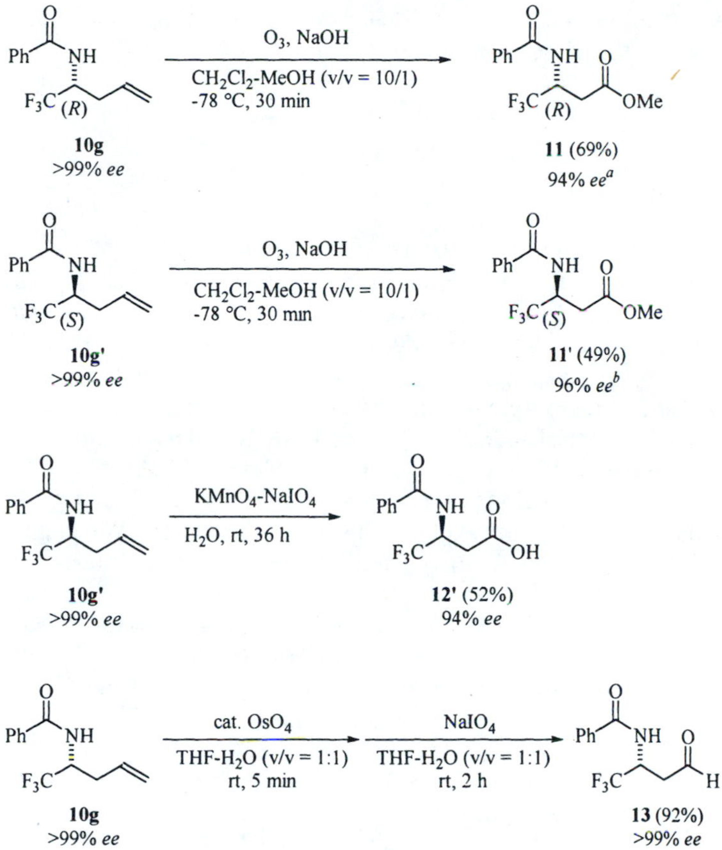

10g (>99% ee) → [O_3, NaOH; CH_2Cl_2-MeOH (v/v = 10/1), -78 °C, 30 min] → 11 (69%), 94% ee[a]

10g' (>99% ee) → [O_3, NaOH; CH_2Cl_2-MeOH (v/v = 10/1), -78 °C, 30 min] → 11' (49%), 96% ee[b]

10g' (>99% ee) → [$KMnO_4$-$NaIO_4$; H_2O, rt, 36 h] → 12' (52%), 94% ee

10g (>99% ee) → [cat. OsO_4; THF-H_2O (v/v = 1:1), rt, 5 min] → [$NaIO_4$; THF-H_2O (v/v = 1:1), rt, 2 h] → 13 (92%), >99% ee

Figure 12. Asymmetric Synthesis of β-Benzoylamino-β-trifluoromethylated Carboxylic Acid Esters, Carboxylic Acid, and Aldehyde

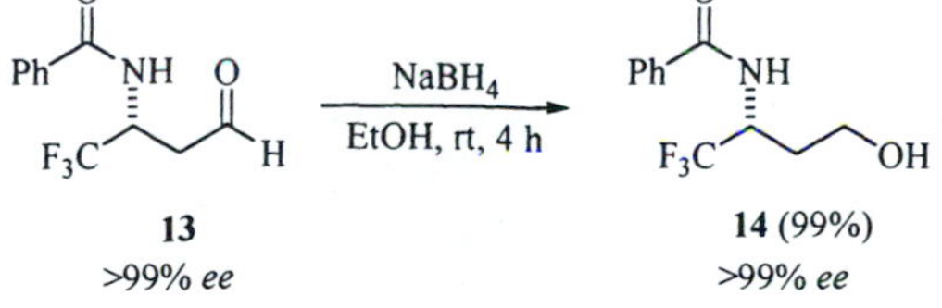

13 (>99% ee) → [$NaBH_4$; EtOH, rt, 4 h] → 14 (99%), >99% ee

Figure 13. Asymmetric Synthesis of γ-Trifluoromethylated γ-Amino Alcohol

Yamanaka, Takashi Ishihara, and Tsutomu Konno of the Kyoto Institute of Technology for the HRMS measurement.

References

1. For recent reviews on asymmetric synthesis of functionalized α-trifluoromethylated chiral amines using chiral auxiliary, see: (a) Bravo, P.; Zanda, M. In *Enantiocontrolled Synthesis of Fluoro-organic Compounds*, Soloshonok, V. A., Ed.; John Wiley & Sons: Chichester, 1999; pp 107-160. (b) Bravo, P.; Bruche, L.; Crucianell, M.; Viani, F.; Zanda, M. In *Asymmetric Fluoroorganic Chemistry: Synthesis, Application, and Future Directions*, Ramachandran, P. V., Ed.; ACS Symposium Series 746, Washington, DC, 1999; pp 98-116. (c) Soloshonok, V. A. In ref. 1b; pp 74-83. (d) Ishii, A.; Kanai, M.; Katsuhara, M. In *Fluorine-Containing Synthons*, Soloshonok, V. A., Ed.; ACS Symposium Series 911, Washington, DC, 2005; pp 368-377. (e) Bonnet-Delpon, D.; Bégué, J.-P.; Crousse, B. In ref. 1d; pp 412-429.
2. For a recent book for fluorinated amino acid, see: *Fluorine-containing Amino Acids : Synthesis and Properties*, Kukhar' V. P.; Soloshonok, V. A., Eds.; John Wiley & Sons: Chichester, 1995.
3. For recent reviews on asymmetric synthesis of functionalized fluorine-containing amino acids see: (a) Soloshonok, V. A. In ref.1a; pp 229-262. (b) Uneyama, K. In ref. 1a; pp 391-418. (c) Soloshonok, V. A. In ref. 1b; pp 74-83. (c) Abouabdellah, A.; Bégué, J.-P.; Bonnet-Delpon, D.; Kornilov, A.; Rodrigues, I.; Nga, T. T. T. In ref. 1b; pp 84-97. (d) Zanda, M.; Bravo, P.; Volonterio, A. In ref. 1b; pp 127-141. (e) Ojima, I.; Inoue, T.; Slaster, J. C.; Lin, S.; Kuduk, S. D.; Chakravarty, S.; Walsh, J. J.; Cresteil, T.; Monsarrat, B.; Pera, P.; Bernacki, R. J. In ref. 1b; pp 158-181. (f) Ojima, I.; Kuznetsova, L., Ungureanu, I. M.; Pepe, A,; Zanardi, I.; Chen, J. In ref. 1c; pp 544-561. (g) Qing, F.-L.; Qiu, X. –L. In ref. 1d; pp 562-571. (h) Fustero, S.; Sanz-Cervera, J. F.; Piera, J.; Sánchez-Roselló, J. F.; Piera, J.; Sánchez-Roselló, M.; Jiménez, D.; Chiva, G. In ref. 1c; pp 593-610.
4. For reviews on the1,2-addition to the carbon-nitrogen double bond see: (a) Denmark, S. E.; Nicaise, O. J. -C. *J. Chem. Soc., Chem. Commun.* **1996**, 999-1004. (b) Enders, D,; Reinhold, U. *Tetrahedron: Asymmetry* **1997**, *8*, 1895-1946. (c) Bloch, R. *Chem. Rev.* **1998**, *98*, 1407-1438. (d) Kobayashi, S.; Ishitani, H. *Chem. Rev.* **1999**, *99*, 1069-1094. (e) Merino, P.; Franco, S,; Merchan, F. L.; Tejero, T. *Synlett* **2000**, 442-454. (f) Alvaro, G.; Savoia, D. *Synlett* **2002**, 651-673.
5. For a recent review, see. Job, A.; Janeck, C. F.; Bettray, W.; Peters, R.; Enders, D. *Tetrahedron* **2002**, *58*, 2253-2329.

6. Enders, D.; Funabiki, K. *Org. Lett.* **2001**, *3*, 1575-1577.
7. Fries, S.; Pytkowicsz, J.; Brigaud, T. *Tetrahedron Lett.* **2005**, *46*, 4761-4764.
8. (a) Bailey, W. F.; Punzalan, E. R. *J. Org. Chem.* **1990**, *55*, 5404-5406. (b) Negishi, E.; Swanson, D. R.; Rousset, C. J. *J. Org. Chem.* **1990**, *55*, 5406-5409.
9. Funabiki, K.; Nagamori, M.; Matsui, M.; Enders, D. *Synthesis* **2002**, 2585-2588.
10. Manabe, K.; Oyamada, H.; Sugita, K.; Kobayashi, S. *J. Org. Chem.* **1999**, *64*, 8054-8057.
11. For the allylation of fluorine-free aldehyde SAMP- or RAMP-hydrazones see: Enders, D.; Schankat, J.; Klatt, M. *Synlett* **1994**, 795-798.
12. For the pioneering work for SmI_2-induced cleavage of nitrogen-nitrogen single bond of hydrazines, see: Souppe, J.; Danon, L.; Namy, J. L.; Kagan, H. B. *J. Organomet. Chem.* **1983**, *250*, 227-236.
13. Enders, D.; Del Signore, G. *Tetrahedron:Asymmetry* **2004**, *15*, 747-751.
14. Enders, D.; Braig, V.; Boudou, M. Raabe, G. *Synthesis* **2004**, 2980-2990.
15. Ding, H.; Friestad, G. K. *Org. Lett.* **2004**, *6*, 637-640.
16. Funabiki, K.; Nagamori, M.; Matsui, M. *J. Fluorine Chem.* **2004**, *125*, 1347-1350.
17. Ref. 11 and Marshall, J. A.; Garofalo, A. W.; Sedrani, C. R. *Synlett* **1992**, 643-645.
18. Juaristi, E.; Escalante, J.; Lamatsche, B.; Seebach, D. *J. Org. Chem.* **1992**, *57*, 2396-2398.
19. Davis, F. A.; Prasad, K. R.; Nolt, M. B.; Wu, Y. *Org. Lett.* **2003**, *5*, 925-927.
20. Davis, S. B.; McKervey, M. A. *Tetrahedron Lett.* **1999**, *40*, 1229-1232.
21. Davis, F. A.; Szewczyk, J. M. *Tetrahedron Lett.* **1998**, *39*, 5951-5954.

Chapter 29

A Novel Strategy for the Synthesis of Trifluoroalanine Oligopeptides

Kenji Uneyama, Yong Guo, Kana Fujiwara, and Yumi Komatsu

Department of Applied Chemistry, Faculty of Engineering, Okayama University, Okayama 700–8630, Japan

2-*N*-(*p*-Methoxyphenyl)(trimethylsilyl)aminoperfluoropropene was prepared in 95% yield by the Mg-promoted C-F bond activation in an Mg-TMSCl-DMF system. The fluorinated enamine is very electrophilic so that it reacts with a variety of α-amino esters in a triethylamine-DMF system at room temperature, affording trifluoroalanine dipeptides in 70-90% yields *via* one-pot process. The reaction proceeds *via* three steps; addition of amino group of amino esters, subsequent dehydrofluorination leading to the corresponding imidoyl fluoride, then acid-catalyzed hydrolysis of the imidoyl fluoride. The protocol is applicable for alanine tripeptides.

The chemistry and science of peptides and their related molecules has been increasingly developed over several decades due to their potential biological activity, and thus, a number of non-natural peptides and peptidomimetics have been designed and synthesized to gain the advantages of increased bioactivity, biostability, and bioselectivity over those of natural peptides.[1] On this basis, highly efficient peptide-bond fomation has been increasingly needed to enable the synthesis of more complex, newly functionalized, and/or structurally new peptides. So far, peptide bonds have been prepared mostly by condensation of amino acids. The rational activation of the carboxyl group of one amino acid followed by condensation with another aminoester,[2] or ring-opening of β-lactams with amino esters[3] has been used.

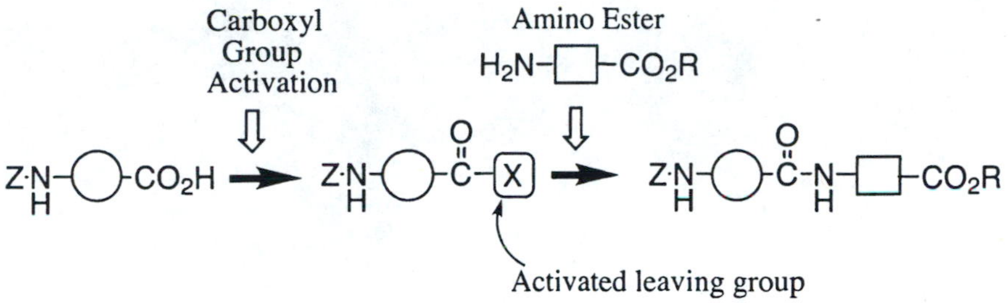

Figure 1. Coventional Method for Peptide Bond Formation

Both enzymatic[4] and chemical[5] activation has been commonly employed for the condensation. One of the enzymatic carboxyl group-

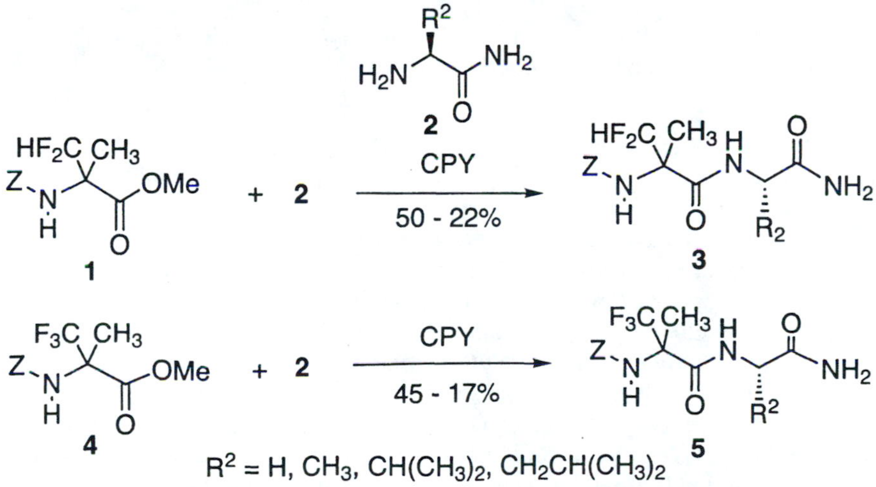

Figure 2. Enzymatic Peptides Formation

activation is shown in Fig. 2, where carboxypeptidase Y (CPY) was employed for the peptide bond formation of α-trifluoromethyl and difluoromethylalanines.[6] One example of the chemical activation of carboxyl group by DCC is shown in Fig. 3, where a rather higher temperature was needed for the peptide bond formation.[7]

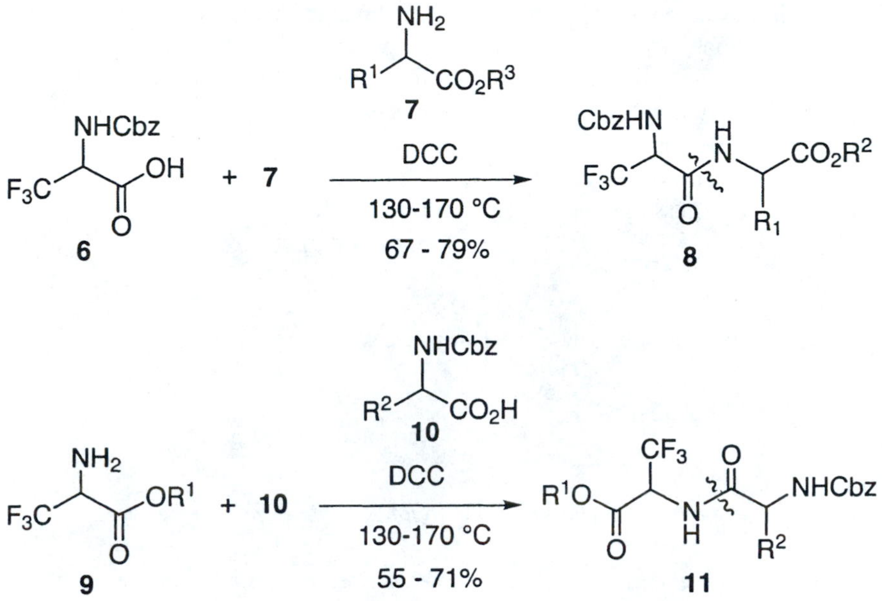

Figure 3. Chemical Activation of Carboxyl Group with DCC for Peptide Bond Formation

Meanwhile, Burger demonstrated Rh-catalyzed insertion of α-ketocarbene to N-H bond of α-aminoamide **13** leading to dipeptide formation *via* non-condensation route[8] (Fig. 4). But, very few methods for non-condensation type peptide synthesis have been known. A novel peptide synthesis method, in which C-N bond formation mechanism is totally different from the conventional condensation of two kinds of amino acids would open a door into the field of structurally unique and highly functionalized peptides, which would create unique biological activity as yet unknown.

In this context, it has been proposed 2-aminoperfluoropropene **15** would be effective synthon for trifluoroalanine which would react smoothly with amino group of amino esters **17**, affording dipeptides **16** (Fig. 5). The concept of this novel peptide-bond formation is shown schematically in Fig. 6.

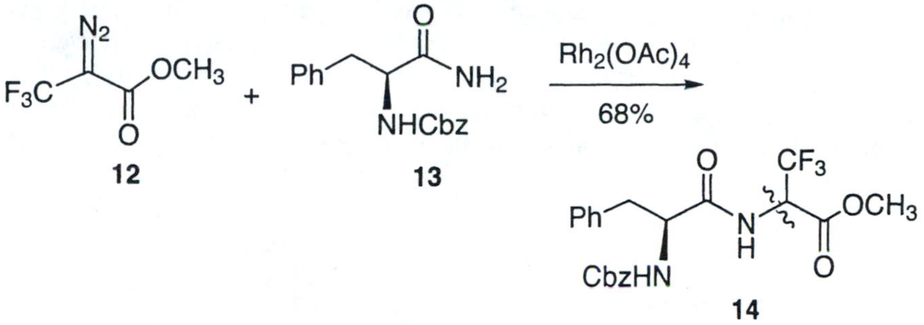

N_2
F_3C OCH_3 O
12
+
O
Ph NH_2
NHCbz
13
$Rh_2(OAc)_4$
68%
O CF_3
Ph N H OCH_3
CbzHN O
14

Figure 4. A New Peptide Formation by Ketocarbene Insertion to Amino Group

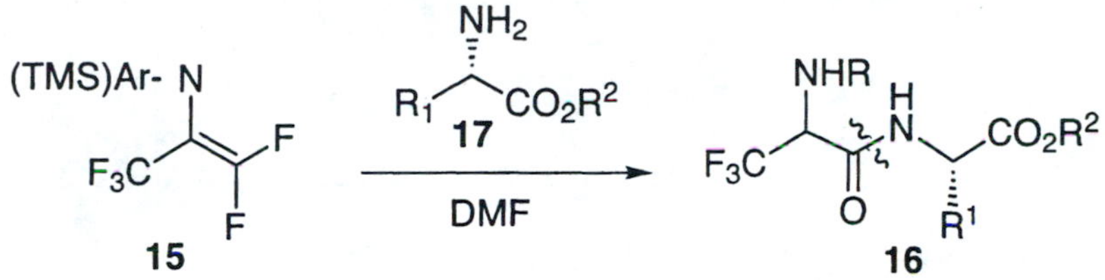

(TMS)Ar- N
F_3C F
F
15
NH_2
R_1 CO_2R^2
17
DMF
NHR H
F_3C N CO_2R^2
O R^1
16

Figure 5. A Novel Concept for Peptide Formation

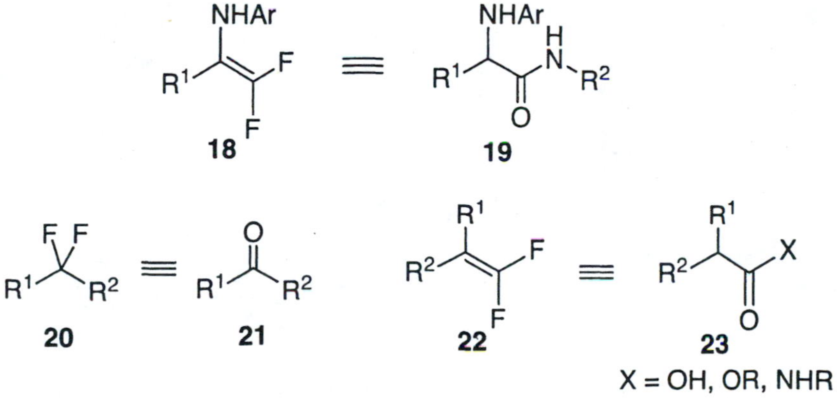

NHAr
R^1 F
F
18
≡
NHAr H
R^1 N R^2
O
19

F F
R^1 R^2
20
≡
O
R^1 R^2
21

R^1
R^2 F
F
22
≡
R^1
R^2 X
O
23
X = OH, OR, NHR

Figure 6. Difluoromethylene Moiety as a Synthon for Carbonyl Group

In principle, difluoromethylene moiety is a synthon for a carbonyl group just as do the dichloromethylene and dibromomethylene groups. Thus, the difluoromethylene group in the alkene **22** would be a synthon for a carboxylic acid, ester[9] and amide[10] since it can be transformed to the carbonyl derivatives on reacting with water, alcohols and amines, respectively. This difluoromethylene-carbonyl group interconversion concept makes it possible to design the difluoroenamine **18** as a synthon for α-aminoacids and peptides **19** as shown in Fig. 6.

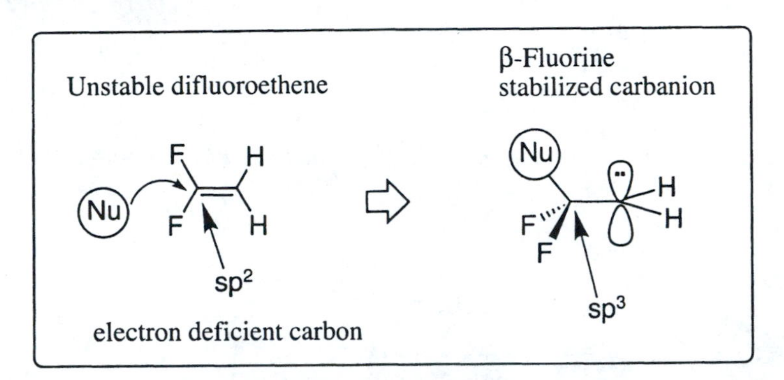

Figure 7. Highly Electrophilic Reactivity of CF_2 Site of Difluoroethene

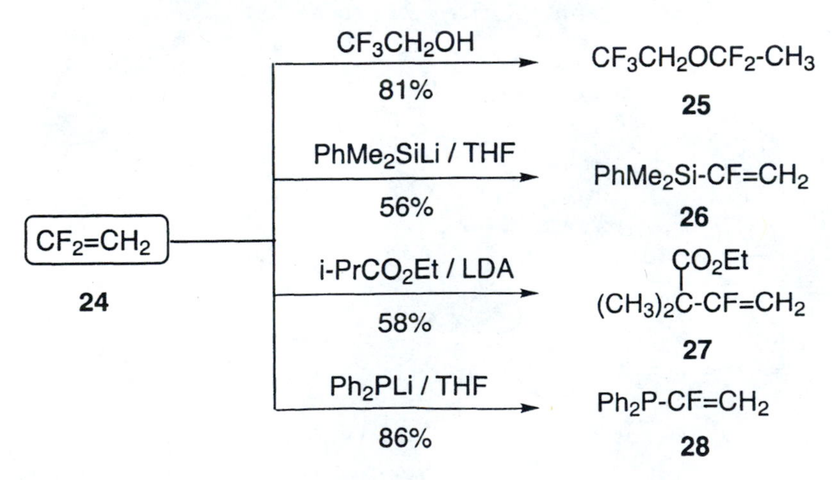

Figure 8. Electrophilic Reactions of Difluoroethene with nucleophiles

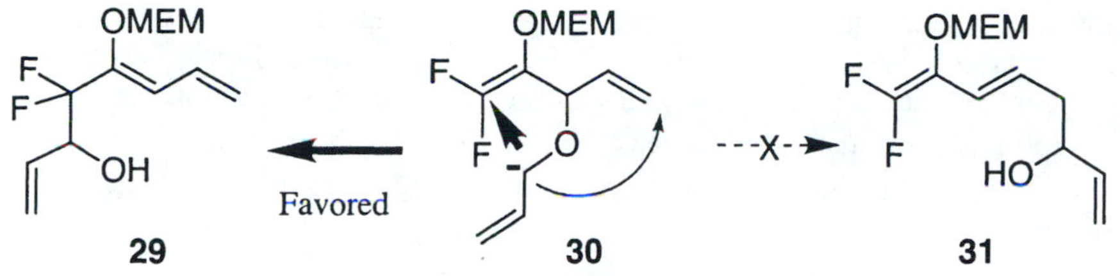

Figure 9. Higher Reactivity toward Nucleophiles at Difluoromethylene Site

The highly electrophilic reactivity of difluoromethylene site of 1,1-difluoroethene is well known.[11] The origin of the high electrophilicity arises from the electron-deficiency at α-carbon and the favorable transformation of sp^2-CF_2 carbon to sp^3-CF_2 carbon, and the stabilization of β-carbanion by the electron-withdrawing CF_2 group (Fig. 7).

Some examples of electrophilic reactions of difluoroethene **24** are shown in Fig. 8. Hydroxyl group,[12] silyl lithium,[13] enolate,[14] and phosphorus anion,[15] all of these nucleophiles attack exactly on α-carbon of **24**, affording an adduct product **25** or products **26**, **27**, and **28** formed *via* addition-elimination process (Fig. 8).

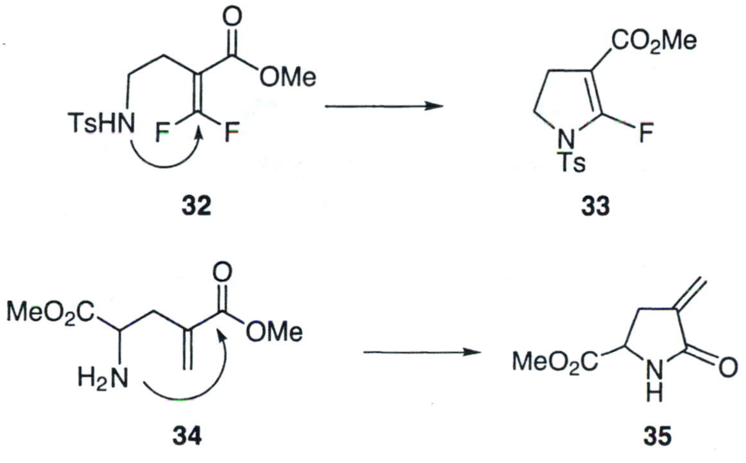

Figure 10. Remarkable Reactivity Difference at β-Carbon between Fluoro and Nonfluoro- α,β-Unsaturated Esters

Figure 9 shows an example of the electrophilicity of CF_2 carbon in the intramolecular reaction. Thus, carbanion **30** attacks exclusively at the difluoromethylene site in [2,3]sigmatropic rearrangement.[16]

Similar preference for nucleophilic attack at the CF_2-site has been also demonstrated by five-membered ring formation shown in Fig.10. Nitrogen-nucleophile attacks preferentially CF_2-carbon rather than the carbonyl carbon in difluorinated α,β-unsaturated ester **32**.[17] On the other hand, reaction can occur at the carbonyl carbon in nonfluorinated α,β-unsaturated ester **34**,[18] leading to lactam **35**.

Results and Discussion

On this basis, 2-aminoperfluoropropene **15** was designed as a precursor for trifluoroalanine peptides **36** (Fig.11). The enamine **15** would be highly electrophilic due to the strong electron-withdrawing CF_3 and CF_2 groups even though amino group at 2-position deactivates the difluoromethylene carbon toward nucleophilic attacks. The enamine **15** was prepared in 95% yield by Mg-promoted C-F bond activation[19] of imine **37**[20] in Mg -TMSCl-DMF system[21] at room temperature.

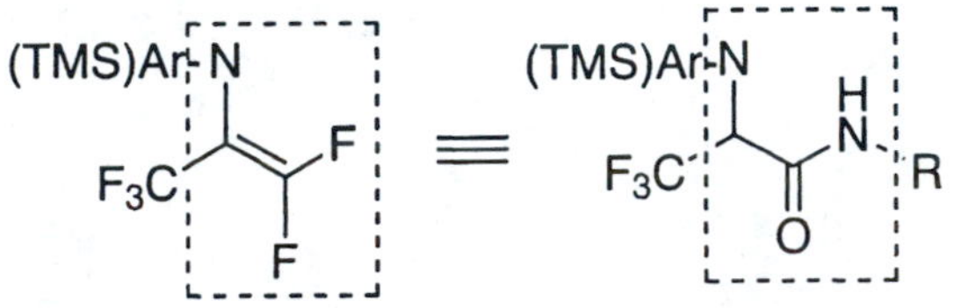

(TMS)Ar-N F3C F F **15** ≡ (TMS)Ar-N F3C H N R O **36**

Figure 11. 2-Aminoperfluoropropene as a Synthon of Trifluoroalanine

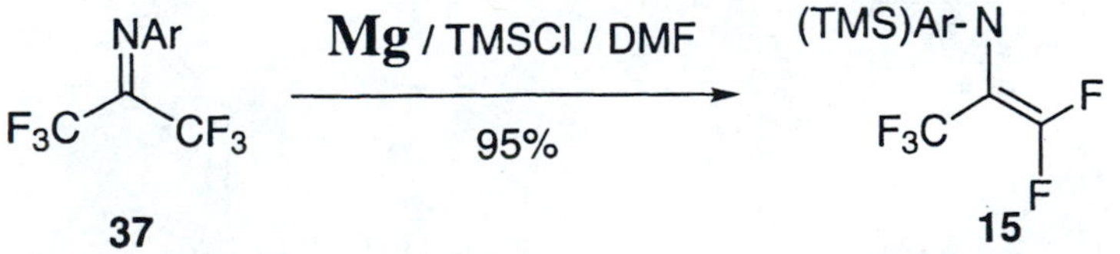

NAr F3C CF3 **37** —Mg / TMSCl / DMF, 95%→ (TMS)Ar-N F3C F F **15**

Figure 12. Mg-Promoted Defluorination of Hexafluoroimine

Enamine **15** was subjected to the reaction with the commercially available amino esters **17**. The enamine **15** is stable enough to be easily handled and can be stored at –5 °C, but it is very reactive even with the less nucleophilic amino group of amino esters due to the strong electron-withdrawing effect of *gem*-difluoromethylene and trifluoromethyl groups. Thus, **15** reacted very smoothly with amino esters in DMF at room temperature within 90 min, affording the desired trifluoroalanine dipeptides **16** as a diastereomeric mixture.[22] The total reaction consists of three steps; addition of amino ester **17** to **15**, subsequent dehydrofluorination from the more acidic N-H bond of **38** to give imidoyl fluoride **39**, and then acid-catalyzed hydrolysis of the imidoyl fluoride **39** to amide **16** (Fig. 13).[23] In some cases the imidoyl fluoride **39** was isolated as an intermediate and characterized spectroscopically.

Table 1 summarizes the results of dipeptide synthesis. Most of the amino esters **17** reacted with **15** smoothly, affording dipeptides **16** in good to excellent yields. Thus, esters of glycine, alanine, phenylalanine and leucine provided the corresponding dipeptide **16** in more than 85% yields. The secondary amino group of proline ester was as reactive to **15** as the primary amino group of glycine, alanine and leucine. Of course, the primary amino group is more reactive than the secondary in tryptophan **41**, where a product **43** with an amide

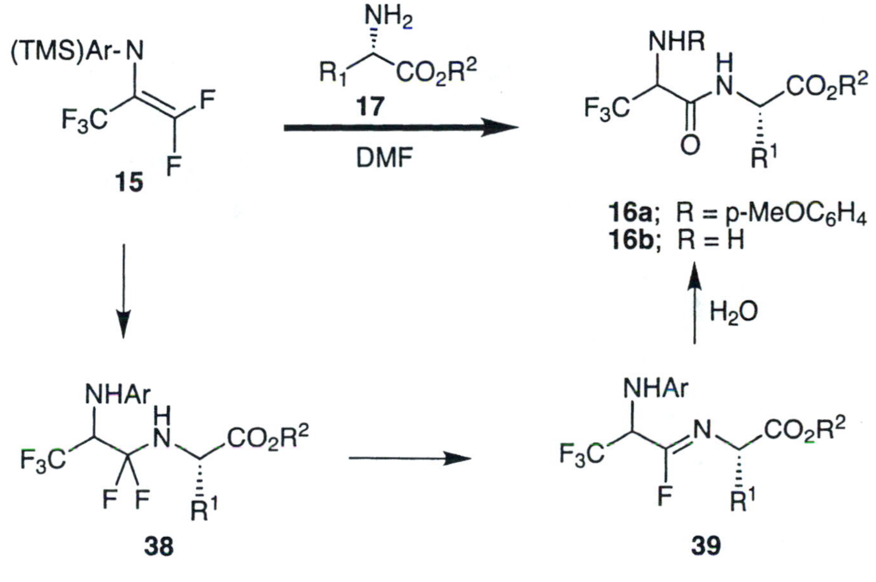

Figure 13. Reaction Mechanism and Intermediates for Peptide Synthesis of ***16***

Table I. Yields of Dipeptides 16 in the Reaction of 15 with 17

Amino ester **17**	R^1	R^2	Yield of **16** (%)		de[a)]
glycine	H	Bn	**16a**	91	-
Alanine	CH_3	Me	**16b**	86	14
Phenyl alanine	$PhCH_2$	Et	**16c**	93	31
Leucine	$(CH_3)_2CHCH_2$	Et	**16d**	85	48
Proline		Me	**16e**	69	36
Serine	$HOCH_2$	Me	**16f**	80	8
Tyrosine	$HOC_6H_4CH_2$	Me	**16g**	68	1
Methionine	$CH_3SCH_2CH_2$	Me	**16h**	88	21
Tryptophan		Me	**16i**	79	19

a) diastereomeric ratios were analyzed by HPLC

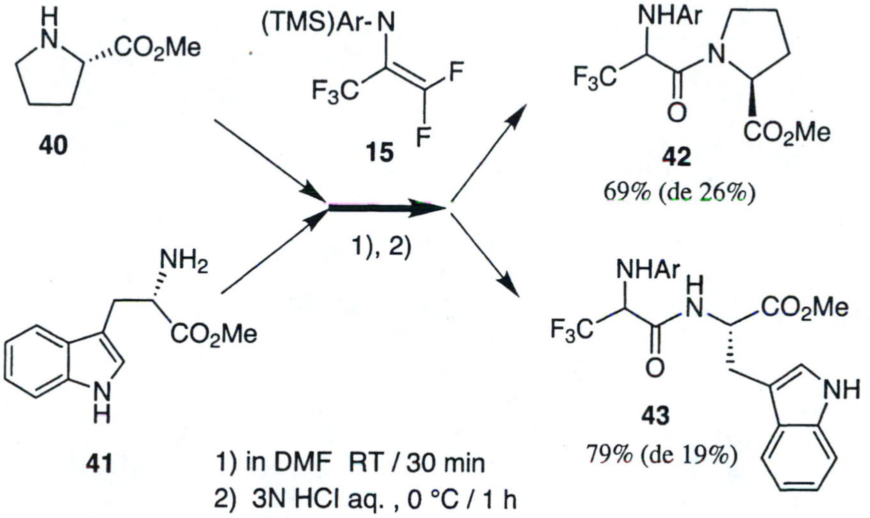

(TMS)Ar-N
F_3C F F
15

40 — H N ...CO2Me

41 — NH2 CO2Me N H

1), 2)

1) in DMF RT / 30 min
2) 3N HCl aq. , 0 °C / 1 h

42
69% (de 26%)

43
79% (de 19%)

Figure 14. Reactions of Proline and Tryptophan Esters

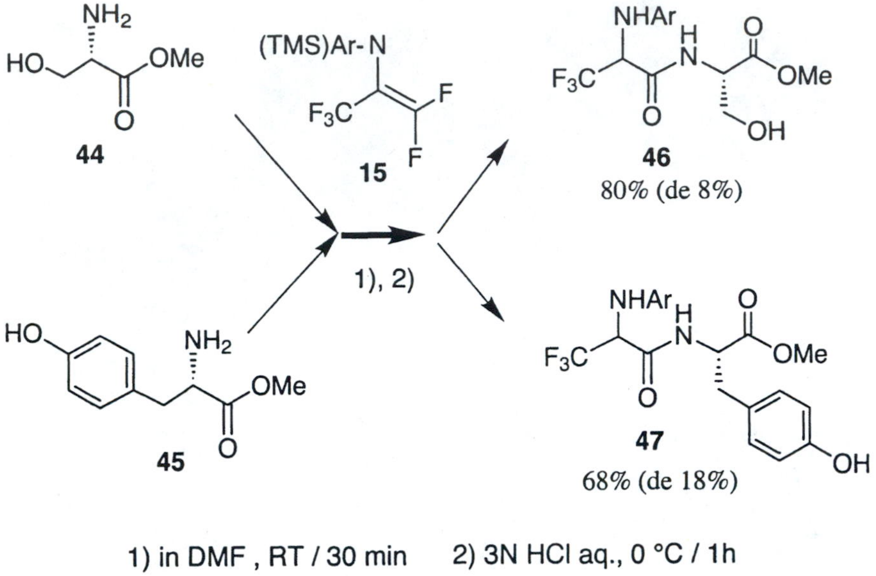

1) in DMF , RT / 30 min   2) 3N HCl aq., 0 °C / 1h

Figure 15. Reactions of Serine and Tyrosine Esters

bond connected with primary amino group in **41** was formed exclusively (Fig. 14).

Interestingly, serine **44** and tyrosine **45** esters reacted exclusively at the amino site rather than the hydroxyl site even when the amino esters were employed without protection of the hydroxyl group (Fig. 15). Both esters provided the desired dipeptides **46** and **47** in 80 and 68% yields, respectively. The addition reaction in the first step is quite chemoselective as shown in **49**. The preferred reactivity of the amino group to **15** over the hydroxyl group is also revealed by the fact that the competitive reaction of **15** with a mixture of equimolar amounts of both benzyl amine and benzyl alcohol provided exclusively benzyl amide **50** of trifluoroalanine in 98% isolated yield and none of the corresponding benzyl ester (Fig. 16).

In relation to the higher chemoselectivity of amino group over hydroxyl group, it is interesting to determine whether the more nucleophilic thiol group can compete with amino group in the reaction of **15** with cysteine. The reaction of **15** with cysteine **51** bearing an unprotected thiol group resulted in the formation of complex miture which might arise from the nucleophilic attack of thiol group to the activated difluoromethylene site. On the other hand, *S*-benzylcysteine **52** provided the desired amide **54** in a reasonable yield (Fig. 17).

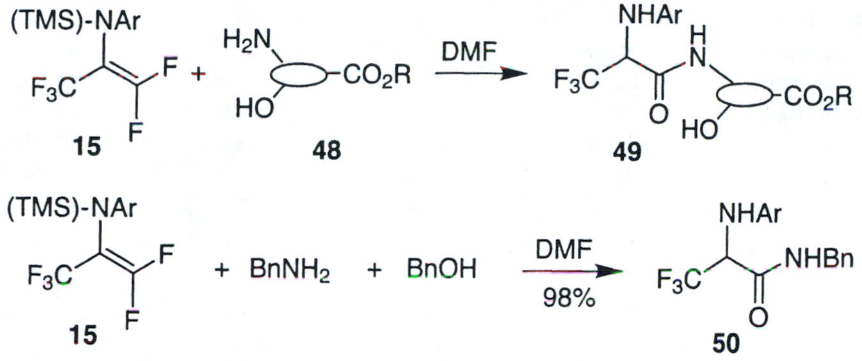

Figure 16. Chemoselective Reaction of ***15*** *with Amine in the Presence of Alcohol*

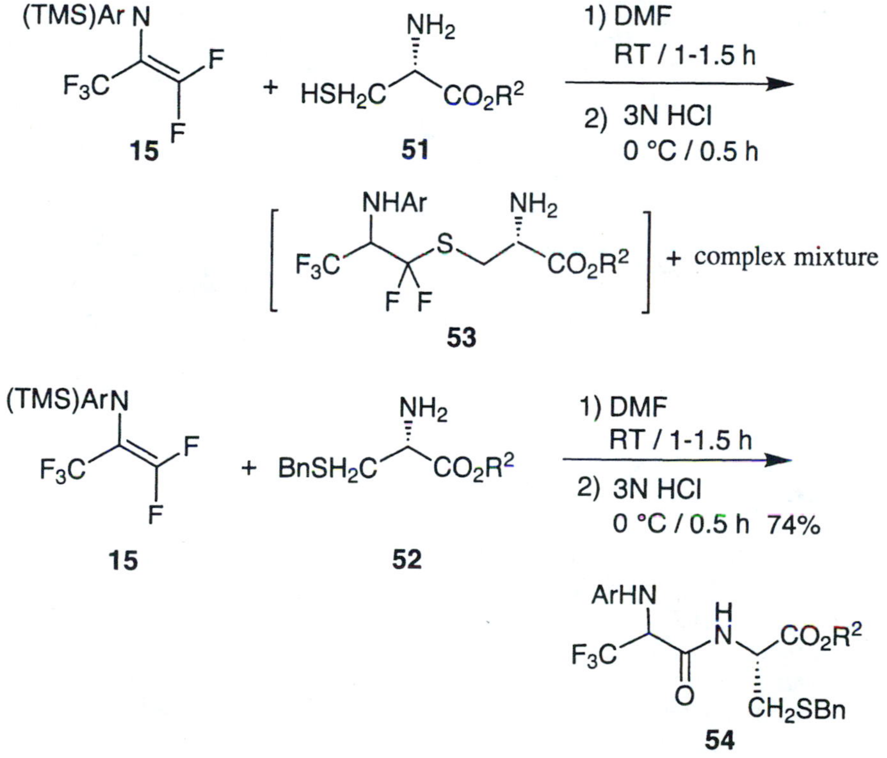

Figure 17. Reaction of Enamine ***15*** *with Cysteine and S-Bn-Cysteine*

The diastereoselectivity of the dipeptides **16** was poor and no improvement of the diastereoselectivity was found with variation of solvents, bases and reaction temperature. The stereochemistry of **16** seemed to be stable under the acid-catalyzed hydrolysis conditions except for phenyl alanine dipeptide where 7 % epimerization of asymmetric carbon of phenylalanine side was observed.

The *N*-protecting group, *p*-methoxyphenyl group, was readily removed by the oxidation of **55** with ammonium cerium nitrate in MeCN-H_2O at room temperature[24] to give the deprotected dipeptides in 80% yield. The diastereoisomers of dipeptides **56** and **57** could be easily separated and isolated as an enantiomerically pure form by column chromatography. The absolute stereochemistry of the major diastereomer **56** (*R*, *S*) was clarified by X-ray crystallographic analysis of the single crystal of its hydrochloric acid salt of **56**. This is the first example of preparation of the enantiomerically pure (*R*,*S*) and (*S*,*S*) trifluoroalanine-phenylalanine dipeptides (Fig. 18).

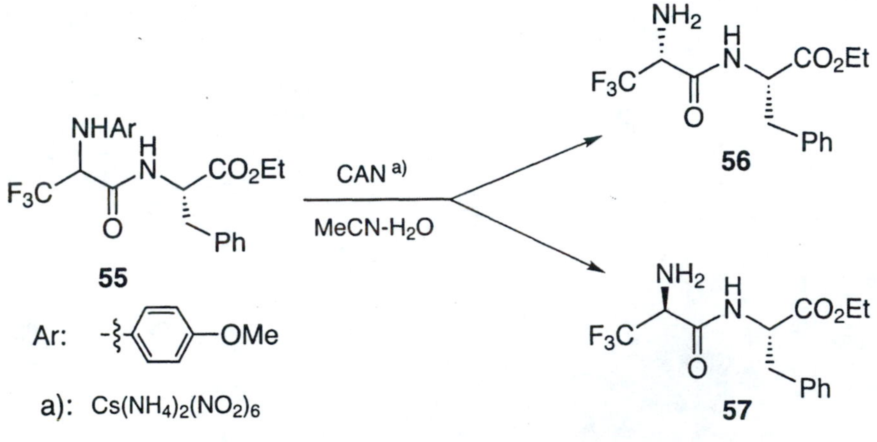

Figure 18. N-Deprotection and Separation of Diastereoisomers of Trifluoroalanine-Phenylalanine Dipeptide

The protocol for trifluoroalanine dipeptide synthesis can be applied to the tripeptide synthesis. Thus, alanine dipeptide **58** was subjected to the peptide formation reaction with **15**. Likewise, the reaction proceeded smoothly and provided the desired tripeptide **59** in a reasonable yield although the stereoselectivity was poor (Fig. 19). The reaction rate (2 hours at room temperature) in tripeptide synthesis seems to be fundamentally as same as that in dipeptide synthesis.[25]

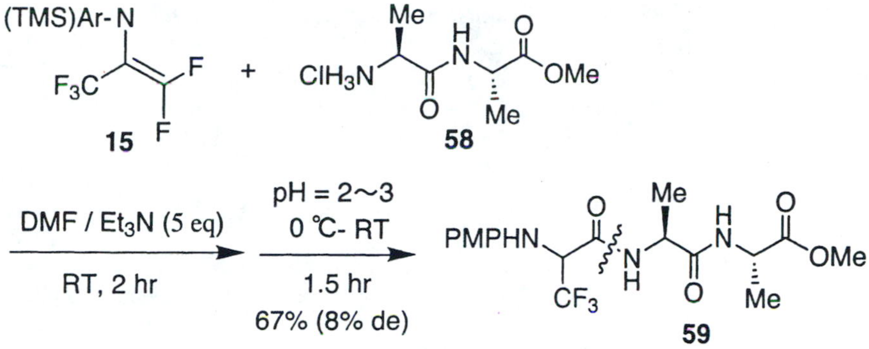

Figure 19. Trifluoroalanine-Alanine-Alanine-Tripeptide

Conclusion

It is suggested that the present protocol can be promising for the construction of peptide skeletons and applicable to the preparation of a variety of trifluoroalanine dipeptides and oligopeptides in an enantiomerically pure form, although 1,4-asymmetric induction of the chiral center of amine esters to the α-carbon atom in trifluoroalanine moiety has remained intractable at this moment.

References

1. (a) Barrett, G. C. Eds.; *Amino Acid Derivatives*, Oxford Univ. Press., Oxford, 1999. (b) Abell, A. Eds.; *Advances in Amino Acid Mimetics and Peptidomimetics*, JAI Press Inc., Greenwich, 1997.
2. (a) Jones, J.; *Amino Acid and Peptide Synthesis*, Oxford Univ. Press., Oxford, 1992. (b) Kusumoto, S. Eds.; *Experimental Chemistry 4th ed.*, No 22, Maruzen, Tokyo, 1990. (c) Hodgson, D. R. W.; Sanderson, J. M. *Chem. Soc. Rev.*, **2004**, *33*, 422.
3. Ojima, I.; Delaloge, F. *Chem. Soc. Revs.*, **1997**, *26*, 377-380.
4. (a) Schellenberger, V.; Jakubke, H. D.; Fachbereich, B. *Angew. Chem.* **1991**, *103*, 1440-1452; (b) Murakami, Y.; Hirata, A. J. Bio Chem., (In Japanese, *Seibutsu Kogaku Kaishi*)**, 1998**, *76*, 238.
5. Bordusa, F.; Dahl, C.; Jakubke, D.-D.; Burger, K.; Koksch, B. *Tetrahedron:Asymmetry* **1999**, *10*, 307.
6. Thust, S.; Koksch, B. *Tetrahedron Lett.*, **2004**, *45*, 1163.
7. Weygand, F.; Steglich, W.; Oettmeier, W. *Chem. Ber.*, **1970**, *103*, 1655.
8. Osipov, S. N.; Sewald, N.; Kolomiets, A. F.; Fokin, A. V.; Burger, K. *Tetrahedron Lett.*, **1996**, *37*, 615.

9. Ishikawa, N.; Takaoka, A.; Iwakiri, H.; Kubota, S.; Kagaruki, S. R. F. *Chem. Lett.*, **1980**, 1107.
10. (a) Bailey, P. D.; Boa, A. N.; Crofts, G. A.; Vandiepen, M.; Helliwell, M.; Gamman, R. E.; Harrison, M. J. *Tetrahedron Lett.*, **1989**, *30*, 7457; (b) England, D. C.; Melby, L. R.; Dietrich, M. A.; Lindsey, Jr. R. V. *J. Am. Chem. Soc.*, **1960**, *82*, 5116.
11. Uneyama, K. "*Organofluorine Chemistry*", **2006**, Blackwell Publishing, Oxford.
12. Murata, J.; Tamura, M.; Sekiya, A.; *Green Chemistry* **2002**, *4*, 60.
13. Hanamoto, T.; Harada, S.; Shindo, K.; Kondo, M.; *Chem. Commun.*, **1999**, 2397.
14. Crouse, G.D.; Webster, J.D.; *J. Org. Chem.*, **1992**, *57*, 6643.
Hanamoto, T.; Shindo, K.; Matsuoka, M.: Kiguchi, Y.; Kondo, M., *J. Chem. Soc., Perkin Trans* 1, **2000**, 103.
16. Garayt, M. R.; Percy, J. M. *Tetrahedron Lett.*, **2001**, *42*, 6377.
17. Ichikawa, J.; Wada, Y.; Fujisawa, M.; Sakoda, K. *Synthesis*, **2002**, 1917.
18. Baldwin, J. E., *J. Chem. Soc., Chem. Commun.*, **1976**, 734.
19. (a) Mae, M.; Amii, H.; Uneyama, K. *Tetrahedron Lett.*, **2000**, *41*, 7893. (b) Amii, H.; Kobayashi, T.; Hatamoto, Y.; Uneyama, K. *Chem. Commun.*, **1999**, 1323.
20. Middleton, W, J.; Krespan, C. G. *J. Org. Chem.*, **1965**, *30*, 1398.
21. Guo, Y.; Fujiwara, K.; Uneyama, K. *Org. Lett.*, **2006**, *8*, 827-829.
22. The absolute stereochemistry (*R*) or (*S*) of trifluoroalanine site of both major and minor diastereomers has not been determined except **16c**.
23. Imidoyl fluoride **39** was isolated as an intermediate which was easily hydrolyzed to the corresponding amide under the hydrolysis conditions.
24. Sakai, T.; Yan, F.; Uneyama, K. *Synlett.*, **1995**, 753.
25. Fujiwara, K.; Komatsu, Y.; Guo, Y.; Uneyama, K. The 86th Annual Meeting of the Chem. Soc. Japan, **2006** March, Funahasi.

Chapter 30

Hyperstable Collagen Based on 4-Fluoroproline Residues

Ronald T. Raines

Departments of Biochemistry and Chemistry, University of Wisconsin at Madison, Madison, WI 53706

Collagen is the most abundant protein in animals. The conformational stability of the collagen triple helix is enhanced by the hydroxyl group of its prevalent (2*S*,4*R*)-4-hydroxyproline residues. For >30 years, the prevailing paradigm had been that this enhanced stability is due to hydrogen bonds mediated by bridging water molecules. We have tested this hypothesis with synthetic collagen triple helices containing (2*S*)-4-fluoroproline residues. The results have unveiled a wealth of stereoelectronic effects that contribute markedly to the stability of collagen (as well as other proteins). This new understanding is leading to hyperstable synthetic collagens for a variety of applications in biotechnology and biomedicine.

Introduction

Each polypeptide chain of collagen is composed of repeats of the sequence: Xaa–Yaa–Gly, where Xaa is often a (2*S*)-proline (Pro) residue and Yaa is often a (2*S*,4*R*)-4-hydroxyproline (Hyp) residue. In natural collagen, three such strands are wound into a tight triple helix in which each strand assumes the conformation of a polyproline type II (PPII) helix. The hydroxyl group of its prevalent Hyp residues increases markedly the conformational stability of the collagen triple helix. For 25 years, the prevailing paradigm had been that the enhanced stability arises from water molecules that form bridges between the hydroxyl group and a main-chain oxygen. Using fluorinated amino acids, we have overturned this paradigm.

Results and Discussion

How does (2*S*,4*R*)-4-hydroxyproline in the Yaa position increase triple helix stability?

Hydroxyl groups can form hydrogen bonds with water, as observed in the structure of crystalline collagen. In addition, the electronegative oxygen in a hydroxyl group is effective at withdrawing electron density by through-bond and through-space interactions.[1] To distinguish between the contributions of hydrogen bonding and inductive effects to collagen stability, we replaced the hydroxyl groups in Hyp residues with fluorine atoms. We chose fluorine because it is the most electronegative atom and thus elicits a large inductive effect, and because organic fluorine does not form hydrogen bonds.[2-7] This latter attribute of fluorine warrants elaboration.

Anionic fluoride forms strong hydrogen bonds. Indeed, the hydrogen bond in gas-phase $[F\cdots H–F]^-$ is the strongest known.[8,9] In contrast to anionic fluoride, organic fluorine is a poor hydrogen bond acceptor. X-ray diffraction analyses[10,11] as well as extensive structure database surveys[12-14] have revealed but few crystalline organofluorine compounds that display short C–F···H–X distances, where X = C, N, or O. In addition, a presumably intimate C–F···H–N interaction does not stabilize DNA double helices.[15] The weakness of the C–F···H–X interaction is likely due to the high charge of the fluorine nucleus, which compacts the surrounding electrons.

The inductive effect of a fluoro group is apparent in our data on the structure[16] and properties[17] of proline derivatives. For example, the nitrogen pK_a of the conjugate acid of (2*S*,4*R*)-4-fluoroproline (FlpOH; 9.23) is lower than that of HypOH (9.68) and ProOH (10.8).[17] The nitrogen of AcFlpOMe is more pyramidal than that of AcHypOMe or AcProOMe.[16] This result indicates that the nitrogen of AcFlpOMe has greater sp^3 character and hence higher electron density. The amide I vibrational mode, which results primarily from the C=O stretching vibration, decreases in the order: AcFlpOMe > AcHypOMe > AcProOMe.[17] The value of $\Delta H^‡$ for amide bond isomerization is smaller for AcFlpOMe than for AcProOMe.[17] Each of these results is consistent with the traditional picture of amide resonance[18] coupled with an inductive effect that increases the bond order in the amide C=O bond and decreases the bond order in the amide C–N bond. These data suggested to us that an inductive effect could contribute to the conformational stability of collagen.

We directly compared the stability conferred to a collagen triple helix by a 4*R* fluoro group and 4*R* hydroxyl group. To do so, we synthesized a collagen-like peptide containing Pro–Flp–Gly units.[19,20] We found that Flp residues allow for triple helix formation. Sedimentation equilibrium experiments with an analytical ultracentrifuge indicated that $(\text{Pro–Flp–Gly})_{10}$ chains form a complex of molecular mass (8.0 ± 0.1) kDa. The expected molecular mass of a $(\text{Pro–Flp–Gly})_{10}$ trimer ($C_{360}H_{480}N_{90}O_{90}F_{30}$) is 8078 Da. The fluorescence of 1-anilinonaphthalene-8-sulfonate,[21] which has affinity for molten globules,[22] was unchanged in the presence of an excess of $(\text{ProFlpGly})_{10}$ trimer. This result suggested that the tertiary structure of the trimer is packed tightly. At low temperature, the circular dichroism (CD) spectrum of the complex formed by $(\text{Pro–Flp–Gly})_{10}$ chains was indistinguishable from that of complexes composed of $(\text{Pro–Hyp–Gly})_{10}$ or $(\text{Pro–Pro–Gly})_{10}$ chains. All three polymers had a CD spectrum with a positive peak at 225 nm and a stronger negative peak at 200–210 nm, which are defining characteristics of a collagen triple helix.[23] The ellipticity at 225 nm of each triple helix decreased in a sigmoidal manner with increasing temperature, which is characteristic of denaturation of the triple helix. This temperature-dependent change in conformational stability was observed in two solvents: 50 mM acetic acid, which stabilizes triple helices by protonating the C-terminal carboxylate groups and thereby eliminating unfavorable Coulombic interactions, and phosphate-buffered saline (PBS), which mimics a physiological environment.

Flp residues enhance triple helix stability. In both 50 mM acetic acid and PBS, the values of T_m and $\Delta\Delta G_m$ for the three triple helices differ dramatically, increasing in the order: $(\text{Pro–Pro–Gly})_{10} < (\text{Pro–Hyp–Gly})_{10} < (\text{Pro–Flp–Gly})_{10}$.[19,20] This order is *inconsistent* with collagen stability arising largely from bridging water molecules, but is consistent with the manifestation of an inductive effect from the electronegative substituent. The stability of the $(\text{Pro–Flp–Gly})_{10}$ triple helix far exceeds that of any untemplated collagen mimic of similar size.

Does (2*S*,4*R*)-4-fluoroproline in the Yaa position endow collagen with hyperstability because of a stereoelectronic effect?

In other words, is the mere presence of an electron-withdrawing group on C^{γ} enough, or does the group have to be in the *R* configuration? To answer this question, we synthesized collagen strands containing (2*S*,4*S*)-4-fluoroproline (flp), which is a diastereomer of Flp. We found that $(\text{Pro–flp–Gly})_7$, unlike $(\text{Pro–Flp–Gly})_7$, does not form a stable triple helix (Table 1).[24] This result provided the first example of a stereoelectronic effect on the conformational stability of a protein. Moreover, the result led us to an explanation for the effect of Flp residues on collagen stability.

We have determined that the remarkable stability of triple helices with

$(Pro–Flp–Gly)_n$ strands derives from the interplay of several factors, all of which arise from the inductive effect of the fluorine atom.[25] First, the gauche effect prescribes a favorable pyrrolidine ring pucker.[17,24] The gauche effect arises when two vicinal carbons bear electronegative substituents. These electronegative substituents prefer to reside *gauche* (60°) to each other so that there is maximum overlap between the σ orbitals of more electropositive substituents, such as hydrogen, and the σ* orbitals of the electronegative substituents. As expected from the manifestation of the gauche effect, the C^{γ}-exo ring pucker is predominant in Hyp residues in the Yaa position of collagen-like peptides,[27] as well as in small-molecule structures of AcHypOMe and AcFlpOMe (Figure 1).[16] The gauche effect between fluoro and amide groups is especially strong.[28,29]

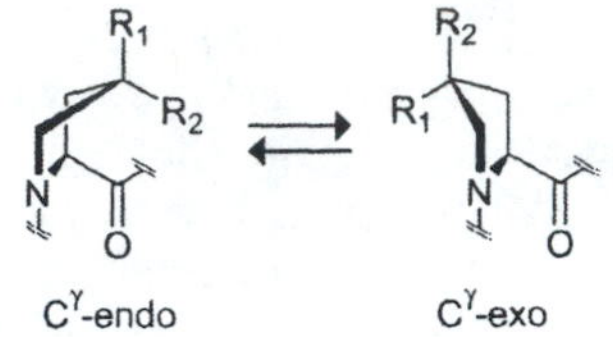

R_1 R_2 R_2 R_1

C^{γ}-endo C^{γ}-exo

Figure 1. Ring conformations of 4-substituted proline residues. The C^{γ}-endo conformation is favored strongly when R_1 = H and R_2 = F (as in flp). The C^{γ}-exo conformation is favored strongly when R_1 = OH (Hyp) or F (Flp) and R_2 = H. The C^{γ}-exo/C^{γ}-endo ratio is near 1:2 when R_1 = R_2 = H (Pro).[26]

Second, the C^{γ}-exo ring pucker preorganizes the main-chain torsion angles of Flp residues. The φ angle correlates with ring pucker, with a C^{γ}-exo pucker giving a high (*i.e.*, less negative) value of φ, and a C^{γ}-endo pucker giving a low value of φ.[26,27] The ψ angle also correlates with ring pucker, as a C^{γ}-exo pucker gives a low value of ψ, and a C^{γ}-endo pucker gives a high value of ψ.[26] The φ and ψ angles in crystalline AcFlpOMe,[16] do not differ significantly from those of residues in the Yaa position of triple-helical collagen.[30]

The ψ angle in AcFlpOMe is not only preorganized for triple helix formation, but also establishes a favorable interaction between a non-bonding electron pair (n) of the amide oxygen (O'_{i-1}) and the π antibonding orbital (π*) of the ester carbon (C'_i). The $O'_{i-1}\cdots C'_i{=}O'_i$ angle in AcFlpOMe is 98°, which is close to the ideal angle for an $n{\rightarrow}\pi^*$ interaction (Figure 2, which is not to be confused with an $n{\rightarrow}\pi^*$ electronic transition).[31-33] Moreover, the $O'_{i-1}\cdots C'_i{=}O'_i$ distance in AcFlpOMe is only 2.76 Å, which

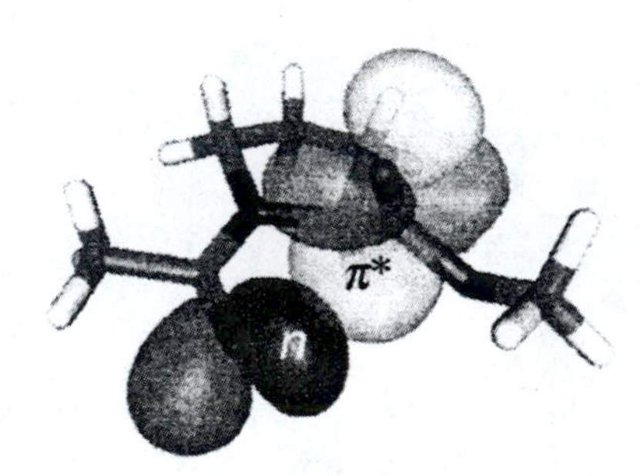

Figure 2. Natural bond orbitals depicting the $n{\rightarrow}\pi^*$ interaction between O'_{i-1} and C' in the trans peptide bond isomer of AcProOMe having C^{γ}-exo ring pucker.[26,34]

predicates a meaningful interaction. Indeed, the ester carbonyl stretching vibration is lower by 6 cm^{-1} in AcFlpOMe than in AcflpOMe, presumably because the $n \rightarrow \pi^*$ interaction decreases the C=O bond order.[24] The $n \rightarrow \pi^*$ interaction stabilizes not only the ideal ψ angle for triple-helix formation, but also the requisite trans peptide bond isomer (ω = 180°) of the Flp peptide bond. In the cis isomer (ω = 0°), C^{α}_{i-1} rather than O'_{i-1} would be proximal to C'_i, and no $n \rightarrow \pi^*$ interaction could occur. These stereoelectronic effects explain why the trans/cis ratio of the amide bond increases as the electronegativity of the substituent in the 4*R* position increases.[17] The reverse trend is true for electronegative 4*S* substituents, which impose a C^{γ}-endo pucker.[24] The association of ω angle with pyrrolidine ring pucker explains the well-known observation that cis prolyl peptide bonds tend to have endo ring puckers in crystalline proteins.[35]

In summary, Flp in the Yaa position stabilizes collagen by a stereoelectronic effect—the gauche effect—that fixes the pyrrolidine ring pucker and thus preorganizes all three main-chain torsion angles: φ, ψ, and ω. Density functional theory (DFT) calculations with the (hybrid) B3LYP method are in gratifying agreement with this explanation and all experimental data.[26] These same arguments apply to the prevalent Hyp residues in natural collagen.

Can triple helix stability be increased by fixing the ring pucker of (2*S*)-proline in the Xaa position?

Having established a link between the C^{γ}-exo ring pucker in the Yaa position and triple helix stability, we next focused our attention on the Xaa position of the collagen triple helix, in which proline residues have C^{γ}-endo pucker.[30] The gauche effect can be used to preorganize this pucker by using proline residues with an electronegative 4*S* substituent. Yet, replacing Pro in the Xaa position of $(\text{Pro–Pro–Gly})_{10}$ with (2*S*,4*S*)-4-hydroxyproline (hyp) is known to produce strands that fail to form triple helices.[36] We suspected that this result could be due to unfavorable steric interactions that develop upon replacing a hydrogen with a hydroxyl group. This suspicion is consistent with molecular modeling of hyp in the Xaa position.[27] Replacing hydrogen with fluorine, on the other hand, typically results in little steric destabilization.[2-7]

To search for a stereoelectronic effect in the Xaa position on collagen stability, we again used a fluoro group as a probe, synthesizing the peptides $(\text{Flp–Pro–Gly})_7$ and $(\text{flp–Pro–Gly})_7$, where Flp and flp refer to the 4*R* and 4*S* diastereomers, respectively. We found that $(\text{flp–Pro–Gly})_7$ but not $(\text{Flp–Pro–Gly})_7$ forms a stable triple helix (Table 1).[37] Moreover, only $(\text{flp–Pro–Gly})_7$ shows the cooperative transition characteristic of triple helix unfolding upon thermal denaturation. The linear decrease in elipticity exhibited by $(\text{Flp–Pro–Gly})_7$ is characteristic of the unfolding of a single polypeptide chain. Sedimentation equilibrium experiments confirm that $(\text{Flp–Pro–Gly})_7$ but not (flp–Pro–Gly)$_7$ is a monomer at 4 °C, whereas both peptides are monomers at 37 °C.

Apparently, stereoelectronic effects can operate adventitiously (or deleteriously) in the Xaa position of collagen. There, flp is able to preorganize the φ and ψ dihedrals as in a triple helix without encountering the steric conflicts that appear to plague hyp in this position.[27] In addition, a 4*S*-substituent in the Xaa position has limited access to solvent, thus making a fluoro group better suited than a hydroxyl group to occupy this position. Altogether, the gain in stability upon replacing hyp with flp in the Xaa position exceeds that of replacing Hyp with Flp in the Yaa position (Table 1).

The conformational stability of a (flp–Pro–Gly)$_7$ triple helix is less than that of a (Pro–Flp–Gly)$_7$ triple helix (Table I). Two factors could contribute to this lower stability. First, Flp in the Yaa position causes favorable preorganization of all three main-chain dihedral angles (φ, ψ, and ω). In the Xaa position, flp increases the probability of ω adopting a cis (ω = 0°) conformation,[24] thus mitigating somewhat the benefit accrued from the preorganization of φ and ψ. Second, a C$^{\gamma}$-endo pucker is already favored in Pro,[7] and flp only increases that preference. In contrast, Flp has the more dramatic effect of reversing the preferred ring pucker, thereby alleviating the entropic penalty of triple-helix

Table I. ***Correlation of ring pucker with collagen triple helix stability.***[24,29,36] ***In a collagen triple helix, proline residues have φ = −73°, ψ = 161° in the Xaa position and φ = −58°, ψ = 152° in the Yaa position.***

	Residue Ring Pucker		Triple Helix T_m	
	C$^{\gamma}$-endo	C$^{\gamma}$-exo	(XaaProGly)$_7$	(ProYaaGly)$_7$
Flp	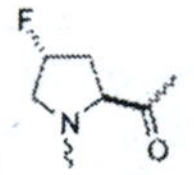 14%	86% ϕ = −55° ψ = 140°	no helix	45 °C
Hyp (natural)			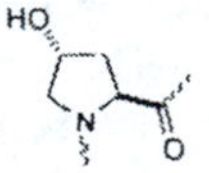 no helix	36 °C
Pro (natural)	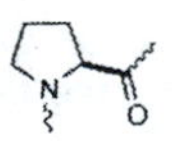 66%	34%	6 °C	6 °C
hyp			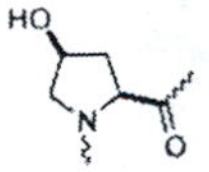 no helix	no helix
flp	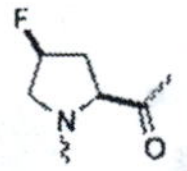 95% ϕ = −76° ψ = 172°	5%	33 °C	no helix

formation to a greater degree. Interestingly, installing both flp in the Xaa position and Flp in the Yaa position does not allow for the formation of a stable triple helix, presumably because of an unfavorable steric interaction between fluoro groups on adjacent strands.[38] Density functional theory calculations indicate that (2*S*,3*S*)-3-fluoroproline (3*S*-flp), like flp, should preorganize the main chain properly for triple-helix formation but without a steric conflict. Synthetic strands containing 3*S*-flp in the Xaa position and Flp in the Yaa position do form a triple helix. This helix is, however, less stable than one with Pro in the Xaa position, presumably because of an unfavorable inductive effect that diminishes the strength of the interstrand 3*S*-FlpC=O···H–NGly hydrogen bond (as we observed with (2*S*,3*S*)-3-hydroxyproline[39]). Thus, other forces can counter the benefits derived from the proper preorganization. Although (Pro–Pro–Gly)$_7$ and (flp–Flp–Gly)$_7$ do not form stable homotrimeric helices, mixtures of these two peptides form stable heterotrimeric helices containing one (Pro–Pro–Gly)$_7$ strand and two (flp–Flp–Gly)$_7$ strands. This stoichiometry can be understood by considering the cross sections of the two possible heterotrimeric helices. This unexpected finding portends the development of a "code" for the self-assembly of determinate triple helices from two or three strands. This code would not be based on hydrogen-bonding patterns, as in the Watson–Crick paradigm for the DNA double helix. Rather, the code would rely on a combination of stereoelectronic effects (Figure 3A) and their steric consequences (Figure 3B).

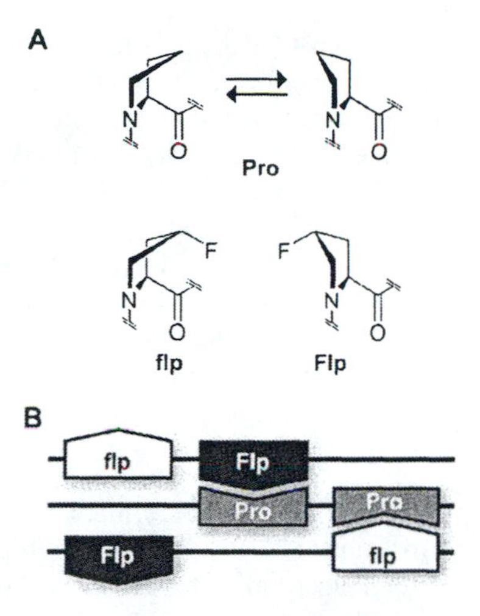

Figure 3. Basis of a code for triple helix formation. (A) Prevalent ring conformations of Pro, flp, and Flp.[26] *(B) Favorable (Flp···Pro and (Pro···flp) and unfavorable (flp···Flp) steric interactions within a triple helix.*[38]

Can the stereoelectronic effects (2*S*)-4-fluoroproline extend beyond collagen?

The PPII helix adopted by collagen strands is a prevalent conformation in both folded and unfolded proteins, and is known to play important roles in a wide variety of biological processes. Polyproline itself can also form a type I (PPI) helix, which has a disparate conformation. Our CD spectral analyses of

$(Pro)_{10}$, $(Hyp)_{10}$, $(Flp)_{10}$, and $(flp)_{10}$ show that 4*R* electron-withdrawing substituents stabilize a PPII helix relative to a PPI helix, even in a solvent that favors the PPI conformation, such as *n*-propanol.[40] The stereochemistry at C^{γ} ordains the relative stability of PPI and PPII helices, as $(flp)_{10}$ forms a mixture of PPI and PPII helices in water and a PPI helix in *n*-propanol. The conformational preferences of $(Pro)_{10}$ are intermediate between those of $(Hyp)_{10}/(Flp)_{10}$ and $(flp)_{10}$. (Interestingly, PPI helices of $(flp)_{10}$ exhibit cold denaturation in *n*-propanol with a value of T_s near 70 °C.) Together, these data show that stereoelectronic effects can have a substantial impact on polyproline conformation and provide a rational means to stabilize a PPI or PPII helix.

Conclusions

Our findings with fluorinated collagen are notable on several fronts. First, they overturn a 25-year old paradigm. Second, they are the first to demonstrate that stereoelectronic effects are critical for the conformational stability of a protein. Finally, they could give rise to hyperstable synthetic collagens for a variety of applications in biotechnology and biomedicine.

Acknowledgement

The author is grateful to Dr. F. W. Kotch for comments on this manuscript. Work on collagen in my laboratory is supported by grant AR44276 (NIH).

References

1. Stock, L. M. *J. Chem. Educ.* **1972**, *49*, 400–404.
2. Welch, J. T.; Eswarakrishnan, S., *Fluorine in Bioorganic Chemistry.* Wiley: New York, 1991.
3. Resnati, G. *Tetrahedron* **1993**, *49*, 9385–9445.
4. Ojima, I.; McCarthy, J. R.; Welch, J. T.; Eds. *Biomedical Frontiers of Fluorine Chemistry.* American Chemical Society: Washington, DC, 1996.
5. O'Hagan, D.; Rzepa, H. S. *Chem. Commun.* **1997**, 645–652.
6. Marsh, E. N. G. *Chem. Biol.* **2000**, *7*, R153–R157.
7. Yoder, N. C.; Kumar, K. *Chem. Soc. Rev.* **2002**, *31*, 335–341.
8. Harrell, S. A.; McDaniel, D. H. *J. Am. Chem. Soc.* **1964**, *86*, 4497.
9. Shan, S.; Loh, S.; Herschlag, D. *Science* **1996**, *272*, 97–101.
10. Murray–Rust, P.; Stallings, W. C.; Monti, C. T.; Preston, R. K.; Glusker, J. P. *J. Am. Chem. Soc.* **1983**, *1983*, 3206–3214.
11. Shimoni, L.; Carrell, H. L.; Glusker, J. P.; Coombs, M. M. *J. Am. Chem. Soc.* **1994**, *116*, 8162–8168.
12. Shimoni, L.; Glusker, J. P. *Struct. Chem.* **1994**, *5*, 383–397.

13. Howard, J. A. K.; Hoy, V. J.; O'Hagan, D.; Smith, G. T. *Tetrahedron* **1996**, *52*, 12613–12622.
14. Dunitz, J. D.; Taylor, R. *Chem. Eur. J.* **1997**, *3*, 89–98.
15. Moran, S.; Ren, R. X.; Kool, E. T. *Proc. Natl. Acad. Sci. U.S.A.* **1997**, *94*, 10506–10511.
16. Panasik, N., Jr.; Eberhardt, E. S.; Edison, A. S.; Powell, D. R.; Raines, R. T. *Int. J. Pept. Protein Res.* **1994**, *44*, 262–269.
17. Eberhardt, E. S.; Panasik, N., Jr.; Raines, R. T. *J. Am. Chem. Soc.* **1996**, *118*, 12261–12266.
18. Pauling, L., *The Nature of the Chemical Bond, 3rd ed.* Cornell University Press: Ithaca, NY, 1960; pp 281–282.
19. Holmgren, S. K.; Taylor, K. M.; Bretscher, L. E.; Raines, R. T. *Nature* **1998**, *392*, 666–667.
20. Holmgren, S. K.; Bretscher, L. E.; Taylor, K. M.; Raines, R. T. *Chem. Biol.* **1999**, *6*, 63–70.
21. Brand, L.; Gohlke, J. R. *Annu. Rev. Biochem.* **1972**, *41*, 843–868.
22. Semisotnov, G. V.; Rodionova, N. A.; Razgulyaev, O. I.; Uversky, V. N.; Gripas', A. F.; Gilmanshin, R. I. *Biopolymers* **1991**, *31*, 119–128.
23. Piez, K. A.; Sherman, M. R. *Biochemistry* **1970**, *9*, 4129–4133.
24. Bretscher, L. E.; Jenkins, C. L.; Taylor, K. M.; DeRider, M. L.; Raines, R. T. *J. Am. Chem. Soc.* **2001**, *123*, 777–778.
25. Jenkins, C. L.; Raines, R. T. *Nat. Prod. Rep.* **2002**, *19*, 49–59.
26. DeRider, M. L.; Wilkens, S. J.; Waddell, M. J.; Bretscher, L. E.; Weinhold, F.; Raines, R. T.; Markley, J. L. *J. Am. Chem. Soc.* **2002**, *124*, 2497–2505.
27. Vitagliano, L.; Berisio, R.; Mazzarella, L.; Zagari, A. *Biopolymers* **2001**, *58*, 459–464.
28. O'Hagan, D.; Bilton, C.; Howard, J. A. K.; Knight, L.; Tozer, D. J. *J. Chem. Soc., Perkin Trans. 2* **2000**, 605–607.
29. Briggs, C. R. S.; O'Hagan, D.; Howard, J. A. K.; Yufit, D. S. *J. Fluorine Chem.* **2003**, *119*, 9–13.
30. Bella, J.; Eaton, M.; Brodsky, B.; Berman, H. M. *Science* **1994**, *266*, 75–81.
31. Bürgi, H. B.; Dunitz, J. D.; Shefter, E. *J. Am. Chem. Soc.* **1973**, *95*, 5065–5067.
32. Bürgi, H. B.; Dunitz, J. D.; Lehn, J. M.; Wipff, G. *Tetrahedron* **1974**, *30*, 1563–1572.
33. Bürgi, H. B.; Lehn, J. M.; Wipff, G. *J. Am. Chem. Soc.* **1974**, *96*, 1965–1966.
34. Hinderaker, M. P.; Raines, R. T. *Protein Sci.* **2003**, *12*, 1188–1194.
35. Milner–White, J. E.; Bell, L. H.; Maccallum, P. H. *J. Mol. Biol.* **1992**, *228*, 725–734.

36. Inouye, K.; Sakakibara, S.; Prockop, D. J. *Biochim. Biophys. Acta* **1976**, *420*, 133–141.
37. Hodges, J. A.; Raines, R. T. *J. Am. Chem. Soc.* **2003**, *125*, 9262–9263.
38. Hodges, J. A.; Raines, R. T. *J. Am. Chem. Soc.* **2005**, *127*, 15923–15932.
39. Jenkins, C. L.; Bretscher, L. E.; Guzei, I. A.; Raines, R. T. *J. Am. Chem. Soc.* **2003**, *125*, 6422–6427.
40. Horng, J. -C.; Raines, R. T. *Protein Sci.* **2006**, *15*, 74–83.

Chapter 31

Fluorinated Amino Acids and Reagents in Protein Design and Biomolecule Separation

He Meng[1], Venkateshwarlu Kalsani[1], and Krishna Kumar[1,2,*]

[1]Department of Chemistry, Tufts University, Medford, MA 02155
[2]Cancer Center, Tufts–New England Medical Center, Boston, MA 02110

The use of fluorinated amino acids and reagents has recently emerged as an important tool in bioorganic chemistry and protein design. The orthogonal phase properties of highly fluorinated compounds have been exploited in the design of protein ensembles that show enhanced chemical and thermal stability. Furthermore, highly fluorinated interfaces exhibit simultaneous hydrophobic and lipophobic behavior thus enabling their use in controlling helix-helix interactions within membranes. The insolubility of long chain fluorinated hydrocarbons has resulted in the development of reagents for use in peptide purification protocols and in proteomics using affinity separation. These applications along with the thermodynamic basis for phase separation of fluorinated materials have been highlighted here.

The use of highly fluorinated compounds has stirred a mini revolution in organic synthesis and combinatorial chemistry. The use of highly fluorinated ("fluorous") solvents has been utilized in reaction acceleration (*1*), catalysis (*2*), combinatorial library synthesis and organic separation methodology (*3,4*). While fluorinated compounds are scarce in biology, on the other hand, phase separation of immiscible domains is ubiquitous, providing the driving force for formation of well-defined structures. Indeed, the structure of globular proteins, formation of the plasma membrane and sub-cellular organelles is essentially driven by the tendency of nonpolar substances to minimize their contacts with water. A new paradigm in supramolecular chemistry has been to make use of materials known to phase separate from water in the design of protein ensembles (*5,6*), for affinity purification in proteomics (*4*), and in the purification of biomolecules (*7*).

The use of fluorous solvents and materials has primarily relied on the use of long chain perfluoroalkyl appendages tagged to the solvent, substrates and/or reagents. Recently, supramolecularly organized fluorinated surfaces have been used to direct protein folding and aggregation in aqueous solutions, and in the context of the nonpolar environments of biological membranes. These protein constructs differ in that the fluorinated groups are incorporated not in a contiguous covalent fashion, but upon folding of the peptide or protein, large expansive fluorinated surfaces are supramolecularly organized and displayed. Overall, this strategy provides the ability to control structural and functional properties of the resulting protein ensembles, and further endows them with extra-biological properties. In addition to use in protein design, fluorinated reagents have gained prominence in the purification of biomolecules synthesized using solid phase methods and in tagging and selective enrichment applications in proteomics.

Phase Separation of Fluorous Materials

Perfluorocarbons phase separate from nonpolar organic solvents and water at room temperature. Hildebrand and Scott's theory of the solubility of nonelectrolytes (*8*) provides an empirical framework for estimating intermolecular interactions and propensity for mutual miscibility. On a simple level, a solubility parameter, δ, determines the extent to which two nonpolar liquids are miscible. The partial molal free energy ($\overline{\Delta F}$) of a component in a mixture of two nonpolar liquids is given by a sum of the entropy of mixing and the heat of mixing where x_1 and x_2 are the respective mole fractions, v_1 and v_2 are the molal volumes, φ_1 and φ_2 are the volume fractions and δ_1 and δ_2 are the solubility parameters (Equations 1 and 2).

$$\overline{\Delta F_1} = RT\ln x_1 + v_1(\delta_1 - \delta_2)^2 \phi_2^2 \quad (1)$$

$$\overline{\Delta F_2} = RT\ln x_2 + v_2(\delta_1 - \delta_2)^2 \phi_1^2 \quad (2)$$

δ is given by $(\Delta E^v/v)^{1/2}$, where ΔE^v is the energy of vaporization of the pure component and v its molal volume at temperature T. When $\delta_1 = \delta_2$, there is no heat of mixing and the two liquids form an ideal solution. When the differences in δ become substantial, phase separation occurs. Small inequalities in δ result in partial miscibility. Fluorous liquids are characterized by low values of δ, signifying exceedingly low propensities for intermolecular interactions, including with other fluorous liquids (*9*). Thus phase separation from organic solvents stems not from "fluorophilicity" but rather from the property of being "omniphobic".

In water, the hydrophobicity of fluorocarbon liquids is sufficient to account for immiscibility (Figure 1). Fluorocarbons are typically more hydrophobic than their hydrocarbon counterparts. The estimated partial molal heat capacities for the process of (aqueous) solution for CF_4 and C_2F_6 (98 and 173 kcal mol^{-1} K^{-1} respectively) are considerably larger than those for the hydrocarbon gases CH_4 and C_2H_6 (50 and 72 kcal mol^{-1} K^{-1}) (*10*). In addition, measured partition coefficients from water to *n*–heptanol for leucine and the unnatural amino acid hexafluoroleucine have apparent free energies of –2.1 kcal/mol and –2.5 kcal/mol respectively, again pointing to the superior hydrophobicity of fluorinated amino acids (*11*). The Hansch hydrophobicity parameters for CF_3 (Π = 1.07) and CH_3 (Π = 0.50) groups affirm this trend (*12*). These properties make fluorinated materials attractive for use in protein design and other applications.

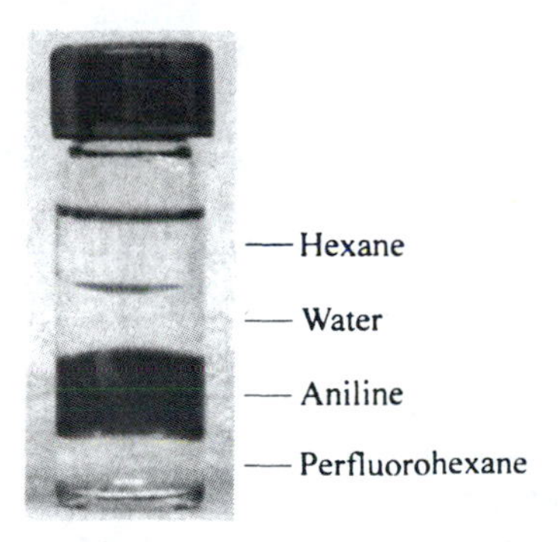

Figure 1. Phase separation of four liquid layers (hexane, water, aniline and f-hexane) at 25 °C. The aniline layer has been false colored to clearly show phase boundaries. (See page 1 of color inserts.)

Protein Design Using Fluorinated Amino Acids

In spite of differential solubility properties and larger size, trifluoromethyl (92 Å^3) groups frequently replace methyl groups (54 Å^3) (*13*) in crystals (solid solutions) and point to a design strategy where such substitution would result in minimal alteration of protein structure. Initial studies from our (*5,14,15*) and David Tirrell's (*16-18*) laboratory clearly established that such substitutions are permissible with concurrent increase in stability of folded conformations adopted by proteins.

Briefly, model peptides based on the coiled coil domain (GCN4-p1) of yeast bZIP transcriptional activator GCN4 were decorated with trifluoroleucine and trifluorovaline (Scheme 1). Coiled coils are characterized by a 4–3 heptad repeat $(abcdefg)_n$ where the *a* and *d* residues are hydrophobic. In helical conformations of pitch 3.5/residues per turn, the hydrophobic residues line one face of each helix providing a dimeric (oligomeric) interface. Residues at the *e* and *g* positions are frequently charged at physiological pH, while the *b*, *c* and *f* positions are more tolerant of a range of side chain functionalities. This protein motif folds into regular helical bundles with 2, 3, 4 and occasionally 5 helices forming part of the ensemble (*19*). In these initial studies, 4,4,4–trifluorovaline and 5,5,5-trifluoroleucine were substituted in place of valine (*a* position) and leucine (*d* position) respectively in GCN4-p1 related peptides.

We envisioned that these substitutions – the more hydrophobic trifluorovaline and trifluoroleucine (over valine and leucine respectively) would lead to more robust and stable structures in aqueous solution. Indeed, even though the fluorinated amino acids had scrambled stereochemistry at the β– (trifluorovaline) and γ– (trifluoroleucine) positions, the aggregate properties of the resulting diastereomeric mixtures could be readily assessed. The fluorinated dimer had a melting temperature of 62 °C that was 15 °C higher than the hydrocarbon coiled coil. Furthermore, the apparent free energy of unfolding, obtained by chaotropic denaturation, was also higher for the fluorinated assembly than the hydrocarbon one by 1.0 kcal/mol (*14*). These increases in thermal and chemical stability are directly attributable to the higher hydrophobicity of trifluoromethyl over methyl groups.

Tirrell and co-workers have also explored similar constructs. They substituted all four leucine residues at the *d* position of GCN4-p1 with trifluoroleucine and saw comparable enhancements in chemical and thermal stability (*16*). Furthermore, the larger 56-amino acid bZIP region when substituted in a similar fashion retained its ability to bind DNA (*16*). Further work from this laboratory showed that incorporation of 8 hexafluoroleucine residues in a 74-residue protein A1 again resulted in elevated thermal (+ 22 °C) and chemical stability (*18*). Recently, an antiparallel four-helix bundle was used as a scaffold where either 2, 4 or 6 layers of core leucine residues were substituted with hexafluoroleucine. Fluorination stabilized the modified structure by 0.3 kcal/mol (per hexafluoroleucine) in the central two layers and by

H1: Ar·HN·RMKQLEDKVEELLSKNAC†LEYEVARLKKLVGE CO·NH$_2$
F1: Ac·HN·RMKQLEDKVEELLSKNASLEYEVARLKKLVGE·CO·NH$_2$

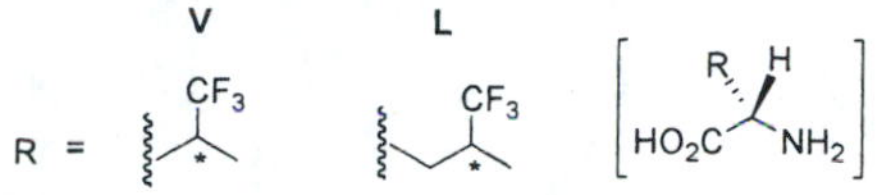

V L

R = CF$_3$ CF$_3$ [R H HO$_2$C NH$_2$]

*Scheme 1. Sequences of GCN4-p1 analogues and structures of fluorinated amino acids used to determine stability of peptide ensembles. Trifluoroleucine (**L**) and trifluorovaline (**V**) were used as mixtures of two diastereomers in these studies. Abbreviations, Ar=4-acetamidobenzoic acid, Ac=acetyl, C†=acetamidocysteine. The asterisk indicates unresolved stereochemistry.*

0.12 kcal/mol for additional layers (*20*). Taken together, these examples illustrate that removal of fluorinated surface area away from water provides a generalizable strategy to stabilize the folded conformation of proteins.

In order to establish that fluorocarbon and hydrocarbon protein-protein interfaces are orthogonal, we prepared disulfide linked peptides **HF**, **HH** and **FF** (Scheme 2). The helical interface in **HH** was composed only of leucine, while that of **FF** only of hexafluoroleucine and **HF** was mixed, where each helix presented a different interface. Disulfide exchange experiments were performed in order to interrogate the thermodynamic preference for homomeric *versus* heteromeric assemblies. When heterodimer **HF** was subjected to equilibrium redox buffer conditions, it disproportionated almost entirely into the two homodimers **HH** and **FF** with <3% of the heterodimer remaining in solution. Control experiments established that the reaction was at equilibrium as the same population resulted irrespective of the starting point of the reaction. The free energy of specificity (ΔGs_{pec}) was calculated to be at least –2.1 kcal/mol in favor of the homodimers (*15,21*). This result set the stage for *de novo* design and structural control of membrane embedded protein components.

Membrane proteins are comprised of two main classes: α–helical bundles and β–barrels (*22*). While binary patterning of the type where the solvent exposed residues are charged or polar, and the interior residues hydrophobic, has provided protein engineers an easy avenue to design globular proteins (*23*), this approach fails for membrane proteins. That is, an "inside out" version of soluble protein poses two problems (*24*) that must be confronted for implementing successful design: (1) addition of many polar and charged residues diminish the partitioning of the protein into the membrane and (2) van der Waals forces play a major role in aggregation of helices in the membrane and tuning differential affinity of side chains for one another in a lipid environment is challenging. But as described earlier, fluorinated interfaces provide an orthogonal hydrophobic surface that is simultaneously lipophobic (*25,26*).

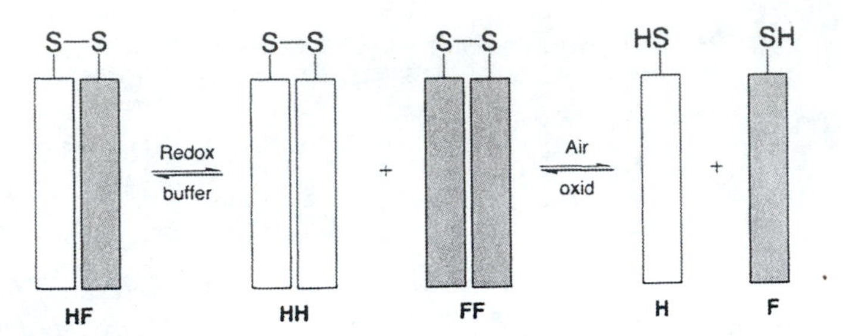

*Scheme 2. Disulfide exchange experiment exploring the preference for homodimer (**HH** and **FF**) versus heterodimer (**HF**) formation. Only 3% or less of the heterodimer is formed (or remained) under equilibrium redox conditions. (See page 1 of color inserts.)*

We envisioned a two-step assembly process (*27*) for insertion and aggregation of transmembrane helices designed with a fluorinated interface allowing for interhelical interactions (Scheme 3). In the first step, hydrophobic peptides would partition into membranes and form α–helices in order to satisfy all hydrogen bonding groups in the nonpolar environment. In the second step, one exposed face decorated with hexafluoroleucine residues would phase separate from lipids to give higher order oligomers.

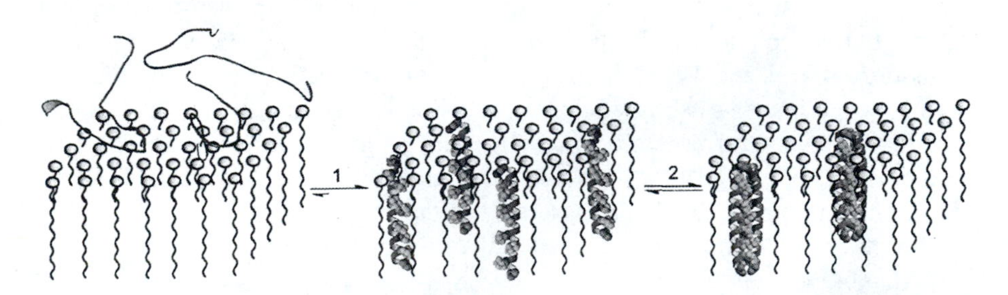

Scheme 3. Schematic diagram of the two-step assembly of membrane embedded protein components. Designed peptides are unstructured in water and partition readily into micelles (step 1). Insertion is followed by formation of α-helices exposing a string of hexafluoroleucine residues (shown in space filling representation) on one face of the helix that fuels the spontaneous assembly of oligomeric structures (step 2). (Reproduced from reference 25. Copyright Proc. Natl. Acad. Sci. 2004.)

Analysis of helix-helix interfaces in extant integral membrane proteins has revealed that packing is reminiscent of the "knobs-into-holes" arrangement observed in soluble coiled coils (*28*). We used a 4-3 hydrophobic repeat to design peptides that would bury 20 residues in the membrane (Scheme 4). This is enough to span the entire bilayer length of biological membranes (~ 30 Å). The putative helix interface region was decorated with either leucine or hexafluoroleucine residues (**TH2** and **TF2** respectively). Degrado and Engelman (*29*) have independently shown that hydrogen bonding driven aggregation of transmembrane helices mediated by side chains of asparagine, glutamine and serine is another method to achieve control in bilayer environments. A second set of peptides was designed and prepared that contained asparagine as a central residue in the interhelical interaction surface but with the rest of the interhelical residues either leucine (**TH1**) or hexafluoroleucine (**TF1**).

A

1 2 .... 6 .... 14 ....

TH1 BQLWIA LLLLIAV NLILLIA LARLKKK KK·$CONH_2$

TH2 BQLWIA LLLLIAV LLILLIA LARLKKK KK·$CONH_2$

TF1 BQLWIA **L**LL**L**IAV N**L**IL**L**IA **L**ARLKKK KKK·$CONH_2$

TF2 BQLWIA **L**LL**L**IAV **L**LIL**L**IA **L**ARLKKK KKK·$CONH_2$

membrane spanning region (3-22)

B

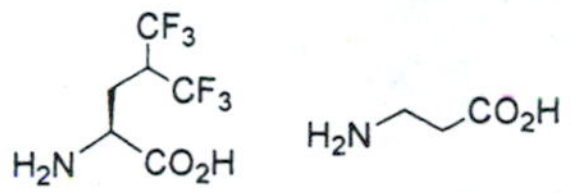

CF_3 CF_3 H_2N CO_2H — **L**

H_2N CO_2H — B

*Scheme 4. (A) Sequences of peptide **TH1**, **TF1**, **TH2** and **TF2**; (B) structure of hexafluoroleucine (**L**) and β-alanine.*

Several techniques were employed to assess the degree of folding and the oligomer number for all peptides in micellar detergents. Circular dichroism spectroscopy in SDS established that all four peptides show signature minima at 222 and 208 nm indicating α–helical structures. Migration of the peptides in SDS/PA gel electrophoresis indicated that **TF1** and **TF2** were retarded to a greater extent than **TH1** and **TH2** respectively suggesting higher oligomer numbers for the fluorinated peptides. Fluorescence Resonance Energy Transfer (FRET) experiments with *N*-terminal labeled peptides (NBD = donor, TAMRA = acceptor) in SDS micelles were also performed. In these experiments, an equimolar mixture of unlabeled and donor peptide was titrated with increasing amounts of acceptor peptide while keeping the nominal peptide (and NBD labeled peptide) concentration constant. Using a model that describes the binomial distribution of donors in the oligomer (*30*) and with the assumption that a single acceptor peptide is sufficient to quench all donor fluorescence in a given assembly, the following dissociation constants were derived: (1) momomerdimer equilibrium: **TH2** (K_d = 5.75 × 10^{-3} MF); **TF2** (K_d = 1.12 × 10^{-3} MF); (2) monomer-trimer equilibrium **TH1** (K_d = 3.42 × 10^{-6} MF^2); **TF1** (K_d = 1.38 × 10^{-6} MF^2) [Expressed in units of peptide/detergent mole fraction (MF)] (*25*). Furthermore, equilibrium analytical ultracentrifugation at 10 μM peptide concentration in the neutral polyether lipid C12E8 revealed **TH2** to be monomeric, **TF2** dimeric while **TH1** and **TF1** were dimeric and tetrameric respectively.

FRET experiments were also carried out in bilayer environments with **TH1** and **TF1**. By varying the ratio of the lipid 1-palmitoyl-2-oleoyl-sn-glycero-3-phosphocholine (POPC) to peptides, and global fitting of FRET data to a monomer-trimer equilibrium, the following dissociation constants could be extracted: **TH1** (K_d = 1.8 × 10^{-5} MF^2); **TF1** (K_d = 0.56 × 10^{-5} MF^2) (*26*). Again the fluorous interface was found to be superior at mediating helix-helix interactions in membranes. In summary, fluorinated interfaces drive the self-association of transmembrane peptides without the aid of hydrogen bonding, and when coupled with hydrogen bonding provide stronger association of membrane embedded protein components.

Fluorous Capping Enabled Purification in Solid Phase Peptide Synthesis

The preparation of peptides and nucleic acids has been facilitated by advances in stepwise synthesis on solid support (*31*). These in turn have allowed the routine use of these biomolecules of diverse sequence and structure resulting in major advances in chemistry, biology and medicine (*32*). Current optimized protocols employed in solid phase synthesis limit the length of peptides to ≈ 50 aa residues (*33*). The main challenge in purification is the removal of deletion sequences, often requiring tedious chromatography and use of solvents that are not environmentally friendly (*33*).

Solid phase synthesis relies on sequential coupling, capping and deprotection steps, all of which must be executed to completion (or near completion). Without this, products lacking one or more amino acid residues accumulate (Scheme 5). In order to alleviate this problem, acetic anhydride is commonly employed immediately after the coupling step to 'cap' any residual amines that result from incomplete coupling. Two problems still plague this general strategy. The capping cocktail typically contains 5–20% v/v of Ac_2O, and repeated use deteriorates the swelling properties of the resin. Furthermore, longer sequences result in multiple acetylated products, making purification using HPLC difficult because of overlapping retention times of the various products.

We and others have introduced a strategy by which fluorous groups can be attached to the growing peptide chain and make the solubility properties of capped or tagged polymers different from the rest (*4,7,34-36*). Here two strategies have been pursued. The first is to 'tag' the final product with a long chain fluorinated compound. This approach necessitates two additional steps, purification to remove unwanted impurities, and then detagging the desired product and a final purification. Nevertheless, Overkleeft and van Boom have shown that such a strategy is a practical and useful way of significantly improving the purity of the desired full-length peptide (*36*). The second approach, one that we have adopted, utilizes a 'capping' step with a perfluoroalkyl group during the course of the synthesis. Final purification is then achieved in a single step, where all deletion products are attached to a fluorous cap and the full length product is not (Scheme 5).

What should the properties of such a capping agent be? Ideally, the reagent should (1) be extremely reactive towards amines; (2) work in the solvents regularly used in peptide synthesis so as not to alter swelling properties of the resin; (3) render the amine unreactive in all subsequent coupling and deprotection steps including final cleavage of the peptide chain from the resin; and (4) be economical so that wide use is possible (*34*).

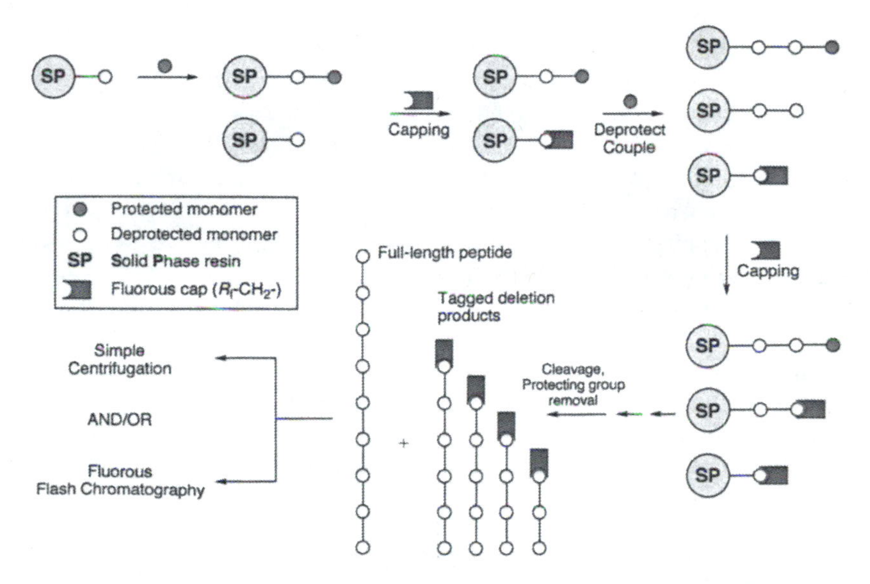

*Scheme 5. Fluorous capping strategy implemented using both t-Boc and Fmoc protocols. Failed amino acid couplings leave an unprotected terminal amino group that is capped with reagent **1** (or **2**) (Scheme 6) in standard peptide synthesis solvents. Side products missing one (or more) residues are therefore capped with a fluorous tag and easily removed at the end of the synthesis by centrifugation or by fluorous chromatography. (Reproduced from reference 38. Copyright Eur. J. Org. Chem. 2006.)*

The fluorous trivalent iodonium salts **1** and **2** (Scheme 6) satisfy all the criteria described above. They are aggressive in their reactivity towards primary amines and optimally suited for the capping reaction. Indeed, in the absence of primary amines, the base commonly employed (2,4,6-collidine) to neutralize the potent acid [bis(trifluoromethylsulfonyl)imide] released during the reaction, is itself consumed. The fluoroalkylated amines are inert in all subsequent steps including the final cleavage reaction (*37*).

In addition, both *t*-Boc and Fmoc protocols for peptide synthesis in both manual and automated modes have proven compatible with fluorous capping reagents. While several peptides have been synthesized using our capping strategy, three examples illustrate the versatility of the reagents. For instance, we synthesized peptide **P1** (analog of adrenocorticotropic hormone 18-39) with deliberate deletions (partial coupling accomplished by using 0.8 equiv of incoming amino acid) introduced at positions shown in bold and marked with an asterisk. In identical syntheses, we capped the deletion products with Ac_2O or using reagent **1** (Scheme 6) (*34*). The synthesis then continued in a normal fashion. At the end of the synthesis both crude peptide samples were subjected to passage through fluorous silica gel using 4:1 H_2O:CH_3CN (1% AcOH) as eluent. The resulting filtrate was then concentrated and analyzed by analytical HPLC. As is seen clearly, the problem of overlapping peaks in HPLC in the case of fluorous capping is substantially diminished and the overall purity is greatly increased. Furthermore, similar experiments were carried out using Fmoc chemistry on peptide **P2** (automated protocol using Advanced Chemtech 348Ω synthesizer) and **P3** (manual synthesis) using reagent **2** (*34*). In the case of **P2**, **P3** and other shorter peptides (of length < 14 residues), the fluorous capped deletion products could be removed by an even simpler method. The crude peptide mixture was dissolved in 1% AcOH and centrifuged at 14,000 rpm for 5 mins. A precipitate resulted, presumably of the fluorous capped deletion products, leaving only the full length desired product in solution. When the same experiment was carried out using Ac_2O capping, no precipitate was visible (*38*). These results suggest that reagents **1** and **2** are robust and efficient capping reagents that can be used with *t*–Boc and Fmoc chemistry including the use of programmable automated synthesizers. We envision that these will find broad use in solid phase peptide, biopolymer and combinatorial chemistry where residual amines require capping.

Conclusions

In summary, we have demonstrated that use of fluorinated amino acids and reagents provide a powerful and unique method to alter the properties of soluble and membrane embedded proteins, and help facilitate biomolecule purification.

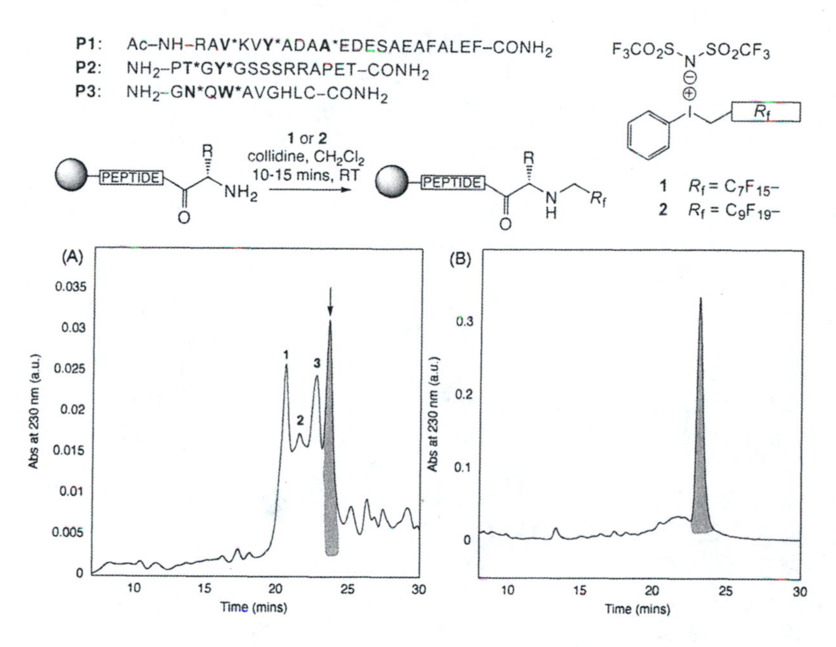

*Scheme 6. Sequences of peptides **P1**, **P2** and **P3**; reactivity of reagents **1** and **2**; (A) HPLC chromatogram of peptide P1 synthesized using Ac_2O capping (peaks 1–3 are acetylated peptides of increasing mass and **P1** is marked with an arrow) and (B) using capping reagent **1**. Samples in (A) and (B) were subjected to filtration using fluorous flash silica gel prior to injection on HPLC.*

The enhanced stabilities of fluorinated proteins allows the construction of robust biomacromolecular assemblies. These advantages also point to new avenues of research. Membrane protein design with increasing levels of sophistication might lead to new solutions in medicine, for example, in modulation of integrin activation, membrane protease inhibition and in control and manipulation of signal transduction pathways. In addition, the chemical synthesis of difficult peptide sequences may be resolved by use of fluorous capping reagents.

Acknowledgment

This work was supported in part by NIH and by a NSF CAREER award. We thank Dr. B. Bilgiçer and Dr. V. Montanari for helpful discussions. The Analytical Ultracentrifugation and Mass Spectrometry facilities at Tufts are supported by the NIH (1S10RR017948) and NSF (CHE-0320783) respectively. K.K. is a DuPont Young Professor.

References

1 Myers, K. E.; Kumar, K. *J. Am. Chem. Soc.* **2000,** *122*, 12025.

2 a) Horvath, I. T.; Rabai, J. *Science* **1994,** *266*, 72. b) Gladysz, J. A. *Chem. Rev.* **2002,** *102*, 3215.

3 a) Studer, A.; Hadida, S.; Ferritto, R.; Kim, S. Y.; Jeger, P.; Wipf, P.; Curran, D. P. *Science* **1997,** *275*, 823. b) de Wolf, E.; van Koten, G.; Deelman, B. J. *Chem. Soc. Rev.* **1999,** *28*, 37. c) Fish, R. H. *Chem. Eur. J.* **1999,** *5*, 1677.

4 Brittain, S. M.; Ficarro, S. B.; Brock, A.; Peters, E. C. *Nat. Biotechnol.* **2005,** *23*, 463.

5 Yoder, N. C.; Kumar, K. *Chem. Soc. Rev.* **2002,** *31*, 335.

6 Jackel, C.; Koksch, B. *Eur. J. Org. Chem.* **2005**, 4483.

7 Miura, T.; Goto, K. T.; Hosaka, D.; Inazu, T. *Angew. Chem., Int. Ed.* **2003,** *42*, 2047.

8 a) Scott, R. L. *J. Am. Chem. Soc.* **1948,** *70*, 4090. b) Hildebrand, J. H.; Cochran, D. R. F. *J. Am. Chem. Soc.* **1949,** *71*, 22.

9 Scott, R. L. *J. Phys. Chem.* **1958,** *62*, 136.

10 a) Wen, W. Y.; Muccitelli, J. A. *J. Solution Chem.* **1979,** *8*, 225. b) Wilhelm, E.; Battino, R.; Wilcock, R. J. *Chem. Rev.* **1977,** *77*, 219.

11 Lee, K. H.; Lee, H. Y.; Slutsky, M. M.; Anderson, J. T.; Marsh, E. N. G. *Biochemistry* **2004,** *43*, 16277.

12 Resnati, G. *Tetrahedron* **1993,** *49*, 9385.

13 Israelachvili, J. N.; Mitchell, D. J.; Ninham, B. W. *Biochim. Biophys. Acta* **1977,** *470*, 185.

14 Bilgicer, B.; Fichera, A.; Kumar, K. *J. Am. Chem. Soc.* **2001,** *123*, 4393.

15 Bilgicer, B.; Xing, X.; Kumar, K. *J. Am. Chem. Soc.* **2001,** *123*, 11815.

16 Tang, Y.; Ghirlanda, G.; Vaidehi, N.; Kua, J.; Mainz, D. T.; Goddard, W. A.; DeGrado, W. F.; Tirrell, D. A. *Biochemistry* **2001,** *40*, 2790.

17 Tang, Y.; Ghirlanda, G.; Petka, W. A.; Nakajima, T.; DeGrado, W. F.; Tirrell, D. A. *Angew. Chem., Int. Ed.* **2001,** *40*, 1494.

18 Tang, Y.; Tirrell, D. A. *J. Am. Chem. Soc.* **2001,** *123*, 11089.

19 Lupas, A. *Trends Biochem.Sci.* **1996,** *21*, 375.

20 Lee, H. Y.; Lee, K. H.; Al-Hashimi, H. M.; Marsh, E. N. G. *J. Am. Chem. Soc.* **2006,** *128*, 337.

21 Bilgicer, B.; Kumar, K. *Tetrahedron* **2002,** *58*, 4105.

22 a) Bowie, J. U. *Proc. Natl. Acad. Sci. U. S. A.* **2004,** *101*, 3995. b) Engelman, D. M.; Chen, Y.; Chin, C. N.; Curran, A. R.; Dixon, A. M.; Dupuy, A. D.; Lee, A. S.; Lehnert, U.; Matthews, E. E.; Reshetnyak, Y. K.; Senes, A.; Popot, J. L. *FEBS Lett.* **2003,** *555*, 122. C) Curran, A. R.; Engelman, D. M. *Curr. Opin. Struct. Biol.* **2003,** *13*, 412.

23 a) Kienker, P. K.; Degrado, W. F.; Lear, J. D. *Proc. Natl. Acad. Sci. U. S. A.* **1994,** *91*, 4859. b) Kamtekar, S.; Schiffer, J. M.; Xiong, H. Y.; Babik, J. M.; Hecht, M. H. *Science* **1993,** *262*, 1680.

24 Bowie, J. U. *Nature* **2005,** *438*, 581.

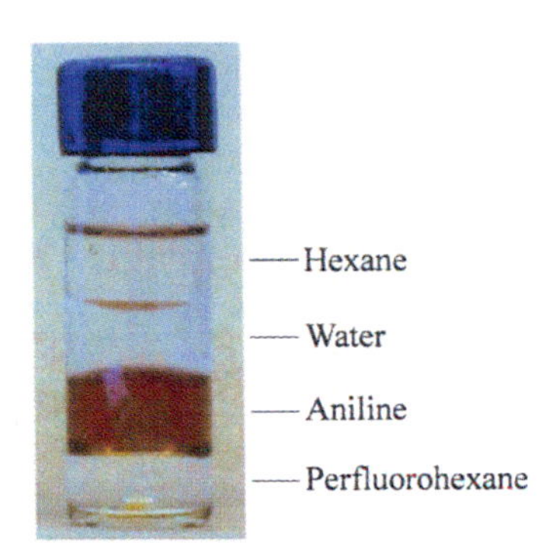

Figure 31.1. Phase separation of four liquid layers (hexane, water, aniline and f-hexane) at 25 °C. The aniline layer has been false colored to clearly show phase boundaries.

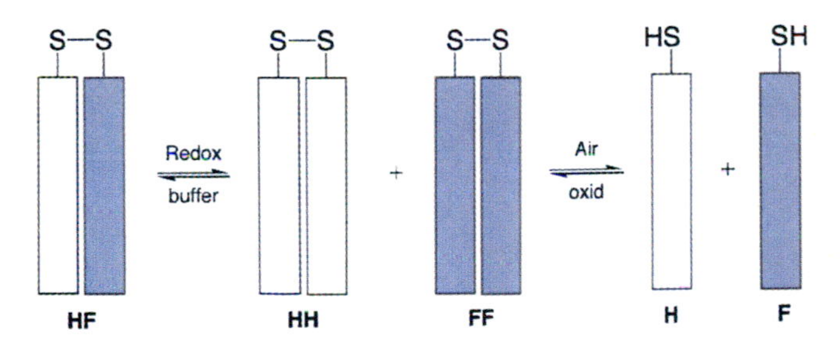

*Scheme 31.2. Disulfide exchange experiment exploring the preference for homodimer (**HH** and **FF**) versus heterodimer (**HF**) formation. Only 3% or less of the heterodimer is formed (or remained) under equilibrium redox conditions.*

25 Bilgicer, B.; Kumar, K. *Proc. Natl. Acad. Sci. U. S. A.* **2004,** *101*, 15324.
26 Naarmann, N.; Bilgicer, B.; Meng, H.; Kumar, K.; Steinem, C. *Angew. Chem., Int. Ed.* **2006,** *45*, 2588.
27 Popot, J. L.; Engelman, D. M. *Annu. Rev. Biochem.* **2000,** *69*, 881.
28 Langosch, D.; Heringa, J. *Proteins* **1998,** *31*, 150.
29 a) Choma, C.; Gratkowski, H.; Lear, J. D.; DeGrado, W. F. *Nat. Struct. Biol.* **2000,** *7*, 161. b) Zhou, F. X.; Cocco, M. J.; Russ, W. P.; Brunger, A. T.; Engelman, D. M. *Nat. Struct. Biol.* **2000,** *7*, 154.
30 Li, M.; Reddy, L. G.; Bennett, R.; Silva, N. D.; Jones, L. R.; Thomas, D. D. *Biophys. J.* **1999,** *76*, 2587.
31 Merrifield, R. B. *J. Am. Chem. Soc.* **1963,** *85*, 2149.
32 a) Dawson, P. E.; Kent, S. B. H. *Annu. Rev. Biochem.* **2000,** *69*, 923. b) Cotton G. J.; Muir, T. W. *Chem. Biol.* **1999,** *6*, R247.
33 Andersson, L.; Blomberg, L.; Flegel, M.; Lepsa, L.; Nilsson, B.; Verlander, M. *Biopolymers* **2000,** *55*, 227.
34 Montanari, V.; Kumar, K. *J. Am. Chem. Soc.* **2004,** *126*, 9528.
35 a) Curran, D. P.; Luo, Z. Y. *J. Am. Chem. Soc.* **1999,** *121*, 9069. b) Pearson, W H.; Berry, D. A.; Stoy, P.; Jung, K. Y.; Sercel, A. D. *J. Org. Chem.* **2005,** *70*, 7114. c) Palmacci, E. R.; Hewitt, M. C.; Seeberger, P. H. *Angew. Chem., Int. E* **2001,** *40*, 4433.
36 a) de Visser, P. C.; van Helden, M.; Filippov, D. V.; van der Marel, G. A.; Drijfhout, J. W.; van Boom, J. H.; Noort, D.; Overkleeft, H. S. *Tetrahedron Let* **2003,** *44*, 9013. b) Filippov, D. V.; van Zoelen, D. J.; Oldfield, S. P.; van der Marel, G. A.; Overkleeft, H. S.; Drijfhout, J. W.; van Boom, J. H. *Biopolymers* **2003,** *71*, 346.
37 Montanari, V.; Kumar, K. *J. Fluorine Chem.* **2006,** in press.
38 Montanari, V.; Kumar, K. *Eur. J. Org. Chem.* **2006**, 874.

Indexes

Author Index

Aceña, José Luis, 54
Afonin, Sergii, 431
Bégué, Jean-Pierre, 337
Berditsch, Marina, 431
Bonnet-Delpon, Danièle, 337
Bruché, Luca, 420
Burton, Donald J., 83
Candiani, Gabriele, 420
Chatterjee, C., 379
Chollet, Constance, 337
Chorki, Fatima, 337
Cottet, Fabrice, 39
Crousse, Benoit, 337
Davis, Benjamin G., 323
del Pozo, Carlos, 54
El Bakkari, Mounir, 271
Enders, Dieter, 447
Fuchigami, Toshio, 69
Fujiwara, Kana, 462
Funabiki, Kazumasa, 141, 447
Fustero, Santos, 54
Gerig, J. T., 379
Grage, Stephan, 431
Grellepois, Fabienne, 337
Grellier, Philippe, 337
Guo, Yong, 462
Hargreaves, Christopher A., 323
Hayakawa, H., 170
Honek, John F., 393
Hovagimyan, G., 379
Huguenot, Florent, 420
Hung, Nguyen Van, 337
Ichikawa, Junji, 155
Ieronimo, Marco, 431
Itoh, Toshiyuki, 352
Itoh, Yoshimitsu, 2
Izawa, Kunisuke, 363
Jagodzinska, Monika, 420
Kalsani, Venkateshwarlu, 487
Kirsch, Peer, 221
Kitamoto, Takamasa, 409
Komatsu, Yumi, 462
Kumar, Krishna, 487
Kuznetsova, Larissa V., 288
Maffezzoni, Raffaella, 420
Magueur, Guillaume, 337
Marubayashi, Shunsuke, 409
Maruyama, Tokumi, 363
Matsui, Masaki, 141, 447
Meng, He, 487
Mikami, Koichi, 2, 190
Moussier, Nathalie, 420
Nagamori, Masashi, 447
Nakamura, Yutaka, 244
Nhan, Doan Hanh, 337
Nhu, Truong Van, 337
Ohashi, Y., 170
Ojima, Iwao, 288
Onishi, Tomoyuki, 363
Ono, T., 170
Ourévitch, Michèle, 337
Petrov, Viacheslav A., 113
Pohl, Nicola L., 261
Qing, Feng-Ling, 305
Qiu, Xiao-Long, 305
Raabe, Gerhard, 447
Raghavanpillai, Anilkumar, 83
Raines, Ronald T., 477
Röschenthaler, Gerd-Volker, 221
Sandford, Graham, 323
Sani, Monica, 420
Sanz-Cervera, Juan F., 54

Schlosser, Manfred, 39
Sinisi, Roberta, 420
Soloshonok, Vadim A., xi
Strandberg, Erik, 431
Sun, Liang, 288
Tajima, Toshiki, 69
Takeuchi, Seiji, 244
Tam, Nguyen Thi Ngoc, 337
Thu, Nguyen Thi Minh, 337
Tonoi, Takayuki, 190
Torii, Takayoshi, 363
Tremouilhac, Pierre, 431
Uekusa, H., 170
Ulrich, Anne S., 431
Uneyama, Kenji, 462
Viani, Fiorenza, 420
Vincent, Jean-Marc, 271
Wadhwani, Parvesh, 431
Wang, Qian, 39
Welch, John T., xi
Yamazaki, Takashi, 409
Yasuda, N., 170
Zanda, Matteo, 420
Zard, Samir Z., 25
Zhang, Wei, 207

Subject Index

A

ACE inhibitors. *See* Angiotensin-converting enzyme
Acquired immunodeficiency syndrome, 363–364
Active matrix liquid crystal displays, super fluorinated materials, 228–229*f*
Acyl group effect on diastereoselective anodic fluorination, 79, 80*f*
Acylfluorodihydronaphthalenes, polysubstituted naphthalene derivatives, 164–165*f*
Agrochemicals and pharmaceuticals, pentafluorosulfanyl-based, 237–238
Alanylglycyldiketopiperazine proton in aqueous fluoroalcohol solutions, cross relaxation rates, 386
Alcohol oxidations, fluorous copper(II)-carboxylate complexes, 277–278*f*
Aldol reactions, trifluoroacetaldehyde ethyl hemiacetal with ketones, 143–149
See also Mukaiyama aldol reaction
Alkane/alkene oxidation, fluorous carboxylate ligand for catalyst recovery, 274–276
Alkenes, fluoro, cationic cyclizations, 155–168
Alkyllithium, 1,2 addition, trifluroacetaldehyde SAMP- or RAMP-hydrazones, 450–455
reaction conditions, 450–451
3-Alkyl-3-methoxycarbonyl-4-thianones, anodic fluorination, 78
cis-2-Alkylsubstituted-1,2-difluoroalkenes, stereospecific synthesis, 93, 95–97f
Amines in reaction optimization, trifluoroacetaldehyde ethyl hemiacetal with ketones, 143–147
Amino acids
cyclic fluorinated, synthesis, 55–60
fluorinated, in biomolecule separations, 494–497
fluorinated, in protein design, 490–494
fluorinated, structure analysis, membrane-active peptides, 431–446
α-Amino hydroxamates, MMP inhibitors, N-alkylated analogs, 422–424
α-Amino hydroxamic matrix metalloproteinases inhibitor, 421–422*f*, 424
2-Aminoperfluoropropene, precursor for trifluoroalanine peptides, 468
Angiotensin-converting enzyme (ACE) inhibitors, 379–380
Angiotensin II fluorine analogs
cross relaxation rates, 386–387
^{1}H{^{19}F} intermolecular NOE spectrum, 386–387*f*
^{1}H NMR spectra, 386–387*f*
Angiotensin II proposed solution conformation, 380–381*f*
Anodic fluorination, highly diastereoselective, 69–82
Anthracene-cyclopropane hybrids, DNA cleavage agents switched by photo irradiation, 355–360

Anticancer agents case study, fluorotaxoids, 288–304
Antimalarial CF_3-substituted artemisinin derivatives, 337–351
Antitumor compounds, fluoroaremisinin dimers, preparation, 346–348
Antiviral fluorinated purine nucleoside, synthesis, 363–378
Arogenate dehydrogenase inhibitor, preparation, 43
Artemisia annua, 338
Artemisinin, 10-trifluoromethyl monomers, dimers and chimeras, 337–351
Artemisinins, first generation, 338–339*f*
cis-1-Arylperfluoroalkenes, stereospecific synthesis, 105–107*t*
cis and *trans*-Arylsubstituted 2,3-difluoroacrylic esters, stereospecific preparation, 98, 99–101*t*
Aryl 3,3,3-trifluoro-2-methoxypropyl sulfides, diastereoselective anodic fluorination, 74–75*f*
Asymmetric anodic fluorination, 69–82
Axial fluorine atom replacement in dielectric anisotropy modification, liquid crystals, 234, 236–237*f*
Aza-lactones, fluorinated. *See* Lactams, lactones and aza-lactones, fluorinated
exo-3-Aza-4-perfluoroalkyltricyclo[4.2.1.0$^{2.5}$]non-3,7-dienes and *exo*-4,4'-fluoroalkyltricyclo[4.2.1.0$^{2.5}$]non-7-enes, polyfluorinated, reactions, 134–136*f*

B

Baeyer-Villiger reaction in nanoflow microreactor, 197, 198*f*–203
compared to batch reaction, 199–201*f*
Benzoylation, trifluoroacetaldehyde SAMP-hydrazines, 455
Bicyclic fluorinated uracil synthesis, ring closing metathesis reactions, 60–61*f*
Bioactive conformations, fluorotaxoids, 296–298*f*
Bioincorporation, fluorinated methionines into proteins, 400–402
Biological chemistry probes, fluorinated methionines, 393–408
Biologically active *gem*-difluorinated cyclopropane compounds, 352–354
Biomolecules
fluorous chemistry for synthesis and purification, 207–220
separations, fluorinated amino acids as reagents, 494–497
Bis-*cis*-1,2-difluoroalkenyl synthons, 104–105
Bistratamide H, syntheses
and C light fluorous protecting group, 252–256
and diastereomer, quasi-racemic mixture, 256–258, 259*f*
heavy fluorous protecting group, 247–252
Bromine rearrangement during fluorination, 3'-α-fluoro-2',3'-dideoxyguanosine, synthesis, 369–373
Bromoiodoarenes, conversion to bromo(trifluoromethyl)arenes, 44–45

C

C_1F_n package delivery option, 44–45
Capping agents, desirable properties, 495
Captopril, ACE inhibitor, 380

Carbohydrate microarrays, fluorous-based, direct formation, 265–268
Carbohydrate synthesis, fluorous-phase, 262–265
fluorous tags for liquid-liquid extractions, 263–264
fluorous tags for solid-phase extractions, 264–265
Carbonyl compounds, fluorinated, cycloaddition to quadricyclane and norbornadiene, 120, 121–125
Carboxylates, fluorous
metal complexes in alcohol oxidations, 277–278
metal complexes in alkane/alkene oxidations, 274–276
reversible phase-switching processes, 280–283
rhodium catalysts, 278–280
Carboxylic acids, fluorous, structures, properties, 272–274
Catalytic metal ligands, fluorous carboxylates, 271–280
Catch-and-release approach, reversible phase-switching process, 280
Cationic cyclizations, fluoro alkenes, 155–168
Cell interactions with fluorinated methionines, 402–405
Chimera for malaria treatment, fluorinated artemisinin derivative and mefloquine, 343–346
3-Chlorophenol, regioexhaustive functionalization, 47–48*f*
Cluster formation based on F–F intermolecular interactions, 183–186*f*
Covalent bi-therapy approach to malaria treatment, 343–344
Cross-metathesis reaction in fluorinated β-amino acid synthesis, 55, 59–60*f*
Cross relaxation rates
alanylglycyldiketopiperazine proton in aqueous fluoroalcohol solutions, 386
angiotensin analogs, 386–387
from experimental values of diffusion coefficients, 384–385*t*
intermolecular interactions, calculations, 383–384
melittin, 387–389
trp-cage, 20-residue peptide, 389–391*f*
Crystal engineering, fluorine-based, hydro-fluoro hybrid compounds, 170–189
Crystal packing motifs in hermaphrodites, 173–179
Cyanobacterium Nostoc spongiaeforme var. tenue, 244
Cyclic compounds, diastereoselective anodic fluorination, 76–81
Cyclic fluorinated amino acids, synthesis, 55–60
Cyclizations, cationic, fluoro alkenes, 155–168
Cyclohexane in xanthate reductions, 32, 33*f*
Cyclopentenones, fluorine-containing, preparation by fluorine-directed Nazarov cyclizations, 156–162

D

Degenerative xanthate transfer, mechanism, 25–27
Diastereoselective anodic fluorination, 69–82
cyclic compounds, 76–81
open-chain compounds,70–75
4,7-Dichloroquinoline, site selective transformation, 49
Didemnum molle, 244
Dielectric anisotropy modification, pentafluorosulfanyl-based liquid crystals, 232–237*f*

See also Physical properties, pentafluorosulfanyl derivatives in liquid crystals
Diffusion coefficients, cross relaxation rate from experimental values, 384–385*t*
gem-Difluorinated cyclopropane compounds, 352–361
biologically active, 352–354
physical properties, 354–355
Difluoro nucleoside analogue, synthesis, 27*f*
1,1-Difluoroalk-1-enes, fluorine-directed Friedel–Crafts-type cyclization, 165–166
cis-1,2-Difluoroalkenes, functionalized, stereospecific preparation, 83–112
1,2-Difluoroalkenyl organometallic synthons, 85–92
cis-1,2-Difluoroalkenyl synthons, stereospecific syntheses, 92–108
gem-Difluorocyclopropane instability under sunlight, 356
Difluoroethene, electrophilic reactions with nucleophiles, 466–468
trans-1,2-Difluoroethenyltrialkyl-silanes, synthons, 90–91*f*
Difluoroethionine, probe in biological chemistry, 394–405
cis-1,2-Difluoro-1-iodostyerenes, stereospecific preparation, 97, 99*t*
cis-1,2-Difluoro-4-oxo-substituted 2-butenoates, stereospecific preparation, 101–102, 103*f*
trans-1,2-Difluoro-1-trialkylsilylethenes photoisomerization to *cis*-1,2-difluoro-1-trialkylsilylethenes, 93
3'-Difluoromethyl-taxoids, synthesis and biological evaluation, 291–294
gem-Difluoromethylated azanucleosides, synthesis, 317–318*f*
gem-Difluoromethylated carbocyclic nucleosides, synthesis, 318–320*f*
gem-Difluoromethylated furanyl (pyranyl) nucleosides, synthesis, 306–313*f*
gem-Difluoromethylated sugar nucleosides, synthesis, 305–322
gem-Difluoromethylated thionucleosides, synthesis, 313–317*f*
cis-1,2-Difluorostyrenes, stereospecific synthesis and reactions, 97–98*t*, 99*t*
3,4-Difluoro-6-substituted-2-pyrones, synthesis, 102, 103*t*
2,2-Difluorovinyl ketones, fluorine-accelerated Friedel–Crafts-type cyclization, 162–164*f*
3'-Difluorovinyl-taxoids, second generation, synthesis and biological evaluation, 294–296
2,2-Difluorovinyl vinyl ketones, fluorine-directed Nazarov-type cyclization, 156–159*f*
Dihydroquinolones and homophthalimides, synthesis, 35, 36*f*
α,ω-Diiodo-perfluoroalkylenes and hexamethylenetetramine, crystal packing, 183–186
1,2-Diol moieties, protecting group effect on diastereoselective anodic fluorination, 71, 72
Direct fluorination as synthetic strategy, 40–44*f*
Disaccharide syntheses, glycosylation reactions, 324–328*f*
trans-1,2-Disubstituted-1,2-difluoroalkenes from corresponding silanes and stannanes, 89–90*f*
1,3-Dithiolanones, diastereoselective anodic fluorination, 76
DNA cleavage agents switched by photo irradiation, anthracene-cyclopropane hybrids, 355–360
Docetaxel. *See* Taxotère
Domino cyclizations, fused polycyclic systems, 164–165*f*

Drug release concept, mechanism-based, 300–301*f*

E

Electrolyte, supporting, effect on diastereoselectivity in anodic fluorination, 71
Electrolytic temperature effect on diastereoselective anodic fluorination, 71, 72*f*
Electrophilic fluorinating reagents, 40–42*f*
Energetic materials. *See* Explosion; Explosives
Environmental Protection Agency (EPA), human health concerns, perfluorooctanoic acid and salts, 274
Enzymatic synthesis with fluorous primers, oligosaccharide synthesis, 215–216
Epothilones, precursor, trifluoromethylated analogue, synthesis, 64–65*f*
Explosion possibility for perchloryl fluoride, 42
Explosives, pentafluorosulfanyl-based, 224–226

F

F–F intermolecular interactions
  basis for cluster formation, 183–186*f*
  in Hermaphrodite molecules, 175, 177–179
^{19}F-labeled amino acids into peptides, incorporation, 436–440
^{19}F NMR label, advantages, 432
^{19}F NMR monitoring, glycosylation reactions, 332–334*f*
^{19}F NMR protein bioincorporation, difluoromethionine and trifluoromethionine, 400–402
^{19}F NMR, solid-state, structure analysis, membrane-active peptides labeled with fluorinated amino acids, 431–446
FDA. *See* Food and Drug Administration
Felkin-Anh model, stereochemical control mechanism, 71–72, 73*f*
Flow rate control in microreactors, 191–193
Fluorinated. *See also* Fluorous
Fluorinated amino acids
  protein design, 490–494
  reagents in protein design and biomolecule separation, 487–499
  structure analysis, membrane-active peptides, 431–446
Fluorinated carbonyl compounds, cycloaddition to quadricyclane and norbornadiene, 120, 121–125
Fluorinated core compounds, functionalization, 45–50
Fluorinated cyclic amino acids, syntheses, 55–60*f*
Fluorinated inhibitors, matrix metalloproteinases, 420–430
Fluorinated lactams, lactones and aza-lactones, synthesis by ring closing metathesis, 62, 63–66
Fluorinated methionines, biological chemistry probes, 393–408
  properties, 396, 398*f*–405
  synthesis, 395–396*f*, 397*f*
Fluorinated nitrogen-containing compounds, cycloaddition to quadricyclane, 125–126
Fluorinated piperidines, synthesis by ring closing metathesis, 61–63*f*
Fluorinated substrates and quadricyclane, 128–129

Fluorinated sugars, 3'-α-fluoro-2',3'-dideoxyguanosine, synthesis, 364–367
Fluorinated sulfur-containing compounds, cycloaddition to quadricyclane, 126–127
Fluorinated uracils, synthesis by ring closing metathesis, 60–61*f*
Fluorination, diastereoselective anodic, 69–82
Fluorine-accelerated. *See* Fluorine-directed
Fluorine-containing cyclopentenones, preparation by fluorine-directed Nazarov cyclizations, 156–162
Fluorine-containing oligopeptides, synthesis and conformational analysis, 409–419
Fluorine contamination, NMR samples, 436
Fluorine-directed
  and accelerated Nazarov cyclization, 1-fluorovinyl vinyl ketones, 160–162
  Friedel–Crafts-type cyclization, 1,1-difluoroalk-1-enes, 165–166
  Nazarov cyclization, 1-(trifluoromethyl)vinyl vinyl ketones, 159–160*f*
  Nazarov-type cyclization, 2,2-difluorovinyl vinyl ketones, 156–159*f*
Fluorine in tumor-targeting anticancer agents, 299–301*f*
Fluoroalanine labeled peptide, 439–440
Fluoroalcohol-water mixtures, solvent-peptide interactions, 379–392
Fluoroalkenes
  cationic cyclizations, 155–168
  Friedel–Crafts-type cyclizations, 162–166
  Nazarov-type cyclizations, 156–162
exo-5(6)-Fluoroalkylnorborn-2-enes, synthesis, 130–132*f*
exo-4,4'-Fluoroalkyltricyclo-[4.2.1.0$^{2.5}$]non-7-enes and *exo*-3-aza-4-perfluoroalkyltricyclo-[4.2.1.0$^{2.5}$]non-3,7-dienes, polyfluorinated, reactions, 134–136*f*
Fluoroaremisinin dimers, antitumor compounds, preparation, 346–348
3'-α-Fluoro-2',3'-dideoxyguanosine, synthesis, 363–378
gem-Fluoromethylated thionucleosides, synthesis, 315–316*f*
3-Fluorophenol, regioexhaustive functionalization, 47
Fluorophenylglycine labeled peptides, 438
4-Fluoroproline residues, hyperstable collagen based on, 477–486
Fluorotaxoid anticancer agents, case study, 288–304
2-Fluoro-1-tetralones, anodic fluorination synthesis, camphanyl enol ester derivatives, 81
Fluorotryptophan labeled peptides, 438
Fluorous. *See also* Fluorinated
Fluorous amide coupling agents, peptide synthesis and separation, 212–213
Fluorous-based carbohydrate microarrays, direct formation, 265–268
Fluorous capping agents in solid-phase synthesis
  oligosaccharide synthesis, 215
  peptide synthesis and separation, 210
Fluorous capping enabled purification in solid phase peptide synthesis, 494–497*f*
Fluorous carboxylates, metal ligands, 271–286
Fluorous carboxylic acids, structures and properties, 272–274

Fluorous chemistry
glycopeptide synthesis, 216–217*f*
oligonucleotide synthesis, 217–218
Fluorous copper(II)-carboxylate complexes in alcohol oxidations, 277–278*f*
Fluorous enrichment of proteomics sample, peptide synthesis and separation, 212
Fluorous manganese-(salen)manganese complexes, catalysts for alkene epoxidations, 275–276
Fluorous materials, phase separation, 488–489*f*
Fluorous microarrays, oligosaccharide synthesis, 216
Fluorous nanoflow microreactor, 190–206
Fluorous phase-switching protocol applications in organic synthesis, 282–284
Fluorous phase synthesis, carbohydrate microarrays, 261–270
Fluorous primers, enzymatic synthesis, oligosaccharide synthesis, 215–216
Fluorous protecting groups in macrolactam marine natural product synthesis, 244–260
Fluorous rhodium-carboxylate catalysts, 278–280
Fluorous solution-phase synthesis, peptide synthesis and separation, 210–211*f*
Fluorous tags
liquid-liquid extractions in carbohydrate synthesis, 263–264
solid-phase extraction in carbohydrate synthesis, 264–265
solid-phase synthesis, peptide synthesis and separation, 209–210
solution-phase synthesis, oligosaccharide synthesis, 213–215*f*
Fluorous trivalent iodonium salts, capping agents, 496, 497*f*
1-Fluorovinyl vinyl ketones, Nazarov cyclization, 160–162
Folding degree assessment, peptides in micellar detergents, 493–494
Food and Drug Administration (FDA) drug approvals, 289, 299, 306
Free energy of unfolding, fluorinated amino acid effects, 490–491
Friedel–Crafts-type cyclizations, fluoro alkenes, 162–166
Functionalization, fluorinated core compounds, 45–50
Fusogenic peptide B18, solid state ^{19}F NMR results, 434*t*, 441

G

Gemcitabine, 306–307*f*
Glutathione-triggered cascade drug release for fluorine-containing tumor-targeting anticancer agents, 300
Glycopeptide synthesis, fluorous chemistry, 216–217*f*
Glycosyl donor development. *See* Polyfluoropyridyl glycosyl donors
Glycosyl donor systems in general use, oligosaccharide synthesis, 326–327
Glycosylation process, stereochemistry, 331–332*f*
Glycosylation reactions with simple disaccharide syntheses from two residues, 324–328*f*
Gramicidin S, PGLa and K3, antimicrobial peptides, solid state ^{19}F NMR results, 434*t*, 441–442
Greenhouse gas properties, SF_5CF_3, 224
Grubbs ruthenium-based catalysts. *See* Ruthenium-based catalysts

Guanosine, 3'-α-fluoro-2',3'-dideoxyguanosine, synthesis with inversion of configuration at the C3'-position of guanosine, 368–369

H

$^1H\{^{19}F\}$ intermolecular NOE spectrum
  angiotensin analog in trifluoroethanol/water, 386–387*f*
  trp-cage peptide and solvent fluorine, in hexafluoroisopropanol/ water, 389–391*f*
^1H NMR chemical shifts, amide and alcoholic protons in hexapeptides and F_3-hexapeptides, 417–418
^1H NMR spectra, angiotensin analogs in trifluoroethanol/water, 386–387*f*
Hansch hydrophobicity parameters, trifluoromethyl group, 489
Hepatitis B and C viruses, 363
Hermaphrodite–new concept in fluorine-based crystal engineering, 170–189
Hexamethylenetetramine and α,ω-diiodo-perfluoroalkylenes, crystal packing, 183–186
Hexapeptides
  and F_3-hexapeptides, spectroscopic comparison, 414–418
  and tri-fluorinated analogs, preparation, 410–414
Hildebrand and Scott's theory, solubility, non-electrolytes, 488–489
Homodimer formation, protected 2,3-diamino-1,1,1,4,4,4-hexafluorobutane, 29–30*f*
Homophthalimides and dihydroquinolones, synthesis, 35, 36*f*
Horner-Wittig reaction, 10,10-difluorocamphene synthesis, 42
Human immunodeficiency virus, 363–364
Hydantoin synthesis assisted by reversible fluorous phase-switching process, 282–284
Hydro-fluoro compounds in fluorine-based crystal engineering, 170–189
Hydro-fluoro hybrids, supramolecular assemblage, 173–176*f*
Hydrogen donor-acceptor functional group in hermaphrodites, 179–183
Hydrogen fluoride content in supporting fluoride salts, diastereoselective anodic fluorination, 78–79
Hydrogen peroxide, oxidizing agent enhancement in Baeyer-Villiger reaction, 197, 198*f*–202
Hyperstable collagen based on 4-fluoroproline residues, 477–486
Hypertension control by ACE inhibitors, 380
Hypervalent sulfur fluorides, functional compounds, 221–243

I

Indolines, fluorinated, alternative synthesis using xanthates, 35, 36*f*
Intermolecular interactions, cross relaxation rate calculations, 383–384
cis-1-Iodoperfluoroalkenes, stereospecific synthesis, 108
Ionic liquids based on pentafluorosulfanyl substituted cation, 226–227*f*

K

Ketones
  asymmetric direct aldol reaction, trifluoroacetaldehyde ethyl

hemiacetal, *L*-proline-catalyzed, 149–152
2,2-difluorovinyl, fluorine-accelerated Friedel–Crafts-type cyclization, 162–164
direct aldol reactions with trifluoroacetaldehyde ethyl hemiacetal, 143–149
vinyl, fluorine-directed Nazarov-type cyclizations, 156–159*f*
Knobs-into-holes packing arrangement, 493
Knorr condensation for trifluoromethylated pyrrole synthesis, 29

L

β-Lactam synthon method, 291
Lactams, lactones and aza-lactones, fluorinated, ring closing metathesis, synthesis, 62, 63–66
Lactones, fluorinated. *See* Lactams, lactones and aza-lactones, fluorinated
Lateral *o*-fluorination, dielectric anisotropy modification, liquid crystals, 233–234, 235*t*
Lewis acid activators in glycosylation reactions, 331–335*f*
Liquid crystals, pentafluorosulfanyl-based, 228–237*f*
Liquid-liquid extractions in carbohydrate synthesis, fluorous tags, 263–264
Lissoclinum bistratum, 244
Lithium enolates
α-CF_3 ketone, structure, 4–5*f*
radical trifluoromethylation, 12–21

M

Macrolactam marine natural products, syntheses, fluorous protecting groups, 244–260
Magic acid in Friedel–Crafts-type cyclization, 1,1-difluoroalk-1-enes, 165–166
Malaria. *See* Antimalarial CF_3-substituted artemisinin derivatives
Marine natural products, macrolactam syntheses, fluorous protecting groups, 244–260
Matrix metalloproteinases, fluorinated inhibitors, 420–430
Melittin
conformation in 35% hexafluoroisopropanol/water, 387–388*f*
cross relaxation rates, 387–389
Membrane-active peptides, solid state NMR results, 441–442
Membrane embedded protein components, two-step assembly, 492
Metal ligands, fluorous carboxylates, 271–286
Metathesis reactions in nitrogen-containing organofluorine compound synthesis, 54–68
Methionine
biochemical role in proteins, 393–394
fluorinated analogs, chemical structures, 395
fluorinated probes, biological chemistry, 393–408
2-Methoxypropyl phenyl sulfide, diastereoselective anodic fluorination, 70, 71
Michael addition to artemisitene for 16-functionalized artemisinin derivatives, 338
Microreactor, fluorous nanoflow, 191–193
Miscibility nonpolar liquids, solubility parameter, 488–489
Mukaiyama aldol reaction in nanoflow microreactor, 193–197, 198*f*
See also Aldol reactions
Mylotarg®, FDA approval, 299

N

N-bromosuccinimide in laboratory-scale vicinal bromofluorination, 42–43
N-chloromethyl-N-fluoro-1,4-diazoniabicyclo[2.2.2]octane bis9tetrafluoroborate), surrogate for perchloryl fluoride, 42
N-fluoroquinuclidinium salts, surrogates for perchloryl fluoride, 42
N-fluorosulfonamides, surrogates for perchloryl fluoride, 42
N-(4-methoxyphenyl)-*F*-2-(2,6-dimethylmorpholino)-propanamide, reverse order-disorder structure in crystal packing, 177–178*f*
N–N single bond cleavage induced by samarium(II) iodide, 456–457
Nanoflow microreactor, fluorous, 190–206
Naphthalene synthesis using xanthate technology, 34–35*f*
Nazarov-type cyclizations, fluoro alkenes, 156–162
Nitrogen-containing compounds, fluorinated, cycloaddition to quadricyclane, 125–126
NMR. *See* ^{19}F NMR; ^{1}H NMR; Nuclear Overhauser effect; Solid state NMR
NOE. *See* Nuclear Overhauser effect
Noncovalent fluorous-fluorous interaction, surface patterning in carbohydrate microarray formation, 265–268
Norbornadiene and quadricyclane, fluorinated carbonyl compound cycloaddition, 120, 121–125
Norbornenes, *exo*-substituted, synthesis, 134
Nuclear Overhauser effect (NOE), 382–384
    angiotensin analog, ^{1}H{^{19}F} intermolecular spectrum, trifluoroethanol/water, 386–387*f*
    computation method reliability, 384–385
    correlation, non-fluorinated and tri-fluorinated hexapeptides, 414–418
    ^{1}H{^{19}F} intermolecular spectrum, trp-cage peptide and solvent fluorines, 389–391*f*
    melittin in hexafluoroisopro-panol/water, 387–389
Nucleophiles, reactions with 10-trifluoromethyl-10-deoxoartemisin allyl bromide, 341–343
Nucleophilic fluorination reagents, 42–44*f*

O

Oligonucleotide synthesis, fluorous chemistry, 217–218
Oligopeptides
    fluorine-containing, synthesis and conformational analysis, 409–419
    trifluoroalanine, synthetic strategies, 462–476
Oligosaccharide synthesis, fluorous chemistry, 213–216*f*
Open-chain compounds, diastereoselective anodic fluorination, 70–75
Organofluorine compounds, synthetic strategies, 39–53
Organometallic approach to functionalized fluoro compounds, toolbox methods, 45–50
Organometallic 1,2-difluoroalkenyl reagents in 1,2-difluoroalkene syntheses, 85–92
exo-3-Oxa-4,4-fluoroalkyltricyclo[4.2.1$0^{2.5}$]non-7-

enes, polyfluorinated, ring opening reactions, 132–133*f*
1,3-Oxathiolanones, diastereoselective
anodic fluorination, 76–77
chemical fluorination, comparison, 77–78
1,3-Oxazolidine from *L*-threonine, diastereoselective anodic fluorination, 79, 80*f*–81

P

Paclitaxel. *See* Taxol
Palladium(0) catalyzed reaction, *cis*-2,3-difluoro-3-stannylacrylic ester with aryl or alkenyl iodides, 100–101*t*
Pentafluorosulfanyl-based
explosives, 224–226
ionic liquids, 226–227*f*
liquid crystals, 228–237*f*
pharmaceuticals and agrochemicals, 237–238
Pentafluorosulfanyl benzene derivatives, synthesis, 222–223
Pentafluorosulfanyl nitramides, oxidizing anions in energetic salts, 225
Pentafluorosulfanyl polyfluoroalkyl imidazolium triflimide salts, densities, 226
Pentafluorosulfanyl polymers, 227–228
Peptide bond synthesis,
trifluoroalanine oligopeptides, 462–476
reaction mechanism and intermediates, 469
Peptide interactions with solvent in fluoroalcohol-water mixtures, 379–392
Peptide synthesis and separation, fluorous chemistry, 209–213
Peptides, membrane-active, structure analysis using fluorinated amino acids, 431–446
Perchloryl fluoride
electrophilic fluorinating reagent, 40–42
safety note, 42
surrogates, 42
trans-Perfluoroalkenylmagnesium reagents, preparation and reactions, 85, 86–87*f*
Perfluorocarbons, phase separation from nonpolar organic solvents and water, 488–489*f*
Perfluorooctanoic acid and salts, human health concerns, 274
Pharmaceuticals and agrochemicals, pentafluorosulfanyl-based, 237–238
Phase separation, fluorous materials, 488–489*f*
Phase-switching, fluorous-phase carbohydrate synthesis, 262–265
Phase-switching process, reversible, using fluorous carboxylates, 280–283
α-Phenylacetates with chiral auxiliaries, diastereoselective anodic fluorination, 70, 71
Phenyllithium and alkyllithiums, 1,2 addition to trifluroacetaldehyde SAMP- or RAMP-hydrazones, 451–453
Photo-irradiation, anthracene-cyclopropane hybrids, DNA cleavage agents, 355–360
Photoisomerization, *trans*-1,2-difluoro-1-trialkylsilylethenes to *cis*-1,2-difluoro-1-trialkylsilylethenes, 93
Physical properties,
pentafluorosulfanyl derivatives in liquid crystals, 229, 231*t*, 232*t*
See also Dielectric anisotropy modification, pentafluorosulfanyl-based liquid crystals

Piperidines, fluorinated, synthesis by ring closing metathesis reactions, 61–63*f*
PM3 calculation in stereo-control mechanism, diastereoselective anodic fluorination, 79–80*f*
Polycyclic systems, fused, domino cyclizations, 164–165*f*
Polyfluorinated *exo*-4,4'-fluoroalkyltricyclo[4.2.1.$0^{2.5}$]non-7-enes and *exo*-3-aza-4-perfluoroalkyltricyclo[4.2.1.$0^{2.5}$]non-3,7-dienes, reactions, 134–136*f*
Polyfluorinated *exo*-3-oxa-4,4-fluoroalkyltricyclo[4.2.$10^{2.5}$]non-7-enes, ring opening reactions, 132–133*f*
Polyfluorinated *exo*-tricyclo[4.2.$10^{2.5}$]non-7-enes, synthesis, 115–120, 121*f*
Polyfluorinated *exo*-tricyclononenes, synthesis using cycloaddition reactions of quadricyclane and norbornadiene, 114–129
Polyfluoropyridine derivatives, reactions, 327–330
Polyfluoropyridyl glycosyl donors, 323–336
Polymers based on pentafluorosulfanyl derivatives, 227–228
Prakash-Ruppert reagent for bromine or iodine displacement, 44
See also Ruppert-Prakash reagent
L-Proline-catalyzed, asymmetric direct aldol reaction, trifluoroacetaldehyde ethyl hemiacetal with ketones, 149–152
2-Propanol in reductive removal of xanthate, 32–34f
Protecting group effect on diastereoselective anodic fluorination, 1,2-diol moieties, 71, 72
Protein conformational stability, stereoelectronic effect, 479–484
Protein design using fluorinated amino acids, 490–494
Proteomics sample, fluorous enrichment, peptide synthesis and separation, 212
Pummerer method, thioethers and primary alcohols oxidation to α-aryl-α,α-difluoroacetaldehydes, 43
Pyridyl tagged procedures, reversible phase-switching processes, 280–284

Q

Quadricyclane and fluorinated substrates, miscellaneous reactions, 128–129
Quadricyclane cycloaddition
fluorinated nitrogen-containing compounds, 125–126
fluorinated sulfur-containing compounds, 126–127
norbornadiene, fluorinated carbonyl compounds, 120, 121–125

R

Radical trifluoromethylation, 5–21
lithium enolates, 12–16
reaction mechanism, titanium ate and lithium enolates, 17–21
titanium ate enolates 5–12
Reformatskii-Claisen reaction, 319
Renin-angiotensin system, 379–380
Retentive fluorination at C3'α-position, 3'-α-fluoro-2',3'-dideoxyguanosine, synthesis, 373–375
Reverse order-disorder structure in crystal packing, *N*-(4-methoxyphenyl)-*F*-2-(2,6-

dimethylmorpholino)-propanamide (MPMPA), 177–178*f*
Reversible hydrocarbon/perfluorocarbon phase-switching, 280–284
Ring-closing metathesis reactions
fluorinated α-amino acid synthesis, 55, 56–58*f*
fluorinated β-amino ester synthesis, 58–59*f*
fluorinated lactams, lactones and aza-lactones, synthesis, 62, 63–66
fluorinated piperidine synthesis, 61–63*f*
fluorinated uracil synthesis, 60–61*f*
Ring-opening reactions
polyfluorinated *exo*-3-oxa-4,4-fluoroalkyltricyclo[4.2.$10^{2.5}$]non-7-enes, 132–133*f*
exo-3-thia-4,4-bis(trifluoromethyl)-tricyclo[4.2.$10^{2.5}$]non-7-ene, 133–134
Ring pucker effect on collagen triple helix stability, 481–483
Ruppert-Prakash reagent in *exo*-norbornen synthesis, 130–131
See also Prakash-Ruppert reagent
Ruthenium-based catalysts, 55
olefin cross-metathesis reaction, artemisinin dimer preparation, 346, 347*f*
unsaturated fluorinated lactams, synthesis, 63–64*f*

S

Safety notes
perchloryl fluoride, 42
perfluorooctanoic acid and salts, toxicity, 274
Salts, supporting fluoride, effect on diastereoselectivity, anodic fluorination, 71
Samarium(II) iodide-induced N–N single bond cleavage, 456–457
Scandium complexes with fluorous bis(perfluoroalkanesulfonyl)-amide ponytails, catalyst
Baeyer-Villiger reaction, 197, 199–202*f*
Mukaiyama aldol reaction, 193–198
Schmidt on crystal engineering, 171
Silyl enol ether in trifluoromethylations, 8, 11–12
Solid-phase extraction in carbohydrate synthesis, fluorous tags, 264–265
Solid-phase synthesis (SPS)
biooligomers, enhancement by fluorous approaches, overview, 208–209
oligosaccharides with fluorous capping agents, 215
peptides with fluorous capping enabled purification, 209–210, 494–497*f*
Solid state ^{19}F NMR measurements, sample preparation, 435–436
Solid state ^{19}F NMR membrane-active peptides, results, 441–442
Solid state ^{19}F NMR structure analysis, 433–435
Solubility, non-electrolytes, Hildebrand and Scott's theory, 488–489
Solubility properties, capped or tagged polymers, fluorous groups, 495–497*f*
Solution-phase synthesis with fluorous tags, oligosaccharide synthesis, 213–215*f*
Solvent-peptide interactions in fluoroalcohol-water mixtures, 379–392
Solvent polarity effect on hexapeptide and F_3-hexapeptide conformations, 414–418
SPS technologies. *See* Solid-phase synthesis
Stereoelectronic effect on protein conformational stability, 479–484

Stereospecific syntheses, *cis*-1,2-difluoroalkenyl synthons, 92–108
Structure-activity relationships, fluorine-containing second-generation taxoids, 290–301
Styrene, rhodium-fluorous carboxylates catalyzed cyclopropanation, 278–280
2-Substituted 4-thiazolidinones, diastereoselective anodic fluorination, 76
Sulfur-containing compounds, fluorinated, cycloaddition to quadricyclane, 126–127
Sulfur fluorides, hypervalent, organic derivatives, 221–243
β-Sulphonyl hydroxamic MMP's inhibitors, 424–425*f*
Super fluorinated materials in active matrix liquid crystal displays, 228–229*f*
Supramolecular assemblage, Hermaphrodite molecules, 173–176*f*
Surface patterning in carbohydrate microarray formation, noncovalent fluorous-fluorous interaction, 265–268
Swern reagent, thioethers and primary alcohols oxidation to α-aryl-α,α-difluoroacetaldehydes, 43

T

Taxoids with fluorine incorporation
bioactive conformations, 296–298*f*
synthesis and biological evaluation, 291–296
Taxol, 289–290, 296–297*f*
Taxotère, 289–290, 296, 297*f*
Tenuecyclamide macrolactams, solid support synthesis, 245–246
Tetralone synthesis using xanthate technology, 34, 35*f*
5,10,15,20-Tetrapyridyl porphyrin, reversible hydrocarbon/perfluorocarbon phase-switching, 280–282
exo-3-Thia-4,4-bis(trifluoromethyl)-tricyclo[4.2.1$0^{2.5}$]non-7-ene, ring opening reactions, 133–134
4'-Thiogemcitabine, synthesis, 313–314*f*
Titanium ate enolates, radical trifluoromethylation, 5–12, 17–21
Titanium enolate of α-CF_3 ketone, direct generation, 3–5*f*
Toolbox methods in functionalization, fluorinated core compounds, 45–50
exo-Tricyclononenes, polyfluorinated, synthesis using cycloaddition reactions of quadricyclane and norbornadiene, 114–129
Trifluoroacetaldehyde, catalytic *in-situ* generation and direct aldol reaction with ketones, 141–154
Trifluoroacetaldehyde ethyl hemiacetal with ketones, direct aldol reactions 143–149
Trifluoroacetaldehyde hemiacetal, synthesis and addition of xanthate derivatives, 30
Trifluoroacetaldehyde hydrazones, preparation, 449–450*f*
Trifluoroacetaldehyde SAMP- or RAMP-hydrazones, 1,2-addition
alkyllithium and phenyllithium, 451–453
prepared allyllithium, 453–455
reaction conditions, 450–451
Trifluoroalanine-alanine-alanine-tripeptide synthesis, 475–475*f*
Trifluoroalanine dipeptide synthesis, 468–474
Trifluoroalanine oligopeptides, synthetic strategy, 462–476

Trifluoroalanine peptides, synthesis, 468–476
Trifluoroalanine-phenylalanine dipeptide, 474
Trifluoromethionine, probe in biological chemistry, 394–405
Trifluoromethyl allyl bromide-artemisinin, reactions with nucleophiles, 341–343
Trifluoromethyl analogs, β-sulphonyl hydroxamic MMP inhibitors, 425–428
10-Trifluoromethyl-10-deoxoartemisin, allyl bromide, 340–343
α-Trifluoromethyl ketone, titanium enolate, direct generation, 3–5*f*
Trifluoromethyl ketones, synthesis by radical pathway, 31, 35, 37*f*
Trifluoromethyl-malic hydroxamate inhibitor of matrix metalloproteinases-2 and -9, 421
Trifluoromethyl radicals, generation and capture, 31–32*f*
3'-Trifluoromethyl-taxoids, synthesis and biological evaluation, 291–294
1-(Trifluoromethyl)-vinyl vinyl ketones, fluorine-directed Nazarov cyclization, 159–160*f*
α-Trifluoromethylated amines, asymmetric synthesis, 455–457
β-Trifluoromethylated β-amino carbonyl compounds, asymmetric synthesis
  1,2-addition to trifluoroacetaldehyde SAMP-or RAMP-hydrazones, 447–461
  via oxidation α-trifluoromethylated homoallyl amines, 457–458, 459*f*
α-Trifluoromethylated homoallyl amines, oxidation, β-trifluoromethylated β-amino carbonyl compounds, 457–458, 459*f*
Trifluromethylated SAMP- or RAMP-hydrazides, SmI_2-induced N–N single bond cleavage, 456–457
Trifluoromethylation, package delivery, 44–45
Trimethyl(trifluoromethyl)silane for bromine or iodine displacement, 44
Triple helix stability, collagen, (2S,4R)-4-hydroxyproline in Yaa position effect, 478–481
Trp-cage peptide and solvent fluorines, $^{1}H\{^{19}F\}$ intermolecular NOE spectrum, in hexafluoroisopropanol/water, 389–391*f*
Tumor-targeting fluorine anticancer agents, 299–301*f*
Tuning reactivity, polyfluoro-pyridyl glycosyl donors, 326

U

Uracils, fluorinated, synthesis by ring-closing metathesis reaction, 60–61*f*

V

Vicinal bromofluorination, laboratory-scale, with *N*-bromosuccinimide, 42–43
Vinyl ketones, fluorine-directed Nazarov-type cyclizations, 156–159*f*

W

Water mixed with fluoro alcohols, solvent-peptide interactions, 379–392

X

X-ray analysis developments, crystal engineering, 172–173
Xanthate radical transfer technology in organofluorine derivatives, syntheses, 25–38

Z

Zinc(II)-dependence proteolytic enzymes. *See* Matrix metalloproteinases